THE THEORY OF VIBRATIONAL SPECTROSCOPY AND ITS APPLICATION TO POLYMERIC MATERIALS

THE THEORY OF VIBRATIONAL SPECTROSCOPY AND ITS APPLICATION TO POLYMERIC MATERIALS

Paul C. Painter
Michael M. Coleman
The Pennsylvania State University
University Park

Jack L. Koenig
Case Western Reserve University
Cleveland

1807 175 YEARS OF PUBLISHING 1982

A WILEY-INTERSCIENCE PUBLICATION
JOHN WILEY & SONS
New York · Chichester · Brisbane · Toronto · Singapore

Library of Congress Cataloging in Publication Data:

Painter, Paul C.
The theory of vibrational spectroscopy and its application to polymeric materials.

"A Wiley-Interscience publication."
Includes index.
1. Polymers and polymerization—Spectra.
2. Vibrational spectra. I. Coleman, Michael M.
II. Koenig, Jack L. III. Title.

QC463.P5P34 547.7′046 81-12969
ISBN 0-471-09346-7 AACR2

Printed in the United States of America

10 9 8 7 6 5 4 3 2 1

For tolerating our idiosyncracies,
this book is dedicated to
Julie Painter, Mary Jane Coleman, and Jeanus Koenig

PREFACE

Next, when I cast mine eyes and see
That brave vibration each way free,
O how that glittering taketh me.
Robert Herrick

I'm picking up good vibrations.
The Beach Boys

Developments in the discipline of vibrational spectroscopy are occurring at a rapid rate owing to the impact of lasers, new detectors, interferometers, and computers. For molecules as complicated as polymers these advances are welcomed with enthusiasm, since polymer spectroscopists need all the help they can get. However, to take full advantage of these developments, the spectroscopist must have available all of the tools necessary for interpretation of the spectra. Although there is certainly no better help than extensive practical experience, it is also true that a fundamental understanding of the vibrational process in polymers is the ultimate foundation of efforts to interpret spectra. Fortunately, the theoretical aspects required for this understanding are available and with modern digital computers, one can take full advantage of the techniques. Unfortunately, except in a few laboratories, the fundamental understanding and techniques do not exist; furthermore, only the reading of a vast literature from a variety of sources enables one to acquire this understanding. What is needed is a monograph that contains all the tools essential for understanding the vibrational spectra of polymers. The authors of this book believe they have provided that compendium. Having taught a number of students the essence of polymer spectroscopy and having been frustrated by the lack of a suitable text, we decided to "take the bull by the horns" and develop such a text. Needless to say, the book represents the view of one group of researchers and will not do justice to all of the work reported. However, our objective is to provide a road map that can be followed by the interested and an atlas of references for those who wish to pursue in depth some aspect of interest to them. It is our hope that our book will be the basis for further improvements. We acknowledge our limitations in aspects of quantum mechanical and field theory of molecular spectroscopy, which have given us the basis for the vibrational analysis. To a great extent we have taken on faith the results and have felt the test is in the laboratory; for the most part, we are satisfied with the results. We welcome new developments and thank the people who will interpret them for us, just as we hope to be able to interpret for you the methods of analyzing spectra of polymer molecules through the vehicle of this book.

PAUL C. PAINTER
MICHAEL M. COLEMAN
JACK L. KOENIG

University Park, Pennsylvania
University Park, Pennsylvania
Cleveland, Ohio

September 1981

ACKNOWLEDGMENTS AND APOLOGIES

At some point during the summer of 1976 we came to the conclusion that there was a need for a book that drew together the various aspects of the theory of vibrational spectroscopy as applied to polymers. Deciding that we were qualified to write such a book was another matter, and we were to some degree inhibited by the prospect of displaying our ignorance to the world.

A person who publishes a book willfully appears before the populace with his pants down.

Edna St. Vincent Millay

However, a project such as this sometimes acquires a life of its own and takes control of an author's existence, a condition not unknown in the ancient world.

An incurable itch for scribbling takes possession of many, and grows inveterate in their insane breasts.

Juvenal (AD 60–140)

We were warned by our wives, friends, acquaintances and the Bible that the writing of such a book would make us vulnerable to attacks by our enemies.

Behold, my desire is, that the Almighty would answer me, and that mine adversary had written a book.

Job xxxi 35.

In addition, scientists are not noted for their ability to write

The only man with anything to say is the man of science, and he can't say it.

James M. Barrie

and apart from the dangers of intemperance

Many contemporary authors drink more than they write.
Maksim Gorki

we were warned that the effort might prove ultimately unsatisfactory.

I have the disease of writing books and feeling ashamed of them afterwards.
Charles de Secondat, Baron de La Brède et Montesquieu

Fortunately, we were encouraged by those authors who took a perverse delight in inflicting their lack of talent upon the world.

The physical business of writing is unpleasant to me, but the psychic satisfaction of discharging bad ideas in worse English makes me forget it. ***H. L. Mencken***

Lacking the arrogance and originality of some authors,

How can anyone do better than that—after all, I wrote it myself.
Molière

we quickly realized that our efforts would be built upon the exertions of some of our more talented colleagues.

This book contains much that is good and new; it's a pity that the good is not new, and the new is not good.
Gotthold Ephraim Lessing

Among the "good" that we recall having a profound influence on our understanding are the texts by Wilson, Decius, and Cross; Steele; and Turrell and reviews and seminal articles by Krimm, Snyder, Schachtschneider, Zerbi, Shimanouchi, and Miyazawa. Perhaps equal in influence are those colleagues with whom we worked and argued: Jim Boerio, Jeanette Grasselli, Marty Hannon, Kurt Holland-Moritz, Bruce Frushour, Dave Tabb, and John Cael, among many others. Those authors and friends whom we have overlooked or whose contributions we have inadvertantly slighted, please accept our humble apologies. Those colleagues who find their original ideas and work reproduced at length we thank with the observation

About the most originality that any writer can hope to achieve honestly is to steal with good judgment.
Josh Billings

In the course of writing this book we have imposed upon our wives and secretaries, so to Julie Painter, Mary Jane Coleman, Jeanus Koenig, Barbara Leach, and Sue Weaver, our humblest apologies and eternal thanks.

Finally, we are under no illusions that this book will be riveting enough to keep readers awake at night, but we are in the company of some of the most prominent authors of antiquity,

Gladstone read Homer for fun, which I thought served him right.
Winston Churchill

and in the final analysis our efforts have been aimed at presenting the essentials of the theory of vibrational spectroscopy by what has proved to be a painful and long sifting and organization of elements that have appeared in various forms at different times. We believe we have learned much from our efforts and hope that we have managed to impart our hard-earned knowledge in an intelligible form

If you wander around in enough confusion, you will soon find enlightement
Digby Diehl

P.C.P.
M.M.C.
J.L.K.

General References

G. Herzberg, "Infrared and Raman Spectra of Polyatomic Molecules," Van Nostrand, New York, 1945.

E. B. Wilson, J. C. Decius, and P. C. Cross, "Molecular Vibrations," McGraw-Hill, New York, 1955.

D. Steele, "Theory of Vibrational Spectroscopy," Saunders, Philadelphia, 1971.

L. A. Woodward, "Introduction to the Theory of Molecular Vibrations and Vibrational Spectroscopy," Oxford (Clarendon Press), New York, 1972.

G. Turrell, "Infrared and Raman Spectra of Crystals," Academic Press, London, 1972.

J. C. Decius and R. M. Hexter, "Molecular Vibrations in Crystals," McGraw-Hill, New York, 1977.

K. Nakamoto, "Infrared and Raman Spectra of Inorganic and Coordination Compounds," 3d edition, Wiley, New York, 1978.

H. Tadakoro, "Structure of Crystalline Polymers," Wiley, New York, 1979.

CONTENTS

THE THEORY OF VIBRATIONAL SPECTROSCOPY AND ITS APPLICATION TO POLYMERIC MATERIALS

1

STATE OF THE ART IN MOLECULAR SPECTROSCOPY

The end of our foundation is the knowledge of causes, and secret motions of things.

Francis Bacon

The Nobel laureate Gerhard Herzberg has called spectroscopy "the science of discovery" (1). Certainly the use of a beam of light to probe the structure of molecules represents a continuing challenge to scientists as the questions asked and the information required become more penetrating and sophisticated. Fortunately, the experimental and theoretical techniques at our disposal have also continued to advance. Periodically, it is desirable to assess where we are and where we have been. As Victor Hugo said, "Let us, while waiting for new monuments, preserve the ancient monuments."

1.1 Historical Background

Spectroscopy can be defined as the study of the interaction of electromagnetic radiation with matter. The nature of this interaction depends on the wavelength or frequency of the radiation, so that regions of the electromagnetic spectrum have become associated with various types of spectroscopy. These frequency ranges are usually named after the most common source of the radiation (e.g., x-rays) or its practical end use (e.g., radio, radar). Infrared rays have wavelengths that range from about 1 μm to 100 μm. They are also commonly expressed by spectroscopists in terms of wavenumbers, which are defined as the reciprocal of the wavelength expressed in centimeters, the range being from 1 to 10,000 cm^{-1}.

Infrared rays were discovered in 1800 by William Herschel (2), who directed a spectrum of the sun's rays from a prism onto a thermometer. From the heating effect, he noted the existence of an invisible radiation beyond the red end of the visible range. These rays obeyed the laws of reflection and refraction of visible light. Subsequently, it was also determined that this radiation could be absorbed by matter and that absorption occurred in the form of several bands localized in discrete frequency intervals. It was not until

1892, however, that Julius (3) made the seminal observation that the type of atoms present and their structural arrangement in the molecule determined the character of infrared absorption. In the early 1900s Coblentz (3), using a spectrometer he had designed and built, painstakingly obtained the infrared spectra point by point of over 100 organic solids, liquids, and gases. An excellent discussion of the history of infrared spectroscopy is detailed in the book edited by Laitinen and Ewing (3).

The basis for the widespread use of infrared spectroscopy was the observation that many chemical groups, such as C=O, absorb in a relatively narrow frequency range, irrespective of the nature of the other functional groups present. Furthermore, within this frequency range the observed frequency can be correlated to specific chemical structures. For example, esters can be differentiated from ketones by the characteristic stretching frequency of the carbonyl group near 1700 cm^{-1}. On the other hand, similar molecules may have significantly different infrared spectra, especially in the region below 1500 cm^{-1}, and the spectral pattern may be likened to a "molecular fingerprint."

In addition to the absorption or emission of infrared radiation, molecular vibrational information can be acquired from the inelastic scattering of light and neutrons. (Neutron scattering will not be discussed in detail in this book.) Light is scattered by a material in several different ways, and the scattering processes are usually named after the scientists who made major contributions to the field. Elastically scattered light that has the same frequency as the incident radiation is known as Rayleigh scattering. Mie scattering is also an elastic scattering process; it is associated with the scattering of large particles. Brillouin scattering, which is in essence a Doppler effect, produces small frequency shifts (≤ 0.1 cm^{-1}) of the scattered light and is usually ignored in molecular vibrational spectroscopy. Raman scattering is an inelastic process in which light exchanges energy with the sample and consequently appears at a different frequency. Sir C. V. Raman (4) first observed the effect that has been given his name, although it was predicted theoretically by Smekal in 1923 (5). This effect is very weak compared to Rayleigh scattering, and Raman spectroscopy has only recently become a routine laboratory tool.

1.2 Advances in Raman Spectroscopy

During the past two decades dramatic changes have occurred in the experimental procedures employed to obtain vibrational spectra. In the 1960s the advent of the laser had an enormous impact on Raman spectroscopy (6). The laser as a light source was quickly adopted by spectroscopists because of the radiation's high power, coherency, and monochromaticity.

In the spontaneous Raman experiment an incident laser beam with frequency ν_0 is directed onto a sample and the scattered radiation is resolved into its frequency components ($\nu_1 = \nu_0 \pm \nu_r$). The strongest emission, usually called a line, is observed at the frequency of the incident radiation and is associated with the elastically scattered light. Symmetrically placed on each side of this emission are the Raman lines. The lines on the low frequency side are termed the Stokes lines (ν_s), while those on the high frequency side are called anti-Stokes. The Raman effect results from the inelastic scattering of photons from a molecule. The energy difference between the incident and scattered photons is the difference in energy between the initial and final states of the molecule. In vibrational spectroscopy, this difference corresponds to the energy of a vibrational transition. Therefore, the energy lost by the incident photon as a result of inelastic scattering excites a vibrational mode in the molecule. The Stokes lines are more intense than the anti-Stokes lines, a consequence of the different populations of molecules in the ground and first excited vibrational states, as described by the Boltzmann distribution.

The principal advantage of lasers over conventional discharge lamps as Raman sources arises from the enhanced power flux density (i.e., average power radiated). For continuous argon-ion lasers, values of power flux density of 10^5 W/cm^2 are typical. In contrast, only relatively small values, of the order of 1 W/cm^2, can be achieved from conventional mercury lamps.

With the laser as a source and the use of sensitive photon-counting techniques for detection, Raman spectroscopy took its place in the laboratory beside the infrared instruments (7, 8). The benefits of the complementary vibrational information available from both Raman and infrared techniques were immediately realized. For the first time, complete infrared and Raman spectra could be obtained. Additionally, Raman spectra could be obtained in a matter of minutes rather than hours or days. Many molecules could be studied with laser sources that previously could not be examined by means of discharge lamps. The highly polarized nature of the Raman source and the small beam size made it possible to obtain Raman polarization results for oriented and crystalline molecules (9). These polarization data are fundamental for mode assignments and structure determination. Concurrently, the development of double and triple monochromators allowed the spectroscopist to readily obtain extremely low frequency data on the Raman effect. These low frequency Raman data have been successfully used to gain information on long-range order in molecules (e.g., the longitudinal acoustic mode) (10) and lattice vibrations. Another major advantage realized concerns the study of molecules in aqueous solution by Raman spectroscopy, since water is a weak Raman scatterer (11). Additionally, the small beam size permits the study of severely limited amounts of sample (12). Thus, the entire field of biochemistry was penetrated by the Raman spectroscopist and Raman spectroscopy is now an accepted biophysical method.

Under normal scattering conditions, only about 1 in 10^8 incident photons emerges as a Raman shifted photon. The inefficiency of Raman scattering arises because typically the molecule is irradiated in a transparent region of its electromagnetic spectrum. The intensity of Raman lines as a function of the frequency of the exciting radiation, ν_0, is predicted by Placzek's theory to have a $(\nu_0 - \nu_{kn})^4$ dependence, where ν_{kn} is the frequency associated with the $k \rightarrow n$

vibrational transition. However, deviations from this theory were first observed by Ornstein and Went (13) from their studies of the 1088 and 288 cm^{-1} Raman lines of calcite. These Raman lines were observed to be intensity enhanced, which was explained on the basis of a correlation with an absorption frequency in the ultraviolet. This effect is known as **resonance Raman scattering** (RRS) (14). Using semiclassical theory, and assuming that the molecules are in a nondegenerate electronic state and that they are all oriented identically in space, Shorygin (15) developed an equation for RRS. In essence, the intensity of the Raman line ν_{kn} depends on the frequency of the exciting radiation ν_0 as follows:

$$I_{kn} = C(\nu_0 \pm \nu_{kn})^4 \left[\frac{\nu_{rk}^2 + \nu_0^2}{(\nu_{rk}^2 - \nu_0^2)^2} \right]^2$$

where r is a single excited electronic level and C is a constant term.

By varying the frequency of the exciting line, the efficiency of the Raman scattering process can be increased by as much as 10^6. Therefore the RRS process can be seen to have great utility in obtaining vibrational information from species that are present in low concentrations. In general, resonance Raman spectroscopy gives information about the vibrational modes, electronic structure, and chemical environment of a molecule (16). With the anticipated improvements in tunable lasers, which will span the whole ultraviolet and visible spectrum, RRS is expected to have even wider applicability (17).

Another method of increasing the sensitivity of the Raman effect is to increase the efficiency of the emission process. The rate of induced emission for a normal absorption-emission process is proportional to the density of the radiation. In the normal Raman scattering process, molecules are excited by light at the frequency ν_0 to some intermediate state and arrive at the Raman-active level with the emission of a photon of frequency ν_s, the Stokes frequency. The frequency difference $\nu_0 - \nu_s$ is equal to the separation of a vibrational energy level, and the conversion from ν_0 to ν_s is an extremely inefficient process (10^{-8}). However, with giant pulses of short duration, which can be achieved by pulsed ruby lasers having electric field strengths of 6×10^6 V/cm or power densities of 4×10^{10} W/cm^2, nonlinear phenomena occur. Thus, the process referred to as **stimulated Raman scattering** (SRS) is feasible (18). Above a certain threshold value of the incident intensity, a gain in the scattering medium at a frequency of $\nu_0 - \nu_s$ is observed that exceeds the losses and becomes amplified. The intensity of this stimulated Raman line may be greater than that of spontaneous emission by a factor of e^{10}–e^{20} and may readily be seen by the naked eye. To date, SRS has been limited to a very few Raman lines of specific molecules and is unlikely to be of great significance to molecular spectroscopists concerned with molecular structure and chemistry.

In the normal Raman effect and SRS we are concerned with emission processes. Conversely, in **inverse Raman scattering** (IRS) the Raman effect is observed in absorption. A sample is irradiated simultaneously with a monochromatic laser of intensity below that necessary to induce SRS (ν_0) and an intense continuum in a frequency range spanning a known Stokes line (ν_s). [Incidentally, this continuum is achieved in a separate experiment through the use of SRS (19).] In IRS we therefore simply monitor the depletion of the laser beam at ν_0 in the presence of the intense source at ν_s. The measurement of absorption rather than emission leads to the phrase "inverse Raman." Theoretically, a 10^6 advantage is expected for IRS compared to ordinary Raman.

With the laser as a source, the principal limitation of Raman spectroscopy is the decrease in signal-to-noise ratio resulting from the occurrence of fluorescence. Whenever the sample is exposed to the laser, absorption of energy can result in fluorescence in the same frequency region as that of the Raman signal. No general technique has been found to eliminate fluorescence, although a number of different schemes have been attempted. The technique showing the greatest potential in this regard is **coherent anti-Stokes Raman scattering** (CARS) spectroscopy (20). In CARS, the output of the fixed laser ν_L and that of a tunable laser ν_t are crossed at a small angle in a sample cell. When the frequency difference ($\nu_L - \nu_t$) between the two laser beams agrees with a Raman frequency, a spatially coherent anti-Stokes signal is observed at $2(\nu_L - \nu_t)$. Since CARS is coherent and on the anti-Stokes frequency side and fluorescence is incoherent and on the Stokes frequency side, the fluorescence is essentially eliminated. With spatial filtering, the fluorescence can be reduced by a factor of about 10^4. Also, the CARS signal is much stronger than the ordinary Raman signal, so the overall rejection of fluorescence relative to ordinary Raman scattering is a factor of 10^9.

1.3 Advances in Infrared Spectroscopy

Just as the laser revolutionized Raman spectroscopy, the discovery of the fast Fourier transform (FFT) algorithm by Cooley and Tukey (21) in 1965 revitalized the field of infrared spectroscopy. The discovery of the FFT and the concurrent introduction of sophisticated minicomputers permitted the development of infrared instruments based on the Michelson interferometer that were applicable over the whole infrared frequency range. Consequently, a new generation of infrared instrumentation, called Fourier transform infrared (FT-IR) spectrophotometers, was born. Incidentally, interferometric methods were known and the optical throughput and multiplex advantages were fully recognized at the turn of the century, but the time, effort, and cost of performing the Fourier transform of the resulting interferograms were prohibitive.

In contrast to dispersive instruments, which use a monochromator and a system of slits to isolate single infrared frequency intervals and then measure each of them sequentially, the scanning interferometer effectively modulates each infrared wavelength at a characteristic frequency and allows all wavelengths to reach the detector during the entire measurement period. The spectra are obtained via a frequency analysis (Fourier transform) of the recorded signal. This multiplexing of the infrared signals results in a significant signal-to-noise enhancement that can reach $N^{1/2}$ where N is the number of spectral elements contained in the spectrum. Furthermore, the optical throughput of the interferometer is substantially greater than that of a dispersive instrument of equal resolution, resulting in another signal-to-noise advantage. The electronics detect and amplify the signal, digitally encode it, and transmit it to the dedicated minicomputer. There the interferogram is Fourier analyzed into a recognizable infrared spectrum. This spectrum has extremely accurate frequency calibration, a high signal-to-noise ratio, and linear photometric response. The theoretical considerations of FT-IR spectrometry have been well documented in the books of Griffiths (22) and Ferraro and Basile (23), and the interested reader is referred to these texts for further details.

Before leaving the subject of FT-IR, however, we draw attention to some important practical ramifications of this instrumentation. At first glance, the necessity of a computer to convert the interferogram to the normal frequency domain via the FFT could be considered a disadvantage, if only from an economic viewpoint. However, the very fact that it is essential to employ a computer resulted in the development of the software necessary to control the spectrometer, undertake signal averaging, and perform mathematical manipulations on the spectral data. Thus data processing techniques have been developed that allow us to extract the maximum possible information from the spectra. Absorbance subtraction makes it possible to eliminate interfering absorbances and magnify the remaining spectral features to the limit of the signal-to-noise ratio. By using the ratio method, the spectra of mixtures can be deconvoluted into their components and from least-squares refinement of the spectra the amount of each component in the mixture can be calculated, along with an indication of the standard deviation of the measurements. Techniques are also available for performing time sequence scanning of the data to extract the spectra of intermediate species. Infrared spectra of the components of complex mixtures can also be obtained "on the fly" by using an FT-IR spectrometer coupled with a gas-liquid chromatograph. Because the high energy throughput of the system and the multiplex advantages, good quality infrared spectra can be obtained from optically dense materials such as a coal. This advantage is also significant for studies of energy-starved systems, such as the measurement of emission spectra that cannot ordinarily be obtained in transmission (e.g., remote samples, coatings). Altogether, the Fourier transform systems represent contemporary infrared spectroscopy at its finest.

1.4 Status of Vibrational Theory in Molecular Spectroscopy

Surprisingly, very little fundamental theory has evolved over the past two decades, but enormous efforts have been directed toward reducing the available theory to useful applications for interpreting the experimental results. The easy availability of large, fast digital computers has radically changed our approach to interpreting spectra. Twenty years ago, a normal coordinate analysis of a small molecule was a slow, demanding, tedious task (24). Now, in a matter of minutes the normal coordinate analysis can be carried out and parametric refinements can be made by comparison with the experimental data. Larger and larger molecules have been subjected to normal coordinate analysis, and today the analysis of polymer molecules is nearly routine (25). At present, different structural models involving configurational and conformational differences can be easily tested. After a suitable structural model has been found, force constant refinement procedures allow insight into the nature of the molecular bonding. Traditionally, only the vibrational frequencies have been calculated, but now the infrared and Raman intensities can also be calculated (26). Although insufficient work has been reported to make the analysis of intensities beyond controversy, it is clear that in the future, methods for the analysis of intensities as well as of frequencies will be available to serve the spectroscopist in the solution of structure problems. Traditionally, only isolated or single molecules have been studied theoretically, but with the larger computers the effects of intermolecular interactions can be evaluated. The complete analysis of three-dimensional crystalline systems can be compared to the results for the single molecule and the experimental data can be evaluated properly for intermolecular forces.

1.5 Summary

It is the opinion of the authors that it is an appropriate time to document the "monuments" of vibrational spectroscopy, particularly as applied to macromolecules (polymers). The literature contains a number of excellent volumes on the methods as applied to "small" molecules, but no text exists that focuses on the important class of molecules known as polymers. However, it is not possible to present the methods and results for polymers without developing some background. Here, we will be as brief as possible and cite additional references for the in-depth study required by the more studious reader. We will present, perhaps for the first time in one place, the assemblage of techniques required for polymer systems. These techniques have been invaluable to us in our spectroscopic research in polymers and it is our belief that without a substantial portion of this material, spectroscopic research involving polymers will be difficult and unrewarding. This is not to imply that good research cannot be carried out without these insights, but such insights certainly make a difficult task easier.

References

1 G. Herzberg, "Infrared and Raman Spectra of Polyatomic Molecules," Van Nostrand, New York, 1945.

2 W. Herschel, *Phil. Trans. Roy. Soc.* **90**, 255 (1800).

3 H. A. Laitinen and G. W. Ewing, "A History of Analytical Chemistry," Division of Analytical Chemistry, American Chemical Society, New York, 1977.

4 C. V. Raman and K. S. Krishnan, *Indian J. Phys.* **2**, 387 (1928).

5 A. Smekal, *Naturwiss.* **11**, 875 (1923).

6 S. P. S. Porto, *J. Opt. Soc. Am.* **56**, 1985 (1966).

7 M. C. Tobin, "Laser Raman Spectroscopy," Wiley, New York, 1971.

8 S. K. Freeman, "Applications of Laser Raman Spectroscopy," Wiley, New York, 1974.

9 T. C. Damen, S. P. S. Porto, and B. Tell, *Phys. Rev.* **142**, 570 (1966).

10 R. F. Schaufelle and T. Shimanouchi, *J. Chem. Phys.* **47**, 3605 (1967).

11 J. L. Koenig and P. L. Sutton, *Biopolymers* **9**, (1970).

12 S. K. Freeman and D. O. Landon, *Anal. Chem.* **41**, 398 (1969).

13 L. S. Ornstein and J. J. Went, *Physica* **2**, 391 (1935).

14 J. A. Koningstein, "Introduction to the Theory of the Raman Effect," D. Reidel, Hingham, Mass., 1972.

15 P. P. Shorygin, *J. Chim. Phys. Physico. Chim. Biol.* **50**, D31 (1953).

16 J. Behringer, in "Molecular Spectroscopy," Vol. 2, R. F. Barrow, D. A. Long, and D. J. Millen, Eds., The Chemical Society, London, 1974, p. 100.

17 J. Behringer, in "Molecular Spectroscopy," Vol. 3, R. F. Barrow, D. A. Long, and D. J. Millen, Eds., The Chemical Society, London, 1975, p. 163.

18 J. B. Grun, A. K. McQuillian, and B. P. Stoicheff, *Phys. Rev.* **180**, 61 (1969).

19 W. J. Jones and B. P. Stoicheff, *Phys. Rev. Lett.* **13**, 657 (1964).

20 W. M. Tolles, J. W. Nibler, J. R. McDonald, and A. B. Harvey, *Appl. Spect.* **31**, 253 (1977).

21 J. W. Cooley and J. W. Tukey, *Math. Comput.* **19**, 297 (1965).

22 P. R. Griffiths, "Chemical Infrared Fourier Transform Spectroscopy," Wiley, New York, 1975.

23 J. R. Ferraro and L. J. Basile, "Fourier Transform Infrared Spectroscopy," Vols. 1 and 2, Academic Press, New York, 1978.

24 E. B. Wilson, Jr., J. C. Decius, and P. C. Cross, "Molecular Vibrations," McGraw-Hill, New York, 1955.

25 F. J. Boerio and J. L. Koenig, *J. Macromol. Sci. Rev.* **C7**, 209 (1972).

26 S. Abbate, M. Gussoni, G. Masetti, and G. Zerbi, *J. Chem. Phys.* **67**, 1519 (1977).

2

ELEMENTARY CLASSICAL MECHANICAL TREATMENT OF MOLECULAR VIBRATIONS

Everything should be made as simple as possible, but not simpler.
A. Einstein

2.1 Background

In the limit of the harmonic approximation the frequencies of the normal modes calculated by classical mechanics are the same as those determined by quantum mechanics. On this basis, the simpler classical mechanical approach will be utilized. There are a number of excellent texts, (1–4) in which the classical and quantum mechanical treatments are given side by side, and these texts should be studied. We intend to present only sufficient background to make it possible to cope with the polymer case. It should be kept in mind that although the frequencies, displacements, and normal modes can be determined with a good deal of confidence by the methods of classical mechanics, many aspects of the spectra of polymers, such as overtones, combinations, and intensities, can be adequately described only by means of more complex treatments.

The classical equations of molecular vibrations become extremely cumbersome for simple polyatomic molecules. Although the equations of motion are now solved by digital computer methods that rapidly diagonalize even very large matrices, a proper selection of molecular coordinates allows a considerable simplification. We will first consider the elementary treatment of simple molecules and then proceed to more advanced treatments.

2.2 The Origin of Vibrational Spectra

A change in the total energy of a molecule occurs upon interaction with electromagnetic radiation. This change is reflected in the observed spectrum of the material. In order to describe this interaction and formulate a useful mathematical model, certain assumptions are usually made. The total energy of a molecule consists of contributions from the rotational, vibrational, electronic, and electromagnetic spin energies. This total energy can be approximated to a sum of the individual components, and any interaction can be treated as a perturbation. The separation of the electronic and nuclear motions, known as the Born-Oppenheimer approximation, depends on the large difference in mass between the electrons and nuclei. Since the former are much lighter, they have relatively greater velocities and their motion can be treated by assuming fixed positions of the nuclei. Conversely, the small (compared to interatomic distance) nuclear oscillations occur in an essentially averaged electron distribution. The change in energy of these nuclear vibrations upon interaction with radiation of suitable frequency is the origin of the vibrational spectrum. This energy change is, of course, quantized. In addition, an absorption band or Raman line nearly always corresponds to discrete vibrational transitions in the ground electronic state. Absorption of higher energy visible or UV light is required to produce changes in electronic energy.

There are requirements for infrared absorption or Raman scattering in addition to a transition from one discrete vibration energy level to another. These will be discussed later. First, we will consider in more detail the form of the molecular vibrations associated with these transitions.

Any vibration of an atom can be resolved into displacements parallel to the x, y, and z axes of a cartesian system. The atom is described as having three degress of freedom. A system of N nuclei therefore has $3N$ degrees of freedom. Once all $3N$ coordinates are fixed, the bond distances and bond angles are likewise fixed.

The number of fundamental vibrational frequencies or **normal modes of vibration** of a molecule is equal to the number of vibrational degrees of freedom. For a nonlinear molecule, however, six of these degrees of freedom correspond to translations and rotations of the molecule as a whole and have zero frequency, leaving $3N-6$ vibrations. For strictly linear molecules, such as carbon dioxide, rotation about the molecular axis does not change the position of the atoms, and only two degrees of freedom are required to describe any rotation. Consequently, linear molecules have $3N-5$ normal vibrations. For an infinite polymer chain only the three translations and one rotation have zero frequency, and there are $3N-4$ degrees of vibrational freedom. Each normal mode consists of vibrations (although not necessarily significant displacements) of *all* the atoms in the system.

Since molecular vibrations cannot be observed directly, a model of the system is required in order to describe these normal modes. The nuclei are considered to be point masses and the forces acting between them springs that obey Hooke's law. The motion of each atom is assumed to be simple harmonic. Even with these assumptions it is intuitively obvious that a system of N atoms is capable of innumerable different complex vibrations, each involving a range of displacements of the various nuclei, rather than a number equal to the vibrational degrees of freedom. In the harmonic approximation, however, any motion of the system can be resolved into a sum of so-called fundamental normal modes of vibration, just as displacements can be represented by components parallel to a set of cartesian coordinates. In a normal vibration each particle carries out a simple harmonic motion of the same frequency, and in general these oscillations are in phase; however, the amplitude may be different from particle to particle. It is these normal modes of vibration that are excited upon infrared absorption or Raman scattering. Naturally, different types of vibrations will have different energies and so absorb or inelastically scatter radiation at different frequencies.

We noted in Section 1.1 that it has been empirically determined that certain functional groups absorb infrared radiation at charcteristic frequencies. Since normal vibrations often involve displacements of all the atoms, it might be expected that the constitution of the rest of the molecule might have a more profound effect than the relatively small frequency shifts often observed for these groups. However, the intensity of an infrared band depends on the displacement of atoms in a particular vibration. All the atoms may be vibrating with the same frequency but the largest displacements from an equilibrium position can be localized in a small group of atoms in the molecule.

It is possible to demonstrate the major factors that influence the frequency of a normal mode by means of a simple model system consisting of the harmonic oscillations of a diatomic molecule. It is convenient to consider the classical solution and then account for the conditions imposed by quantum mechanics.

2.3 The Vibrations of a One-Dimensional Diatomic Molecule

Consider two point masses m_1 and m_2 connected by a Hookean spring as shown diagrammatically in Figure 2.1. For this one-dimensional model, the z axis is coincident with the molecular axis and movements of the atoms in the x and y directions are not allowed. Let us define the displacement of the two atoms relative to each other by $z=(z_1'-z_2')-(z_1-z_2)$ where z_1-z_2 is the equilibrium separation between m_1 and m_2 and $(z_1'-z_2')$ is the distance after a finite extension or compression. Assuming that the spring obeys Hooke's law, we have as the exerted force $-fz$ where f is the force constant or ("stiffness") of the spring. It is also convenient at this stage to define m_r, the reduced mass, equal to $(m_1m_2)/(m_1+m_2)$.

Since force is defined as mass times acceleration, for a conservative field (i.e., no frictional forces) the following equation holds:

$$-fz=m_r\ddot{z} \tag{2.1}$$

where $\ddot{z}=d^2z/dt^2$. Hence

$$m_r\ddot{z}+fz=0 \tag{2.2}$$

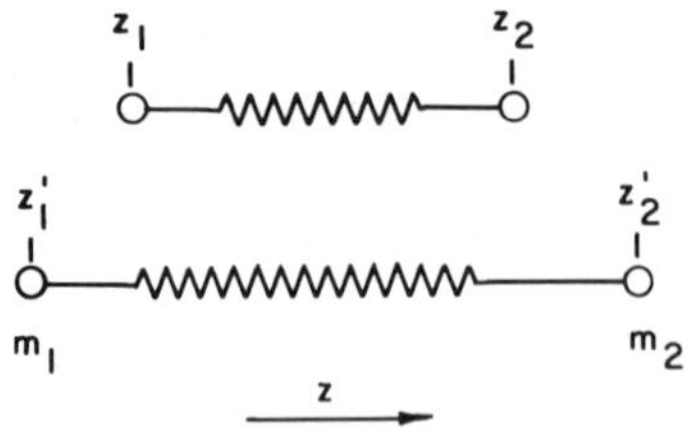

Figure 2.1. The coordinates of a model diatomic molecule.

This is a second-order differential equation that has a periodic solution of the form

$$z = A\cos(2\pi\nu t + \varepsilon) \tag{2.3}$$

where A is an amplitude, ε is a phase angle, and ν is the vibrating frequency. Differentiating twice with respect to time leads to

$$\ddot{z} = -4\pi^2\nu^2 z \tag{2.4}$$

Substituting back into equation (2.2) we obtain

$$\left(-4\pi^2\nu^2 m_r + f\right)z = 0 \tag{2.5}$$

Assuming that $z \neq 0$, we obtain the familiar equation

$$\nu = \frac{1}{2\pi}\sqrt{\frac{f}{m_r}} \tag{2.6}$$

This is the basic vibrational equation for the harmonic oscillator. Although the model considered is extremely simple, it does demonstrate that the vibrational frequency depends inversely on mass and directly on the force constant. Consider as an example the isolated stretching vibrations of the bond C—X, where X can be a substituent such as hydrogen, chlorine, or oxygen. The chemical bond can be assumed to be the focus of the forces acting between the atoms, that is, the spring. Then as the mass of a substituent is increased, equation (2.6) indicates a decrease in frequency. In fact, the C—H stretching modes absorb near 2900 cm^{-1}, while C–Cl stretching vibrations occur near 600 cm^{-1}. Conversely, increasing the force constant between atoms, say by formation of a double bond, increases the frequency. The C—O stretching vibrations are found near 1100 cm^{-1}, while C=O frequencies are characteristically observed at about 1700 cm^{-1}.

For more complex vibrating systems the vibrational frequencies naturally depend on the type of motion and the geometry in addition to the mass of the atoms and the forces acting between them.

The absorption of infrared radiation or inelastic scattering of light (Raman effect) changes the vibrational energy of a system. The total energy E of a

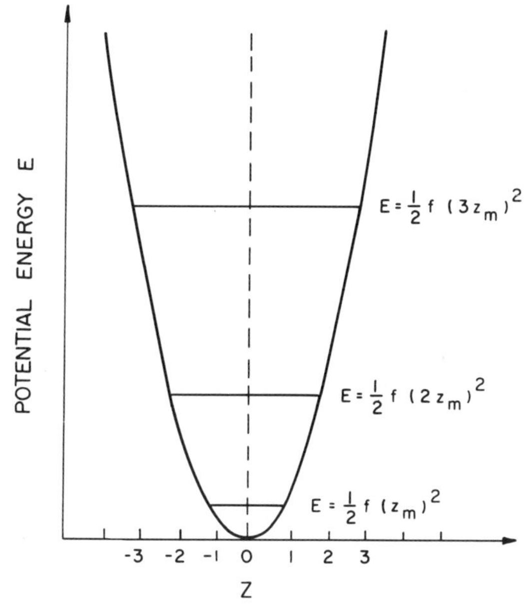

Figure 2.2. The potential energy (E) for a harmonic diatomic oscillator as a function of displacement z.

classical harmonic oscillator, in a conservative field in which there are no frictional losses, is constant and equal to the sum of the potential and kinetic energies at any point in the oscillation. The potential energy (PE) is given by

$$\mathrm{PE} = \int_0^z fz\,dz = \tfrac{1}{2} fz^2 \tag{2.7}$$

A plot of PE as a function of displacement z is a parabola, and a typical curve is shown in Figure 2.2. At the maximum displacement (z_m) in the vibration there is no movement for an instant in time, and the kinetic energy is zero. Hence, the total energy is given by

$$E = \tfrac{1}{2} f z_m^2$$

In the classical solution with z being continuous, E is continuously variable from 0 to $\frac{1}{2} f z_m^2$. However, the quantum mechanical solution requires discrete quantized energy levels rather than a continuum. We will not repeat the quantum mechanical derivation here, since it is covered in many elementary texts (1–4). Briefly, the Schrödinger equation is set up in order to obtain the vibrational energy levels. This equation is more readily solved by a change in the coordinate system from the usual cartesian to a set of normal coordinates (see Section 2.8). As a result of using normal coordinates, the energy problem reduces to a sum of the energies of different harmonic oscillators; so the total energy of a system can be expressed as the sum of the energies of the individual normal modes. In general

$$E = \sum_i E_i = \sum_i \left(n_i + \tfrac{1}{2}\right) h\nu_i \tag{2.8}$$

where ν_i is the classical vibrational frequency for the ith normal mode and n_i is the vibrational quantum number. Additionally, quantum mechanical selection rules for the harmonic oscillator only allow $n_i = \pm 1$. For our simple example

$$E_i = \left(n_i + \tfrac{1}{2}\right) h\nu_i$$

$$\Delta E = \left(n + 1 + \tfrac{1}{2}\right) h\nu_i - \left(n + \tfrac{1}{2}\right) h\nu_i = h\nu_i \tag{2.9}$$

where ΔE_i is the energy change for absorption or emission of radiation.

This is an extremely important result since it allows the calculation of vibrational frequencies by the standard methods of classical mechanics, provided that the harmonic approximation holds.

Figure 2.3 illustrates the absorption of various energy quanta. A molecule in the ground state E_0 can absorb an energy quantum $h\nu$ to reach the first excited state E_1. The subsequent absorption of another quantum $h\nu$ would allow the second excited state E_2 to be reached. Similarly, the absorption of $2h\nu$ would allow the second excited state to be reached directly. For a complex system, quanta of $h\nu_i + h\nu_j$ could be absorbed, where ν_i and ν_j are the frequencies of different normal modes. These processes are the basis for the observation of overtone and combination bands, respectively. For a harmonic oscillator, however, a fundamental condition of quantum mechanics is that only transitions between adjacent levels are observed.

Combination and overtone bands are observed, although only with relatively weak intensities. These observations reflect the limitations of our theoretical model. Molecular vibrations are not strictly harmonic, and the potential energy curve is closer to that shown in Figure 2.4 than to the parabola of Figure 2.2. Low energy transitions approximate the harmonic model closely,

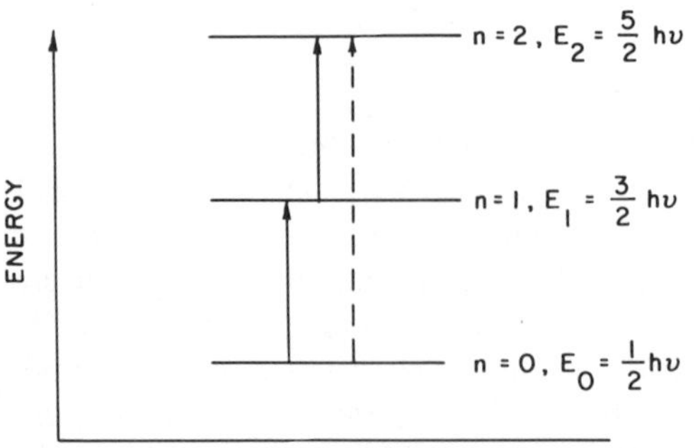

Figure 2.3. The energy levels of a diatomic harmonic oscillator.

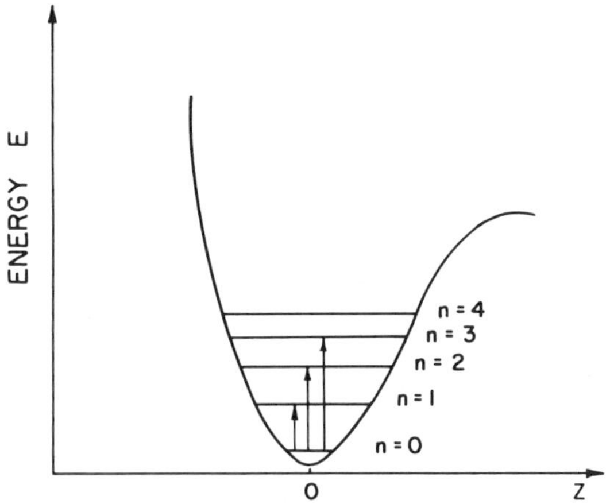

Figure 2.4. Energy levels for an anharmonic oscillator.

while the higher energy levels, including the second, are more accurately given by

$$E = h\nu\left[\left(n+\tfrac{1}{2}\right) - x\left(n+\tfrac{1}{2}\right)^2\right] \tag{2.10}$$

where x is known as the anharmonicity constant and is usually a small fraction of unity. In most cases the anharmonicity is small but sufficient for overtone and combination bands to appear. Their relative intensity is weak because of the low number of transitions of this type that occur.

2.4 Vibrational Selection Rules

We have discussed one basic and inherent condition for infrared absorption and Raman scattering, namely, that the frequency of the absorbed radiation or the frequency shift of the scattered light must correspond to the frequency of a normal mode of vibration and hence to a transition between vibrational energy levels. An additional restraint imposed by the assumption of simple harmonic motion is that only **fundamentals**, or transitions between adjacent energy levels, occur. Since most molecules are in the ground state under ambient conditions, most observed vibrational frequencies arise when the quantum number changes from 0 to 1. A second fundamental requirement is that there must be some mode of interaction between the impinging radiation and the molecule. Even if infrared radiation with the same frequency as a fundamental normal vibration is incident on the sample, it will be absorbed only under certain conditions. The rules determining optical activity, whether Raman or infrared, are known as **selection rules**. An understanding of the selection rules can be reached only through the methods or theories capable of successfully describing the interaction. It is easier to obtain a physical picture of these interactions by first considering the classical interpretation. We will then state the equivalent quantum mechanical description without deriving the appropriate equations.

2.5 Conditions for Infrared Absorption

Infrared absorption is simply described by classical electromagnetic theory; an oscillating dipole is an emitter or absorber of radiation. Consequently, the periodic variation of the dipole moment of a vibrating molecule results in the absorption or emission of radiation of the same frequency as that of the oscillation of the dipole moment. The intensity of the absorption or emission is proportional to the square of the change in the dipole moment. If we, for now, leave the constant of proportionality undefined, then we can write,

$$I_k = C|\mu'|^2$$

where $\mu' = \partial\mu/\partial x$, μ is the dipole moment, and x is the displacement coordinate.

The requirement of a change of the dipole moment with molecular vibration is fundamental, and it is simple to illustrate that certain normal modes do not result in such a change (e.g., the stretching of a homonuclear diatomic molecule). Consider the two in-plane stretching vibrations of CO_2 shown in Figure 2.5. The molecular dipole moment of the symmetric unperturbed molecule is zero. In the totally symmetric vibration (Figure 2.5*a*), the two oxygen atoms move in phase successively from and toward the carbon atom. The symmetry of the molecule is maintained in this vibration and there is no net change in dipole moment. Hence, no interaction with infrared radiation occurs for this motion. Conversely, in the asymmetric stretching vibration (Figure 2.5*b*) the symmetry of the molecule is perturbed and there is a change in the net dipole moment, allowing infrared absorption.

Infrared selection rules are expressed quantum mechanically by the following conditions:

$$\int \psi_i \mu \psi_f \, d\tau \neq 0 \tag{2.11}$$

This integral is called the transition moment integral and its square is a

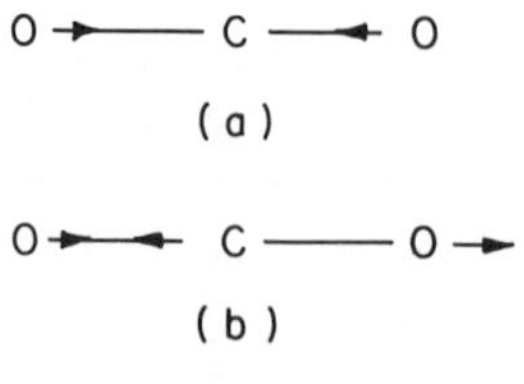

Figure 2.5. The in-plane stretching modes of CO_2 (*a*) symmetric stretching, 1387 cm^{-1}; (*b*) antisymmetric stretching, 2350 cm^{-1}.

measure of the probability of the transition occurring; μ is the dipole moment vector and ψ_i and ψ_f are wave functions describing the initial and final states of the molecule. The vector μ can be expressed as the sum of three components μ_x, μ_y, and μ_z in a cartesian system, so that for a given mode of vibration all that is required is that any one of these components has a nonzero value. Note that the magnitude of the integral need not be calculated in order to determine optical activity; we need to know only whether or not its value is zero.

The selection rules for the symmetric and antisymmetric stretching modes of CO_2 were determined above by inspection. For this simple molecule it is easy to see that the symmetric stretching mode does not result in a net change in the dipole moment of the molecule. It is intuitively clear that this behavior is due to the symmetry of the CO_2 molecule. We will discuss symmetry in detail later, but it is worth noting that the number of optically active normal modes can be predicted from symmetry considerations alone. All molecules can be classified into groups according to the symmetry elements (mirrors, rotation axes, etc.) they possess. Each normal mode belongs to a particular symmetry species. Each transition moment integral has a clearly defined behavior with respect to products of these symmetry species. Consequently, the vanishing or nonvanishing of the integrals is the same for all transitions between states of two particular symmetry classes. Nonvanishing integrals occur when the excited state belongs to the same symmetry species as one component of the dipole moment. Thus the optical activity of each class of vibrations is predetermined by its symmetry species, and the infrared activity of the normal modes can be determined solely from a knowledge of the symmetry of the molecule in question. The symmetry species of the normal mode also determines the dichroic behavior of the infrared absorption bands for oriented molecules.

2.6 Conditions for Raman Scattering

As opposed to infrared absorption, the Raman effect is not concerned with the *intrinsic* or permanent dipole moment of the molecule. For Raman scattering to occur, the electric field of the light must *induce* a dipole moment by a change in the polarizability of the molecule. This change in polarizability can be physically (and simplistically) pictured as a change in the shape of the electron cloud surrounding the molecule. If the electric field of the incident light is imagined to act as a condenser plate, then any charge will distort the electron cloud of the molecule and transmit energy to it. This process is classically described as the induction of a variable dipole moment in the molecule by the electric field of the light, which then oscillates with the frequency of the incident radiation and therefore emits radiation in all directions. The intensity of this radiation can be written

$$I = \frac{16\pi^4 \nu^4}{3c^2} |P|^2 \tag{2.12}$$

where P is the induced dipole moment, given by

$$P=\alpha\cdot E \tag{2.13}$$

where α is the molecular polarizability and E is the electric vector of the incident light. Since

$$E=E_0\cos(2\pi\nu_0 t) \tag{2.14}$$

then

$$P=\alpha\cdot E_0\cos(2\pi\nu_0 t) \tag{2.15}$$

For a normal vibrational mode of frequency ν_k, the polarizability will execute periodic motion of the form

$$\alpha=\alpha_0+\alpha_k\cos(2\pi\nu_k t+\phi_k) \tag{2.16}$$

where ϕ_k is a phase factor, α_0 a constant, and α_k is the maximum change in α. Substituting in equation (2.13), we obtain the induced dipole moment

$$P=\left[\alpha_0+\alpha_k\cos(2\pi\nu_k t+\phi_k)\right]\left[E_0\cos(2\pi\nu_0 t)\right] \tag{2.17}$$

that is,

$$P=\alpha_0 E_0\cos(2\pi\nu_0 t)+\tfrac{1}{2}\alpha_k E_0\{\cos[2\pi(\nu_0+\nu_k)t+\phi_k]+\cos[2\pi(\nu_0-\nu_k)t-\phi_k]\} \tag{2.18}$$

Consequently, when the molecule vibrates with a normal frequency of ν_k the induced dipole oscillates not only with frequency ν_0 (Rayleigh scattering) but with frequencies ν_0+v_k and ν_0-v_k, the anti-Stokes and Stokes lines of the Raman spectrum.

As with the case of infrared absorption, the classical description permits a broad understanding of the nature of the interaction but fails to account for many details of the Raman effect. For example, classical Raman theory predicts that the Stokes and anti-Stokes lines will be of equal intensity [substitute equation (2.18) into (2.12)], which, as the Raman spectrum of CCl_4 shown in Figure 2.6 demonstrates, is not the case. However, if we consider the quantum mechanical description, according to which a molecule with discrete energy levels interacts with photons of light, then Rayleigh scattering corresponds to an elastic collision, whereas Raman scattering is an inelastic process. Inelastic collisions can result in the photon losing a quantum of vibrational energy (Stokes lines) or gaining such a quantum (anti-Stokes lines). But in a collection of molecules at room temperature, the majority are in the ground vibrational state and only a small fraction exist in higher vibrational energy

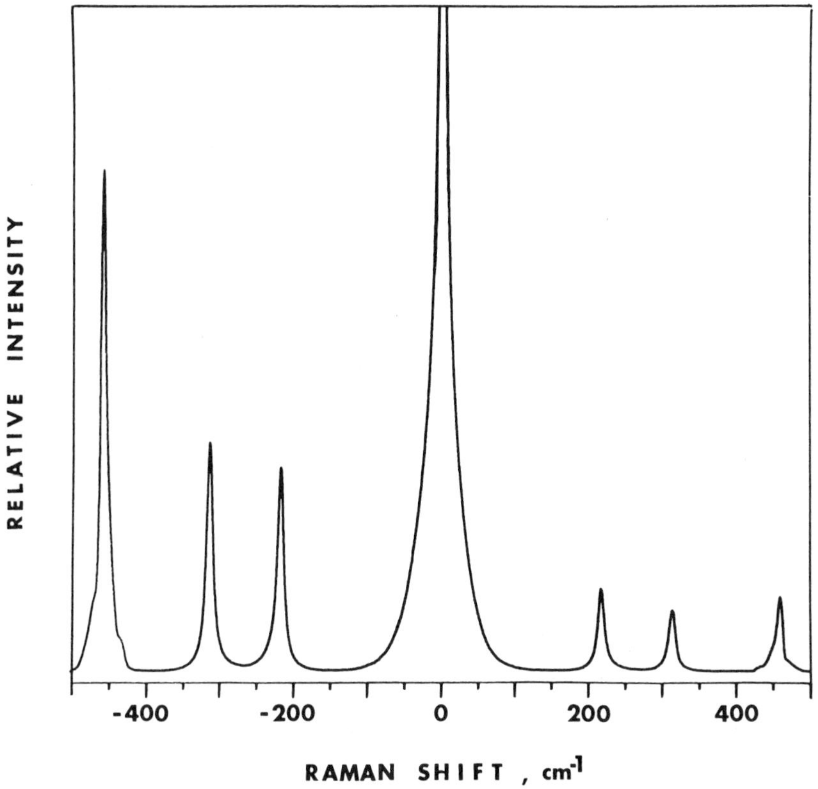

Figure 2.6. The Raman spectrum of carbon tetrachloride showing both the Stokes and anti-Stokes portions of the spectrum.

levels. Consequently, the number of scattering processes in which the molecules acquire vibrational energy from the photons greatly outnumbers the processes in which the molecules give up energy, which explains the greater intensity of the Stokes lines.

The quantum theory of Raman scattering is complex. The quantum field treatment is much more general and expressions for Raman intensities have been derived by using a second-order perturbation theory. A review of the theories of Raman intensities has been given by Tang and Albrecht (5). Under conditions that are usually fulfilled by most molecules, the quantum mechanical analogue of the polarizability can be used:

$$\alpha_{nm} = \int \psi_n \alpha \psi_m \, d\tau \tag{2.19}$$

This treatment was developed by Placsek and in its developed form simplifies spectral analysis by allowing the experimental determination of the symmetry species of Raman-active modes. The selection rules arising from the harmonic approximation (only transitions between adjacent energy levels are allowed) are again imposed, in addition to those determined by symmetry.

Since both the induced dipole moment P and the electric field of the light E are vector quantities, α_{ik} is a tensor defined by an array of nine components as

$$P_x = \alpha_{xx}E_x + \alpha_{xy}E_y + \alpha_{xz}E_z$$

$$P_y = \alpha_{yx}E_x + \alpha_{yy}E_y + \alpha_{yz}E_z$$

$$P_z = \alpha_{zx}E_x + \alpha_{zy}E_y + \alpha_{zz}E_z \tag{2.20}$$

The tensor is symmetric; that is,

$$\alpha_{xy} = \alpha_{yx}, \qquad \alpha_{xz} = \alpha_{zx}, \qquad \alpha_{yz} = \alpha_{zy}$$

so that the requirement or selection rule for Raman scattering is that at least one of the six integrals

$$\int \psi_n \alpha_{ik} \psi_m \, d\tau$$

be totally symmetric. In an analogous fashion to the infrared selection rules, a fundamental is allowed in Raman scattering only if it belongs to the same symmetry species as a component of the polarizability tensor.

2.7 Elementary Treatment of the Vibrations of Small Molecules

The equations of motion for a molecule are derived from expressions of the kinetic energy T and potential energy V. T is given by the familiar equation

$$T = \tfrac{1}{2}mv^2 \tag{2.21}$$

where m is the mass and v the velocity. Consequently, for a molecule consisting of N atoms this equation becomes

$$2T = \sum_{\alpha=1}^{N} m_\alpha \left[\left(\frac{d\Delta x_\alpha}{dt} \right)^2 + \left(\frac{d\Delta y_\alpha}{dt} \right)^2 + \left(\frac{d\Delta z_\alpha}{dt} \right)^2 \right] \tag{2.22}$$

where Δx_α, Δy_α, and Δz_α are the respective cartesian displacements of atom α from its equilibrium position. Since we will always be working in terms of displacements when determining molecular vibrations, Δx_1, Δy_1, Δz_1, $\Delta x_2 \cdots \Delta z_n$ will be replaced by a general coordinate $x_i(x_1, x_2, \ldots, x_{3N})$, where the Δ has been dropped for convenience. Hence in this new notation equation

(2.22) becomes

$$2T = \sum_{i=1}^{3N} m_i(\dot{x}_i)^2 \tag{2.23}$$

where $\dot{x}_i$ is dx_i/dt.

The notation of the equations of motion is simplified by using **mass-adjusted cartesian displacement coordinates** $q_1, \ldots, q_n$, defined as

$$q_i = (m_i)^{1/2} x_i \tag{2.24}$$

so that

$$2T = \sum_{i=1}^{3N} \dot{q}_i^2 \tag{2.25}$$

All that is known about the potential energy V is that it must be some function of the displacement coordinates. It is therefore assumed that displacements are small, allowing a Taylor series expansion of the potential energy in displacement coordinates as

$$2V = 2V_0 + 2\sum_{i=1}^{3N} \left(\frac{\partial V}{\partial q_i}\right)_0 q_i + \sum_{ij=1}^{3N} \left(\frac{\partial^2 V}{\partial q_i \partial q_j}\right)_0 q_i q_j + \text{higher terms} \tag{2.26}$$

Putting

$$f_i = \left(\frac{\partial V}{\partial q_i}\right)_0, \qquad f_{ij} = \left(\frac{\partial^2 V}{\partial q_i \partial q_j}\right)_0$$

we then have

$$2V = 2V_0 + 2\sum_{i=1}^{3N} f_i q_i + \sum_{ij=1}^{3N} f_{ij} q_i q_j + \text{higher terms} \tag{2.27}$$

Since we are concerned only with the change in potential energy with displacment, the potential energy at equilibrium V_0 is arbitrary and can be put equal to zero. Furthermore, since the equilibrium configuration is by definition at a potential energy minimum, it follows that

$$f_i = \left(\frac{\partial V}{\partial q_i}\right)_0 = 0$$

Finally, for small displacements the higher-order terms are neglected, leaving

$$2V=\sum_{ij=1}^{3N} f_{ij}q_iq_j \tag{2.28}$$

The f_{ij} coefficients are called force constants, since they represent the forces acting to restore a displaced atom to its equilibrium for displacements q_i and q_j. In this equation the force constants are mass adjusted. If F_{ij} are the nonadjusted force constants

$$f_{ij}=\frac{F_{ij}}{\sqrt{m_im_j}}$$

Truncating the potential energy to this quadratic term corresponds to the harmonic approximations that the displacements are small and forces between atoms are elastic. The equations of motion for a vibrating molecule can now be obtained from the expressions for T and V. It is convenient to write Newton's second law of motion in the Lagrange form, since this form is independent of the coordinate system used:

$$\frac{d}{dt}\left(\frac{\partial T}{\partial \dot{q}_j}\right)+\left(\frac{\partial V}{\partial q_j}\right)=0 \tag{2.29}$$

Substituting equations (2.5) and (2.8) into (2.9) yields

$$\ddot{q}_j+\sum_{i=1}^{3N} f_{ij}q_i=0, \qquad j=1,2,\ldots,3N$$

As in the diatomic harmonic oscillator discussed earlier, we will assume that this set of $3N$ simultaneous equations has periodic solutions of the form

$$q_i=A_i\cos(\lambda^{1/2}t+\varepsilon) \tag{2.30}$$

in which λ is equal to $4\pi^2c^2\nu^2$ where ν is the frequency of the vibration in cm^{-1} and c is the velocity of light; ε is a phase factor and A_i is the amplitude of vibration of atom i. Using equations (2.29) and (2.30), we obtain

$$\sum_{i=1}^{3N}(f_{ij}-\delta_{ij}\lambda)A_i=0, \qquad j=1,2,\ldots,3N \tag{2.31}$$

where δ_{ij} is the Kronecker delta; $\delta_{ij}=1$ when $i=j$ and $\delta_{ij}=0$ otherwise. This equation has the trivial solution $A_i=0$. For $A_i\neq 0$, equation (2.31) holds only if the coefficients of A_i are zero; therefore

$$|f_{ij} - \delta_{ij}\lambda| = 0 \tag{2.32}$$

This is the **secular equation**, written in complete notation as

$$\begin{vmatrix} f_{11}-\lambda & f_{12} & f_{13} & \cdots & f_{1,3N} \\ f_{21} & f_{22}-\lambda & f_{23} & \cdots & f_{2,3N} \\ f_{31} & f_{32} & f_{33}-\lambda & \cdots & f_{3,3N} \\ \vdots & & & & \\ f_{3N,1} & f_{3N,2} & f_{3N,3} & \cdots & f_{3N,3N}-\lambda \end{vmatrix} = 0 \tag{2.33}$$

The values of λ obtained as solutions of the secular equation are known as **eigenvalues**. There are $3N$ values of λ, but six of these are zero. The zero roots correspond to the three translations and three rotations of the molecule as a whole.

By substituting calculated values of λ back into equation (2.31) we obtain the eigenvectors, which give the amplitudes of vibration for each coordinate. However, it can be shown from the theory of linear systems of equations that it is not possible to determine all $3N$ values of A_i, but only their ratios. Nevertheless, the relative amplitudes, which give us a "picture" of the normal modes, can be determined. This will be illustrated in more detail in the following section.

2.8 Vibrational Frequencies and Displacements for a One-Dimensional Carbon Dioxide Molecule

As an example we will consider the one-dimensional vibrations of the carbon dioxide molecule. This model system is shown schematically in Figure 2.7; x_1, x_2, and x_3 are the cartesian displacement coordinates for the one dimension. If it is assumed that the forces f_{ij} between the atoms behave like Hookean springs, the potential energy V is given by

$$2V = \sum_{ij=1}^{3} f_{ij} x_i x_j$$

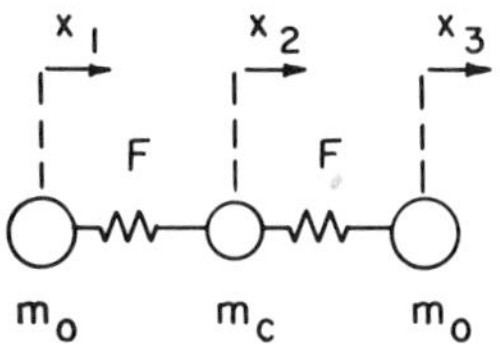

Figure 2.7. One-dimensional displacements of carbon dioxide molecule.

where x is a displacement from the equilibrium position. Consequently, for CO_2 we obtain

$$2V = F(x_2 - x_1)^2 + F(x_3 - x_2)^2 + k(x_2 - x_1)(x_3 - x_2)$$
$$= F(x_1^2 + 2x_2^2 + x_3^2 - 2x_1x_2 - 2x_2x_3) + k(x_2x_3 - x_1x_3 - x_2^2 + x_1x_2)$$

where $f_{12} = f_{23} = F$ and $f_{13} = k$. The kinetic energy can be written in the form

$$T = \tfrac{1}{2}mv^2$$

so that

$$2T = m_O\dot{x}_1^2 + m_C\dot{x}_2^2 + m_O\dot{x}_3^2$$

Converting from cartesian displacement coordinates to mass-adjusted cartesian displacements, we have

$$2T = \dot{q}_1^2 + \dot{q}_2^2 + \dot{q}_3^2$$

$$2V = \frac{F}{m_O}q_1^2 + \frac{(2F-k)}{m_C}q_2^2 + \frac{F}{m_O}q_3^2$$
$$- \frac{(2F-k)}{\sqrt{m_O m_C}}q_1q_2 - \frac{(2F-k)}{\sqrt{m_O m_C}}q_2q_3 - \frac{k}{m_O}q_1q_3$$

Using the Lagrange equation of motion

$$\frac{d}{dt}\left(\frac{\partial T}{\partial \dot{q}_k}\right) + \frac{\partial V}{\partial q_k} = 0$$

we can generate the following three equations:

$$\frac{d}{dt}\left(\frac{\partial T}{\partial \dot{q}_1}\right)_{2,3} + \left(\frac{\partial V}{\partial q_1}\right)_{2,3} = 0$$

$$\frac{d}{dt}\left(\frac{\partial T}{\partial \dot{q}_2}\right)_{1,3} + \left(\frac{\partial V}{\partial q_2}\right)_{1,3} = 0$$

$$\frac{d}{dt}\left(\frac{\partial T}{\partial \dot{q}_3}\right)_{1,2} + \left(\frac{\partial V}{\partial q_3}\right)_{1,2} = 0$$

Differentiating the kinetic energy equation yields

$$\left(\frac{\partial T}{\partial \dot{q}_1}\right)_{2,3} = \frac{(2)\dot{q}_1}{(2)} = \dot{q}_1$$

Similarly,

$$\left(\frac{\partial T}{\partial \dot{q}_2}\right)_{1,3} = \dot{q}_2, \qquad \left(\frac{\partial T}{\partial \dot{q}_3}\right)_{1,2} = \dot{q}_3$$

It then follows that

$$\frac{d}{dt}\left(\frac{\partial T}{\partial \dot{q}_1}\right) = \ddot{q}_1 \qquad \text{etc.}$$

Turning our attention to the potential energy V, we find that

$$\left(\frac{\partial V}{\partial q_1}\right)_{2,3} = \frac{F}{m_{\mathrm{O}}}q_1 - \frac{(2F-k)}{2\sqrt{m_{\mathrm{O}}m_{\mathrm{C}}}}q_2 - \frac{k}{2m_{\mathrm{O}}}q_3$$

$$\left(\frac{\partial V}{\partial q_2}\right)_{1,3} = -\frac{(2F-k)}{2\sqrt{m_{\mathrm{O}}m_{\mathrm{C}}}}q_1 + \frac{(2F-k)}{m_{\mathrm{C}}}q_2 - \frac{(2F-k)}{2\sqrt{m_{\mathrm{O}}m_{\mathrm{C}}}}q_3$$

$$\left(\frac{\partial V}{\partial q_3}\right)_{1,2} = -\frac{k}{2m_{\mathrm{O}}}q_1 - \frac{(2F-k)}{2\sqrt{m_{\mathrm{O}}m_{\mathrm{C}}}}q_2 + \frac{F}{m_{\mathrm{O}}}q_3$$

Consequently, the equations of motion are

$$\ddot{q}_1 + \frac{F}{m_{\mathrm{O}}}q_1 - \frac{(2F-k)}{2\sqrt{m_{\mathrm{O}}m_{\mathrm{C}}}}q_2 - \frac{k}{2m_{\mathrm{O}}}q_3 = 0$$

$$\ddot{q}_2 - \frac{(2F-k)}{2\sqrt{m_{\mathrm{O}}m_{\mathrm{C}}}}q_1 + \frac{(2F-k)}{m_{\mathrm{C}}}q_2 - \frac{(2F-k)}{2\sqrt{m_{\mathrm{O}}m_{\mathrm{C}}}}q_3 = 0$$

$$\ddot{q}_3 - \frac{k}{2m_{\mathrm{O}}}q_1 - \frac{(2F-k)}{2\sqrt{m_{\mathrm{O}}m_{\mathrm{C}}}}q_2 + \frac{F}{m_{\mathrm{O}}}q_3 = 0$$

One possible solution of these equations is

$$q_i = A_i\cos(\lambda^{1/2}t + \varepsilon)$$

Substituting yields

$$\left(\frac{F}{m_{\mathrm{O}}} - \lambda\right)A_1 - \frac{(2F-k)}{2\sqrt{m_{\mathrm{O}}m_{\mathrm{C}}}}A_2 - \frac{k}{2m_{\mathrm{O}}}A_3 = 0$$

$$-\frac{(2F-k)}{2\sqrt{m_{\mathrm{O}}m_{\mathrm{C}}}}A_1 + \left[\frac{(2F-k)}{m_{\mathrm{C}}} - \lambda\right]A_2 - \frac{(2F-k)}{2\sqrt{m_{\mathrm{O}}m_{\mathrm{C}}}}A_3 = 0$$

$$-\frac{k}{2m_{\mathrm{O}}}A_1 - \frac{(2F-k)}{2\sqrt{m_{\mathrm{O}}m_{\mathrm{C}}}}A_2 + \left(\frac{F}{m_{\mathrm{O}}} - \lambda\right)A_3 = 0$$

These equations have nontrivial solutions when the secular equation, written next as a determinant, is zero:

$$\begin{vmatrix} \frac{F}{m_O}-\lambda & -\frac{(2F-k)}{2\sqrt{m_O m_C}} & -\frac{k}{2m_O} \\ -\frac{(2F-k)}{2\sqrt{m_O m_C}} & \frac{(2F-k)}{m_C}-\lambda & -\frac{(2F-k)}{2\sqrt{m_O m_C}} \\ -\frac{k}{2m_O} & -\frac{(2F-k)}{2\sqrt{m_O m_C}} & \frac{F}{m_O}-\lambda \end{vmatrix} = 0 \tag{2.34}$$

The secular determinant gives a cubic equation in λ. The roots or eigenvalues are

$$\lambda_1 = \frac{1}{2m_O}(2F+k), \qquad \lambda_2 = \frac{1}{2}(2F-k)\left(\frac{1}{m_O}+\frac{2}{m_C}\right),$$

$$\lambda_3 = 0 \tag{2.35}$$

These eigenvalues correspond to the frequencies of the normal modes of vibration of one-dimensional CO_2. Now, if we assume that $F=15.98$ mdyn/Å and $k=2.45$ mdyn/Å, the frequencies in wavenumbers (cm^{-1}) can be calculated from λ by substituting into

$$\begin{aligned} \nu_k &= \frac{1}{2\pi c}\sqrt{\lambda_k} \\ &= 1302.83\sqrt{\lambda_k} \end{aligned} \tag{2.36}$$

which yields values of 1351 and 2396 cm^{-1}. By separate substitution of these values into the secular equations, the relative values of the amplitudes or eigenvectors A_{ij} corresponding to each frequency can be derived. For λ_1

$$\left[\frac{F}{m_O}-\frac{(2F+k)}{2m_O}\right]A_{11} - \frac{(2F-k)}{2\sqrt{m_O m_C}}A_{12} - \frac{k}{2m_O}A_{13} = 0 \tag{2.37}$$

$$-\frac{(2F-k)}{2\sqrt{m_O m_C}}A_{11} + \left[\frac{(2F-k)}{m_C}-\frac{(2F+k)}{2m_O}\right]A_{12} - \frac{(2F-k)}{2\sqrt{m_O m_C}}A_{13} = 0 \tag{2.38}$$

$$-\frac{k}{2m_O}A_{11} - \frac{(2F+k)}{2\sqrt{m_O m_C}}A_{12} + \left[\frac{F}{m_O}-\frac{(2F+k)}{2m_O}\right]A_{13} = 0 \tag{2.39}$$

where A_{ij} refers to the ith eigenvalue and the jth atom. Solution for A_{12} yields

$$A_{12} = 0$$

Substituting in equation (2.38), we get

$$A_{11} = -A_{13}$$

Hence, if we arbitrarily set $A_{11} = -1$, we have determined the form of the normal vibration to be

$$\begin{array}{ccc} A_{11} & A_{12} & A_{13} \\ \leftarrow\bullet & \cdot & \bullet\rightarrow \end{array}$$

which is the symmetric stretching mode. Similarly, for λ_2 it can be shown that

$$A_{21} = A_{23} = -\frac{1}{2}\left(\frac{m_C}{m_O}\right)^{1/2} A_{22}$$

$$\begin{array}{ccc} A_{21} & A_{22} & A_{23} \\ \bullet\rightarrow & \leftarrow\bullet & \bullet\rightarrow \end{array}$$

which is the asymmetric stretching mode. For $\lambda_3 = 0$ we obtain $A_{31} = A_{32} = A_{33}$, which is a translation of the molecule as a whole.

As noted above, a unique solution for the amplitudes cannot be found. However, an important solution is obtained by introducing a normalized ratio, defined as

$$l_{ij} = \frac{A_{ij}}{|\Sigma_i (A_{ij})^2|^{1/2}} \tag{2.40}$$

These amplitudes are considered to be normalized to unity in the sense that

$$\sum_i l_{ij}^2 = 1 \tag{2.41}$$

We can now construct an eigenvector matrix **L** that consists of the normalized elements l_{ij}. The first column of this matrix, corresponding to the first eigenvalue, will be given by the three quantities

$$l_{11}=\frac{A_{11}}{\left(A_{11}^2+A_{12}^2+A_{13}^2\right)^{1/2}}$$

$$l_{12}=\frac{A_{12}}{\left(A_{11}^2+A_{12}^2+A_{13}^2\right)^{1/2}}$$

$$l_{13}=\frac{A_{13}}{\left(A_{11}^2+A_{12}^2+A_{13}^2\right)^{1/2}} \tag{2.42}$$

Since the ratio of amplitude for this mode is given by

$$A_{11}:A_{12}:A_{13}=1:0:-1 \tag{2.43}$$

we obtain

$$l_{11}=\frac{1}{(1+0+1)}\frac{1}{(1+0+1)^{1/2}}=\frac{1}{2^{1/2}}$$

$$l_{12}=\frac{0}{(1+0+1)}\frac{0}{(1+0+1)^{1/2}}=0 \tag{2.44}$$

$$l_{13}=\frac{-1}{(1+0+1)}\frac{-1}{(1+0+1)^{1/2}}=-\left(\frac{1}{2}\right)^{1/2}$$

In a similar fashion the remaining columns of the $\mathbf{L}_q$ matrix can be constructed by using the appropriate l_{ij} elements; they are

$$l_{21}=l_{23}=\left[\frac{m_{\mathrm{C}}}{2(m_{\mathrm{C}}+2m_{\mathrm{O}})}\right]^{1/2}$$

$$l_{22}=-\left(\frac{2m_{\mathrm{O}}}{m_{\mathrm{C}}+2m_{\mathrm{O}}}\right)^{1/2} \tag{2.45}$$

and for the null mode

$$l_{31}=l_{33}=\left(\frac{m_{\mathrm{O}}}{m_{\mathrm{C}}+2m_{\mathrm{O}}}\right)^{1/2}, \qquad l_{32}=\left(\frac{m_{\mathrm{C}}}{m_{\mathrm{C}}+2m_{\mathrm{O}}}\right)^{1/2} \tag{2.46}$$

We can see that these l factors are a set of coefficients describing how the mass-weighted cartesian displacement coordinates take part in the normal mode of vibration. As such they can be used to generate the normal coordinates, Q_i, as follows:

$$q_1 = l_{11}Q_1 + l_{12}Q_2 + l_{13}Q_3$$

$$q_2 = l_{21}Q_1 + l_{22}Q_2 + l_{23}Q_3$$

$$q_3 = l_{31}Q_1 + l_{32}Q_2 + l_{33}Q_3 \tag{2.47}$$

The desirability of using normal coordinates rests on the resulting simple and useful form of the potential and kinetic energies. For the kinetic energy, for example,

$$2T = \dot{q}_1^2 + \dot{q}_2^2 + \dot{q}_3^2$$

$$= (l_{11}\dot{Q}_1 + l_{12}\dot{Q}_2 + l_{13}\dot{Q}_3)^2 + (l_{21}\dot{Q}_1 + l_{22}\dot{Q}_2 + l_{23}\dot{Q}_3)^2$$

$$+ (l_{31}\dot{Q}_1 + l_{32}\dot{Q}_2 + l_{33}\dot{Q}_3)^2$$

Substitution of the l values leads to

$$2T = \dot{Q}_1^2 + \dot{Q}_2^2 + \dot{Q}_3^2 \tag{2.48}$$

Thus the simple representation given by mass-adjusted coordinates is retained in normal coordinate notation. This is a general property of the normal coordinate basis and applies to all molecules.

$$2T = \Sigma \dot{Q}_1^2 \tag{2.49}$$

In general, for the potential energy, we similarly go from q coordinates to normal coordinates and we find that

$$2V = \frac{(2F+k)}{2m_O} Q_1^2 + \frac{1}{2}(2F-k)\left(\frac{1}{m_O} + \frac{2}{m_C}\right) Q_2^2$$

which according to equation (2.35) reduces to

$$2V = \lambda_1 Q_1^2 + \lambda_2 Q_2^2 + \lambda_3 Q_3^2 \tag{2.50}$$

and for the general molecular system

$$2V = \sum_i \lambda_i Q_i^2 \tag{2.51}$$

in the normal coordinate basis. We see that the ith normal coordinate has the frequency associated with λ_i and is not mixed with any of the other modes.

We can solve the equation (2.47) for the normal coordinates Q in terms of the mass-adjusted coordinates q as follows:

$$Q_1 = \frac{1}{\sqrt{2}} q_1 - \frac{1}{\sqrt{2}} q_3$$

$$Q_2 = \left[\frac{m_C}{2(m_C + 2m_0)}\right]^{1/2} q_1 - \left(\frac{2m_0}{m_C + 2m_0}\right)^{1/2} q^2 + \left[\frac{m_C}{2(m_C + 2m_0)}\right]^{1/2}$$

$$Q_3 = \left(\frac{m_0}{m_C + 2m_0}\right)^{1/2} q_1 + \left(\frac{m_C}{m_C + 2m_0}\right)^{1/2} q_2 + \left(\frac{m_0}{m_C + 2m_0}\right) q_3$$

These coefficients are useful for calculating a number of parameters. As an example we will consider a simplified treatment of infrared and Raman intensities. We will examine more complex expressions for these terms in Chapter 9.

2.9 Infrared Intensities of One-Dimensional Carbon Dioxide

We noted in Section 2.5 that infrared absorption depends on the change in dipole moment with molecular vibration. Since the displacements of atoms in a normal mode can be directly expressed in terms of the corresponding normal coordinate, it is convenient to express infrared band intensities for the CO_2 molecule as

$$I_k = C\left|\frac{\partial \mu_x}{\partial Q_k}\right|^2$$

and, changing to mass-adjusted cartesian coordinates,

$$I_k = C\left|\sum_i \frac{\partial \mu_x}{\partial q_i}\frac{\partial q_i}{\partial Q_k}\right|^2$$

$$= C\left|\sum_i \frac{\partial \mu}{\partial q_i} l_{ik}\right|^2$$

Because of the symmetry of CO_2, the total dipole moment is zero. However, the dipole moment of each C—O bond has a finite value. We can express this condition as

$$\mu_x = \mu_{12}^x + \mu_{23}^x = 0$$

so that

$$-\mu_{12}^{x}=+\mu_{23}^{x}$$

A positive change in the coordinate of oxygen atom 1, x_1, will result in a compression of the C—O bond and a negative change in the dipole moment (since the dipole moment of the C—O bond is the product of the charge on each atom multiplied by the distance or bond length between them). Consequently, we have a negative bond dipole moment, $-\mu_{12}^{x}$ and a negative change in dipole moment, so that

$$-\frac{\partial(-\mu_{12}^{x})}{\partial q_1}=+\mu'_x$$

and there is an increase in the total dipole, therefore a positive μ'. When the position of the carbon atom, or displacement x_2, changes, the effective change in dipole moment is zero, since an increase in one bond results in an equivalent loss for the other; therefore

$$\frac{\partial\mu_x}{\partial q_2}=0$$

Similarly, for a displacement x_3, the dipole μ_{23}^{x} is positive and the change in dipole is positive (since there is an increase in the length of the C—O bond), so

$$\frac{\partial\mu_{23}^{x}}{\partial q_3}=\mu'_x$$

We can now calculate the relative infrared intensities. For λ_1 (1356 cm^{-1}) $l_{11}=-l_{13}=1/\sqrt{2}$ and $l_{12}=0$, so

$$I_{\lambda_1}=C\left|l_{11}\frac{\partial\mu_x}{\partial q_i}+l_{12}\frac{\partial\mu_x}{\partial q_2}+l_{13}\frac{\partial\mu_z}{\partial q_3}\right|^2$$

Substituting, we obtain

$$I_{\lambda_1}=C\left|\frac{1}{\sqrt{2}}\mu'_x+(0)(0)+\left(-\frac{1}{\sqrt{2}}\right)\mu'_x\right|^2$$

$$=0$$

Consequently, the infrared intensity of this symmetric stretching mode is zero; that is, it is infrared inactive.

For λ_2 (2396 cm^{-1})

$$l_{21}=\left[\frac{m_c}{2(m_c+2m_0)}\right]^{1/2}=l_{23}$$

$$l_{22}=-\left(\frac{2m_0}{m_c+2m_0}\right)^{1/2}$$

The infrared intensity is given by

$$I_{\lambda_2}=C\left|l_{21}\frac{\partial\mu_x}{\partial q_1}+l_{22}\frac{\partial\mu_x}{\partial q_2}+l_{23}\frac{\partial\mu_x}{\partial q_3}\right|^2$$

$$=C\left|\left[\frac{m_c}{2(m_c+2m_0)}\right]^{1/2}\mu'_x-\left(\frac{2m_0}{m_c+2m_0}\right)(0)+\left[\frac{m_c}{2(m_c+2m_0)}\right]^{1/2}\mu'_x\right|^2$$

$$=C\left[\frac{m_c}{2(m_c+2m_0)}\right]|\mu'_x|^2$$

So the antisymmetric stretching mode has an intensity proportional to the square of the change in dipole moment of the C—O bond with extension. As we will see in Chapter 7, this μ' is termed an electro-optical parameter and is assumed to be characteristic of each type of chemical bond.

2.10 Intensities of Raman Lines for One-dimensional Carbon Dioxide

In order to determine the relative Raman intensities, we can rewrite equation (2.12) as

$$I_k=K\left|\frac{\partial\alpha}{\partial Q_k}\right|^2$$

where I_k is the intensity of Raman scattered radiation of the kth normal mode and K is a constant of proportionality. This equation can be written

$$I_k=K\left|\sum_i\frac{\partial\alpha}{\partial x_i}\frac{\partial x_i}{\partial Q_k}\right|$$

$$=K\sum_i l_{k_i}\frac{\partial\alpha}{\partial x_i}$$

The polarizability is a function of each atomic displacement:

$$\alpha=\alpha_1+\alpha_2+\alpha_3$$

so

$$\frac{\partial\alpha}{\partial x_1}=\frac{\partial\alpha_1}{\partial x_1},\qquad \frac{\partial\alpha}{\partial x_2}=\frac{\partial\alpha_2}{\partial x_2},\qquad \text{and}\qquad \frac{\partial\alpha}{\partial x_3}=\frac{\partial\alpha_3}{\partial x_3}$$

The effect of displacements x_1 and x_3 on the polarization must be of opposite sign and equal, so

$$\frac{\partial\alpha_1}{\partial x_1}=-\frac{\partial\alpha_3}{\partial x_3}=\alpha'$$

and

$$\frac{\partial\alpha_2}{\partial x_2}=0$$

so

$$I_k=K\left|l_{k_1}\frac{\partial\alpha}{\partial x_1}+l_{k_2}\frac{\partial\alpha}{\partial x_2}+l_{k_3}\frac{\partial\alpha}{\partial x_3}\right|^2$$

For λ_1

$$l_{11}=-l_{13}=\frac{1}{\sqrt{2}}l_{12}=0$$

$$I_{\lambda_1}=K\left|\frac{1}{\sqrt{2}}\alpha'-\left(\frac{1}{\sqrt{2}}\right)(-\alpha')\right|^2$$

$$=\frac{K}{2}|\alpha'|^2$$

so the intensity is proportional to the change in polarizability with extension of the C—O bond of CO_2. The α' is again termed an electro-optical parameter characteristic of the chemical bond. For x_2, $I_{\lambda_2}=0$ and the λ_2 mode is Raman inactive. Observe that λ_1 is Raman active but infrared inactive, whereas λ_2 is infrared active but Raman inactive. This result is an example of the mutual exclusion principle that applies to molecules with a center of symmetry. (See Chapter 4.)

2.11 Elementary Treatment Using Matrices

The vibrational problem is much easier to describe by means of matrix algebra. The relevant aspects of matrix theory are discussed in Woodward (2), and a familiarity with the subject is assumed. In order to review the notation, however, we will reexamine the elementary treatment presented above in terms of matrix theory and illustrate the simplifications introduced by the use of normal coordinates.

The displacement coordinates are expressed as a column matrix **X**.

$$\mathbf{X}=\begin{bmatrix} x_1 \\ x_2 \\ \vdots \\ x_{3N} \end{bmatrix} \tag{2.52}$$

The transpose of **X**, $\mathbf{X}^T$, is found by exchanging rows and columns, so that the transpose of the column matrix **X** is a row matrix $\mathbf{X}^T$:

$$\mathbf{X}^T=[x_1 \quad x_2 \cdots x_{3N}]$$

The force constants can be represented as a symmetric square $(3N\times 3N)$ matrix $\mathbf{F}_x$, where x refers to the force constants in cartesian coordinates.

$$\mathbf{F}_x=\begin{bmatrix} F_{11} & F_{12} & \cdots & F_{ij3N} \\ F_{21} & F_{22} & & \\ \vdots & & & \\ & & & \vdots \\ F_{3N,1} & & \cdots & F_{3N,3N} \end{bmatrix} \tag{2.53}$$

The matrix is symmetric since $F_{21}=F_{12}$, $F_{13}=F_{31}$, and so on. The potential energy written in the quadratic form is

$$2V=\sum_{i,j=1}^{3N} F_{ij}x_ix_j=\mathbf{X}^T\mathbf{F}\mathbf{X} \tag{2.54}$$

Similarly, for the kinetic energy we have

$$2T=\sum_{i=1}^{3N} m_i\dot{x}_i^2=\dot{\mathbf{X}}^T\mathbf{M}\dot{\mathbf{X}} \tag{2.55}$$

where

$$\mathbf{M}=\begin{bmatrix} m_1 & & \\ & m_2 & \\ & & m_{3N} \end{bmatrix} \tag{2.56}$$

is a diagonal matrix.

The linear transformation to mass-adjusted coordinates is accomplished by the following matrix equation:

$$\mathbf{q}=\mathbf{M}^{1/2}\mathbf{X}$$

or

$$\mathbf{X}=\mathbf{M}^{-1/2}\mathbf{q} \tag{2.57}$$

where $\mathbf{q}$ is now a column matrix:

$$\mathbf{q}=\begin{bmatrix} q_1 \\ \vdots \\ q_{3N} \end{bmatrix} \tag{2.58}$$

The kinetic and potential energies are then written

$$2T=\dot{\mathbf{q}}^T\dot{\mathbf{q}} \tag{2.59}$$

$$2V=\mathbf{q}^T\mathbf{F}_q\mathbf{q} \tag{2.60}$$

where $\mathbf{F}_q$ is the force constant matrix in mass-adjusted coordinates given by

$$\mathbf{F}_q=(\mathbf{M}^{-1/2})^T\mathbf{F}_x\mathbf{M}^{-1/2} \tag{2.61}$$

The Lagrange equation in matrix form is then

$$\ddot{\mathbf{q}}+\mathbf{F}_q\mathbf{q}=0 \tag{2.62}$$

and the secular determinant is determined in the same way as in the elementary treatment given above. Although for molecules such as CO_2 solutions can be obtained by simple expansion of the determinant, this approach would obviously be difficult to apply to more complex polyatomic molecules. However, the solution of the Lagrange equation would be trivial if $\mathbf{F}_q$ could be obtained in diagonal form. It was shown in Section 2.8 that this form involves the normal coordinates. By solving the equation of motion in normal coordinates and comparing the solution with that obtained in terms of cartesian displacements, it can be shown that the coefficients of the linear transformation from normal coordinates to mass-adjusted cartesian displacement coordinates correspond to the normalized ratios l_{ij} defined by equation (2.40). If $\mathbf{L}_q$ is the matrix representing l_{ij}, then we can write

$$\mathbf{Q}=\mathbf{L}_q^{-1}\mathbf{q} \tag{2.63}$$

and

$$\mathbf{q}=\mathbf{L}_q\mathbf{Q} \tag{2.64}$$

where **q** and **Q** are both column matrices. Using CO_2 as an example, we demonstrated that with the linear transformation between **q** and **Q** defined by the **L** matrix, the form of the kinetic and potential energies is greatly simplified, with

$$2T=\dot{\mathbf{Q}}^T\dot{\mathbf{Q}} \tag{2.65}$$

and

$$2V=\mathbf{Q}^T\mathbf{\Lambda}\mathbf{Q} \tag{2.66}$$

where $\mathbf{\Lambda}$ is a diagonal matrix whose elements are the frequency parameters

$$\lambda_k=4\pi^2c^2\nu_k^2 \tag{2.67}$$

The potential energy expression does not involve any cross products of normal coordinates, unlike the corresponding equation in terms of cartesian displacements coordinates. Unfortunately, this particular simplification of the secular equation requires a prior knowledge of the quantities we are seeking to determine, namely, the normal coordinates. Nevertheless, it illustrates that the vibrational problem can be considerably simplified by an appropriate choice of coordinate systems. A system of internal coordinates and symmetry coordinates is particularly useful and will be explored later.

References

1 E. B. Wilson, Jr., J. C. Decius, and P.C. Cross, "Molecular Vibrations," McGraw-Hill, New York, 1955.
2 L. A. Woodward, "Introduction to the Theory of Molecular Vibrations and Vibrational Spectroscopy," Oxford (Clarendon Press), New York, 1972.
3 D. Steele, "Theory of Vibrational Spectroscopy," Saunders, Philadelphia, 1971.
4 S. Califano, "Vibrational States," Wiley-Interscience, New York, 1976.
5 J. Tang and A. C. Albrecht, in "Raman Spectroscopy," Vol. 2, H. A. Szymanski, Ed., Plenum Press, New York, 1974, p. 33.

3

THE VIBRATIONAL PROBLEM IN INTERNAL COORDINATES

Science is always wrong: it never solves a problem without creating ten more.

Bernard Shaw

3.1 Advantages of Internal Coordinates

In Chapter 2 it was demonstrated that we can represent the kinetic and potential energies in terms of the $3N$ cartesian displacement coordinates as

$$2T=\dot{\mathbf{X}}^T\mathbf{M}\dot{\mathbf{X}}, \qquad 2V=\mathbf{X}^T\mathbf{F}_x\mathbf{X}$$

The cartesian coordinate system is desirable from the point of view of defining the kinetic energy in that **M** is a diagonal matrix of the masses of the nuclei and as a result the kinetic energy of the molecular system is simple to calculate. In addition, the interaction effects of rotation and vibration are given proper treatment using the cartesian coordinate basis. However, the descriptions of the force constants, except for linear molecules like CO_2, are difficult to interpret physically, because the force constants in cartesian coordinates are projections on the laboratory frame of the forces involved with the chemical bonds. The number of possible orientations of the molecule within the frame makes impossible the transfer of the force field from a bond in one molecule to a similar or identical bond in another molecule.

These difficulties can be overcome by using a set of internal coordinates that describe the bond distortions, such as bond stretchings, angle deformations, and torsions. The internal coordinates are the same regardless of the position and orientation of the molecule in space. Furthermore, for a nonlinear molecule $3N-6$ coordinates are required rather than $3N$. The remaining six coordinates are required to describe the position and configuration of the molecule in the laboratory coordinate system. These coordinates do not contribute to the potential energy and result in zero roots in the solution to the vibrational problem. The size of the secular equation in internal coordinates is smaller and hence easier to solve. Nevertheless, the principal advantage of internal coordinates is the representation of the potential energy or force constant matrix in terms of bond stiffness and resistance to bond angle deformations, which make these constants physically comprehensible. Potentially, force constants in internal coordinates associated with a given chemical bond or group are transferable from one molecule to another. The difficulties of internal coordinates arise from the algebra involved in transforming the kinetic energy, easily described in cartesian coordinates, into terms of the internal coordinates. Using an internal coordinate description, one is also neglecting the interactions of the rotational and vibrational energies of the molecule. Fortunately, the transformation of the kinetic energy into internal coordinates can be easily accomplished on the computer by using suitable programs. These programs require the definition of bond distances and angles for the generation of the kinetic energy matrix. Alternatively, one can perform the reverse transformation on the force constant matrix in internal coordinates and still retain the use of cartesian coordinates. This description of the vibrational problem has particular advantages in the vibrational analysis of polymers and in the calculation of infrared and Raman intensities.

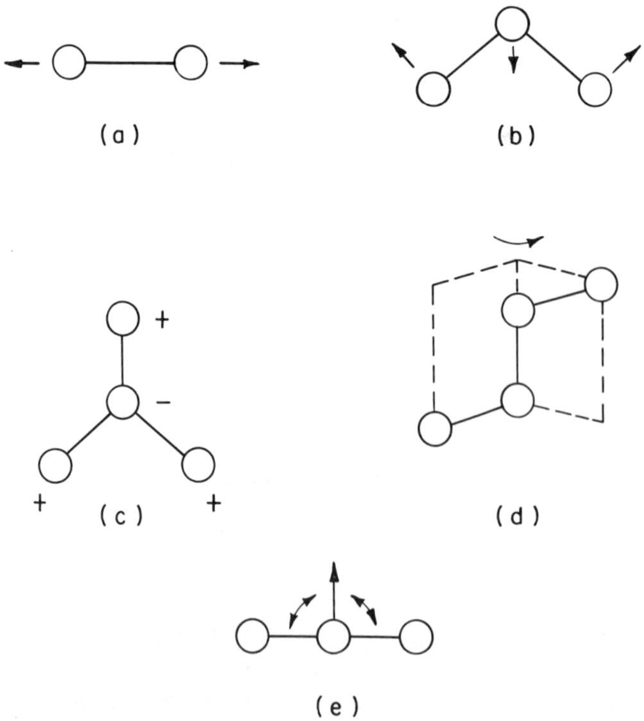

Figure 3.1. Definition of internal coordinates: (*a*) bond stretching; (*b*) valence angle bending; (*c*) out-of-plane bending or angle between a bond and a plane defined by two bonds; (*d*) torsion; (*e*) linear valence angle bending.

The most commonly used internal coordinates are illustrated in Figure 3.1 and consist of the following five types:

1 Bond stretching coordinate (Figure 3.1*a*) represents a change in the length of a chemical bond.
2 In-plane bending coordinate (Figure 3.1*b*) represents a change in the angle between two chemical bonds having one atom in common.
3 Out-of-plane bending coordinate (Figure 3.1*c*) represents a variation in the angle defined by two bonds with one atom in common and a third bond connected to the common atom.
4 Torsion coordinate (Figure 3.1*d*) represents a variation in the dihedral angle between the planes determined by three consecutive bonds connecting four atoms.
5 Linear bend coordinate (Figure 3.1*e*) represents the bending of a linear three-atom bond.

In order to illustrate the use of internal coordinates, we will again consider solutions to the one-dimensional CO_2 molecule. It is illuminating to consider first the nonmatrix solution in order to illustrate the general form of the equations. We will then derive a general form of the secular equation using matrix notation.

3.2 Vibrational Analysis of the One-Dimensional Carbon Dioxide Molecule

The internal coordinates of the simple one-dimensional CO_2 molecule are the extensions of the bond distances r_{12} and r_{23} shown in Figure 3.2. These coordinates can be expressed in terms of the cartesian coordinates defined in Chapter 2 as

$$r_{12}=x_2-x_1 \tag{3.1}$$

$$r_{23}=x_3-x_2 \tag{3.2}$$

The potential energy can now be expressed very simply as

$$2V=f\left(r_{12}^2+r_{23}^2\right)+kr_{12}r_{23} \tag{3.3}$$

where f is a force constant describing a bond stretch and k is an interaction

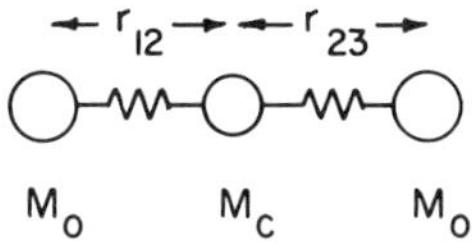

Figure 3.2. Internal coordinate definitions for the one-dimensional CO_2 molecule.

force constant. The kinetic energy expression in cartesian coordinates is

$$2T = m_O\dot{x}_1^2 + m_C\dot{x}_2^2 + m_O\dot{x}_3^2 \tag{3.4}$$

It is important to note that this expression includes the translational kinetic energy, which we must eliminate in order to use the internal coordinate system. We therefore introduce the condition that the linear momentum of the system **as a whole** is zero. That is,

$$m_O\dot{x}_1 + m_C\dot{x}_2 + m_O\dot{x}_3 = 0 \tag{3.5}$$

We can now obtain the vibrational kinetic energy in terms of the internal coordinates r_{12} and r_{23}.

$$\dot{r}_{12} = \dot{x}_2 - \dot{x}_1 \tag{3.6}$$

$$\dot{r}_{23} = \dot{x}_3 - \dot{x}_2 \tag{3.7}$$

We now use equations (3.5), (3.6), and (3.7) to express $\dot{x}_1$, $\dot{x}_2$, and $\dot{x}_3$ in terms of $\dot{r}_{12}$ and $\dot{r}_{23}$ as

$$\dot{x}_1 = -\left(\frac{m_C + m_O}{m_C + 2m_C}\right)\dot{r}_{12} - \left(\frac{m_O}{m_C + 2m_O}\right)\dot{r}_{23} \tag{3.8}$$

$$\dot{x}_2 = \left(\frac{m_O}{mm_C + 2m_O}\right)\dot{r}_{12} - \left(\frac{m_O}{m_C + 2m_O}\right)\dot{r}_{23} \tag{3.9}$$

$$\dot{x}_3 = \left(\frac{m_O}{m_C + 2m_O}\right)\dot{r}_{12} + \left(\frac{m_C + m_O}{m_C + 2m_O}\right)\dot{r}_{23} \tag{3.10}$$

By using equation (3.5) we have excluded the contribution of the translational kinetic energy. We substitute into equation (3.4) in order to obtain an expression for T, which is now purely the vibrational kinetic energy.

$$2T = \left[\frac{m_O(m_C + m_O)}{m_C + 2m_O}\right]\dot{r}_{12}^2 + \left(\frac{2m_O^2}{m_C + 2m_O}\right)\dot{r}_{12}\cdot\dot{r}_{23} + \left[\frac{m_O(m_C + m_O)}{m_C + 2m_O}\right]\dot{r}_{23}^2 \tag{3.11}$$

It is obvious that the expression for kinetic energy is much more complicated in internal coordinates than cartesian coordinates [cf. equation (3.4)].

We now use the Lagrange equation expressed in terms of internal coordinates:

$$\frac{d}{dt}\left(\frac{\partial T}{\partial \dot{r}_{ij}}\right) + \frac{\partial V}{\partial r_{ij}} = 0 \tag{3.12}$$

By taking derivatives, making the appropriate substitution, and, as in Chapter 2, assuming periodic solutions of the form

$$r_{12}=A_1\cos(\lambda^{1/2}t+\varepsilon) \tag{3.13}$$

$$r_{23}=A_2\cos(\lambda^{1/2}t+\varepsilon) \tag{3.14}$$

the amplitude equations we obtain are

$$\left[f-\frac{m_{\mathrm{O}}(m_{\mathrm{C}}+m_{\mathrm{O}})\lambda}{m_{\mathrm{C}}+2m_{\mathrm{O}}}\right]A_1+\left[k-\left(\frac{m_{\mathrm{O}}^2\lambda}{m_{\mathrm{C}}+2m_{\mathrm{O}}}\right)\right]A_2=0 \tag{3.15}$$

$$\left(k-\frac{m_{\mathrm{O}}^2\lambda}{m_{\mathrm{C}}+2m_{\mathrm{O}}}\right)A_1+\left[f-\frac{m_{\mathrm{O}}(m_{\mathrm{C}}+m_{\mathrm{O}})\lambda}{m_{\mathrm{C}}+2m_{\mathrm{O}}}\right]A_2=0 \tag{3.16}$$

Expanding the secular equation (in determinant form) gives a quadratic equation, compared to the cubic form obtained by using a mass-weighted coordinate system. The roots of this quadratic equation are

$$\lambda_a=\frac{f+k}{m_{\mathrm{O}}} \tag{3.17}$$

$$\lambda_b=\left(\frac{m_{\mathrm{C}}+2m_{\mathrm{O}}}{m_{\mathrm{C}}m_{\mathrm{O}}}\right)(f-k) \tag{3.18}$$

Using equations (3.15) and (3.16) and substituting the eigenvalues, we can also determine the ratio of amplitudes in each of the two vibrational modes as

$$A_{1a}=A_{2a} \tag{3.19}$$

$$A_{1b}=-A_{2b} \tag{3.20}$$

These results are obviously the same as those obtained previously. We see that in internal coordinates the expressions for the potential energy are simple but the kinetic energy terms are more complicated. Fortunately, we can use computers to calculate the kinetic energy from the molecular geometry. We now have the potential energy in a physically meaningful form, which enables us to generate "data banks" of force constants.

3.3 General Matrix Solution

Initially, we will follow the general approach of the preceding section by first expressing the equations of motion and the secular equation in terms of the internal coordinates.

Let the linear transformation between cartesian (x) and internal (R) coordinates be given by

$$R_t = \sum_{k=1}^{3N} B_{tk} x_k \tag{3.21}$$

where $t=1,\ldots,3N-6$; $k=1,\ldots,3N$.

In matrix form this is simply

$$\mathbf{R} = \mathbf{BX} \tag{3.22}$$

We can write a similar relationship between internal coordinates R_t and mass-adjusted coordinates q_k as

$$R_t = \sum_{k=1}^{3N} D_{tk} q_k \tag{3.23}$$

or in matrix form

$$\mathbf{R} = \mathbf{Dq} \tag{3.24}$$

Note that since $\mathbf{q}=\mathbf{M}^{1/2}\mathbf{X}$ by definition, we can relate the $\mathbf{B}$ and $\mathbf{D}$ matrices by the expression $\mathbf{D}=\mathbf{B}(\mathbf{M}^{-1/2})$.

The kinetic energy in terms of mass-weighted cartesian coordinates is given by $2T=\Sigma_i \dot{q}^2$. In matrix form this relation is written

$$2T = \dot{\mathbf{q}}^T\dot{\mathbf{q}} \tag{3.25}$$

We wish eventually to express the kinetic and potential energy in terms of the internal coordinates $\mathbf{R}$. In order to do this we now deviate from the development used in the nonmatrix method by defining a quantity p, which is the momentum conjugate to q, as

$$p_j = \frac{\partial T}{\partial \dot{q}_j} = \dot{q}_j \tag{3.26}$$

so that the kinetic energy can now be written

$$2T = \mathbf{p}^T\mathbf{p} \tag{3.27}$$

We can now proceed to write the kinetic energy T as a function of the velocities in the internal coordinates; applying the rules for partial differentiation, we obtain

$$p_j = \frac{\partial T}{\partial \dot{q}_j} = \sum_{t=1}^{3N-6} \frac{\partial T}{\partial \dot{R}_j}\frac{\partial \dot{R}_t}{\partial \dot{q}_j} \tag{3.28}$$

However, in a fashion analogous to that in which we defined p_j as the momentum conjugate to the mass-adjusted cartesian coordinate q_j, we can define P_t as the momentum conjugate to the internal coordinates R_t:

$$P_t = \frac{\partial T}{\partial \dot{R}_t} \tag{3.29}$$

Although the rationale for the introduction of these momentum conjugates is not immediately apparent, they eventually allow the kinetic energy T to be more easily evaluated.

An expression for $\partial \dot{R}_t / \partial \dot{q}_j$ can be obtained by differentiating equation (3.23) to give

$$\dot{R}_t = \sum_{k=1}^{3N} D_{tk}\dot{q}_k, \qquad t=1,\ldots,3N-6 \tag{3.30}$$

Therefore

$$\frac{\partial \dot{R}_t}{\partial \dot{q}_k} = D_{tk} \tag{3.31}$$

We can now substitute equations (3.29) and (3.31) into (3.28) to obtain

$$p_j = \sum_{t=1}^{3N-6} P_t D_{tj} \tag{3.32}$$

In matrix form this equation is written

$$\mathbf{p}^T = \mathbf{P}^T\mathbf{D} \tag{3.33}$$

It is necessary to write this equation using $\mathbf{p}^T$ and $\mathbf{P}^T$ because of the laws of matrix multiplication. Since $\mathbf{D}$ is a $(3N-6)\times 3N$ matrix, we can multiply it by $\mathbf{P}^T$, which is a $1\times(3N-6)$ matrix, to obtain a $1\times 3N$ matrix that corresponds to $\mathbf{p}^T$. From equation (3.33) we have

$$\mathbf{p} = \mathbf{D}^T\mathbf{P} \tag{3.34}$$

Substituting (3.33) and (3.34) into the equation for kinetic energy (3.27), we obtain

$$2T = \mathbf{p}^T\mathbf{p} = \mathbf{P}^T\mathbf{D}\mathbf{D}^T\mathbf{P} \tag{3.35}$$

We now define the matrix $\mathbf{G}$ as

$$\mathbf{G} = \mathbf{D}\mathbf{D}^T \tag{3.36}$$

From equation (3.24) we can relate **G** to the **B** and **M** matrices as

$$\mathbf{G} = \left[\mathbf{B}(\mathbf{M}^{-1/2})\right]\left[\mathbf{B}(\mathbf{M}^{-1/2})\right]^T$$

$$= \mathbf{B}(\mathbf{M}^{-1/2})(\mathbf{M}^{-1/2})^T\mathbf{B}^T$$

$$\mathbf{G} = \mathbf{B}(\mathbf{M}^{-1})\mathbf{B}^T \tag{3.37}$$

The **B** matrix is readily determined from a given equilibrium geometry by using trigonometric methods (1). Consequently, there is considerable advantage in expressing the kinetic energy T in terms of **G**. As we shall see later, in the final form of the secular equation the momentum conjugate terms are eliminated, allowing a simple determination of T from **G**.

From equations (3.35) and (3.36)

$$2T = \mathbf{P}^T\mathbf{G}\mathbf{P} \tag{3.38}$$

From equation (3.29), by application of Hamilton's equations of motion

$$\dot{R}_t = \frac{\partial T}{\partial P_t}$$

differentiating equation (3.38) gives

$$\dot{\mathbf{R}} = \mathbf{G}\mathbf{P} \tag{3.39}$$

which can be solved for **P** as long as $|\mathbf{G}| \neq 0$. If $|\mathbf{G}| \neq 0$, then $\mathbf{G}^{-1}$ exists:

$$\mathbf{P} = \mathbf{G}^{-1}\dot{\mathbf{R}} \tag{3.40}$$

We can now write the kinetic energy in terms of the velocities of internal coordinates, which is a much more useful expression for our purposes.

$$2T = \mathbf{P}^T\mathbf{G}\mathbf{P}$$

$$= (\mathbf{G}^{-1}\dot{\mathbf{R}})^T\mathbf{G}(\mathbf{G}^{-1}\dot{\mathbf{R}})$$

$$= \dot{\mathbf{R}}^T(\mathbf{G}^{-1})^T\mathbf{G}\cdot\mathbf{G}^{-1}\dot{\mathbf{R}}$$

$$= \dot{\mathbf{R}}^T(\mathbf{G}^{-1})^T\dot{\mathbf{R}} \tag{3.41}$$

However, **G** is a symmetric matrix; therefore $\mathbf{G}^{-1}$ is a symmetric matrix, and from matrix theory $(\mathbf{G}^{-1})^T = \mathbf{G}^{-1}$. Therefore

$$2T = \dot{\mathbf{R}}^T\mathbf{G}^{-1}\dot{\mathbf{R}} \tag{3.42}$$

The expression for potential energy can be written in a straightforward manner if the force constant matrix is expressed in terms of the internal coordinates ($\mathbf{F}_r$):

$$2V = \mathbf{R}^T \mathbf{F}_r \mathbf{R} \tag{3.43}$$

Having obtained the kinetic and potential energies in the required form as functions of the internal coordinates, we can now obtain the secular equation as before by writing the equations of motion in the Lagrange form:

$$\frac{d}{dt}\left(\frac{\partial T}{\partial \dot{R}_t}\right) + \left(\frac{\partial V}{\partial R_t}\right) = 0 \tag{3.44}$$

Thus

$$\mathbf{G}^{-1}\ddot{\mathbf{R}} + \mathbf{F}_r\mathbf{R} = \mathbf{0} \tag{3.45}$$

or

$$\ddot{\mathbf{R}} + \mathbf{G}\mathbf{F}_r\mathbf{R} = \mathbf{0} \tag{3.46}$$

Therefore, assuming periodic solutions of the form

$$\mathbf{R} = \mathbf{L}_r \cos(\lambda^{1/2} t + \varepsilon)$$

we have

$$(\mathbf{G}\mathbf{F}_r)\mathbf{L}_r = \boldsymbol{\lambda}\mathbf{L}_r$$

or

$$(\mathbf{G}\mathbf{F}_r - \boldsymbol{\lambda}\mathbf{E})\mathbf{L}_r = \mathbf{0} \tag{3.47}$$

and the secular determinant becomes

$$|\mathbf{G}\mathbf{F}_r - \boldsymbol{\lambda}\mathbf{E}| = \mathbf{0} \tag{3.48}$$

where **E** is the unit matrix. The final form of the expression yields the product **GF** first described by Wilson, and this method is referred to as the Wilson **GF** method. The eigenvector matrix $\mathbf{L}_r$ is a function of the displacements of the internal coordinates and can therefore be related to the eigenvectors in mass-adjusted cartesian displacement coordinates ($\mathbf{L}_q$), defined in Chapter 2, by the appropriate transformation (see the next section).

3.4 General Approach to Normal Coordinate Analysis

So far, the examples used in the development of the theory of molecular vibrations have taken the perspective of calculating the frequencies of the normal modes from the force constants between atoms of a given mass in simple molecules. Once these frequencies or eigenvalues of the secular equation have been calculated, the eigenvectors or relative displacements in each of these vibrations can be found. Various coordinate systems have been introduced in order to simplify the solution of the secular determinant. It was shown in Chapter 2 that the introduction of normal coordinates resulted in particularly simple expressions for the kinetic and potential energy. Actually, neither the force constants of the normal coordinates are known *a priori*, but are in fact the quantities we wish to determine from a set of experimentally observed frequencies. As noted earlier, each normal coordinate is a description of the displacements in a normal mode of vibration and is related to the mass-adjusted cartesian coordinates by the matrix $\mathbf{L}_q$:

$$\mathbf{q}=\mathbf{L}_q\mathbf{Q}$$

The transformation from internal to normal coordinates can therefore be written as

$$\mathbf{R}=\mathbf{BX}=\mathbf{BM}^{-1/2}\mathbf{q}=\mathbf{BM}^{-1/2}\mathbf{L}_q\mathbf{Q}=\mathbf{L}_r\mathbf{Q} \tag{3.49}$$

where

$$\mathbf{L}_r=\mathbf{BM}^{-1/2}\mathbf{L}_q \tag{3.50}$$

Consequently, the coefficients connecting the internal coordinates **R** with the normal coordinates **Q** are related by a constant factor to the normalized ratios of amplitudes of vibration in each normal mode (see Chapter 2). In the same way as the coefficients of the matrix $\mathbf{L}_q$ resulted in a simplification of the potential and kinetic energies expressed in cartesian coordinates, the transformation matrix $\mathbf{L}_r$ accomplishes a corresponding reduction of V and T expressed in terms of internal coordinates:

$$2V=\mathbf{R}^T\mathbf{F}_r\mathbf{R}=\mathbf{Q}^T\mathbf{L}_r^T\mathbf{F}_r\mathbf{L}\mathbf{Q} \tag{3.51}$$

$$2T=\dot{\mathbf{R}}^T\mathbf{G}^{-1}\dot{\mathbf{R}}=\dot{\mathbf{Q}}^T\mathbf{L}_r^T\mathbf{G}^{-1}\mathbf{L}_r\mathbf{Q} \tag{3.52}$$

However, it was shown in Chapter 2 that

$$2V=\mathbf{Q}^T\mathbf{\Lambda}\mathbf{Q} \tag{3.53}$$

and

$$2T=\dot{\mathbf{Q}}^T\mathbf{F}\dot{\mathbf{Q}} \tag{3.54}$$

where $\mathbf{\Lambda}$ is a diagonal matrix of the eigenvalues and $\mathbf{E}$ is the unit matrix. Consequently,

$$\mathbf{L}_r^T \mathbf{F}_r \mathbf{L}_r = \mathbf{\Lambda} \tag{3.55}$$

and

$$\mathbf{L}_r^T \mathbf{G}^{-1} \mathbf{L}_r = \mathbf{E} \tag{3.56}$$

Therefore, solving (3.56) for $\mathbf{L}_r^T = \mathbf{L}_r^{-1}\mathbf{G}$ and substituting in (3.55), we obtain

$$\mathbf{L}_r^{-1} \mathbf{G} \mathbf{F}_r \mathbf{L}_r = \mathbf{\Lambda} \tag{3.57}$$

or

$$\mathbf{G} \mathbf{F}_r \mathbf{L}_r = \mathbf{L}_r \mathbf{\Lambda} \tag{3.58}$$

Equation (3.57) is a set of simultaneous equations that can be used to calculate the transformation $\mathbf{L}_r$ and hence the displacements in each normal mode. Equation (3.58) in terms of components has the form

$$\sum_t \left[(GF_r)_{tt'} - \delta_{tt'} \lambda_k \right] (L_r)_{t'k} = 0, \qquad k = 1, 2, 3, \ldots, 3N-6 \tag{3.59}$$

or in matrix form

$$\mathbf{G} \mathbf{F}_r (\mathbf{L}_r)_k = \lambda_k (\mathbf{L}_r)_k \tag{3.60}$$

Note that

$$\mathbf{\Lambda} = \begin{bmatrix} \lambda_1 & & & & \\ & \lambda_2 & & & \\ & & \lambda_3 & & \\ & & & \ddots & \\ & & & & \lambda_{3N-6} \end{bmatrix}$$

and the $(\mathbf{L}_r)_k$'s are column vectors such that

$$\mathbf{L}_r = [l_1, l_2, l_3, \ldots, l_{3N-6}] \tag{3.61}$$

For nontrivial solution ($\mathbf{L}_r \neq 0$) of (3.60) it is required that;

$$|\mathbf{G}\mathbf{F}_r - \mathbf{E}\lambda_k| = \mathbf{0} \tag{3.62}$$

which is the secular determinant in internal coordinates.

To summarize, the transformation matrix $\mathbf{L}_r$ is the matrix whose columns are the eigenvectors of $\mathbf{GF}_r$. $\mathbf{\Lambda}$ contains the $3N-6$ eigenvalues of $\mathbf{GF}_r$ on its diagonal and l_k is the eigenvector associated with λ_k.

3.5 Redefining the Secular Equation in Mass-Adjusted Cartesian Coordinates

With the secular equation in internal coordinates written as in equation (3.62) the determinant is not symmetric. A symmetric secular equation is easier to solve on the computer through well-established mathematical matrix diagonalization routines. As the equation now stands, it would be necessary to diagonalize the $\mathbf{G}$ and $\mathbf{F}$ matrices separately before multiplication. To obtain a symmetric matrix in the secular equation, we will examine the eigenvalues and eigenvectors of the matrix product $\mathbf{D}^T\mathbf{F}_r\mathbf{D}$, which is symmetric. (You will recall that $\mathbf{D}$ was defined as $\mathbf{BM}^{-1/2}$.)

Let $\mathbf{L}_x$ and $\mathbf{\Lambda}'$ be the eigenvector and eigenvalue matrices of $\mathbf{D}^T\mathbf{F}_r\mathbf{D}$. ($\mathbf{L}_x$ is the matrix whose columns are eigenvectors and $\mathbf{\Lambda}'$ is a diagonal matrix containing the eigenvalues.) Thus

$$(\mathbf{D}^T\mathbf{F}_r\mathbf{D})\mathbf{L}_x = \mathbf{L}_x\mathbf{\Lambda}' \tag{3.63}$$

Then, premultiplying by $\mathbf{D}$, we obtain

$$\mathbf{DD}^T\mathbf{F}_r\mathbf{DL}_x = \mathbf{DL}_x\mathbf{\Lambda}' \tag{3.64}$$

Since $\mathbf{G} = \mathbf{DD}^T$, (3.64) is equivalent to

$$(\mathbf{GF}_r)\mathbf{DL}_x = \mathbf{DL}_x\mathbf{\Lambda}' \tag{3.65}$$

However, a matrix has a unique set of eigenvalues. Consequently, from comparison of equations (3.58) and (3.65), $\mathbf{\Lambda} = \mathbf{\Lambda}'$ and $\mathbf{L}_r$ is related to $\mathbf{DL}_x$. These matrices can be equated once they have been normalized (presuming each eigenvector is unique).

The matrix $\mathbf{L}_r$ is a $(3N-6)\times(3N-6)$ matrix, whereas $\mathbf{D}^T\mathbf{FD}$ is a $3N\times 3N$ matrix. Consequently, $\mathbf{L}_x$ is a $3N\times(3N-6)$ matrix (number of cartesian coordinates $\times$ number of nonzero frequencies) and is the eigenvector matrix in terms of cartesian displacement coordinates. The equation

$$|\mathbf{D}^T\mathbf{F}_r\mathbf{D} - \lambda\mathbf{E}| = 0 \tag{3.66}$$

is the vibrational secular equation in mass-adjusted cartesian coordinates with the force constant matrix defined in internal coordinates. This form reintroduces the zero frequencies associated with translation and rotation but is more easily solved by computer methods (see Chapter 7). The rationale for the detailed examination of the use of both mass-adjusted cartesian coordinates and internal coordinates should now be clear.

As noted earlier, the secular determinant expressed in the form of equation (3.66) is symmetric, making diagonalization easier and saving considerable computing time. The method has the additional advantages of allowing the solution of larger matrices on computers with limited memory storage and it eliminates the need to remove redundant coordinates. In fact, as many redundant coordinates as desired may be incorporated without affecting the results of the calculations or increasing the dimensions of the secular determinant from $3N \times 3N$. This form of the equation also simplifies the calculation of infrared and Raman intensities. Finally, in the $\mathbf{GF}_r$ form [equation (3.60)] the order of the secular determinant is increased by the introduction of extraneous zero roots. The problem of redundant coordinates will be discussed more fully in Chapter 5.

3.6 Construction of the B Matrix

The solution of the secular determinant written either as equation (3.60) or equation (3.66) requires the construction of the **B** matrix, which relates internal coordinates to cartesian displacement coordinates. This is a straightforward process once the molecular geometry is known. In the next section we will present an example in which the construction is performed simply by inspection. For most polyatomic molecules, however, the calculations are tedious but fortunately can be performed quickly by using standard computer programs. The trigonometry involved is discussed in detail in Wilson, Decius, and Cross (1) and other standard texts (2, 3).

3.7 Example: Two-dimensional Carbon Dioxide Molecule

We will again consider the solution of the linear CO_2 molecule, this time in two dimensions using the Wilson **GF** method. The internal coordinates are as defined in Figure 3.3 and are the displacements of the bond distances r_{12} and r_{23} and the deformation α of the bond angle. So the matrix **R** of the internal coordinates is

$$\mathbf{R} = \begin{Bmatrix} r_{12} \\ r_{23} \\ \alpha \end{Bmatrix}$$

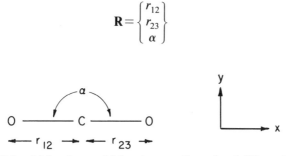

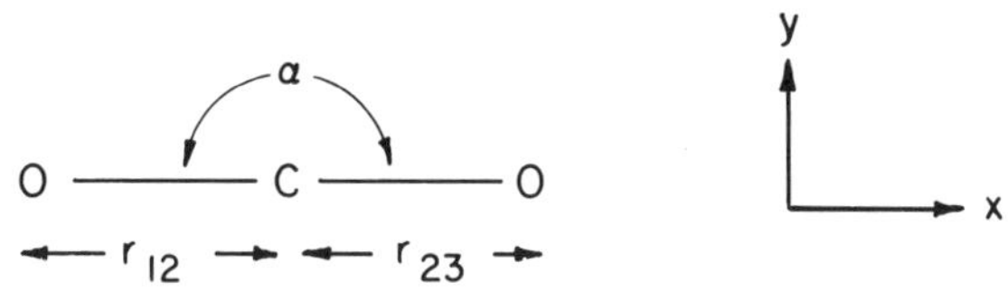

Figure 3.3. Molecular model for the two-dimensional CO_2 molecule.

The transformation between the two coordinate systems is

$$\mathbf{R}=\mathbf{BX}$$

The relationship between the bond extensions and the cartesian displacements can be simply written as

$$r_{12}=x_2-x_1$$

$$r_{23}=x_3-x_2$$

The relationship between bond angle bending and the cartesian displacements is not so intuitively obvious. We will simply state the relationship as

$$r\alpha=y_1+y_3-2y_2$$

The angle bend coordinate α is multiplied by r to give all coordinates the dimensions of length. The **B** matrix can now be written by inspection of these three equations:

$$\mathbf{B}=\begin{matrix} x_1 & x_2 & x_3 & y_1 & y_2 & y_3 \end{matrix}$$

$$\mathbf{B}=\begin{bmatrix} -1 & 1 & 0 & 0 & 0 & 0 \\ 0 & -1 & 1 & 0 & 0 & 0 \\ 0 & 0 & 0 & 1 & -2 & 1 \end{bmatrix}$$

The **G** matrix can be calculated from

$$\mathbf{G}=\mathbf{BM}^{-1}\mathbf{B}^T=\mathbf{DD}^T$$

where

$$\mathbf{D}=\qquad \mathbf{B} \qquad\qquad \mathbf{M}^{-1/2}$$

$$\mathbf{D}=\begin{bmatrix} -1 & 1 & 0 & 0 & 0 & 0 \\ 0 & -1 & 1 & 0 & 0 & 0 \\ 0 & 0 & 0 & 1 & -2 & 1 \end{bmatrix}$$

$$\times\begin{bmatrix} m_O^{-1/2} & 0 & 0 & 0 & 0 & 0 \\ 0 & m_C^{-1/2} & 0 & 0 & 0 & 0 \\ 0 & 0 & m_O^{-1/2} & 0 & 0 & 0 \\ 0 & 0 & 0 & m_O^{-1/2} & 0 & 0 \\ 0 & 0 & 0 & 0 & m_C^{-1/2} & 0 \\ 0 & 0 & 0 & 0 & 0 & m_O^{-1/2} \end{bmatrix}$$

Therefore,

$$\mathbf{D}=\begin{bmatrix} -m_{\mathrm{O}}^{-1/2} & m_{\mathrm{C}}^{-1/2} & 0 & 0 & 0 & 0 \\ 0 & -m_{\mathrm{C}}^{-1/2} & m_{\mathrm{O}}^{-1/2} & 0 & 0 & 0 \\ 0 & 0 & 0 & m_{\mathrm{O}}^{-1/2} & -2m_{\mathrm{C}}^{-1/2} & m_{\mathrm{O}}^{-1/2} \end{bmatrix}$$

Therefore,

$$\mathbf{D}^{T}=\begin{bmatrix} -m_{\mathrm{O}}^{-1/2} & 0 & 0 \\ m_{\mathrm{C}}^{-1/2} & -m_{\mathrm{C}}^{-1/2} & 0 \\ 0 & m_{\mathrm{O}}^{-1/2} & 0 \\ 0 & 0 & m_{\mathrm{O}}^{-1/2} \\ 0 & 0 & -2m_{\mathrm{C}}^{-1/2} \\ 0 & 0 & m_{\mathrm{O}}^{-1/2} \end{bmatrix}$$

Thus, $\mathbf{G}=\mathbf{D}\mathbf{D}^{T}$:

$$\mathbf{G}=\begin{bmatrix} \frac{1}{m_{\mathrm{O}}}+\frac{1}{m_{\mathrm{C}}} & \frac{-1}{m_{\mathrm{C}}} & 0 \\ \frac{-1}{m_{\mathrm{C}}} & \frac{1}{m_{\mathrm{O}}}+\frac{1}{m_{\mathrm{C}}} & 0 \\ 0 & 0 & \frac{2}{m_{\mathrm{O}}}+\frac{4}{m_{\mathrm{C}}} \end{bmatrix}$$

The $\mathbf{F}_r$ matrix can be written by inspection:

$$2V=fr_{12}^2+fr_{23}^2+kr_{12}r_{23}+f_{\alpha}\alpha^2+k_{r\alpha}r_{12}\alpha+k_{r\alpha}r_{23}\alpha$$

$$F_r=\begin{Bmatrix} f & k & 0 \\ k & f & 0 \\ 0 & 0 & f_{\alpha} \end{Bmatrix}$$

where $k_{r\alpha}=0$ because of the orthogonal nature of the coordinates.

$$\mathbf{GF}=\begin{bmatrix} f\left(\frac{1}{m_{\mathrm{O}}}+\frac{1}{m_{\mathrm{C}}}\right)-\frac{1}{m_{\mathrm{C}}}k & -\frac{f}{m_{\mathrm{C}}}+k\left(\frac{1}{m_{\mathrm{O}}}+\frac{1}{m_{\mathrm{C}}}\right) & 0 \\ -\frac{f}{m_{\mathrm{C}}}+k\left(\frac{1}{m_{\mathrm{O}}}+\frac{1}{m_{\mathrm{C}}}\right) & f\left(\frac{1}{m_{\mathrm{O}}}+\frac{1}{m_{\mathrm{C}}}\right)-\frac{1}{m_{\mathrm{C}}}k & 0 \\ 0 & 0 & \left(\frac{2}{m_{\mathrm{O}}}+\frac{4}{m_{\mathrm{C}}}\right)f_2 \end{bmatrix}$$

and the secular determinant becomes

$$\begin{vmatrix} f\left(\dfrac{1}{m_O}+\dfrac{1}{m_C}\right)-\dfrac{1}{m_C}k-\lambda & -\dfrac{f}{m_C}+k\left(\dfrac{1}{m_O}+\dfrac{1}{m_C}\right) & 0 \\ -\dfrac{f}{m_C}+k\left(\dfrac{1}{m_O}+\dfrac{1}{m_C}\right) & f\left(\dfrac{1}{m_O}+\dfrac{1}{m_C}\right)-\dfrac{1}{m_C}k-\lambda & 0 \\ 0 & 0 & \left(\dfrac{2}{m_O}+\dfrac{4}{m_C}\right)f_2-\lambda \end{vmatrix}=0$$

The solutions are

$$\lambda_1=\frac{1}{m_O}(f_r+k), \qquad 1387\ \text{cm}^{-1}$$

$$\lambda_2=\left(\frac{1}{m_O}+\frac{2}{m_C}\right)(f_r-k), \qquad 2350\ \text{cm}^{-1}$$

$$\lambda_3=\left(\frac{2}{m_O}+\frac{4}{m_C}\right)f_\alpha, \qquad 667\ \text{cm}^{-1}$$

Going from one to two dimensions introduced the additional frequency due to the bending mode. Extending the problem from two to three dimensions introduces another frequency, namely, bending in the third dimension. Since it is physically impossible to distinguish in which dimension perpendicular to the chemical bonds the molecule is vibrating, the new frequency is identical to λ_3 and is called a degenerate mode. With $|\mathbf{GF}_s-\lambda\mathbf{E}|\mathbf{L}_r=0$ the eigenvectors are

$$\lambda_1:\ L_{11}=+L_{12} \qquad \begin{array}{ccc} L_{11} & & L_{12} \\ \leftarrow & \cdot & \rightarrow \end{array}$$

$$\lambda_2:\ L_{21}=-L_{22} \qquad \begin{array}{ccc} L_{21} & & L_{22} \\ \rightarrow & \leftarrow & \rightarrow \end{array}$$

$$\lambda_3:\ L_{33}\ (L_{31}=L_{32}=0) \qquad \begin{array}{ccc} & L_{33} & \\ \uparrow & \downarrow & \uparrow \end{array}$$

Two stretching frequencies are observed for CO_2, at 1387 cm^{-1} (symmetric stretching$=\lambda_1$) and 2350 cm^{-1} (antisymmetric stretching$=\lambda_2$), and the bending mode (λ_3) is observed at 667 cm^{-1}.

3.8 Potential Energy Distribution

In determining the eigenvector matrix $\mathbf{L}_r$ we have constructed a visual description of the normal coordinates. It is often desirable also to determine the contribution of each internal coordinate to the potential energy of a given normal coordinate. The potential energy is given by

$$2V = \mathbf{Q}^T \boldsymbol{\Lambda} \mathbf{Q} = \mathbf{Q}^T \mathbf{L}_r^T \mathbf{F} \mathbf{L}_r \mathbf{Q}$$

If this is written for one normal coordinate whose frequency is λ_n, we obtain

$$\lambda_n = \sum_{i,j} L_{ni}^T F_{ij} L_{jn} = \sum_{i,j} F_{ij} L_{in} L_{jn}$$

In general, the value of $F_{ij} L_{in} L_{jn}$ is large when $i=j$. Thus the $F_{ii} L_{in}^2$ terms are the most important in determining the distribution of the potential energy and the ratios of the $F_{il} L_{in}^2$ terms provide a measure of the relative contribution of each internal coordinate q_i to the normal coordinate Q_{Ni}. Therefore, one can write

$$\mathrm{POT}_{ij} = \frac{F_{ii} L_{ij}^2}{\Sigma_i F_{ii} L_{ij}^2}$$

where POT_{ij} are elements of the potential energy distribution matrix. The summation in the denominator serves to normalize the matrix elements. Many normal coordinates have an overwhelming contribution from a given "internal" coordinate, so we tend to label the normal coordinate as a particular kind of vibration. In many molecules, modes occur with "characteristic" frequencies, and we often call them "C—H stretch," "H—C—H bend," and so on. These band assignments are possible if the appropriate element of the potential energy matrix makes up most of the potential energy of the normal coordinate. Because these normal coordinates are dominated by these internal coordinates, the normal coordinates take on the characteristics of group vibrations and have nearly the same representation regardless of the remaining skeleton of the molecule. Consequently, the infrared and Raman spectra often reflect frequencies characteristic of a particular internal coordinate or chemical group.

References

1 E. B. Wilson, Jr., J. C. Decius, and P. C. Cross, "Molecular Vibrations," McGraw-Hill, New York, 1955.

2 L. A. Woodward, "Introduction to the Theory of Molecular Vibrations and Vibrational Spectroscopy," Oxford (Clarendon Press), New York, 1972.

3 D. Steele, "Theory of Vibrational Spectroscopy," Saunders, Philadelphia, 1971.

4 THE APPLICATION OF SYMMETRY TO VIBRATIONAL SPECTROSCOPY

Tyger! Tyger! burning bright
In the forests of the night
What immortal hand or eye
could frame thy fearful symmetry?
William Blake

In the development of the theory of molecular vibrations presented in the preceding three chapters we hinted at the central role of symmetry in determining many features of the vibrational spectrum. Molecular symmetry, through the application of group theory, is in fact of general use in determining many molecular properties. A French mathematician, Evariste Galois, formulated group theory in a letter to a friend the night before he was shot to death in a duel over a woman (1). More recently, a mathematical group with more than 10^{54} members, a number known as the Monster and representing a set of rotations in a space of 196,883 dimensions, was constructed by R. L. Griess, Jr. (reported in ref. 1). Fortunately, in applying group theory to vibrational spectroscopy we can confine our discussion to the mundane world of three dimensions. We will first consider the symmetry operations that are used to describe molecular symmetry and then proceed to a brief consideration of the relevant aspects of group theory that allow us to determine and classify the number of active normal modes of small molecules.

4.1 Symmetry Elements and Operations

When a symmetry operation is performed on a molecule, it can result in an interchange of atoms of the same type but does not change the configuration of the molecule. Consider the water molecule shown in Figure 4.1. A rotation of 180° about the z axis interchanges the position of the two hydrogen atoms, H_a and H_b, but since these atoms are equivalent, the new orientation is

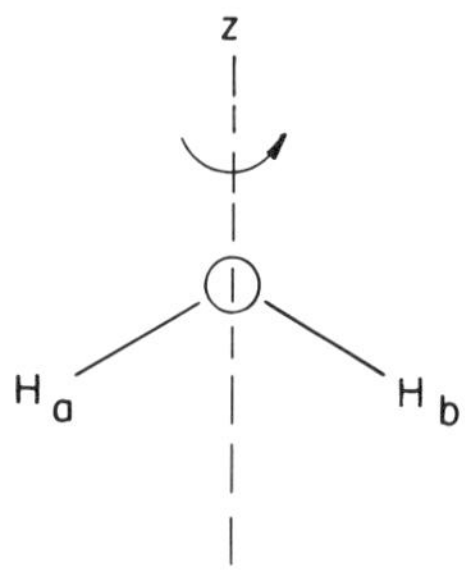

Figure 4.1. The twofold rotation axis of a molecule of water.

indistinguishable from the original and can be superimposed upon it. The symmetry operation in this example is the rotation about the z axis, while the symmetry element is the rotational axis itself.

By definition, a symmetry operation leaves the center of gravity of a molecule unchanged. There are five basic symmetry operations. The symmetry elements associated with each of these operations are:

1 Rotational axes
2 Center of symmetry or inversion center
3 Mirror plane
4 Inversion axis
5 Identity element

We will now describe these in more detail.

1 ***Rotation About a Symmetry Axis*** We have considered one example of rotational axis above. In general, if θ represents the angle through which the molecule must be rotated in order to obtain a new but indistinguishable configuration, then the molecule is said to have a $360°/\theta$-fold rotational axis. The water molecule has a 360/180 or twofold axis, written as C_2.

Some molecules have a number of superimposed rotational axes. Consider benzene in Figure 4.2. The z axis contains C_2, C_3, and C_6 rotational axes. Both the x and y axes are C_2 rotational axes. It is conventional to make the axis of highest order the z axis in simple molecules. For polymers, the axis of the polymer chain is generally taken to be the z axis.

Linear molecules such as CO_2 can be rotated about the lengthwise axis (passing through the center of the atoms) by any angle, so that this axis is, in effect, C_∞.

2 ***Inversion at a Center of Symmetry*** If a molecule has a center such that a line drawn from any atom in one half of the molecule through this point encounters an equivalent atom equidistant from this point in the other half, the molecule possesses a center of symmetry or inversion designated **i**. The origin of the cartesian coordinate system defined in Figure 4.2 for benzene is a center of symmetry.

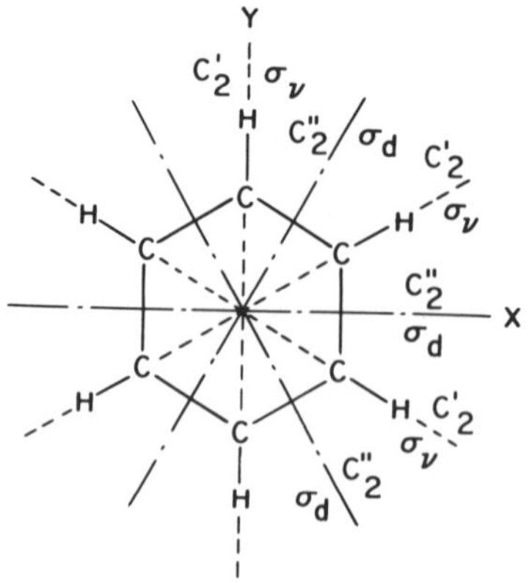

Figure 4.2. The symmetry elements of a molecule of benzene. The z axis (a C_6 and S_6 symmetry axis) is perpendicular to the plane of the figure, which is the σ_h symmetry plane.

3 ***Reflection at a Mirror Plane*** A molecule can be bisected by a plane, such that the geometric arrangement of atoms on one side of a plane is a mirror image of the other side. This symmetry operation is designated $\boldsymbol{\sigma}$. Again using benzene as an example, it can be seen from Figure 4.2 that the xy, xz, and yz planes are all mirror planes, designated $\sigma(xy)$, $\sigma(xz)$, and $\sigma(yz)$, respectively. In molecules were the z axis is considered vertical, $\sigma(xz)$ or $\sigma(yz)$ (or both) is designated σ_v. A mirror plane perpendicular to the principal rotation axis is designated σ_h.

4 ***Inversion Axis*** If a molecule is rotated by $360°/\theta$ and the resulting configuration is reflected at a mirror plane perpendicular to this axis to obtain an orientation superimposable on the original, the molecule is said to have a rotation-reflection or inversion axis, designated $\mathbf{S}_p$, where $p = 360/\theta$. An example is *trans*-dichloroethylene, illustrated in Figure 4.3.

If a molecule has a center of symmetry **i**, it must also possess an $\mathbf{S_2}$ **axis**, so that when one is specified the other is implied and need not be explicitly stated.

5 ***The Identity Symmetry Element*** There is a symmetry operation possessed by every molecule, called the identity, given the symbol E (or in some texts I). It is the operation in which the molecule remains in the original position. Although this operation appears trivial, it is mathematically useful in the development and application of group theory.

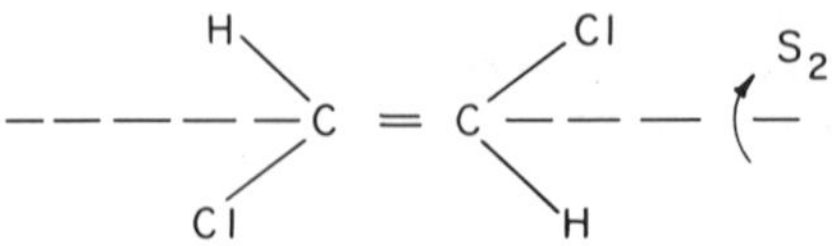

Figure 4.3. The inversion axis of a molecule of *trans*-dichloroethylene.

4.2 Point Groups

Consider the molecule of water shown in Figure 4.4. By inspection it can be determined that this molecule has four symmetry elements, E, $C_2(z)$, $\sigma(xz)$, and $\sigma(yz)$. All of the corresponding symmetry operations leave one point, the center of gravity of the molecule, unchanged, and such a group of operations is therefore known as a **point group**. There are only a limited number of such point groups and every molecule can be assigned to one of them, depending on the symmetry elements it possesses. The number of independent operations in a group is called the **order** of the group. We will discuss the different types of groups and the rules for classifying a given molecule later.

Point groups are also groups in the mathematical sense. Consequently, the set of symmetry operations constituting a point group must have the following four properties:

1 The product of any two operations in the group and the square of each element must be an element in the group; that is, the set is closed.

2 The associative law must hold; that is,

$$A(BC)=(AB)C$$

where A, B, and C are symmetry operations.

3 The set contains the identity operation E, such that

$$AE=EA=A$$

4 Each operation A has an inverse A^{-1}, also a member of the group, where

$$AA^{-1}=A^{-1}A=E$$

Rule 1 is of particular significance in molecular spectroscopy. It can be illustrated by considering the water molecule shown in Figure 4.4. The operation $\sigma(xz)$ followed by $\sigma(yz)$, or $\sigma(xz)\sigma(yz)$, is equivalent to $C_2(z)$. In this example the order of multiplication does not matter, that is, $AB=C$ and $BA=C$, and the multiplication is commutative. This is not generally true for all groups. Groups for which it is true are called **Abelian**.

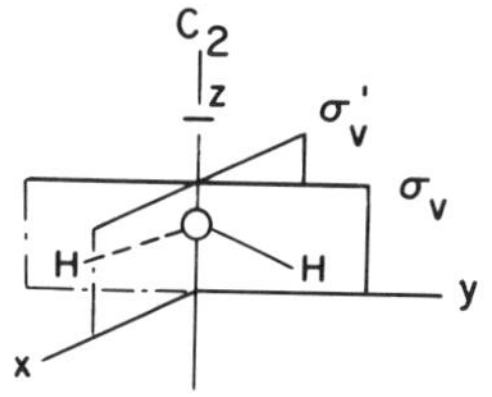

Figure 4.4. The symmetry elements of a molecule of water.

Table 4.1 Multiplication Table for the Point Group Consisting of the Symmetry Operations of the Water Molecule

	E	$C_2(z)$	$\sigma_v(xz)$	$\sigma_v(yz)$
E	E	$C_2(z)$	$\sigma_v(xz)$	$\sigma_v(yz)$
$C_2(z)$	$C_2(z)$	E	$\sigma_v(yz)$	$\sigma_v(xz)$
$\sigma_v(xz)$	$\sigma_v(xz)$	$\sigma_v(yz)$	E	$C_2(z)$
$\sigma_v(yz)$	$\sigma_v(yz)$	$\sigma_v(xz)$	$C_2(z)$	E

Since the multiplication of any two operations of a set is also a member of the set, we can construct a multiplication table for the group. Using the four symmetry elements applicable to the water molecule, Table 4.1 can be derived.

4.3 Classes of Symmetry Operations

The definition of classes of symmetry operations in a particular point group will be of use in applying group theory to the determination of the number of normal modes of vibration. Consider the ammonia molecule NH_3 shown in planar projection

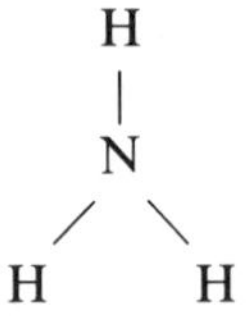

A counterclockwise rotation C_3' produces the same orientation as do two successive clockwise C_3 rotations, denoted C_3^2. As a result, C_3' is not specified, and the complete set of symmetry operations is E, C_3^1, C_3^2, σ_v, σ_v', and σ_v''. The mirror planes σ_v pass vertically through the chemical bonds of the molecule. They can be transformed into one another by the operations C_3^1 and C_3^2, as

Table 4.2 Multiplication Table for Symmetry Operations of NH_3 $EC_3^1C_3^2\sigma_v\sigma_v'\sigma_v''$

	E	C_3^1	C_3^2	σ_v	σ_v'	σ_v''
E	E	C_3^1	C_3^2	σ_v	σ_v'	σ_v''
C_3^1	C_3^1	C_3^2	E	σ_v''	σ_v	σ_v'
C_3^2	C_3^1	E	C_3^1	σ_v'	σ_v''	σ_v
σ_v	σ_v	σ_v'	σ_v''	E	C_3^1	C_3^2
σ_v'	σ_v'	σ_v''	σ_v	C_3^2	E	C_3^1
σ_v''	σ_v''	σ_v	σ_v'	C_3^1	C_3^2	E

shown in Table 4.2. Consequently, C_3 and C_3^1 are similar symmetry operations; differing only in the direction of rotation, they are said to belong to one **class**.

4.4 Equivalent Atoms and Subgroups

Point groups can often be divided into subgroups. For example, referring back to Table 4.1, the multiplication table for the symmetry operations of H_2O, we can see that the first two columns and rows

	E	$C_2(z)$
E	E	$C_2(z)$
$C_2(z)$	$C_2(z)$	E

also obey the rules for a mathematical group. This **subgroup** of the molecular point group has the additional property of not altering the position of the oxygen atom. For complex molecules such as ethane, as shown in Figure 4.5, subgroups can leave groups of atoms in their original position. The staggered conformation of ethane is specified by the symmetry operations I, C_3^1, C_3^2, C_2, C_2', C_2'', **i**, S_6^5, σ_v, σ_v', and σ_v''. The operations I, C_3^1, C_3^2, σ_v, σ_v', and σ_v'' form a subgroup that does not alter the position of either carbon atom. As we shall see later, the concept of subgroups is extremely useful in relating the vibrations of an isolated polymer chain to those in an ordered crystal.

In contrast to the definition of subgroups that leave certain atoms or groups of atoms unchanged is the equally important concept of symmetric

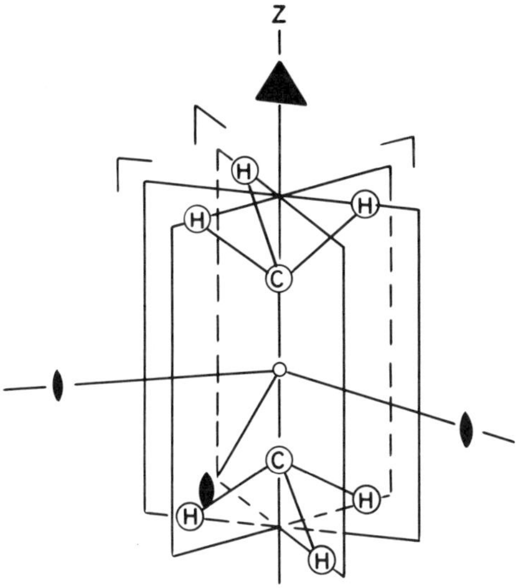

Figure 4.5. The symmetry elements of the staggered conformation of ethane.

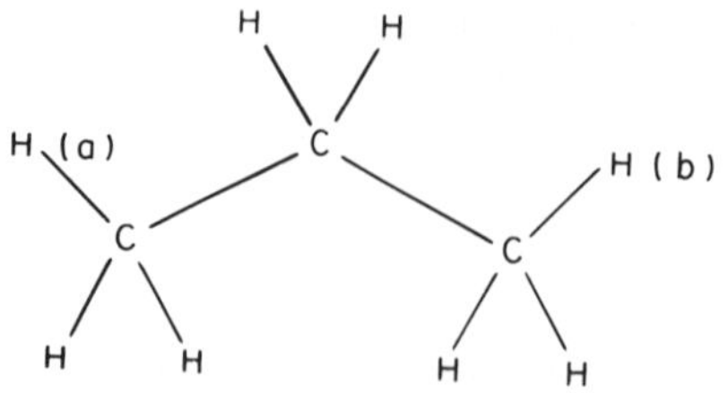

Figure 4.6. The structure of the propane molecule.

equivalence. In ethane all six of the hydrogen atoms are said to be equivalent, since by starting with any one of them, all the others can be generated by application of the symmetry operations of the group. However, atoms of the same type are not always equivalent. For example, in propane, shown in Figure 4.6, the methyl hydrogens (a) and (b) belong to an equivalent set, the methylene hydrogens to a second, and the remaining methyl hydrogens to a third. We will use this concept of equivalent atoms in Chapter 6 to determine symmetry coordinates.

4.5 Classification of Point Groups

There are two major systems of notation for point groups. The Schonfliess system based on the $C, \sigma, \mathbf{i}$ notation is used in vibrational spectroscopy, while the Hermann-Mauguin system is used by crystallographers. We will not discuss the latter. As noted earlier, every molecule can be assigned to a point group, depending on the symmetry elements it possesses. Even though there is theoretically an infinite number of such point groups, they can be classified into just a few types.

Type 1: *No rotational axes.*

(a) Point group C_1: this group has no symmetry elements.
(b) Point group C_s contains a single mirror plane σ.
(c) Point group C_i contains only a center of inversion $\mathbf{i}$.

Type 2: *Only one axis of rotation.*

(a) Point group C_p contains a single rotational axis, $p > 1$.
(b) Point groups S_p have more symmetry elements than their C_p counterparts; for example, S_2 has C_2 and $\mathbf{i}$.
(c) Point groups C_{pv} have the symmetry elements C_p and p vertical mirror planes (σ_v) intersecting the axis.
(d) Point groups C_{ph} have the symmetry element C_p and at right angles to it a horizontal mirror plane σ_h.

Type 3: *One p-fold axis and p twofold axes.* Point groups D_p, D_{ph}, and D_{pd}, with h designating a horizontal mirror plane (as above) and d diagonal mirror

Table 4.3 Preliminary Identification of Point Groups

	C_n	C_{nh}	C_{nv}	D_n	D_{nh}	D_{nd}
σ_h	√			√		
σ_v or σ_d			√		√	√
C_2				√	√	√

planes. Depending on the specific point group, the addition of σ_h and σ_d implies other symmetry elements.

Type 4: *More than one axis higher than twofold.* Tetrahedral and octahedral molecules belong to point groups of this type. They are not usually encountered in the spectroscopy of polymers.

A given molecule can be classified into its appropriate point group in a systematic manner. We will briefly outline an approach where the point group is tentatively identified by means of certain characteristic symmetry element combinations. A more systematic and elegant approach is described by Harris and Bertolucci (2). However, as we will see, the point group of most polymers can be determined by inspection.

The point groups of linear molecules are easily recognized and belong to either $C_{\infty v}$ or $D_{\infty h}$. The latter point group applies to linear molecules with a center of symmetry. Similarly, tetrahedral and octahedral molecules can be recognized by inspection. The point group of a nonlinear or noncubic molecule can often be determined from just the first row of the character table. First, the principal symmetry elements must be identified.

1. Highest-order or principal rotational axis
2. Symmetry planes
3. A twofold axis at right angles to the principal axis

The point group can then be tentatively identified from Table 4.3.

Once a tentative identification has been made, the other symmetry elements in the group listed in the character table can be applied as a test for correctness.

4.6 Introduction to Character Tables

For expository purposes we illustrate the use of character tables in vibrational spectroscopy before discussing their formal mathematical derivation. Consider displacements of the atoms of water parallel to the set of cartesian coordinate axes defined previously in Figure 4.4. As noted earlier, H_2O belongs to the

Table 4.4 Rudimentary Character Table for C_{2v}

	E	$C_2(z)$	$\sigma_v(xz)$	$\sigma_v(yz)$	Designation	
Translations parallel to						
z	+1	+1	+1	+1	A_1	Γ^1
x	+1	−1	+1	−1	B_1	Γ^3
y	+1	−1	−1	+1	B_2	Γ^4
Rotation about						
z	+1	+1	−1	−1	A_2	Γ^2
x	+1	−1	+1	−1	B_1	Γ^3
y	+1	−1	−1	+1	B_2	Γ^4

point group C_{2v} and has four symmetry operations, E, $C_2(z)$, $\sigma_v(xz)$, and $\sigma_v(yz)$. If we take as an example displacements parallel to the y direction, reflection by $\sigma_v(xz)$ produces a mirror image in which the direction of motion is reversed. This motion is said to be antisymmetric with respect to the symmetry operation $\sigma_v(xz)$. Conversely, the operation $\sigma_v(yz)$ does not change the direction of the displacements of the molecule, and the vibration is said to be symmetric with respect to this symmetry operation. If we now consider translations parallel to our set of cartesian axes and represent symmetric behavior by +1 and antisymmetric behavior by −1, we can construct Table 4.4, which is called a **character table**.

Rotational motion can be characterized in a similar fashion and is included in Table 4.4. It can be seen that rotations about the x and y axes behave in the same way as translations parallel to the y and x directions, respectively. However, rotations about the z axis are distinct in their behavior with respect to the symmetry operations. We have therefore defined four different behavior patterns, or what we will call **symmetry species**. Since E must always be represented by +1 and only two of the remaining operations are independent (the third can be generated by multiplication of the other two), there are only 2^2 ways to assign +1 and −1 to two independent vibrations. Consequently, there are only the four symmetry species derived earlier. We can now write the character table in the more usual form shown in Table 4.5

Table 4.5 Character Table for Point Group C_{2v}

C_{2v}	E	$C_2(z)$	$\sigma_v(xz)$	$\sigma_v(yz)$	
A_1	1	1	1	1	T_z
A_2	1	1	−1	−1	R_z
B_1	1	−1	1	−1	T_x, R_y
B_2	1	−1	−1	1	T_y, R_x

In the last two columns of Table 4.4 we have listed two common notations for symmetry species. In the first, those species that are symmetric with respect to the rotational axis are designated *A*, those that are antisymmetric, *B*. For molecules with a number of rotational axes, the one of highest order determines the *A* or *B* symbol. If several axes of the highest order occur, *A* refers to species symmetric to all of them. In many point groups there are several species of *A* and *B*, distinguished (as earlier) by subscripts or primes. In the special case where a center of symmetry is present, the subscripts *g* and *u* are used to represent behavior symmetric and antisymmetric, respectively, to the center of symmetry.

The alternate system of notation, not often used in polymer work, labels the species with the symbol Γ by numeric superscripts starting with *A*.

In addition to species of the types *A* and *B*, *E* and *F* symmetry species are commonly encountered. We will discuss these degenerate species later.

The characters listed in Tables 4.4 and 4.5 have a deeper significance than is apparent from this preliminary discussion. To understand the origin of the characters, we have to consider matrix representations of symmetry operations.

4.7 Matrix Representations of Symmetry Operations

In the preceding section we chose the **characters** $+1$ and -1 to represent symmetric and antisymmetric behavior under the symmetry operations of the point group C_{2v}. In the simple example above the choice was arbitrary. Actually, the characters of the character table are derived precisely from the use of matrices to represent symmetry transformations. Consider displacements parallel to a system of cartesian coordinates x, y, z to be our basis vectors. Let x', y', and z' be the displacement coordinates after a particular symmetry operation. Under the identity operation the coordinates remain unchanged, so that the symmetry transformation matrix is

$$[\mathbf{x}'] = \mathbf{D}[\mathbf{E}] \quad [\mathbf{x}]$$

$$\begin{bmatrix} x' \\ y' \\ z' \end{bmatrix} = \begin{bmatrix} 1 & 0 & 0 \\ 0 & 1 & 0 \\ 0 & 0 & 1 \end{bmatrix} \begin{bmatrix} x \\ y \\ z \end{bmatrix} \tag{4.1}$$

However, a reflection in a mirror parallel to the *xy* plane, $\sigma(xy)$, changes the sign of the *z* displacement coordinate, so the symmetry transformation matrix $\mathbf{D}[\boldsymbol{\sigma}(\mathbf{xy})]$ is given by

$$[\mathbf{x}'] = \mathbf{D}[\boldsymbol{\sigma}(\mathbf{xy})] \quad [\mathbf{x}]$$

$$\begin{bmatrix} x' \\ y' \\ z' \end{bmatrix} = \begin{bmatrix} 1 & 0 & 0 \\ 0 & 1 & 0 \\ 0 & 0 & -1 \end{bmatrix} \begin{bmatrix} x \\ y \\ z \end{bmatrix} \tag{4.2}$$

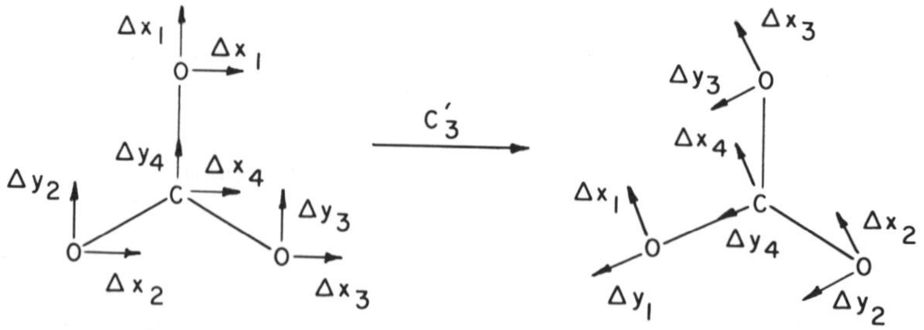

Figure 4.7. The effect of the symmetry operation C_3' on the cartesian displacement coordinates of CO_3^{2-}.

Rotations around, for example, the z axis leave the z coordinate unchanged but transform and mix the x and y coordinates, depending on the angle of rotation θ. Consider, as an example, the carbonate ion CO_3^{2-}. The symmetry operation C_3' (i.e., a rotation of $-120°$) results in the coordinate transformation illustrated in Figure 4.7. The coordinates of atom 1 are transformed into the combination of atom 2 coordinates given by

$$[\mathbf{x}'] = \mathbf{D}[\mathbf{C}(\boldsymbol{\theta})] \quad [\mathbf{x}]$$

$$\begin{bmatrix} x_1' \\ y_1' \\ z_1' \end{bmatrix} = \begin{bmatrix} \cos\theta & -\sin\theta & 0 \\ \sin\theta & \cos\theta & 0 \\ 0 & 0 & 1 \end{bmatrix} \begin{bmatrix} x_2 \\ y_2 \\ z_2 \end{bmatrix} \tag{4.3}$$

where $\theta = -120°$. The overall transformation of the coordinates of the molecule is illustrated in Figure 4.8. Other symmetry operations can be represented in the same way.

It can be shown that the matrix representations of symmetry operations of a particular point group satisfy the requirements of a mathematical group, listed in Section 4.2. It is also evident that we can construct an infinite number of such representations based on different choices of basis coordinate systems. For instance, the orientation of the x, y, and z axes could have been chosen differently; alternatively, internal coordinates rather than cartesian coordinates could have been chosen as the basis set. However, if any new basis set of coordinates is equal to a linear combination of the original set, then the two sets are **equivalent**. An important property of equivalent matrices is that they have the same **trace** or **character**. The trace is the sum of the diagonal elements of the matrix. For example, the trace of the matrix shown in Figure 4.8 is zero.

The infinite various matrix representations of any symmetry operation can be shown to be what are known as **reducible representations**. If we write the coordinate transformation expressed by, for example, the matrix shown in Figure 4.8 in the general form

$$[\mathbf{x}] \stackrel{R}{\rightarrow} [\mathbf{x}']$$

R	Δx_1	Δy_1	Δz_1	Δx_2	Δy_2	Δz_2	Δx_3	Δy_3	Δz_3	Δx_4	Δy_4	Δz_4
$\Delta x_1'$	0	0	0	$-\frac{1}{2}$	$\frac{\sqrt{3}}{2}$	0	0	0	0	0	0	0
$\Delta y_1'$	0	0	0	$-\frac{\sqrt{3}}{2}$	$-\frac{1}{2}$	0	0	0	0	0	0	0
$\Delta z_1'$	0	0	0	0	0	1	0	0	0	0	0	0
$\Delta x_2'$	0	0	0	0	0	0	$-\frac{1}{2}$	$\frac{\sqrt{3}}{2}$	0	0	0	0
$\Delta y_2'$	0	0	0	0	0	0	$-\frac{\sqrt{3}}{2}$	$-\frac{1}{2}$	0	0	0	0
$\Delta z_2'$	0	0	0	0	0	0	0	0	1	0	0	0
$\Delta x_3'$	$-\frac{1}{2}$	$\frac{\sqrt{3}}{2}$	0	0	0	0	0	0	0	0	0	0
$\Delta y_3'$	$-\frac{\sqrt{3}}{2}$	$-\frac{1}{2}$	0	0	0	0	0	0	0	0	0	0
$\Delta z_3'$	0	0	1	0	0	0	0	0	0	0	0	0
$\Delta x_4'$	0	0	0	0	0	0	0	0	0	$-\frac{1}{2}$	$\frac{\sqrt{3}}{2}$	0
$\Delta y_4'$	0	0	0	0	0	0	0	0	0	$-\frac{\sqrt{3}}{2}$	$-\frac{1}{2}$	0
$\Delta z_4'$	0	0	0	0	0	0	0	0	0	0	0	1

Figure 4.8. The form of the transformation matrix describing the operation C_3' on CO_3^{2-}.

where R is the symmetry operation, then if R_{11}, $R_{12}, R_{3N,3N}$ are the elements of the transformation matrix,

$$x_1' = R_{11}x_1 + R_{12}x_2 + \cdots + R_{1,3N}x_{3N}$$

$$x_2' = R_{21}x_1 + R_{22}x_2 + \cdots + R_{2,3N}x_{3N}$$

$$\vdots \quad \vdots$$

$$x_{2N}' = R_{3N,1x_1} + R_{3N,2x_2} + \cdots + R_{3N,3N}x_{3N}$$

It is possible to choose a new coordinate system η such that this transformation is reduced to a simpler form. This form is diagonal; that is,

$$\begin{bmatrix} \eta_1' \\ \eta_2' \\ \vdots \\ \eta_{3N}' \end{bmatrix} \begin{matrix} = \\ = \\ \\ = \end{matrix} \begin{bmatrix} R_{11}\eta_1 & & 0 \\ & R_{22}\eta_2 & \\ & & \\ 0 & & R_{3N}\eta_{3N} \end{bmatrix} \tag{4.4}$$

It has achieved a maximum simplicity or is **completed reduced**. It is not always possible to achieve by a single coordinate transformation a completely diagonal form where none of the coordinates η_i mix with η_j. More often, a block diagonal form, illustrated in Figure 4.9 and consisting of sets of nonmixing coordinates, is the completely reduced representation.

To summarize, the symmetry operations of a point group can be represented by matrices describing coordinate transformations. In general these

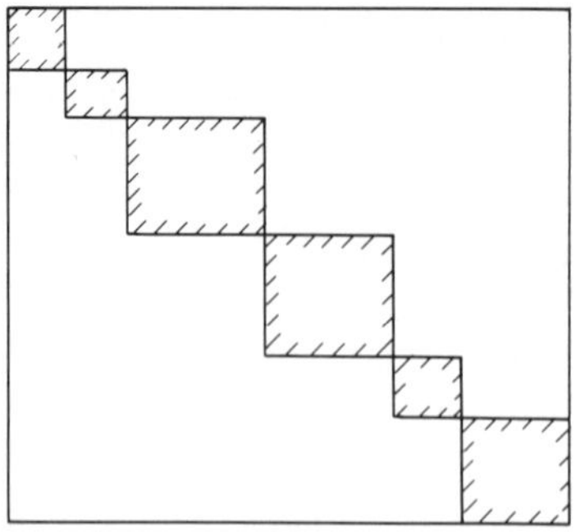

Figure 4.9. The block diagonal form of the reduced transformation matrix.

matrices are **reducible representations** in that they can be reduced to a set of 1×1 matrices or to a set of matrices that cannot be reduced further, that is, the **irreducible representations**. (Mathematically, this reduction is accomplished by similarity transformations.) At this point, the relevance of irreducible representations to molecular spectroscopy is probably not clear. In the next two sections, however, we will demonstrate that the normal coordinates of any molecule are an irreducible representation of the molecular point group. This fact allows us to classify the normal modes of vibration according to symmetry species, an extremely powerful tool in the analysis of the vibrational spectrum.

4.8 The Symmetry Properties of Normal Coordinates

We will now consider the transformation matrices that represent the effect of symmetry operations on the set of normal coordinates of a molecule. We again take CO_3^{2-} as our example; the normal modes of this molecule, calculated from the procedures outlined in Chapters 2 and 3, are illustrated in Figure 4.10.

The rotation C_3' transforms normal mode Q_1 into itself; that is,

$$Q_1 \xrightarrow{C_3'} Q_1$$

Similarly, for Q_2

$$Q_2 \xrightarrow{C_3'} Q_2$$

However, Q_3 is transformed into a linear combination of Q_3 and Q_4, as is Q_4.

$$Q_3 \xrightarrow{C_3'} (-\tfrac{1}{2})Q_3 - (\tfrac{1}{2})3^{1/2}Q_4$$

$$Q_4 \xrightarrow{C_3'} (+\tfrac{1}{2})3^{1/2}Q_3 - (\tfrac{1}{2})Q_4$$

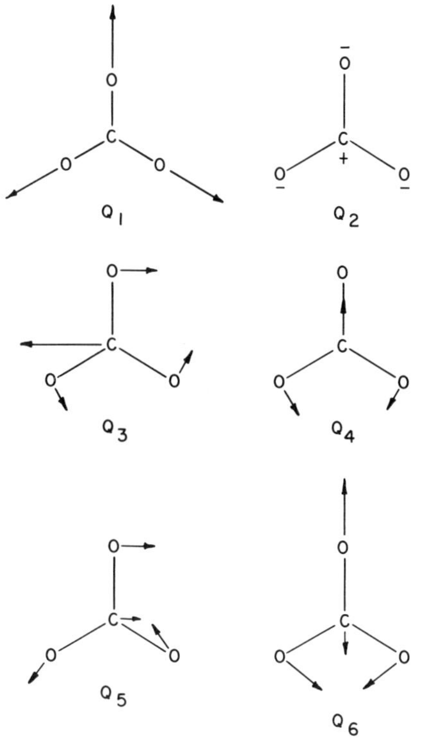

Figure 4.10. The normal modes of CO_3^{2-}.

This is a general result. A symmetry operation of a molecule will transform a member of a degenerate set of vibrations into a linear combination of the members of the degenerate set. Normal coordinates Q_5 and Q_6 also form a degenerate set, so

$$Q_5 \xrightarrow{c_3'} (-\tfrac{1}{2})Q_5 - (\tfrac{1}{2})3^{1/2}Q_6$$

$$Q_6 \xrightarrow{C_3'} (+\tfrac{1}{2})3^{1/2}Q_5 - (\tfrac{1}{2})Q_6$$

The representation of the symmetry operation C_3' in terms of normal coordinates can therefore be written

$$\mathbf{D}[\mathbf{C}_3^1] = \left|\begin{array}{cccccc} 1 & & & & & \\ & 1 & & & & \\ & & (-\tfrac{1}{2}) & (-\tfrac{1}{2})3^{1/2} & & \\ & & (+\tfrac{1}{2})3^{1/2} & (-\tfrac{1}{2}) & & \\ & & & & (-\tfrac{1}{2}) & (-\tfrac{1}{2})3^{1/2} \\ & & & & (+\tfrac{1}{2})3^{1/2} & (-\tfrac{1}{2}) \end{array}\right|$$

This representation is completely reduced. It cannot be broken down into smaller nonmixing sets of coordinates. The transformation matrices that represent the effect of the other symmetry operations of CO_3^{2-} upon the normal coordinates have this same property. Consequently, the matrix representations of each symmetry operation, such as C_3', can be written as the sum of a number of submatrices, each of which describes the transformation of all the nonmixing sets of normal coordinates. If we work with the trace or character of these submatrices, we can write C_3' as

	Trace of C_3'
Q_1	1
Q_2	1
Q_3, Q_4	−1
Q_5, Q_6	−1

The matrix describing the transformation of Q_1 is one-dimensional and its character is therefore the single element +1. The doubly degenerate modes are represented by two-dimensional matrices. Normal mode Q_1 forms a separate set under all the symmetry operations, since it does not mix with any other coordinate. Together, Q_3 and Q_4 form a separate set. Writing the characters of the matrices that describe the effect of all the symmetry operations of CO_3^{2-} (point group D_{3h}) acting on the nonmixing sets of normal coordinates, we obtain Table 4.6. Each row of this table is a representation of the transformations of a nonmixing set of normal coordinates under all of the symmetry operations of this group. Since no change in coordinates can reduce such a representation any further, it is an **irreducible representation**.

We have used normal coordinates to illustrate the general concept of irreducible representations. For a given point group there are only a definite small number of such irreducible representations, which can be presented in tabular form by giving the characters for each symmetry operation. We have also seen that for the molecule CO_3^{2-} each normal coordinate corresponds to one of the irreducible representations of the point group. However, there is often more than one normal mode in each irreducible representation. This can be seen from Table 4.6, where the characters of the transformation of Q_3, Q_4 and of Q_5, Q_6 are the same. In addition, with normal coordinates we did not derive the complete character table for point group D_{3h}, shown in Table 4.7.

Table 4.6 Representations of the Normal Coordinates of CO_3^{2-}

	E	C'	C_3^2	σ_v	σ_v'	σ_v''	σ_h	S_3'	S_3^2	C_2	C_2'	C_2''
Q_1	1	1	1	1	1	1	1	1	1	1	1	1
Q_2	1	1	1	1	1	1	−1	−1	−1	−1	−1	−1
Q_3, Q_4	2	−1	−1	0	0	0	2	−1	−1	0	0	0
Q_5, Q_6	2	−1	−1	0	0	0	2	−1	−1	0	0	0

Table 4.7 Characters of the Irreducible Representations of D_{3h}

D_{3h}	E	C_3^1	C_3^2	C_2	C_2'	C_2''	σ_h	S_3^1	S_3^5	σ_v	σ_v'	σ_v''
$\Gamma^1 = A_1'$	1	1	1	1	1	1	1	1	1	1	1	1
$\Gamma^2 = A_1''$	1	1	1	1	1	1	−1	−1	−1	−1	−1	−1
$\Gamma^3 = A_2'$	1	1	1	−1	−1	−1	1	1	1	−1	−1	−1
$\Gamma^4 = A_2''$	1	1	1	−1	−1	−1	−1	−1	−1	1	1	1
$\Gamma^5 = E'$	2	−1	−1	0	0	0	2	−1	−1	0	0	0
$\Gamma^6 = E''$	2	−1	−1	0	0	0	−2	1	1	0	0	0

Character tables for all the point groups have been derived and have been published in many standard texts (2,3). We think it is redundant to reproduce these tables again.

An important property of irreducible representations is that they can be used to describe any reducible representation of the group. Consider the set of matrices

$$\mathbf{G}[\mathbf{E}], \mathbf{G}[A_1], \ldots, \mathbf{G}(A_n)$$

that describe the transformations under the symmetry operations of a group. By applying the appropriate similarity transformation, all the matrices $\mathbf{G}[\mathbf{R}]$ can be reduced to the form

$$\mathbf{G}[\mathbf{R}] = \begin{bmatrix} \underbrace{D_1[R] \;\; \ddots \;\; D_1[R]}_{n_1 \text{ times}} & & & 0 \\ & \underbrace{D_2[R] \;\; \ddots \;\; D_2[R]}_{n_2 \text{ times}} & & \\ & & \ddots & \\ 0 & & & \underbrace{D_k[R] \;\; \ddots \;\; D_k[R]}_{n_k \text{ times}} \end{bmatrix}$$

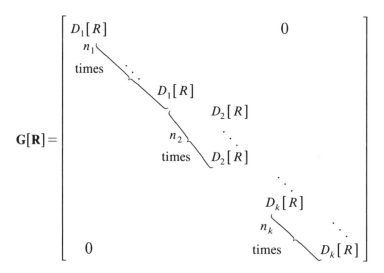

where $D_1[R], \ldots$ are the k irreducible representations of the symmetry group and n_i is the number of times $D_i[R]$ is contained in **G**[R]. If, for example, we had used cartesian displacement coordinates as a basis for **G**[R], then n_i could be shown to be equal to the number of normal modes in each irreducible representation or symmetry species. This is because the normal modes form a basis for the irreducible representations $D[R]$ of the group.

It is important to note that each irreducible representation is orthogonal to the others. For example, multiplying and adding the appropriate characters of species A_1 and A_2 of point group C_{2v} (Table 4.5), we obtain

$$A_1 \cdot A_2 = 1 \cdot 1 + 1 \cdot 1 + 1 \cdot (-1) + 1 \cdot (-1) = 0$$

In addition, each symmetry species is normalized to the **order** of the group, or the number of symmetry species.

$$A_1 A_1 = (1)^2 + (1)^2 + (1)^2 + (1)^2 = 4$$

$$A_2 A_2 = (1)^2 + (1)^2 + (-1)^2 + (-1)^2 = 4, \qquad \text{etc.}$$

These two properties can be more generally stated as

$$\frac{1}{g} \sum_{j=1}^{k} g_j \chi_j^{(\gamma')} \chi_j^{(\gamma)} = \delta_{\gamma'\gamma} \tag{4.5}$$

where g_j is the number of operations in the jth symmetry class, g is the total number of symmetry classes in the point group (which is equal to the order of the group), $\chi^{(\gamma)}$ is the character of the γth representation of an operations, and $\delta_{\gamma'\gamma}$ is the Kronecker delta. As we shall demonstrate in the next section, equation (4.5) can be applied to the determination of the number of normal modes in each irreducible representation or symmetry species.

4.9 The Calculation of the Number of Normal Modes in Each Symmetry Species

As mentioned earlier, any reducible representation of the group can be expressed as a linear integer combination of the orthogonal irreducible representations. In other words, the cartesian displacement coordinates can be used as a basis set for a reducible representation. It will be recalled that these reducible representations $\boldsymbol{\Gamma}$ are the matrices used to describe the coordinate transformations of the symmetry operations. These matrices can be expressed as an integer sum of the irreducible representations $\boldsymbol{\Gamma}^{(\gamma)}$.

$$\boldsymbol{\Gamma} = \sum n^{(\gamma)} \boldsymbol{\Gamma}^{(\gamma)} \tag{4.6}$$

The number of times $\boldsymbol{\Gamma}^{(\gamma)}$ appears in $\boldsymbol{\Gamma}$ is the number of normal modes $n^{(\gamma)}$ in each symmetry species; $n^{(\gamma)}$ can be calculated by considering the characters of the irreducible representations. Let χ_R be the character of a symmetry operation R in a reducible representation $\boldsymbol{\Gamma}$. It is independent of the choice of basis coordinates and equal to the sum of the diagonal elements in $\boldsymbol{\Gamma}$. It must therefore also be equal to the sum of the traces of the individual submatrices $\mathbf{\Gamma}^{(\gamma)}$, each multiplied by $n^{(\gamma)}$, the number of times each $\mathbf{\Gamma}^{(\gamma)}$ appears in γ.

$$\chi_j = \sum_{j=1}^{k} n^{(\gamma)}\chi_j^{(\gamma)}, \qquad j=1,2,k \tag{4.7}$$

The sum is over all k classes of the point group.

Multiplication of both sides of this equation by $\chi_j^{(\gamma')}$, weighted by the factor g_j, followed by summation over the class index j yields

$$\sum_{j=1}^{k} g_j\chi_j^{(\gamma')}\chi_j = \sum_{j=1}^{k} g_j\chi_j^{(\gamma')}\left[\sum_{j=1}^{k} n^{(\gamma)}\chi_j^{(\gamma')}\right]$$

$$= \sum_{j=1}^{k} n^{(\gamma)}\left[\sum_{j=1}^{k} g_j\chi_j^{(\gamma')}\chi_j^{(\gamma)}\right] \tag{4.8}$$

However, with the orthogonality relationship, equation (4.5), the quantity in brackets is equal to $g\delta_{\gamma'\gamma}$, so that

$$n^{(\gamma)} = \frac{1}{g}\sum_{j=1}^{k} g_j\chi_j^{(\gamma)}\chi_j \tag{4.9}$$

Consequently, the number of normal modes in each symmetry species can be calculated by using the character table of the point group to obtain the $\chi_j^{(\gamma)}$'s and obtaining the χ_j's by inspection of the original reducible representation. We will illustrate the application of this method using water as an example.

Example: Classification of the Normal Modes of Water

Consider the nine cartesian displacement coordinates of water, shown in Figure 4.11. We have $3N=9$ basis vectors and require a 9×9 matrix to describe each symmetry operation in the point group (C_{2v}). However, it is not required to write out the entire matrix, since we are *interested* only in *the trace or character* of each, in order to use equation 4.9. If under any symmetry operation a basis vector, say x_1, is transformed into itself, a $+1$ is found on the

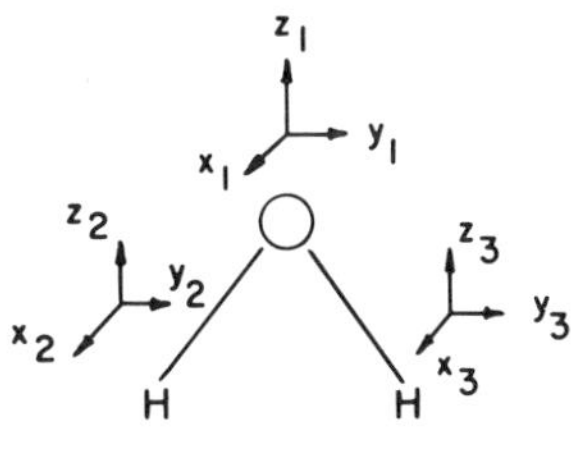

Figure 4.11. Definition of the cartesian displacement coordinates of water.

diagonal. If x_1 goes to $-x_1$, a -1 will result. Finally, if x_1 is transformed into the corresponding coordinate of another atom (e.g., x_2), a zero will result. Consequently, using the operation C_2 as an example, we can calculate the character of the transformation matrix as follows:

Coordinate Transformation under C_2	Term on Diagonal of Transformation Matrix
$x_1 \rightarrow -x_1$	-1
$y_1 \rightarrow -y_1$	-1
$z_1 \rightarrow z_1$	$+1$
$x_2 \rightarrow -x_3$	0
$y_2 \rightarrow -y_3$	0
$z_2 \rightarrow z_3$	0
$x_3 \rightarrow -x_2$	0
$y_3 \rightarrow -y_2$	0
$z_3 \rightarrow z_3$	0
	Sum $= -1 =$ character χ of C_2

By this procedure the characters of the reducible representation $\boldsymbol{\Gamma}_{3N}$ can be determined to be

C_{2v}	E	C_2	$\sigma_v(xz)$	$\sigma_v(yz)$
Γ_{3N}	9	-1	3	1

We can now apply equation (4.9) to determine the sum of the irreducible representations that correspond to this reducible representation. $\chi_j^{(\gamma)}$ can be read directly from the character table for C_{2v} given in Table 4.5. Then $\Sigma_{j=1}^{k} g_j \chi_j^{(\gamma)} \chi_j$, which can be written as the product $\Gamma_{3N}\chi_{A_1}$ is equal to

$$\Gamma_{3N}\chi_{A_1} = 9 \cdot 1 + 1 \cdot (-1) + 1 \cdot 3 + 1 \cdot 1 = 12$$

This result has to be normalized by the order of the group, that is, multiplied by $1/g = 1/4$. Consequently, there are $12/4 = 3$ normal modes in the A_1 symmetry species. Similarly,

$$\Gamma_{3N} = 3A_1 + A_2 + 3B_1 + 2B_2$$

Because we have used cartesian displacement coordinates as the basis for our reducible representation, this expression contains the translations and rotations. The quantities are listed in most character tables and they can be easily subtracted:

$$\begin{aligned} \Gamma_{3N} &= 3A_1 + A_2 + 3B_1 + 2B_2 \\ \Gamma_{t,r} &= \underline{\ \ A_1 + A_2 + 2B_1 + 2B_2} \\ \Gamma_{vib} &= 2A_1 \qquad\quad + \mathrm{B}_1 \end{aligned}$$

Thus there will be $2A_1 + 1B_1$ vibrational modes.

This method can be considerably simplified by deriving the reducible representation from the number of unshifted atoms. Atoms whose coordinates are converted into the coordinates of other atoms contribute only zeros to the trace or character (see above for water). Consequently, if we could obtain a characteristic factor corresponding to the trace of the transformation matrix of each symmetry operation acting on a single atom, and then multiply this factor by the number of times it appears in the transformation matrix for the whole molecule (i.e., by the number of unshifted atoms), we could obtain the reducible character for each symmetry operation. This characteristic factor can be derived from the general equation for symmetry transformations:

$$R_j = \begin{bmatrix} \cos\phi_j & -\sin\phi_j & 0 \\ \sin\phi_j & \cos\phi_j & 0 \\ 0 & 0 & \pm 1 \end{bmatrix} \tag{4.10}$$

Here $\chi_j = m_j(2\cos\phi_j \pm 1)$ where m_j is the number of unshifted atoms and ϕ_j is the angle of rotation for the particular symmetry operation. For the operation E, the characteristic factor $(2\cos\phi_j \pm 1)$ is equal to $(2\cos 0° + 1) = 1$; for C_2, $(2\cos 180 + 1) = -1$, and so on. We can construct Table 4.8, listing this characteristic factor for the most commonly encountered symmetry operations.

Consequently, for water the reducible representation can be determined very easily by constructing the following table:

C_{2v}	E	C_2	$\sigma_v(xz)$	$\sigma_v(yz)$
Factor (from Table 4.8)	3	-1	1	1
Number of unshifted atoms	3	1	3	1
Γ_{3N}	9	-1	3	1

This result is identical to that previously obtained by the more laborious route.

Table 4.8 Characteristic Factor For Commonly Encountered Symmetry Operations

Symmetry Operation	Characteristic Factor
E	3
C_2	-1
C_3	0
C_4	1
C_6	2
$S_1 = \sigma$	1
$S_2 = i$	-3
S_3	-2
S_4	-1
S_6	0

4.10 The Classification of Normal Modes Using an Internal Coordinate Representation

When analyzing the vibrational spectrum of a molecule it is often important to obtain a rough idea of the types of internal coordinates that can contribute to the vibrational modes in a particular symmetry species. This can often be achieved by using internal coordinates as the reducible representation. It should be kept in mind, however, that the results of this type of analysis are only approximate, since the mixing of coordinates (e.g., bond stretch with an angle bend) in a given symmetry species is not taken into account. A normal coordinate analysis is necessary for a more complete understanding. Nevertheless, for symmetry species containing vibrational modes of widely different frequency this type of analysis can be extremely useful.

As an example of the classification of the normal modes we will again use water. The internal coordinates of this molecule are defined in Figure 4.12. The reducible representation in internal coordinates can be determined by considering which coordinates are unshifted under each symmetry operation.

C_{2v}	E	C_2	$\sigma_v(xz)$	$\sigma_v(yz)$
Number of unshifted stretches, r	2	0	2	0
Number of unshifted bends, α	1	1	1	1

The assignment of normal modes corresponding to stretching and bending modes to the appropriate symmetry species Γ_r and Γ_α can then be made by using the character table and equation (4.9); for example,

$$\Gamma_r(A_1) = \tfrac{1}{4}[2(1) + 0(1) + 2(1) + 0(1)] = 1A_1$$

so that

$$\Gamma_r = A_1 + B_1 \quad \text{stretching modes}$$

$$\Gamma_\alpha = A_1 \quad \text{bending modes}$$

For simple molecules such as water the form of the normal modes can also be generated. The O—H stretching modes are classified as symmetric (symmetry species A_1) and antisymmetric (B_1) because one is symmetric to the C_2 symmetric operation whereas the other is not. Consequently, if we draw the symmetric stretching of each bond as in Figure 4.13*a* and sum the displacement at the oxygen atom, we obtain the A_1 mode shown in the figure.

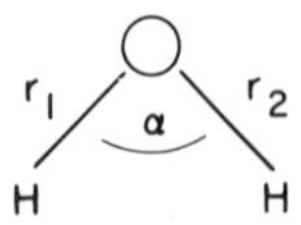

Figure 4.12. Definition of the internal coordinates of water.

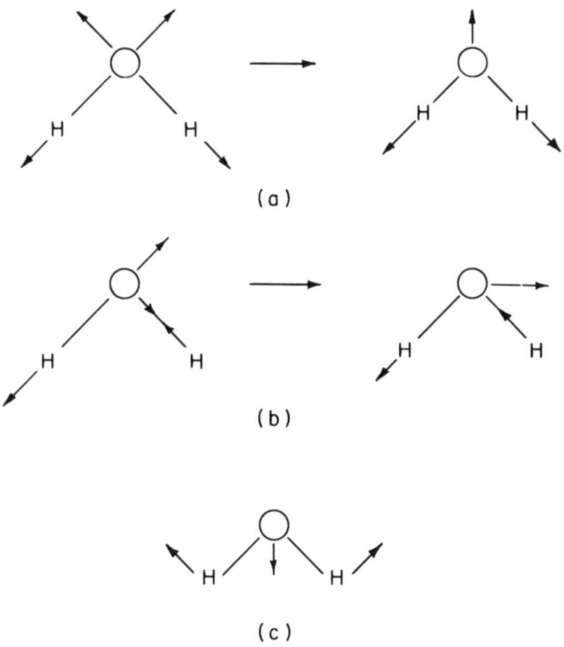

Figure 4.13. The normal modes of water (approximate). (*a*) Symmetric stretching mode. (*b*) Antisymmetric stretching mode. (*c*) Angle bending mode.

Conversely, the antisymmetric stretching mode is constructed from the stretch of one O—H bond and the compression of the other, as illustrated in Figure 4.13*b*. Finally, the bending mode has A_1 symmetry (totally symmetric) and, from our definition of bending coordinates, must have the form illustrated in Figure 4.13*c*. Of course, it must be kept in mind that this is only an approximate treatment, and that mixing between the symmetric stretching and bending modes could result in the form of the normal coordinates being different. For water, however, this approximation turns out to be a good one; for certain other molecules having the structure XY_2 (e.g., Cl_2O) the modes are mixed and cannot truly be classified separately as pure stretching or bending. It must be kept in mind that modes in different symmetry species (A_1 and B_1 for water) are not permitted to mix.

In this example we have in fact generated (in picture form) what are known as **symmetry coordinates**. In simple molecules such symmetry coordinates approximate the normal modes and in many cases correspond to the normal coordinates. We will consider a method for deriving symmetry coordinates and illustrate their use in factoring the secular equation in the following chapter. First, we will discuss briefly a general problem involved in using internal coordinates as reducible representations, the problem of **redundancy**. Redundancy is a factor we will refer to often later in this text and is the potential source of numerous difficulties.

4.11 The Problem of Redundant Coordinates

The problem of redundancy can be illustrated by reference to the chloroform molecule, shown in Figure 4.14. We have defined 10 internal coordinates to describe $3N-6=9$ normal modes. These internal coordinates are: r, r_1, r_2, r_3, α_{12}, α_{13}, α_{23}, β_1, β_2, and β_3. If we use the methods described in Section 4.10 to assign the normal modes to the various symmetry species (point group C_{3v}), we obtain

(a) With cartesian coordinates as a basis set

$$\Gamma_{\text{vib}}=3A_1+3E$$

(b) With internal coordinates as a basis set

$$\Gamma_{\text{vib}}=4A_1+3E$$

(Keep in mind that the E modes are doubly degenerate.)

The internal coordinate system results in one fictitious degree of freedom in symmetry species A_1. This redundancy is a consequence of defining six interbond angles when only five are independent; that is,

$$\alpha_{12}+\alpha_{23}+\alpha_{13}+\beta_1+\beta_2+\beta_3=0$$

Ignoring one of the interbond angles is not allowed and would lead to a masking of the contribution of the neglected angle to the vibrational energy. The nature of this redundancy introduces a variability in the mathematical description and an ambiguity in the calculation of the force field. The redundancy condition can be solved by defining appropriate symmetry coordinates (see Chapter 5). In addition, with modern computer methods of solving the vibrational problem by diagonalization of matrices, it is not necessary to

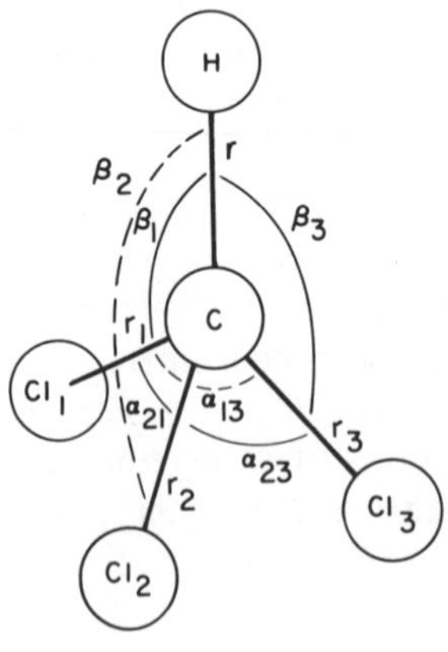

Figure 4.14. Defintion of the internal coordinates of chloroform.

remove the redundancy condition explicitly. The extra coordinate simply results in a zero root. However, when internal coordinate relationships exist such that the number of internal coordinates exceeds the number of degrees of vibrational freedom, then certain force constants can be indeterminate and only appropriate combinations can be calculated. This is not generally a problem in the normal coordinate analysis of polymers, where in any case we have the situation of the force field being vastly underdetermined by the number of observed frequencies. In this situation various force constants are to some degree assumed, approximated, or neglected. Nonetheless, if elements of a force field are being built up from a vibrational analysis of appropriate low molecular weight materials this could be a problem. In addition, there will be a correlation between values of valence force constants if there is a redundant coordinate. Consequently, careful attention to redundancies is often required. We will consider this problem further in Section 6.8.

4.12 Selection Rules

We have determined a method of classifying the normal modes according to symmetry species. If we could predict which symmetry species were Raman and/or infrared active for a given molecular model, it would be possible to specify the spectral activity according to the symmetry and hence structure of the molecule. In fact, this information has been determined for each point group and is usually listed as part of the character table. We will only outline the derivation of the selection rules here.

Consider the intensity of an infrared band, determined by the transition moment integrals

$$\int \psi_i \mu_x \psi_f, \qquad \int \psi_i \mu_y \psi_f, \qquad \int \psi_i \mu_z \psi_f$$

where $\mu_k = \partial\mu/\partial Q_R$. The transition moment integrals specify the change in dipole moment upon the vibrational transition $i \rightarrow f$. We need to know only whether these integrals are zero or not in order to predict optical activity. This can be determined from the symmetry properties of the integral. As an illustration, consider a function $y(x)$ where y is symmetric in x. One such function, $y = x^2$, is illustrated in Figure 4.15*a*.

This is an even function and the integral $\int y\,dx$ is nonzero. However, for an odd function, such as $y = x^3$ illustrated in Figure 4.15*b*, the integral $\int y\,dx$ is zero. In a similar fashion the components of the transition moment integral vanish wherever the integrand is antisymmetric with respect to any symmetry element. In other words, for a normal mode to be allowed, the transition moment integral has to belong to the totally symmetric representation, that is, species A, A_1, A', or A_g of the appropriate point group. (This is not to say that the normal mode has this symmetry.) To determine if this is the case we need

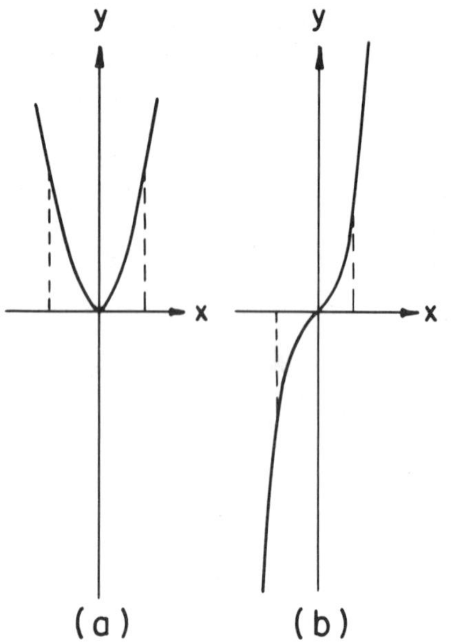

Figure 4.15. Illustration of typical even (a) and odd (b) functions.

to know the symmetry properties of the wave functions ψ_i, ψ_f and the dipole moment derivatives μ_x, μ_y, and μ_z.

All observed fundamental transitions are between the ground and first vibrational energy levels. Since the ground state (ψ_i) has all of the symmetry of the undistorted molecule, it is totally symmetric. Consequently, for the transition moment integral to be totally symmetric, the product of the symmetry species of ψ_f with that of any of the components of μ must be totally symmetric, that is, by the orthogonality requirement they must belong to the same symmetry species. It can be shown that ψ_f has the same symmetry as the corresponding normal coordinate. The dipole moment vector has components μ_x, μ_y, and μ_z, and it is a property of such vectors that they transform in the same way as the translations T_x, T_y, and T_z. Consequently, only those normal modes appearing in the same symmetry species as the translations T_x, T_y, and T_z are **infrared active**.

Raman selection rules are derived in a similar manner by determining the symmetry species to which each of the six elements of the polarizability tensor α_{ij} belongs (2). This is not an easy matter, and the derivation will not be given here. A brief discussion of Raman selection rules is given in Chapter 7 as part of the discussion of intensities. Here we will take the easy way out and note that vibrations are Raman active if they belong to the same symmetry species as any one of the elements of the polarizability tensor, given in the character table.

It is important to note that the selection rules only predict which modes are allowed. Allowed modes can have extremely weak intensities in the Raman or infrared spectra and thus not be observed.

4.13 Example: Determination of the Number of Active Modes of the Molecule AB_4

Consider a molecule of the form

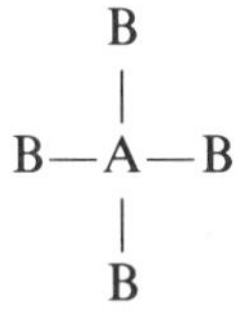

This molecule belongs to the point group D_{4h}, whose character table is given in Table. 4.9. The normal modes can be classified by the method outlined earlier:

D_{4h}	E	$2C_4$	C_2	$2C_2'$	$2C_2''$	i	$2S_4$	σ_h	$2\sigma_v$	$2\sigma_d$
Factor	3	1	−1	−1	−1	−3	−1	1	1	1
No. of unchanged atoms	5	1	1	3	1	1	1	5	3	1
Γ_{3N}	15	1	−1	−3	−1	−3	−1	5	3	1

Using this reducible representation, Γ_{3N}, and the character table, we have

$$\Gamma = A_{1g} + A_{2g} + B_{1g} + B_{2g} + E_g + 2A_{2u} + B_{2u} + 3E_u$$

Subtracting the translations and rotations

$$\Gamma_{\mathrm{TR}} = A_{2g} + E_g + A_{2u} + E_u$$

(note that this is a total of six, since the E modes are double degenerate); then

$$\Gamma_{\mathrm{vib}} = A_{1g} + B_{1g} + B_{2g} + A_{2u} + B_{2u} + 2E_u$$

Table 4.9 Character Table for Point Group D_{4h}

D_{4h}	E	$2C_4$	C_2	$2C_2'$	$2C_2''$	i	$2S_4$	σ_h	$2\sigma_v$	$2\sigma_d$		
A_{1g}	1	1	1	1	1	1	1	1	1	1		$\alpha_{xx}+\alpha_{yy}, \alpha_{sz}$
A_{2g}	1	1	1	−1	−1	1	1	1	−1	−1	R_z	
B_{1g}	1	−1	1	1	−1	1	−1	1	1	−1		$\alpha_{xx}-\alpha_{yy}$
B_{2g}	1	−1	1	−1	1	1	−1	1	−1	1		α_{xy}
E_g	2	0	−2	0	0	2	0	−2	0	0	(R_x, R_y)	$(\alpha_{yz}, \alpha_{zx})$
A_{1u}	1	1	1	1	1	−1	−1	−1	−1	−1		
A_{2u}	1	1	1	−1	−1	−1	−1	−1	1	1	T_z	
B_{1y}	1	−1	1	1	−1	−1	1	−1	−1	1		
B_{2u}	1	−1	1	−1	1	−1	1	−1	1	−1		
E_u	2	0	−2	0	0	−2	0	2	0	0	(T_x, T_y)	

The active infrared modes are those in the same symmetry species as the translations x, y, z, that is, A_{2u} and E_u; while the active Raman modes belong to the same symmetry species as the elements of the polarizability tensor, that is, A_{1g}, B_{1g}, and B_{2g}. Consequently there are three active Raman modes

$$A_{1g}+B_{1g}+B_{2g}$$

and three active infrared modes

$$A_{2u}+2E_{2u}$$

4.14 Symmetry Considerations and Infrared Dichroism

The absorption of infrared radiation by a molecule can be given in terms of the absorbance A according to the formula

$$A=\log_{10}\left(\frac{I_0}{I}\right)(\mathbf{M}\cdot\mathbf{E})^2 \tag{4.11}$$

where I_0 and I are the incident and transmitted intensities of the absorbing frequency, $\mathbf{M}$ is the transition moment vector of the normal mode,

$$\mathbf{M}=\frac{\partial\mu}{\partial Q_r}$$

and $\mathbf{E}$ is the electric field vector of the incident beam at the absorbing frequency. (We will discuss infrared and Raman intensities later. In this chapter we will simply state the intensity relationships.) At normal incidence to the sample, the electric field vector can be linearly polarized in two mutually perpendicular directions. For a given normal mode of vibration, the transition moment vector has a definite orientation in the molecule. Consequently, if the angle between $\mathbf{M}$ and $\mathbf{E}$ is θ, the absorbance is proportional to $\cos^2\theta$, as illustrated in Figure 4.16.

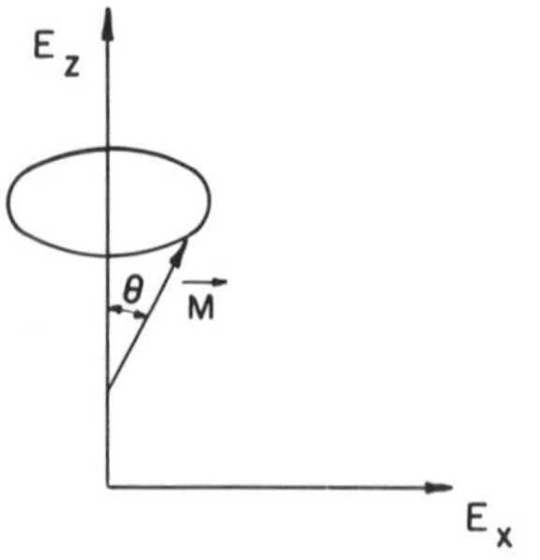

Figure 4.16. The geometry of the infrared dichroism experiment.

For gases and liquids, the absorbance is independent of the polarization of the incident beam because the random orientation of the molecules produces a random orientation of the resultant transmission moment vectors, and no preferred direction results in the sample. In the ordered solid state, for example, a single crystal, the molecules are fixed and the direction of the transition moment vector of each molecule has the same orientation in space. The absorbance of a given infrared band will then change, depending on (a) the direction of the transition moment vector of the particular normal mode with respect to the molecular axis, and (b) the polarization of the electric vector of the incident radiation. A maximum absorbance will occur when the electric vector of the polarized light is parallel to the direction of the transition moment ($\theta=0$) and no absorption will occur when the two vectors are perpendicular ($\theta=90°$). As an example, let us assume we have a single crystal of a substance that we can align perfectly ($\theta=0$) with the laboratory fixed spectrometer axis. That is, the x, y, z axes of the molecule correspond to the X, Y, Z axes in the laboratory or spectrometer framework. For this simple example, the infrared dichroic behavior can be predicted from the previously developed symmetry considerations. Infrared activity occurs only for those vibrational modes belonging to the same symmetry species as the components of the molecular dipole moment vector. The symmetry of each of these x, y, and z components is normally given in the character tables. Therefore, a mode that is optically active for the xth component of the molecular dipole moment has a maximum absorbance when the linearly polarized light vector is parallel to the x axis of the crystal, which for our example is parallel to the X axis of the laboratory. When the electric vector of the polarized light is perpendicular to the x axis, the modes having activity arising only from the x-axis component of the dipole moment will **not** absorb any light even though the modes are optically active. These active modes are "extinguished" by the geometry of the sample relative to the incident radiation. Corresponding statements apply for the modes in the symmetry species corresponding to the Y and Z directions. It is obvious that given an appropriate experimental arrangement and known orientation of the sample, infrared dichroic measurements can be used to assign the observed vibrational modes to the appropriate symmetry species. This experimental result is of great importance in assigning bands for polymers and other complicated molecules that can be prepared with anisotropic orientation. In a drawn polymer, the macromolecular chains are preferentially oriented in the direction of strain. When measurements are made with the electric vector parallel or perpendicular to the preferred direction, a dichroic ratio R

$$R=\frac{A_{\parallel}}{A_{\perp}} \tag{4.12}$$

can be measured where $A_{\parallel}$ is the absorbance for linearly polarized light.

4.15 Symmetry Considerations and Raman Polarization

When a laser beam with electric vector **E** is incident on a molecule, an electric dipole **P** is induced according to the relation

$$\mathbf{P}=\boldsymbol{\alpha}E \tag{4.13}$$

where $\boldsymbol{\alpha}$, the polarizability of the molecule, is a second-order symmetric tensor. In general, the induced dipole is not necessarily in the direction of the incident electric vector, so that equation (4.13) has the complex form

$$\begin{aligned} P_X &= \alpha_{XX}E_X+\alpha_{XY}E_Y+\alpha_{XZ}E_Z \\ P_Y &= \alpha_{YX}E_X+\alpha_{YY}E_Y+\alpha_{YZ}E_Z \\ P_Z &= \alpha_{ZX}E_Z+\alpha_{ZY}E_Y+\alpha_{ZZ}E_Z \end{aligned} \tag{4.14}$$

where the quantities α'_{FF} are independent of the components of the electric vector, but are dependent on the orientation of the molecule relative to the space-fixed axes X, Y, Z. The intensity of the Raman scattered light is proportional to the square of the magnitude of $\overline{P}$.

Let is consider the excitation of the Raman effect by a laser, which supplies linearly polarized light. The geometry of the Raman experiment is shown in Figure 4.17. The path of the exciting line is along Z, and can be polarized parallel to X or parallel to Y. The scattered radiation may be viewed at right angles to the sample, say in the X direction, in which case it may have components parallel to Y and Z. A polarized Raman spectrum is denoted by four symbols, two inside the parentheses and two outside, for example, $A(ba)C$. The symbols inside the parentheses are, left to right, the polarization

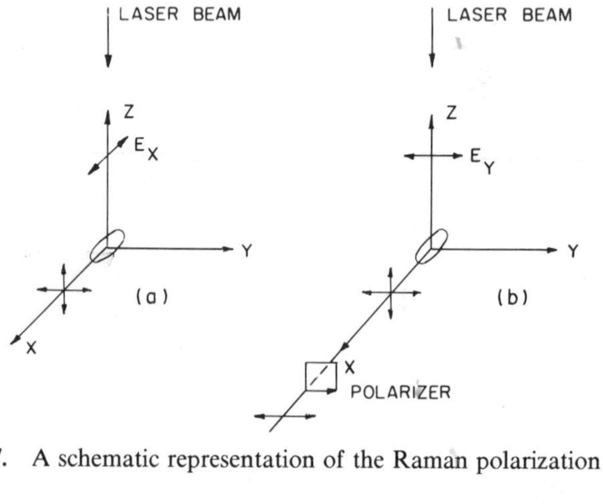

Figure 4.17. A schematic representation of the Raman polarization experiment.

directions of the incident and scattered light, while the ones to the left and the right outside the parentheses are the propagation directions of the electric vector of the incident and scattered light, respectively. Thus, for the experimental arrangement shown in Figure 4.17 we can obtain spectra with components $Z(ff')X$ with $f=x$ or y and $f'=y$ or z. If we observe the *total* scattered radiation in the X direction, as illustrated in Figure 4.17*a*, then the intensity will be proportional to the square of the induced dipole moments in the Y and Z directions, P_Y^2 and P_Z^2. Substituting these conditions ($E_Y=E_Z=0$; $P_X=0$) into equation (4.14), we obtain

$$\begin{vmatrix} 0 \\ P_Y \\ P_Z \end{vmatrix} = \begin{vmatrix} \alpha_{XX} & \alpha_{XY} & \alpha_{XZ} \\ \alpha_{YX} & \alpha_{YY} & \alpha_{YZ} \\ \alpha_{ZX} & \alpha_{ZY} & \alpha_{ZZ} \end{vmatrix} \begin{vmatrix} E_X \\ 0 \\ 0 \end{vmatrix}$$

Consequently the intensity of light collected in the X direction per unit solid angle is given by

$$I_T(\text{obs}\parallel) = C\left(P_Y^2 + P_Z^2\right) = C\left(\alpha_{YX}^2 + \alpha_{ZX}^2\right)E_X^2$$

where the symbol C represents a constant of proportionality and $I_T(\text{obs}\parallel)$ means that we are observing the total scattered light in a direction parallel to that of the electric vector of the incident beam.

Naturally, we can also have the condition of incident light propagating along the Z axis but with the electric vector parallel to the Y axis, as illustrated in Figure 4.17*b*. However, instead of collecting the total scattered light we can insert a polarizer beyond the sample and collect the part that is polarized in the direction of, for example, E_Y. We can use the symbol $I_\parallel(\text{obs}\perp)$ to designate this experimental geometry, or in terms of our space-fixed axes, $I[Z(yy)X]$. This term is given by

$$I[Z(yy)X] = C\alpha_{YY}^2 E_Y^2$$

Substituting for C and using the relation

$$I_0 = \frac{c}{8\pi}E^2$$

where c is the velocity of light, we obtain

$$I[Z(yy)X] = \frac{16\pi^4\nu^4}{c^4} I_0\left(\alpha_{YY}^2\right)$$

In a similar manner, the intensity of Raman scattered light will be related to various components of the polarizability tensor, depending on the direction of the electric vector of the incident light and the direction in which the scattered light is collected. For a perfectly oriented system—for example, a single crystal—it is possible to align perfectly the molecular axis (g) with the space-fixed axes (F). There are six ways of identifying the two axis systems and since a pair of polarized spectra is obtained for each location, we obtain a total of 12 spectra. Usually the directions of the crystallographic axes can be recognized and the crystal placed in the sample holder in such a way that the edges of the cube are parallel to the space-fixed axes X, Y, and Z. For single crystals, an X-ray goniometer head can be used for the Raman experiment and the X-ray pattern can be used to establish the direction of the crystal axes. The anisotropy ($\alpha_{ff'}$) of scattering by a vibration of any given symmetry can be predicted by standard group theoretical methods and are generally given in the character tables.

The polarized Raman spectra of triclinic n-paraffins will serve as an example (6). The Raman polarization spectra of a triclinic single crystal of n-$C_{24}H_{50}$ are shown in Figure 4.18. The scattering geometries are also shown in this figure. Here a means parallel to the crystallographic a axis, b' is perpendicular to it on the film surface, and c' is perpendicular to the film surface. Only those vibrations of the methylene units that are most in phase (see Chapter 10)

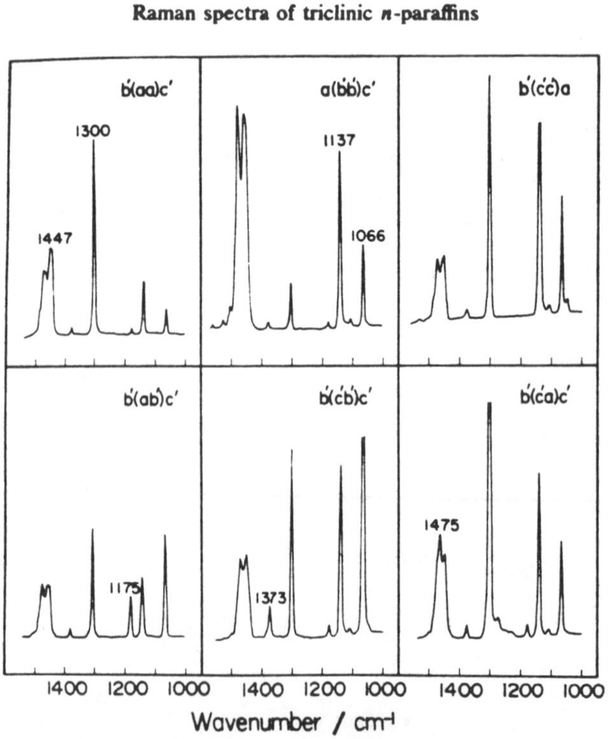

Figure 4.18. Polarized Raman spectra (1000–1500 cm^{-1}) of n-$C_{24}H_{50}$ at room temperature. (Reprinted by permission from ref. 6.)

Table 4.10 Observed Raman Frequencies and Polarizations of Single-Crystal $n-C_{24}H_{50}$ in the Range of 1000–1500 cm^{-1}

Frequency (cm^{-1})	Polarization[a]	Mode[b]	Species[c]
1066	$b'c'$	v(CC)	B_{1g}
1137	$c'c'$	v(CC)	A_g
1175	ab'	$r(CH_2)$	B_{3g}
1300	ac'	$t(CH_2)$	B_{2g}
1373	$b'c'$	$w(CH_2)$	B_{1g}
1447	$b'b'$	$\delta(CH_2)$	A_g
1475	$b'b'$	$\delta d(CH_3)$	—

[a]Polarization giving the most intense scattering.
[b]v denotes stretching, r, rocking, t, twisting, w, wagging, δ scissoring, and d degenerate.
[c]Symmetry species in the factor group approximation (D_{2h}).

with one another will exhibit appreciable intensity, so the observed Raman bands coincide with those of the D_{2h} factor group of polyethlene. The observed polarizations of the single crystal of $n—C_{24}H_{50}$ in the range of 1000–1500 cm^{-1} are consistent with the assignments given in Table 4.10.

An aspect of the Raman spectrum that differs fundamentally from the infrared is the ability to observe band polarization in liquids and gases where the molecules are randomly oriented. This is because the squares of the polarizability components average to different values over all angles. In order to derive an appropriate set of equations for such a system we have to introduce a second reference system (x, y, z) fixed to the molecule and rotating with it. It is possible to express the result (see refs. 3 and 4) in terms of two quantities associated with the tensor α, the **mean value** $\bar{\alpha}$, and the **anisotropy** γ, defined as follows:

$$\bar{\alpha}=\tfrac{1}{3}(\alpha_{xx}+\alpha_{yy}+\alpha_{zz}) \tag{4.15}$$

$$\gamma^2=\tfrac{1}{2}\Big[(\alpha_{xx}-\alpha_{yy})^2+(\alpha_{yy}-\alpha_{zz})^2 + (\alpha_{zz}-\alpha_{xx})^2+6\big(\alpha_{xy}^2+\alpha_{yz}^2+\alpha_{zx}^2\big)\Big] \tag{4.16}$$

Then for 90° scattering we can define an experimentally measurable quantity ρ, known as the depolarization ratio, and representing the ratio of intensities scattered perpendicular and parallel to the direction of the electric vector, E. The polarization ratio ρ is given by

$$\rho=\frac{3\gamma^2}{45\bar{\alpha}^2+4\gamma^2} \tag{4.17}$$

and can vary between the limits

$$0 \leqslant \rho \leqslant \tfrac{3}{4}$$

since $\bar{\alpha}$ can be equal to zero. The cases in which $\bar{\alpha}$ is equal to zero can be obtained by symmetry considerations. In fact, $\bar{\alpha}$ is different from zero only for totally symmetric normal modes. Consequently, for totally symmetric modes, $\rho < \frac{3}{4}$, and such Raman lines are said to be **polarized**. For nontotally symmetric vibrations, $\rho = \frac{3}{4}$ and the Raman line is said to be **depolarized**.

Finally, when we consider polymers it is possible to obtain samples that are uniaxially oriented. In other words, one of the axes can be oriented but the other two are assumed to be random. A particular example is for a bundle of oriented polymer fibers. The z component of the tensor, by convention defined parallel to the fiber axis, can be determined, but the x and y components are randomly distributed. It is possible to place the partially oriented sample in the three different alignments shown in Figure 4.19 and designated A, B, and C, corresponding to alignment of the z axis parallel to the Y, X, and Z space-fixed axes, respectively. The Raman scattering activities for such partially oriented systems have been derived by Snyder (7).

If we consider the sample orientation A, the transformation matrix relating axis systems F and g that allows us to express the elements α_{FF} in terms of elements α_{gg}, based on principal axes of polarizability, is

$$(A) = \begin{Bmatrix} -\cos\phi & \sin\phi & 0 \\ 0 & 0 & 1 \\ \sin\phi & \cos\phi & 0 \end{Bmatrix}$$

were ϕ is the angle between Z and Y. The matrices $T(B)$ and $T(C)$ are similar except for the appropriate permutation of rows and columns.

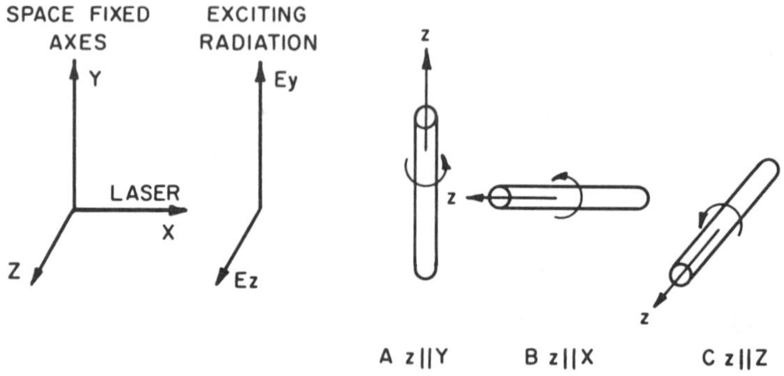

Figure 4.19. The geometry of polarized Raman studies of uniaxially oriented samples.

It is required to average over ϕ to determine the polarizability elements referred to space-fixed axes. The results are

$$\alpha(F)_A^2 = \begin{Bmatrix} A_1 & A_3 & A_2 \\ A_3 & A_4 & A_3 \\ A_2 & A_3 & A_1 \end{Bmatrix}$$

$$\alpha(F)_B^2 = \begin{Bmatrix} A_4 & A_3 & A_3 \\ A_3 & A_1 & A_2 \\ A_3 & A_2 & A_1 \end{Bmatrix}$$

$$\alpha(F)_C^2 = \begin{Bmatrix} A_1 & A_2 & A_3 \\ A_2 & A_1 & A_3 \\ A_3 & A_2 & A_4 \end{Bmatrix}$$

where

$$A_1 = (1/8)(2a^2 + b^2) \qquad \text{and} \qquad a^2 = (\alpha_{xx} + \alpha_{yy})^2$$

$$A_2 = (1/8)b^2 \qquad\qquad b^2 = (\alpha_{xx} - \alpha_{yy})^2 + 4\alpha_{xy}^2$$

$$A_3 = (1/2)c^2 \qquad\qquad c^2 = \alpha_{yz}^2 + \alpha_{zx}^2$$

$$A_4 = d^2 \qquad\qquad d^2 = \alpha_{zz}^2$$

Table 4.11 Raman Scattering Activities for Right-Angle Viewing of Partially Oriented Polypropylene

Point Group Type	Species	Polarization of Excitation	Polarization of Raman Radiation[a]					
			A		B		C	
			X	Y	X	Y	X	Y
C_3	A	Y	0	d^2	0	$\frac{a^2}{4}$	0	$\frac{a^2}{4}$
		Z	0	0	0	0	0	0
	E	Y	$\frac{c^2}{2}$	0	$\frac{c^2}{2}$	$\frac{b^2}{8}$	$\frac{b^2}{8}$	$\frac{b^2}{8}$
		Z	$\frac{b^2}{8}$	$\frac{c^2}{2}$	$\frac{c^2}{2}$	$\frac{b^2}{8}$	$\frac{c^2}{2}$	$\frac{c^2}{2}$

[a] Right-angle viewing.

For the uniaxially oriented case, we have four independent terms A_1, A_2, A_3, and A_4, which is two less than the perfectly oriented case, but two more than the random case. Although we would like to measure the absolute values of A_1, A_2, A_3, and A_4, this is usually not feasible. But we can establish the symmetry of the vibration from the intensity changes observed when either the sample orientation or the direction of polarization of the excitation is changed.

For optically homogeneous samples, the symmetry species of any molecular vibration can be uniquely determined. For example, for fibers of polypropylene the vibrational modes may be separated under point group C_3 into species A and E (see Chapter 11). With Z polarization, the A modes are extinguished but the E modes are not. The complete Raman activities for all three cases are show in Table 4.11.

References

1 J. Friendly, reported in *The New York Times*, Sunday, June 22, 1980, p. E11.

2 D. C. Harris and M. D. Bertolucci, "Symmetry and Spectroscopy," Oxford University Press, New York, 1978.

3 E. B. Wilson, J. C. Decius, and P. C. Cross, "Molecular Vibrations," McGraw-Hill, New York, 1955.

4 S. Califano, "Vibrational States," Wiley, London, 1976.

5 B. Jasse and J. L. Koerig, *J. Macromol. Sci.* **C17** (1), 61 (1979).

6 M. Kobayshi, T. Kobayshi, T. Uesaka, and H. Tadokoro, *Spectrochim. Acta* **35A**, 1277 (1979).

7 R. G. Snyder, *J. Mol. Spect.* **37**, 353 (1971).

5

THE VIBRATIONAL PROBLEM IN SYMMETRY COORDINATES

Mankind always sets itself only such problems as it can solve; since, looking at the matter more closely it will always be found that the task itself arises only when the material conditions for its solution already exist or are at least in the process of formation.

Karl Marx

The use of normal coordinates as a basis set considerably simplifies the vibrational problem by diagonalizing the secular equation. However, we must first solve the secular equation to obtain these normal coordinates. In this chapter we will discuss the use of the symmetry of a molecule to obtain a set of **symmetry coordinates**, which can also considerably simplify the form of the secular equation. In special cases the secular equation is completely diagonalized and the symmetry coordinates therefore correspond to normal coordinates. In order to appreciate the simplifications introduced, consider the vibrations of the one-dimensional CO_2 molecule expressed in terms of internal coordinates, as shown previously in Figure 3.2.

We will discuss solutions in terms of symmetry coordinates, S. Let

$$s_1 = (1/\sqrt{2})(r_{12} + r_{23}) \tag{5.1}$$

$$s_2 = (1/\sqrt{2})(r_{12} - r_{23}) \tag{5.2}$$

The methods used to construct these symmetry coordinates will be given later.

We now need to transform the **F** and **G** matrices from the internal coordinate system **R** to the symmetry coordinate system **S**, in the matrix form:

$$\mathbf{S} = \mathbf{U}\mathbf{R} \tag{5.3}$$

where

$$\mathbf{U}=\begin{bmatrix} \frac{1}{\sqrt{2}} & \frac{1}{\sqrt{2}} \\ \frac{1}{\sqrt{2}} & \frac{-1}{\sqrt{2}} \end{bmatrix} \tag{5.4}$$

It should be noted that when the symmetry coordinates are normalized, as in this case, the **U** matrix is orthogonal; that is,

$$\mathbf{U}^T=\mathbf{U}^{-1} \qquad \text{and} \qquad \mathbf{U}\mathbf{U}^T=\mathbf{E} \tag{5.5}$$

The potential and kinetic energies expressions derived in Chapter 3 can now be written

$$2T=\dot{\mathbf{R}}^T\mathbf{G}^{-1}\dot{\mathbf{R}}=\dot{\mathbf{S}}^T\mathbf{U}\mathbf{G}^{-1}\mathbf{U}^T\dot{\mathbf{S}} \tag{5.6}$$

$$2V=\mathbf{R}^T\mathbf{F}_r\mathbf{R}=\mathbf{S}^T\mathbf{U}\mathbf{F}_r\mathbf{U}^T\mathbf{S} \tag{5.7}$$

The **G** and **F** matrices can therefore be defined in terms of symmetry coordinates:

$$\mathbf{G}_S^{-1}=\mathbf{U}\mathbf{G}^{-1}\mathbf{U}^T \tag{5.8}$$

$$\mathbf{G}_S=\mathbf{U}\mathbf{G}\mathbf{U}^T \tag{5.9}$$

and

$$\mathbf{F}_S=\mathbf{U}\mathbf{F}_r\mathbf{U}^T \tag{5.10}$$

and using equations (5.6) and (5.7) as before, we can derive the secular determinant in the following form:

$$|\mathbf{G}_S\mathbf{F}_S-\lambda\mathbf{E}|=0 \tag{5.11}$$

Similarly, it can be shown that the eigenvectors $\mathbf{L}_S$ of this equation are given by

$$\mathbf{L}_S=\mathbf{U}\mathbf{L}_r \tag{5.12}$$

The **G** and **F** matrices for our model system, CO_2, were derived in Chapter 2:

$$\mathbf{F}_r = \begin{vmatrix} f & k \\ k & f \end{vmatrix} \tag{5.13}$$

$$\mathbf{G} = \begin{bmatrix} \dfrac{1}{m_O} + \dfrac{1}{m_C} & -\dfrac{1}{m_C} \\ -\dfrac{1}{m_C} & \dfrac{1}{m_O} + \dfrac{1}{m_C} \end{bmatrix} \tag{5.14}$$

so

$$\mathbf{F}_S = \mathbf{U}\mathbf{F}_r\mathbf{U}^T = \begin{Bmatrix} f+k & 0 \\ 0 & f-k \end{Bmatrix} \tag{5.15}$$

$$\mathbf{G}_S = \mathbf{U}\mathbf{G}\mathbf{U}^T = \begin{bmatrix} \dfrac{1}{m_O} & 0 \\ 0 & \dfrac{1}{2m_O} + \dfrac{1}{m_C} \end{bmatrix} \tag{5.16}$$

The $\mathbf{G}_S\mathbf{F}_S$ matrix is therefore

$$\mathbf{G}_S\mathbf{F}_S = \begin{Bmatrix} \dfrac{1}{m_O}(f+k) & 0 \\ 0 & \left(\dfrac{1}{2m_O} + \dfrac{1}{m_C}\right)(f-k) \end{Bmatrix} \tag{5.17}$$

and the secular equation is

$$\begin{Bmatrix} \left(\dfrac{f+k}{m_O}\right) - \lambda & 0 \\ 0 & \left(\dfrac{1}{2m_O} + \dfrac{1}{m_C}\right)(f-k) - \lambda \end{Bmatrix} = 0 \tag{5.18}$$

so that the two roots, λ_1 and λ_2, can be derived directly as

$$\lambda_1 = \frac{(f+k)}{m_O}, \qquad \lambda_2 = (f-k)\left(\frac{1}{2m_O} + \frac{1}{m_C}\right)$$

In this example the symmetry coordinates completely diagonalized the **G** and **F** matrices, so that the $\mathbf{G}_S\mathbf{F}_S$ matrix was also diagonal. For more complex molecules the simplification is not so great, and the **G** and **F** matrices are usually reduced to a block diagonal form, as illustrated schematically in Figure 4.10. Nevertheless, this is still a major reduction in the dimensions of the problem.

Each of these blocks corresponds to a symmetry species of the molecular point group. This can be demonstrated by considering the equations for the kinetic and potential energies in nonmatrix form, as presented in Chapter 3. Cross terms of the type $\dot{r}_{12}\dot{r}_{23}$ occur in the kinetic energy expression which would transform to $\dot{s}_1\dot{s}_2$ in terms of symmetry coordinates. If $\dot{s}_1$ and $\dot{s}_2$ belonged to different symmetry species, there would be at least one symmetry operation for which one of these coordinates would not change in sign while the other would; that is, the product $\dot{s}_1\dot{s}_2$ would change in sign. However, the potential and kinetic energy must be invariant with respect to all symmetry operations. Consequently, such cross terms between different species give rise to zeros in the $\mathbf{G}_S$ and $\mathbf{F}_S$ matrices, and the secular determinant assumes the characteristic block diagonal form, with each block corresponding to one species. The determinant can now be written as the product of factors corresponding to each symmetry species.

There are several useful types of symmetry coordinates. We will consider the two most frequently used, known as internal symmetry coordinates and external symmetry coordinates. The former consist of linear combinations of internal coordinates; the latter are constructed from linear combinations of cartesian displacement coordinates. In this chapter we will consider only the generation of the symmetry coordinates of simple molecules. The methods are the same when applied to polymers, as will be shown in Chapter 11.

5.1 The Construction of Internal Symmetry Coordinates

In Section 5.1 we used a set of internal symmetry coordinates to simplify the form of the secular equation of CO_2. These nondegenerate symmetry coordinates can be easily derived from the characters of a given symmetry species (1). We will not repeat the derivation of the formula but simply state the result:

$$S^{(A)} = N \sum_{p} \chi_p^{(A)} \cdot PR_1 \tag{5.19}$$

where $S^{(A)}$ is the symmetry coordinate (belonging to symmetry species A), N is a normalization factor, $\chi_p^{(A)}$ is the character of symmetry operation P in symmetry species A, PR_1 is the coordinate to which the displacement R_1 is transferred by the symmetry operation P, and N is given by

$$\frac{1}{\sqrt{a^2}} \tag{5.20}$$

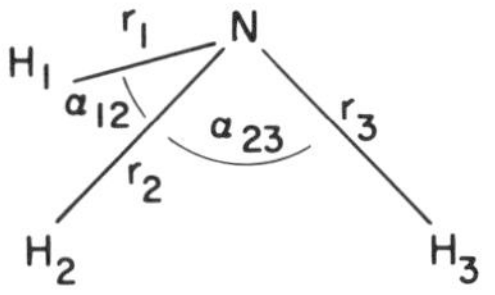

Figure 5.1. Definition of the internal coordinates of ammonia.

where a^2 is the sum of the squared coefficients. Degenerate coordinates can be generated by a similar rule. We will illustrate the construction of both types of coordinates using ammonia as an example.

The internal coordinates are defined as in Figure 5.1. By the methods of Chapter 5 it can be shown that this molecule belongs to the point group C_{3v} and has $2A_1$ modes and $2E$ modes. We therefore have to construct symmetry coordinates for each of these symmetry species.

The first step is the determination of the changes in internal coordinates that take place upon a given symmetry operation. These are listed in Table 5.1.

The character table together with equation (5.19) can now be used to generate the symmetry coordinates. Symmetry coordinates are constructed out of equivalent internal coordinates, that is, those that are exchanged by the symmetry operations of the point group. Consequently, it can be seen from Table 5.1 that we can construct a maximum of two symmetry coordinates for each symmetry species, one from the bond stretching coordinates and one from the angle bending coordinates, since any single bond stretching or bond angle bending coordinate will generate all the other bond stretches and bends under the symmetry operations of the group.

For the A_1 modes,

$$S_1^{(A')} = N[(1)r_1 + (1)r_2 + (1)r_3 + (1)r_1 + (1)r_2 + (1)r_3]$$
$$= N\{2r_1 + 2r_2 + 2r_3\}$$

Normalizing, we obtain

$$N = (2^2 + 2^2 + 2^2)^{-1/2} = \frac{1}{2\sqrt{3}}$$

Table 5.1 The Transformation of the Internal Coordinates of NH_3 under the Symmetry Operations of the Point Group C_{3v}

	I	C_3^1	C_3^2	σ_v	σ_v'	σ_v''
r_1	r_1	r_2	r_3	r_1	r_2	r_3
r_2	r_2	r_3	r_1	r_3	r_1	r_2
r_3	r_3	r_1	r_2	r_2	r_3	r_1
α_{12}	α_{12}	α_{23}	α_{13}	α_{13}	α_{12}	α_{23}
α_{13}	α_{13}	α_{12}	α_{23}	α_{12}	α_{23}	α_{13}
α_{23}	α_{23}	α_{13}	α_{12}	α_{23}	α_{13}	α_{12}

Thus,

$$S_1^{(A_1)} = \frac{1}{\sqrt{3}}(r_1 + r_2 + r_3)$$

Similarly,

$$S_2^{(A_1)} = \frac{1}{\sqrt{3}}(\alpha_{12} + \alpha_{13} + \alpha_{23})$$

[Note that if there were only one symmetry coordinate (and hence one normal mode) in this symmetry species, we would have determined either S_1 or $S_2 = 0$.]

For the E symmetry species

$$S_1^{(E)} = N[(2)r_1 + (-1)r_2 + (-1)r_3 + (0)r_1 + (0)r_2 + (0)r_3]$$

$$S_1^{(E)} = \frac{1}{\sqrt{6}}(2r_1 - r_2 - r_3)$$

Similarly,

$$S_2^{(E)} = \frac{1}{\sqrt{6}}(-\alpha_{12} - \alpha_{13} + 2\alpha_{23})$$

So the **U** matrix, defined by

$$\mathbf{S} = \mathbf{U}\mathbf{R}$$

is described by

$$\mathbf{U} = \begin{bmatrix} \frac{1}{\sqrt{3}} & \frac{1}{\sqrt{3}} & \frac{1}{\sqrt{3}} & 0 & 0 & 0 \\ 0 & 0 & 0 & \frac{1}{\sqrt{3}} & \frac{1}{\sqrt{3}} & \frac{1}{\sqrt{3}} \\ \frac{2}{\sqrt{6}} & \frac{-1}{\sqrt{6}} & \frac{-1}{\sqrt{6}} & & & \\ & & & \frac{-1}{\sqrt{6}} & \frac{-1}{\sqrt{6}} & \frac{2}{\sqrt{6}} \end{bmatrix}$$

The reduction in the size of the vibrational problem can be demonstrated further by setting up the **GF** matrix in order to calculate force constants from observed frequencies. Recalling that

$$\mathbf{G}_S = \mathbf{U}\mathbf{G}\mathbf{U}^T, \qquad \mathbf{F}_S = \mathbf{U}\mathbf{F}\mathbf{U}^T$$

where **F** has the form

$$\mathbf{F} = \begin{bmatrix} k & k' & k' & & & \\ k' & k & k' & & 0 & \\ k' & k' & k & & & \\ & & & f & f' & f' \\ & 0 & & f' & f & f' \\ & & & f' & f' & f \end{bmatrix}$$

where k' is the interaction force constant between adjacent stretching coordinates and f' is the interaction force constant between adjacent bending coordinates, we can obtain

$$\mathbf{F}_S = \begin{bmatrix} 2k'+k & 0 & 0 & 0 \\ 0 & 2f'+f & 0 & 0 \\ 0 & 0 & k-k' & 0 \\ 0 & 0 & 0 & f-f' \end{bmatrix}$$

Partitioning the diagonal elements, we form

$$\mathbf{F}_S^{(A_1)} = \begin{bmatrix} 2k'+k & 0 \\ 0 & 2f'+f \end{bmatrix}$$

$$\mathbf{F}_S^{(E)} = \begin{bmatrix} k-k' & 0 \\ 0 & f-f' \end{bmatrix}$$

The same partitioning can be performed for the **G** matrix, resulting in $\mathbf{G}_S^{(A_1)}$ and $\mathbf{G}_S^{(E)}$.

For $\mathbf{G}_S^{(A_1)}$ one ultimately obtains

$$\mathbf{G}_S^{(A_1)} = \begin{bmatrix} \dfrac{1}{M_{\mathrm{H}}} + \dfrac{1}{3M_{\mathrm{N}}} & \dfrac{-2\sqrt{2}}{3M_{\mathrm{N}}} \\ \dfrac{-2\sqrt{2}}{3M_{\mathrm{N}}} & \dfrac{8}{3M_{\mathrm{N}}} + \dfrac{1}{8M_{\mathrm{H}}} \end{bmatrix} = \begin{bmatrix} 1.0238 & -0.0673 \\ -0.0673 & 0.3154 \end{bmatrix}$$

From $\mathbf{F}_S^{(A_1)}$ let $a = 2k'+k$, $b = 2f'+f$, from which a and b can be determined from the observed frequencies; for A_1

$$\nu_1 = 3330\ \mathrm{cm}^{-1}, \qquad \lambda_1 = 6.530$$

$$\nu_2 = 950\ \mathrm{cm}^{-1}, \qquad \lambda_2 = 0.531$$

The $\mathbf{G}_S^{(A_1)}\mathbf{F}_S^{(A_1)}$ product can be written as

$$\mathbf{G}_S^{(A_1)}\mathbf{F}_S^{(A_1)} = \begin{bmatrix} 1.0238 & -0.0673 \\ -0.0673 & 0.3154 \end{bmatrix} \begin{bmatrix} a & 0 \\ 0 & b \end{bmatrix}$$

$$= \begin{bmatrix} 1.0238a & -0.0673b \\ -0.0673a & 0.3154b \end{bmatrix}$$

We now require

$$|\mathbf{G}_S\mathbf{F}_S - \lambda E| = \mathbf{0}$$

or

$$\begin{vmatrix} 1.0238a-\lambda & -0.0673b \\ -0.0673a & 0.3154-\lambda \end{vmatrix} = 0$$

which gives

$$1.0238a + 0.3154b = \lambda_1 + \lambda_2 = 7.061$$

$$0.3184ab = \lambda_1\lambda_2 = 3.467$$

from which

$$a = 6.369, \qquad b = 1.710$$

Therefore,

$$2k' + k = 6.369 \times 10^5 \text{ dyn/cm}$$

$$2f' + f = 1.710 \times 10^5 \text{dyn/cm}$$

Using the same procedure for the E species, we can derive additional equations for calculating the valence force constants.

Although the construction of degenerate E symmetry species coordinates was trivial for the example chosen, NH_3, this is usually not the case and care has to be taken. We will consider this point in a little more detail before proceeding to a corresponding discussion of external symmetry coordinates.

5.2 Symmetry Coordinates for Degenerate Species

For degenerate symmetry species we can construct symmetry coordinates in exactly the same way as above, provided that only one internal coordinate set contributes to a given symmetry species. This is obviously not the case for NH_3, where we have internal coordinate sets r and α. In such cases the first symmetry coordinate s_1 is constructed in the same way, but the others must use coordinates or sets of coordinates that have the same symmetry properties as s_1. In addition, all symmetry coordinates must be orthogonal. Constructing such degenerate internal symmetry coordinates is trivial for NH_3, since the angle bending coordinate α, say α_{23}, opposite the bond stretching coordinate r_1 has the same symmetry properties as r_1 and is clearly orthogonal. However, for

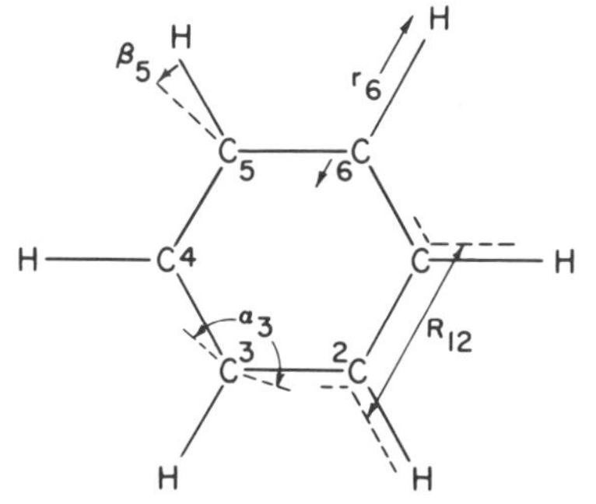

Figure 5.2. Definition of the internal coordinates of benzene.

a molecule such as benzene, whose in-plane internal coordinates are illustrated in Figure 5.2, the choice of coordinates for construction of the E_{1u} symmetry coordinates is more complicated. For example, if we chose the C—H stretching coordinate r_i to construct the first symmetry coordinate, construction of symmetry coordinates involving the ring angle bend α will be equivalent, since these coordinates transform under the symmetry operations of the group in the same way. However, C—C stretching coordinates R and hydrogen bending coordinates β transform in a different manner. Consequently, we have to choose appropriate combinations of the various R (and of the various β) coordinates as our basis set. One such combination is $R_1 + R_6$ (similarly, $\beta_2 - \beta_6$).

Each set of coordinates has to be constructed orthogonal to all other sets, so that there are no cross terms involving different blocks in the symmetry-reduced **G** and **F** matrices. A special case involves redundant coordinates. When solving the vibrational problem using internal symmetry coordinates the redundancy is best coped with by expressing the redundant condition as one of the symmetry coordinates. The other symmetry coordinates in the symmetry species can then be constructed orthogonal to this coordinate.

The construction of degenerate and orthogonal symmetry coordinates is best illustrated by an example. Since the method is the same for both internal and external symmetry coordinates, we will return to this problem in more detail in the following section.

5.3 The Construction of External Symmetry Coordinates

External symmetry coordinates consist of linear combinations of cartesian displacement coordinates. Equation (5.19) can be used in the same way as earlier to construct nondegenerate coordinates. We will illustrate a method for constructing degenerate coordinates by using the x and y coordinates of the hydrogen atoms of NH_3 as an example. These coordinates are illustrated in Figure 5.3.

When we use cartesian displacement coordinates we do not remove the translations and rotations. We therefore also construct symmetry coordinates

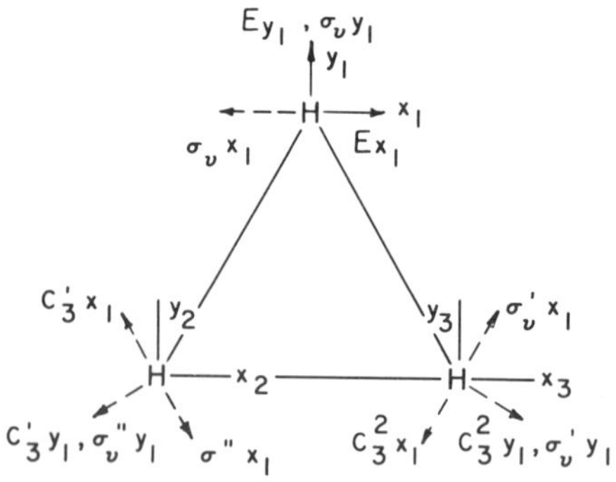

Figure 5.3. The effect of symmetry operations on the cartesian displacement coordinates of the hydrogen atoms of a molecule of NH_3.

for symmetry species A_2. The eigenvalues corresponding to translations and rotations will, of course, be zero roots of the secular equation. Consider Table 5.2, which shows the changes in the x and y coordinates that take place under the symmetry operations of the molecular point group. Since all of the x and y coordinates of the H atoms of the molecules have been generated by the symmetry operations, we do not have to consider them further in order to derive a complete set of symmetry coordinates. The character table of point group C_{3v} can be used with equation (5.19) to generate the external symmetry coordinates.

For example, for symmetry species A_1 using x_1 as a basis

$$S^{A_1}=N\left[(1)x_1+(1)\left(\left(-\frac{1}{2}\right)x_2+\left(\frac{\sqrt{3}}{2}\right)y_2\right)+(1)\left(\left(-\frac{1}{2}\right)x_3-\left(\frac{\sqrt{3}}{2}\right)y_3\right)\right.$$

$$\left.-(1)x_1+(1)\left(\left(\frac{1}{2}\right)x_3+\left(\frac{\sqrt{3}}{2}\right)y_3\right)+(1)\left(\left(\frac{1}{2}\right)x_2-\left(\frac{\sqrt{3}}{2}\right)y_2\right)\right]=0$$

That is, there is no external symmetry coordinate based on the combination of cartesian coordinates generated from x_1 in Table 5.2.

However, using the coordinates generated from y_1 we obtain

$$S^{A_1}=\frac{1}{\sqrt{3}}\left(y_1-\left(\frac{1}{2}\right)y_2-\left(\frac{\sqrt{3}}{2}\right)x_2-\left(\frac{1}{2}\right)y_3+\left(\frac{\sqrt{3}}{2}x_3\right)\right)$$

Table 5.2 Transformation of Cartesian Coordinates under the Symmetry Operations of the Point Group C_{3v}

	I	C_3^1	C_3^2	σ_v	σ_v'	σ_v''
x_1	x_1	$\left(-\frac{x_2}{2}+\frac{\sqrt{3}\,y_2}{2}\right)$	$\left(-\frac{x_3}{2}-\frac{\sqrt{3}\,y_3}{2}\right)$	$-x_1$	$\frac{x_3}{2}+\frac{\sqrt{3}\,y_3}{2}$	$\frac{x_2}{2}-\frac{\sqrt{3}\,y_2}{2}$
y_1	y_1	$\left(\frac{\sqrt{3}\,x_2}{2}-\frac{y_2}{2}\right)$	$\left(\frac{\sqrt{3}\,x_3}{2}-\frac{y_3}{2}\right)$	y_1	$-\frac{\sqrt{3}}{2}x_2-\frac{y_2}{2}$	$\frac{\sqrt{3}\,x_3}{2}-\frac{y_3}{2}$

To generate symmetry coordinates of the E modes we can start with, say, y_1 and generate

$$S_1^{(E)} = \frac{1}{\sqrt{6}}\left(2y_1 + \left(\frac{1}{2}y_2 + \left(\frac{\sqrt{3}}{2}\right)x_2 + \left(\frac{1}{2}\right)y_3 - \left(\frac{\sqrt{3}}{2}\right)x_3\right)\right.$$

Then, as noted in Section 5.2, the second degenerate pair should be generated by a coordinate or combination of coordinates that has the same symmetry properties as y_1. From Figure 5.3 it can be seen that (y_2+y_3) or (x_2-x_3) fulfills this requirement. If (y_2+y_3) is used as a coordinate, S' is obtained, which is independent of $S_1^{(E)}$ but not orthogonal to it.

$$S' = N\left[y_1 + \left(\frac{5}{2}\right)y_2 + \left(\frac{5}{2}\right)y_3 - \left(\frac{\sqrt{3}}{2}\right)x_2 + \left(\frac{\sqrt{3}}{2}\right)x_3\right]$$

An orthogonal coordinate can be constructed from a linear combination of $S_1^{(E)}$ and S' as

$$S' + k\left(S_1^{(E)}\right) = S_2^{(E)} \tag{5.21}$$

where k is a constant. Consequently,

$$S_2^{(E)} = N(y_1 + 2ky_1) + \left(\left(\frac{5}{2}\right)y_2 + \left(\frac{k}{2}\right)y_2\right) + \cdots.$$

The factor k can now be determined from the orthogonality condition. That is, the sum of the corresponding coefficients of $S_1^{(E)}$ and $S_2^{(E)}$ multiplied together must be zero:

$$(y_1 + 2ky_1)2y_1 + \left(\left(\frac{5}{2}\right)y_2 + \left(\frac{k}{2}\right)y_2\right)\left(\frac{1}{2}\right)y_2 + \cdots = 0$$

Once k has been determined $S_2^{(E)}$ can be set up and normalized. So,

$$S_2^{(E)} = \frac{1}{2\sqrt{6}}\left(3y_2 + 3y_3 - \sqrt{3}\,x_2 + \sqrt{3}\,x_3\right)$$

We will apply this procedure to the vibrational analysis of benzene, discussed in Chapter 8.

References

1 E. B. Wilson, Jr., J. C. Decius, and P. C. Cross, "Molecular Vibrations," McGraw-Hill, New York, 1955.

2 L. A. Woodward, "Introduction to the Theory of Molecular Vibrations and Vibrational Spectroscopy," Oxford (Clarendon Press), New York, 1972.

3 D. Steele, "Theory of Vibrational Spectroscopy," Saunders, Philadelphia, 1971.

6

THE MOLECULAR FORCE FIELD

Though the Life Force supplies us with its own purpose, it has no other brains to work with than those it has painfully and imperfectly evolved in our heads.

George Bernard Shaw

In the preceding chapters we have dealt only briefly with the potential function or force field. In the initial simple treatment (Chapter 2) the force constants were defined in terms of cartesian displacement coordinates. Subsequently, it was noted that force constants defined in terms of internal displacement coordinates were not only more convenient but also physically more meaningful. In this chapter we will consider the molecular force field in greater detail. In Chapter 7 we can then proceed to describe methods for determining force constants and can discuss the assumptions and simplifications required in such calculations.

Implicit in our derivations of the secular equation is the assumption that the vibrational frequencies (eigenvalues) and the form of the normal modes (eigenvectors) can be determined from a knowledge of the atomic masses, the geometry of their arrangement, and the forces acting between them. However, it is the vibrational frequencies that are known (from experimental observation), while the force field of most polyatomic molecules remains an elusive quantity. Normal coordinate analysis is primarily a problem involving the determination of appropriate force constants from the observed frequencies. The form of the normal modes can then be calculated and visualized in terms of cartesian displacement coordinates or the potential energy distribution (PED).

There are a number of difficulties associated with the determination of a suitable force field. The most critical of these is that, in general, the force constants outnumber the observed frequencies. We will elaborate this point in the next section, but first let us reconsider the general form of the potential energy function V. For a diatomic molecule V is a function of the interatomic distance R. Figure 6.1 illustrates the general form of this function for such a molecule. The potential energy tends asymptotically to approach infinity as R

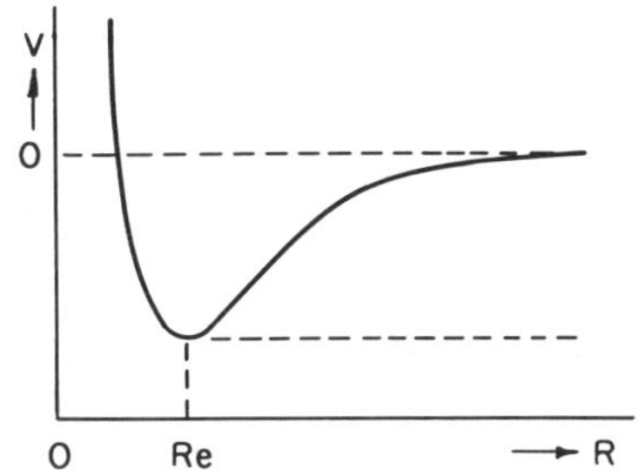

Figure 6.1. The potential energy curve of a diatomic molecule.

approaches zero, takes a minimum at an equilibrium structure where R is denoted R_e, and then increases to a constant value as R becomes large.

It is customary to determine expressions for V by a Taylor's series expansion in appropriate coordinates. In Chapter 2 we used cartesian displacement coordinates. Here we will use internal displacement coordinates r, so that $r=R-R_e$ for the bond stretching of the simple diatomic molecule. Then for the general case of a molecule with N atoms

$$2V=2V_0+2\sum_{i=1}^{3N}\left(\frac{\partial V}{\partial r}\right)_0 r_i+\sum_{i,j=1}^{3N}\left(\frac{\partial^2 V}{\partial r_i\,\partial r_j}\right)_0 r_i r_j + \sum_{i,j,k=1}^{3N}\left(\frac{\partial^3 V}{\partial r_i\,\partial r_j\,\partial r_k}\right)_0 r_i r_j r_k + \text{higher terms} \qquad (6.1)$$

The reference V_0 from which the vibrational potential energies are measured can be put equal to zero. At equilibrium the potential energy is at a minimum, so that $\partial V/\partial r$ is also equal to zero. The cubic term, together with the quartic force constant and other higher terms, is generally small compared to the quadratic force constant $\partial^2 V/\partial r_i\,\partial r_j$, and so is usually neglected. It should be noted that the anharmonic terms (cubic and higher) can become significant for certain vibrations and significantly influence the selection rules and intensities.

Even though we have (by making various assumptions) obtained a mathematical expression for the potential energy, it tells us very little about the origin and nature of the forces between the atoms of a molecule. The simplest case of a diatomic molecule has been discussed by Shimanouchi (1). The attractive energy associated with the valence electrons can be approximately separated from the repulsive energy, associated with the nuclei and the electrons localized at these nuclei. This can be written

$$V(R)=V^{\mathrm{REP}}(R)-V^{\mathrm{ATT}}(R) \qquad (6.2)$$

and is illustrated in Figure 6.2 where V^{REP} and V^{ATT} are, respectively, the repulsive and attractive components of the potential energy. At the equilibrium

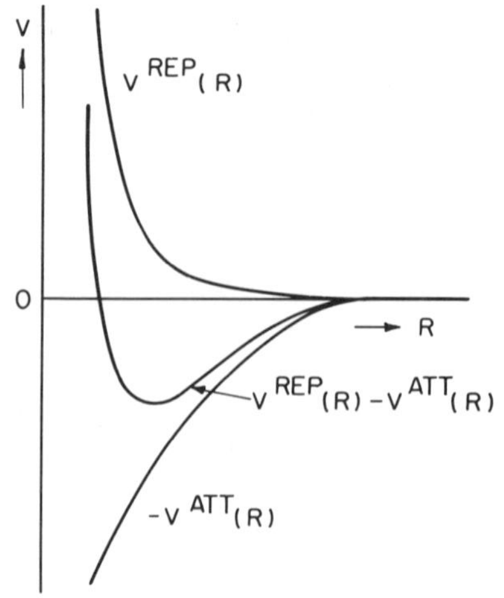

Figure 6.2. Plot of repulsive and attractive potentials.

distance R_e it is required that the attractive and repulsive energies balance:

$$\left(\frac{\partial V^{\mathrm{REP}}}{\partial R}\right)_{\mathrm{e}} = \left(\frac{\partial V^{\mathrm{ATT}}}{\partial R}\right)_{\mathrm{e}} \tag{6.3}$$

As the distance between atoms becomes less than R_e, the repulsive term dominates, so that $\partial V^{\mathrm{REP}}/\partial R$ increases more than $\partial V^{\mathrm{ATT}}/\partial R$. Furthermore, as the interatomic distance becomes greater than R_e, $\partial V^{\mathrm{ATT}}/\partial R$ decreases more slowly than $\partial V^{\mathrm{REP}}/\partial R$, since the attractive force dominates. Consequently, $\partial^2 V^{\mathrm{REP}}/\partial R^2$ is larger than $\partial^2 V^{\mathrm{ATT}}/\partial R^2$ at the equilibrium point and the quadratic force constant is more dependent on the repulsive term. This argument can be generalized to the interactions between two atoms forming a bond in a polyatomic molecule (1). The repulsive term dominates in determining the force constant, but by similar arguments the attractive term is more significant in determining the molecular energy involving light atoms, such as hydrogen, where the magnitude of the displacements is greater than for normal modes involving heavier atoms. Even though the harmonic approximation is still adequate, care has to be taken when corefining frequencies from deuterated molecules, since the anharmonic terms for the two isotopes can be significantly different because of the large difference in mass.

6.1 The General Quadratic Potential Function

In the general form of the harmonic approximation the potential function can be written in terms of cartesian displacement coordinates as

$$2V = \sum_{i,j=1}^{3N} f_{ij} q_i q_j \tag{6.4}$$

The force constant matrix consists of $(3N)^2$ terms. However, symmetry requires that $f_{ij}=f_{ji}$, so that this number is reduced. In addition, the choice of cartesian displacement as basis coordinates is not the most convenient or meaningful, since there are only $3N-6$ independent coordinates for nonlinear molecules. However, the coordinate set chosen should be related in a simple manner to the cartesian coordinates because the kinetic energy term is usually based on this coordinate system.

The coordinates chosen should also be as close as possible to the normal coordinates, so that the off-diagonal force constants f_{ij} are smaller than the diagonal terms f_{ii}. Band assignments and the description of the normal modes then become simpler. Furthermore, we shall see that for complex molecules the number of force constants can greatly exceed the number of observed frequencies. If certain off-diagonal terms can to a first approximation be set equal to zero, the problem may be sufficiently simplified to allow a viable solution. Force constants expressed in terms of internal coordinates satisfy these conditions very well. These coordinates are usually not only independent of one another, but in some cases are an approximate description of the normal coordinates. Many vibrational modes can be assigned to predominantly bond stretching, angle bending, and torsional vibrations of specific bonds in a given molecule. This is not only intuitively satisfying from a chemical viewpoint, but also allows an easy description and visualization of a particular vibrational mode. In addition, the occurrence of group frequency correlations suggests that force constants in internal coordinates may be transferable.

There is one major problem with the description of the potential energy in terms of internal coordinates—the problem of redundancy, which was discussed in Chapter 5. Even though methods for selecting a set of internal coordinates in which the redundancy condition is removed have been described, with modern computer methods of solving the secular equation it is not always worthwhile. In fact, it can be advantageous to calculate the corresponding zero root (frequency) as a check on the round-off errors in the calculation. We will return to the problem of redundant coordinates in Section 6.8.

As mentioned above, the number of force constants in the general quadratic potential function is equal to the elements on or above the diagonal of the force constant matrix, which has $3N-6$ independent elements on each side (assuming that any redundant coordinates are removed). This number is equal to

$$1+2+3+\cdots+(3N-7)+(3N-6)=(\tfrac{1}{2})(3N-6)(3N-5)$$

However, this applies only to molecules with no symmetry. Consider, as an example, the water molecule, which has C_{2v} symmetry. The internal coordinate set consists of the two bond stretching coordinates r_1 and r_2 and the valence angle bend α. The force constant matrix can be written

$$2V=\begin{vmatrix} f_{11} & f_{12} & f_{13} \\ & f_{22} & f_{23} \\ & & f_{33} \end{vmatrix} \tag{6.5}$$

where f_{11}, f_{22} are the bond stretching force constants and f_{33} is the angle bending force constant. However, f_{11} and f_{22} are required to be equivalent by symmetry, as are f_{13} and f_{23}. Consequently, there are only four independent force constants to be determined instead of six. Such relationships occur in more complex molecules and can be reduced to a general principle (1). If a symmetry operation exists that sends the coordinate product $r_t r_{t'}$ into $r_{t''} r_{t'''}$, then $f_{tt'} = f_{t''t'''}$. Furthermore, in specific cases certain interaction force constants are required to be zero by symmetry. We will defer discussion of this point until Section 6.9.

Turning now to a consideration of the block diagonal form of the force constant matrix $\mathbf{F}_s$ expressed in symmetry coordinates, we note that if there are n symmetry coordinates in a particular symmetry species, the number of force constants in the corresponding $\mathbf{F}$ matrix block is

$$n+(n-1)+(n-2)+\cdots+1=\left(\frac{n}{2}\right)(n+1)$$

Obviously, the number of force constants in the general potential function is still impractically large, so that a general force field for only certain small molecules (e.g., asymmetric ABC triangular molecules) can be determined. Even for simple molecules additional data sources are often required. In terms of calculating force fields applicable to polymers, the two most important sources of additional data are obtained from studying isotopes and oligomers. We will defer our discussion of this source of additional frequencies, the use of chain molecules of finite length (e.g., paraffins in calculating force constants for polyethylene), to Chapters 10 and 13.

6.2 The Isotope Effect

In using isotopes of a particular molecule as a source of additional frequency data, it is assumed that the potential energy and the geometry of the molecule remain unchanged by substitution. However, the $\mathbf{G}$ matrix takes on different values as a result of the change in mass of various atoms. The most frequently encountered isotopes in spectroscopy are deuterated molecules. These isotopes are also extremely useful in making band assignments because they involve a large percentage change in mass of specific atoms. Consequently, vibrational modes involving large displacements of the substituted hydrogen atom are subject to a much greater frequency shift than modes in which the displacement of this atom are small.

The use of deuterium isotopes in calculating force constants can be illustrated by reference to the spectra of HCN and DCN, as discussed by Shimanouchi (1). The secular equation for HCN can be considered as two separate blocks, corresponding to in-plane and out-of-plane vibrations. Consider the in-plane stretching modes. If the force constants f_{ii}, f_{jj}, and f_{ij} are the C—H and C≡N stretching and interaction force constants, respectively, then the secular determinant is

$$\begin{bmatrix} g_{ii} & g_{ij} \\ g_{ij} & g_{jj} \end{bmatrix} \begin{bmatrix} f_{ii} & f_{ij} \\ f_{ij} & f_{jj} \end{bmatrix} = \begin{bmatrix} \lambda_a & 0 \\ 0 & \lambda_b \end{bmatrix} \tag{6.6}$$

which can be expanded to give the following two equations:

$$f_{ii}g_{ii} + f_{jj}g_{jj} + 2f_{ij}g_{ij} = \lambda_a + \lambda_b \tag{6.7}$$

$$\left(f_{ii}f_{jj} - f_{ij}^2\right)\left(g_{ii}g_{jj} - g_{ij}^2\right) = \lambda_a\lambda_b \tag{6.8}$$

With three unknown force constants these two equation cannot be solved. However, a relationship between f_{ii} and f_{ij} and between f_{jj} and f_{ij} can be determined (1). The former is plotted in Figure 6.3. A corresponding relationship can be determined from the two observed frequencies of DCN, and this is also plotted in Figure 6.3. It should be noted that there is not a unique solution for f_{ii} and f_{ij}, but two possible solutions corresponding to the intersection of the curves at points *A* and *B*. Shimanouchi (1) rejected point *B* because the values of the force constants did not correspond to those of similar molecules.

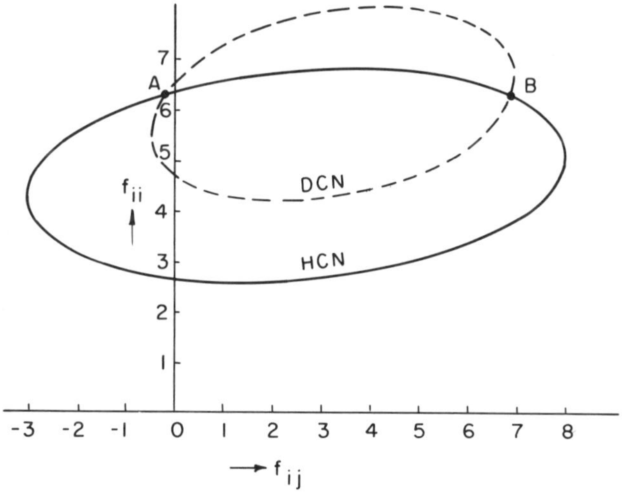

Figure 6.3. A plot of the f_{ii}, f_{ij} relationship for HCN and DCN (units of mdyn/Å). (Reprinted by permission from ref. 1.)

In addition, one would expect that the interaction force constant should be approximately an order of magnitude less than the diagonal force constant, a consideration that also indicates point A to be the more reasonable solution. However, the important point is that even for a simple molecule the use of isotopes does not necessarily lead to a unique solution of the vibrational problem. A certain degree of chemical intuition and a "feel" for the relative magnitude of certain force constants is involved. In this case, the inclusion of data from additional isotopes (e.g., ^{13}C and or ^{15}N) should theoretically allow the determination of a unique solution. However, such isotopes often result in only small frequency shifts, so that extremely precise frequency measurements are essential. In addition, other more fundamental problems can lead to uncertainties in the determination of the frequencies of the fundamental modes when there are only small isotopic frequency shifts; the most common of these is probably Fermi resonance (see Section 6.9). McKean and Duncan (2) have discussed the problems associated with the use of isotopic substitution in force field calculations. They have shown that in certain cases asymmetric isotopic substitution can be used to determine a unique solution.

6.3 Relationships Between the Vibrational Frequencies of Isotopes

The usefulness of observing frequency shifts between isotopes in making band assignments is enhanced by the application of certain rules governing these shifts. Consider the totally symmetric C—H stretching vibration of methane, CH_4, observed at 2914 cm^{-1}. This mode is observed at 2085 cm^{-1} in the completely deuterated molecule CD_4. In the limiting case of a vibration in which only the hydrogen atoms are moving, the frequency should shift a factor of $1/\sqrt{2}$ upon deuteration, since this is the square root of the ratio of the masses of the isotopically substituted atoms. In fact, the ratio of the two observed frequencies (ν_{CD_4}/ν_{CH_4}) is equal to 0.715, compared to the expected value of 0.707. The influence of anharmonic terms leads to such small discrepancies, particularly for vibrations involving hydrogen, where the amplitudes of vibration are relatively large.

The generalized form of this relationship is known as the Redlich-Teller product rule. This rule is derived in Wilson et al. (3) and relates the ratio of the product of the frequencies of two isotopic molecules in the same symmetry species as

$$\prod_{i=1}^{3N-6} \frac{\nu_i'}{\nu_i} = \prod_{j=1}^{3N} \left[\left(\frac{m_j}{m_j'} \right)^{k_j} \left(\frac{M'}{M} \right)^{t} \left(\frac{I_x'}{I_x} \right)^{\delta x} \left(\frac{I_y'}{I_y} \right)^{\delta y} \left(\frac{I_z'}{I_z} \right)^{\delta z} \right]^{1/2} \tag{6.9}$$

where ν_i refers to the normal frequencies appearing in symmetry species i, m_j is the mass of atoms belonging to the same atomic group, M is the mass of the whole molecule, and I_x, I_y, I_z refer to the moment of inertia with respect to the

x, y, z axes, respectively. The factor k_j is the number of degrees of freedom of the atomic group in question in the given symmetry species, t is the number of translations belonging to the given symmetry species, and the factors $\delta x, \delta y, \delta z$ are either 0 or 1, depending on whether there is a rotation about the appropriate axis in the given symmetry species. For degenerate vibrations in t and δ only one direction has to be considered. The factors involving M and I are included to account for the effect of translations and rotations (3). The use of this rule is best illustrated by an example, which we will defer to Chapter 9, where we consider the vibrational analysis of benzene in detail. In cases where isotopic substitution lowers the molecular symmetry, the rule has to be applied as if both molecules had this lower symmetry. In addition to the Redlich-Teller product rule, there is a rule relating the sum of the squares of the frequencies of isotopic molecules. The basis for this rule is that the sum of the squares of the frequencies is a linear function of the reciprocal masses of the atoms. Consequently, if several isotopic systems can be geometrically superimposed with appropriate signs in such a way that the atoms vanish at all positions, then the corresponding linear combinations of the sums of the squares of the frequencies should also vanish. If

$$\sigma = \sum_i \lambda_i = 4\pi^2 \sum_i \nu_i^2 \tag{6.10}$$

then for the water molecule

$$\sigma(\text{HOD}) + \sigma(\text{DOH}) - \sigma(\text{HOH}) - \sigma(\text{DOD}) = 0 \tag{6.11}$$

or for isotopes of ammonia

$$\sigma(\text{N}^{14}\text{H}_3) + \sigma(^{15}\text{ND}_3) - \sigma(^{14}\text{ND}_3) - \sigma(^{15}\text{NH}_3) = 0 \tag{6.12}$$

In the latter example all the molecules have the same symmetry, so that the sum rule applies separately to each factor of the secular equation. In order to use factoring when the isotopic molecules have different symmetry the rule has to be applied independently to the frequencies of the subgroup common to all the molecules. For the example of isotopes of water used above, this subgroup consists of the identity and mirror plane operations, constituting the subgroup C_s.

6.4 Additional Sources of Experimental Data for the Determination of Force Constants

As mentioned above, for polymeric materials additional frequency data can be obtained by using oligomers, if available. In certain cases elements of the force constant matrix of a polymer can be determined by solving the secular equation for an appropriate small molecule. For example, the force field for the phenyl group of polystyrene can be determined from simple monosubstituted alkyl benzenes. There are often additional sources of information concerning the force constants of such small molecules. These data are obtained from measurements of Coriolis coupling constants, centrifugal distortion constants, and mean amplitudes of vibration. A detailed discussion of these effects is largely outside the scope of this book, and we will only outline the origin of these constants.

The Coriolis vibrational-rotational interaction constant is due to the coupling between the two components of a doubly degenerate vibration. It can be obtained from the rotational fine structure of vibrational bands. An interaction between the components of degenerate vibrations when the molecule undergoes rotational motion results in an increase in the separation of the two levels by the rotational quantum number. The Coriolis interaction coefficient is determined by the off-diagonal force constants and can yield information that is not obtainable from the vibrational frequencies.

Centrifugal distortion coefficients are not so widely used as Coriolis constants, but have a similar application. They are obtained from microwave and electron spectra and are determined by the coefficients of the higher-order terms in the rotational energy levels. From appropriate measurements the centrifugal distortion constant can be correlated with the stretching force constants.

Mean square amplitudes of vibration can be determined from electron diffraction experiments. It is possible to demonstrate that such measurements are related to the force constants by a simple argument; the amplitude of vibration is larger for bonds with smaller force constants. However, the mean square amplitude is not very sensitive to changes in the force constants, so the results of such calculations can be uncertain.

6.5 Simplified Force Fields

It has been shown that a general force field, even in the quadratic approximation, cannot always be determined from experimental data. This is principally because there are insufficient observable vibrational frequencies relative to the number of force constants to be determined. Consequently, it is usually necessary to assume a model force field by making certain approximations. The ultimate test of such model force fields is, of course, their ability to reproduce independent experimental data. For example, a general force field determined from the vibrational frequencies of *n*-paraffins should be applicable to the determination of the normal frequencies of polyethylene. In this section we will discuss some simplified force fields and then proceed to more complex model systems that have been shown to be reasonably valid and useful in their application.

In elementary classical treatments of molecular vibrations it is assumed that the forces holding the atoms in their equilibrium position act as Hookean springs along the lines joining pairs of atoms. If we connect every pair of atoms in the molecule by such a force, whether or not these atoms are also linked by a chemical bond, then we obtain a central force field. This approximation is useful for molecules whose atoms are bonded solely by ionic interactions. In addition, the force constant matrix consists only of diagonal elements. Unfortunately, this coordinate system has to consist of the complete set of interatomic distances. The central force field has a number of deficiencies and is now little used (1). For example, in the case of CO_2 the molecule is linear and a small bond angle bend does not significantly alter any of the interatomic distances. Consequently, a zero frequency would be calculated for the bending mode using the central forces assumption, since no force resisting this motion is generated.

This difficulty with describing the bending of linear molecules is not apparent if we use another simplified force field, valence forces. The valence force field is defined in terms of the forces resisting stretching, bending, or torsion of chemical bonds. Interaction force constants or forces between nonbonded atoms are not considered in this simple approximation. For ammonia, the potential function in the valence field approximation can be simply written as

$$2V = f_r\left(r_1^2 + r_2^2 + r_3^2\right) + f_\alpha\left(\alpha_{13}^2 + \alpha_{23}^2 + \alpha_{31}^2\right) \tag{6.13}$$

where r_i is the extension of the bond and α_{ij} is the distortion of the angle between bonds i and j. [It should be noted that in many calculations $(r^0)^2 f_\alpha \alpha^2$ is used instead of $f_\alpha \alpha^2$ in defining the potential energy, where r^0 is the equilibrium length of the bond defining α; this definition gives the bending force constant the same dimensions as the stretching force constant.] By assuming a simple force field of this type we can often obtain more observed frequencies than defined force constants, so that the extra frequencies can be used as a check on the calculated force field. In this rough approximation it has been noted (1) that deviations of observed from calculated frequencies can be of the order of $\pm 10\%$. Despite the obvious oversimplification, the simple valence force field can still be useful in assigning observed infrared bands and Raman lines to modes of vibration involving specific bond stretching or angle bending coordinates. In addition, the force constants determined have been found to be approximately characteristic of the type of bond involved. For example, the C=C stretching force constant is roughly the same in whatever molecule it is found. This observation is the basis for transferring force constant values from one molecule to another of similar chemical structure. However, the vibrational spectrum is sensitive to differences in the local environment of chemical bonds, and this should be accounted for by the model force field. Consequently, the most useful models are those that attempt to strike a balance between the simple force fields that ignore all interaction terms and the general quadratic force field, which includes all interactions and is generally indeterminant.

6.6 Modified Simple Force Fields

The simple force fields discussed earlier can be modified in a number of ways to give a better fit between observed and calculated frequencies. This improvement can be achieved by introducing the interaction force constants that seem physically most necessary. In essence, such interaction force constants usually measure the effect of a particular bond distortion on the force constant describing an adjacent bond distortion. A good example is the stretch-stretch interaction of carbon dioxide, as described in Wilson et al. (3). Consider CO_2 as a mixture of the following forms:

O=C=O	$^{+}$O≡C—O^{-}	$^{-}$O—C≡O^{+}
a b	a b	a b
I	**II**	**III**

then the stretching of bond (a) in structure **I** will result in structure **III** becoming more favorable. The lengthening of bond (a) and subsequent shortening of bond (b) will tend to give them more single and triple bond character, respectively. Accordingly, the stretching force constant of bond (b) will increase, since the bond is stiffer. The stretching interaction force constant between the two bonds is positive, since the extension of one leads to a stiffening of the other. There are two stretching modes, the symmetric and antisymmetric vibrations. From the observed frequencies we can calculate two different force constants in the simple valence force field approximation, 16.8 mdyn/Å for the symmetric mode and 14.2 mdyn/Å for the antisymmetric mode. However, by introducing the interaction term, a diagonal force constant of 15.5 mdyn/Å can be calculated together with an interaction constant of 1.3 mdyn/Å. Note that the interaction force constant is an order of magnitude less than the diagonal term.

In the case of complex polyatomic molecules it can become extremely difficult to decide which interaction terms should be employed and which can be set equal to zero. Usually, there is a degree of physical reasoning or intuition that can be employed. For instance, interactions between nonconjugated bonds with no common nucleus are expected to be very small. For polymers, only interactions between adjacent monomer units are usually considered. However, the necessarily arbitrary nature of some assumptions makes a comparison of published force fields difficult. Nevertheless, particularly in the case of polymers, the use of the valence force field has been highly successful. By using overlay methods and calculating and refining force constant values (by an iterative least-squares method) for several similar molecules simultaneously, a good, transferable force field for hydrocarbons has been determined. We will discuss this force field in detail in Chapter 12. At this time, however, we reemphasize that despite the use of such overlay techniques the force field remains a model and as such is subject to revision. For example, the classic and seminal work of Snyder and Schachtschneider (4) in determining a force field for hydrocarbon polymers used an incorrect assignment for the B_{2g} wagging mode. As subsequently pointed out by Snyder (5):

> Although there are abundant illustrations in the literature of the dangers of "predicting" or "confirming" vibrational assignments using computer calculated frequencies, the present example is singularly spectacular. What is so surprising at first sight is the ease with which it is possible to fit precisely the B_{2g} frequency to an "observed" value either at 1415 cm^{-1} or at 1381 cm^{-1}, while in either case the other 185 calculated frequencies are maintained in equally excellent agreement with the observed values. This reflects in part the fact that the B_{2g} mode occupies a position on the wagging dispersion curve, which is considerably isolated from that segment well established by other wagging fundamentals. Thus an extrapolation of this curve to the limit $\phi=\pi$ is certain to involve a greater error than does the calculation of a frequency which falls on a known part.

The uncertainty inherent in a model force field where certain interaction constants are set equal to zero has persuaded many authors to favor a different type of potential function. It is generally known as the Urey-Bradley or Urey-Bradley-Shimanouchi (UBS) force field. Urey and Bradley first suggested the use of this field, but much of the development and application was accomplished by Shimanouchi. Essentially, the UBS force field is a mixed potential function, employing the principal bond stretching and bond angle bending diagonal force constants of the simple valence force field but then adding central force terms, namely, repulsions between nonbonded atoms (1). For example, for a nonlinear XY_2 molecule the potential energy is expressed as

$$2V=f_{xy}(\Delta r_1^2+\Delta r_2^2)+f_\alpha(\Delta\alpha)^2+f_{y\cdots y}\Delta q^2+F_r'(\Delta r_1+\Delta r_2)+F_q'q+F_\alpha'\Delta\alpha \tag{6.14}$$

where Δq is the change in distance between the nonbonded pair of nuclei, Y, from the equilibrium value. The linear F' terms are nonzero because the coordinates Δr, $\Delta\alpha$, and Δq are not independent. In using this type of potential function it is necessary to eliminate the redundant coordinates. Furthermore, the redundancy conditions must be obtained to the second order so that the linear F' terms are properly accounted for (1). The potential energy finally takes the form

$$\begin{aligned}2V=&\sum_i 2\Delta r_i(F_r'r_i+F_q'q_{ij}S_{ij})+2\Delta\alpha\left(r_1r_2F_\alpha'+q_{12}(t_{12}t_{21}r_1r_2)^{1/2}F_q'\right)\\&+\sum_i\Delta r_i^2(f_{xy}+t_{ij}^2F_q'+S_{ij}^2f_{yy})+r_1r_2(\Delta\alpha)^2(f_\alpha-S_{12}S_{21}F_q'+t_{12}t_{21}f_{yy})\\&+2\Delta r_1\Delta r_2(-t_{12}t_{21}F_q'+S_{12}S_{21}f_{yy})+\sum_{i=1,2}2r_j\Delta r_i\Delta\alpha(t_{ij}s_{ji}F_q'+S_{ij}t_{ji}f_{yy})\end{aligned} \tag{6.15}$$

where

$$S_{ij} = \frac{r_i - r_j \cos\alpha}{q_{ij}} \qquad \text{and} \qquad t_{ij} = \frac{r \sin\alpha}{q_{ij}} \tag{6.16}$$

UBS force fields have largely been adopted by the Japanese schools. However, the pure UBS force field is unsatisfactory and it has been found necessary to introduce additional valence-type interactions (1, 4), thus dissipating the rigid character of the UBS force field. In the case of *n*-paraffins Schachtschneider and Snyder (4) determined that the simple Urey-Bradley force field gives only a crude approximation to the observed frequencies. In order to improve the fit between observed and calculated frequencies, it has been necessary to modify it with interaction constants to the extent that it is almost equivalent to the valence force field. However, it should be pointed out that through familiarity the authors believe that they are biased in favor of the valence force field. Zerbi (6) has noted that little attention has been given to the problem of the differences in the description of the normal modes of a given molecule as determined by a valence force field or a modified UBS potential function.

In determining the normal modes of polymers, valence force fields and UBS potential functions have been preeminent. For completeness, however, two other model systems should be mentioned. The Heath-Linnet (see ref. 7) orbital force field attempts to explain differences in deformation force constants in terms of electronic rehybridization phenomena. Basically, the sp^3, sp^2, or *sp* orbitals are fixed to each atom and the molecular deformation is ascribed to changes in the overlapping orbitals forming chemical bonds. Mills presented a hybrid-bond orbital force field (see discussion in refs. 1 and 7) where these types of arguments were put in a more quantitative form through the correlation of certain off-diagonal force constants. Shimanouchi (1) has pointed out that even though such models can give well-conditioned force constants, this does not mean that the initial assumption is correct. Nevertheless, the theoretical basis for a correlation between interaction force constants and electronic charge movements appears sound.

6.7 Local Symmetry Force Constants and Force Constant Transferability

One method for obtaining sufficient data points (observed frequencies) to solve the vibrational problem is to assume the transferability of force constants between similar molecules. Even though the force field describing a particular functional group is influenced by the nature of the neighboring atoms, it seems reasonable to assume that the force constants of, for example, the methyl group in $CH_3CH_2CH_3$ can be transferred to the methyl group of molecules such as $CH_3CH_2CH_2CH_3$, CH_3CH_2Cl, or CH_3CH_2OH. In this way the number of observed frequencies can be increased dramatically. However, it must be kept in mind that not all of these additional data are truly independent. The characteristic asymmetric methyl bending mode observed near 1370 cm^{-1} will appear at or near this frequency in a whole series of *n*-paraffins and therefore provides no additional data points. In contrast, the progression of observed frequencies of many of the CH_2 bending modes of the *n*-paraffins does provide new data.

Snyder and Schachtschneider (8) used this approach to obtain a set of valence force constants for saturated hydrocarbons. However, Barnes and Fanconi (9) have pointed out that in, for example, the force field describing the $—CH_2—$ unit, there is a redundant bending coordinate about the carbon atom that leads to a degree of correlation between the force constants. This correlation is minimized by use of local symmetry force constants. Furthermore, in cases where there are redundant coordinates there can also be redundant valence force constants that cannot be determined. All that can be calculated is the local symmetry force constant. We will consider these points in more detail in the following section and in Chapter 13. First we will consider in a little more detail the nature of local symmetry force constants. It was shown in Chapter 5 that symmetry coordinates can be used to factor the force constant matrix $\mathbf{F}_r$ into block diagonal form as

$$\mathbf{F}_s = \mathbf{U}\mathbf{F}_r\mathbf{U}^T \tag{6.17}$$

However, even for molecules with little or no overall symmetry, local symmetry force constants can be utilized, since they take advantage of the local symmetry of a particular group in a molecule. For example, in $CH_3CH_2CH_2Cl$ the CH_3 and CH_2 groups have C_{3v} and C_{2v} local symmetry, respectively. The force constants are expressed in terms of the local symmetry coordinate describing each type of unit. For the methylene unit, $—CH_2—$, the use of local symmetry coordinates combines the valence force constants to give methylene rocking, twisting, wagging, and bending coordinates.

6.8 Coordinate Redundancies and Indeterminate Force Constants

In certain cases it is not possible to determine individual valence force constants, but only certain combinations. Consider as an example formaldehyde, shown in Figure 6.4. The sum of the three in-plane angle deformations

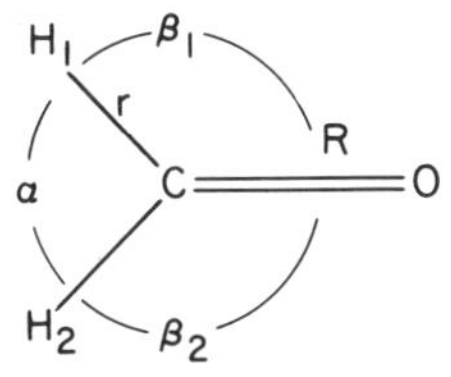

Figure 6.4. Definition of the internal coordinates of formaldehyde.

must be zero, so that

$$\delta H_1CH_2 = -\delta H_1CO - \delta H_2CO$$

or

$$\alpha = -\beta_1 - \beta_2$$

The symmetry coordinates for the A_1 symmetry species can be determined to be

$$S_1 = \Delta R, \qquad S_2 = \frac{1}{\sqrt{2}}(\Delta r_1 + \Delta r_2)$$

$$S_3 = \frac{1}{\sqrt{6}}(2\alpha - \beta_1 - \beta_2) \qquad \text{or} \qquad S_3 = \frac{1}{\sqrt{2}}(\beta_1 + \beta_2) \tag{6.18}$$

where R is the C=O bond length and r_1 and r_2 refer to the C—H bonds. The force constants for these internal coordinates can now be written

i.c.→ ↓	R	r_1	r_2	α	β_1	β_2
R	f_R	$f(rR)$	$f(rR)$	$f(R\alpha)$	$f(R\beta)$	$f(R\beta)$
r_1		f_r	$f(rr)$	$f(r\alpha)$	$f(r\beta)$	$f(r\beta')$
r_2			f_r	$f(r\alpha)$	$f(r\beta')$	$f(r\beta)$
α				f_α	$f(\alpha\beta)$	$f(\alpha\beta)$
β_1	symmetric				f_β	$f(\beta\beta)$
β_2						f_β

where the off-diagonal interaction terms are simply noted in parentheses. Applying the symmetry coordinates to this matrix to obtain the $\mathbf{F}_s$ block for the A_1 species, we have

$$\mathbf{F}^{A_1} = \begin{bmatrix} f_R & \sqrt{2}\,f(rR) & \frac{2}{\sqrt{6}}[f(R\alpha) - f(R\beta)] \\ & f_r + f(rr) & \frac{1}{\sqrt{3}}[2f(r\alpha) - f(r\beta) - f(r\beta')] \\ \text{symmetric} & & \frac{1}{3}[2f_\alpha + f_\beta + f(\beta\beta) - 4f(\alpha\beta)] \end{bmatrix} \tag{6.19}$$

Similarly, for the B_1 mode we have symmetry coordinates

$$S_4 = \frac{1}{\sqrt{2}}(\Delta r_1 - \Delta r_2), \qquad S_5 = \frac{1}{\sqrt{2}}(\beta_1 - \beta_2) \tag{6.20}$$

and a block of symmetry force constants for the β_1 mode:

$$\mathbf{F}^{\beta_1} = \begin{bmatrix} f_r & -f(rr) & f(r\beta)-f(r\beta') \\ & & f_\beta - f(\beta\beta) \end{bmatrix} \tag{6.21}$$

From an examination of equations (6.19) and (6.20) it can be seen that the symmetry of this molecule is such that all of the nine valence force constants involving α and β cannot be independently determined. The only determinable combinations are $[f(R\alpha)-f(R\beta)]$, $f(r\beta)-f(r\beta')$, $f_\beta-f(\beta\beta)$, $f_\alpha+(\beta\beta)-2(\alpha\beta)$, and $(r\alpha)-(r\beta')$. Inclusion of vibrational data for completely isotopically substituted molecules would only improve the precision with which these combinations could be determined. In certain cases, however, these symmetry blocks can be "broken down" by partial isotopic substitution, which can lower the symmetry of the molecule. Even so, the new vibrational data obtained in this way might not be sufficiently different (i.e., the frequencies might shift only slightly) to allow a good determination of force constant values. In molecules of this type the problem is usually solved by arbitrarily fixing the values of certain force constants [e.g., $f(\beta\beta)$ and $f(\alpha\beta)$] to zero.

It should be noted that this problem can be obscured when the secular equation is set up in the cartesian representation. For molecules with a high degree of symmetry, internal symmetry coordinates should be used initially to set up the $\mathbf{F}_s$ matrix; it can then be examined to ascertain which force constant combinations can be independently determined.

6.9 Anharmonicity; Combinations, Overtones, and Fermi Resonance

We have assumed that only quadratic terms in the potential function need to be considered, but this approximation is valid only for small displacements. If this assumption were true, infinite energy would be required to break a valence bond. However, for the purposes of normal coordinate analysis the frequencies of the fundamentals appearing in the vibrational spectrum can be successfully calculated by using this approach. But in order to account for a number of the weak or even medium-intensity secondary bands that appear in the spectrum it is necessary to consider either simultaneous changes in two or more vibrational quantum numbers, which give **combination bands**, or lines; or multiple changes in one vibrational quantum number, which result in **overtones**. Although the selection rule imposed by the harmonic approximation (i.e., $\Delta n = \pm 1$) no longer applies, those based on the nonzero values of the transition moment integrals under the symmetry operations of the molecular point group are still valid.

In order to determine whether a combination or overtone mode is allowed, it is necessary to determine the symmetry of its normal mode. We will only state the results here and refer to other texts for the derivation of the appropriate equations (3). The symmetry species of a multiple transition is related through group theory to those of the fundamental modes of which it is composed. The representation formed by the wave functions of a combination level can be shown to be a direct product of the representations formed by the fundamentals involved. This can be written

$$\chi_R^{\text{prod}} = \chi_R^{\Gamma_1} \cdot \chi_R^{\Gamma_2} \tag{6.22}$$

where Γ_1 and Γ_2 are the symmetry species of the two fundamentals. This allows us to determine the symmetry species of the combination, hence whether it is allowed by the general selection rules.

It has been observed in many spectra that when an overtone or combination band is close in frequency (and hence in energy) to a fundamental of the same symmetry species, it can be considerably enhanced in intensity. This phenomenon is known as Fermi resonance. When two energy levels are close to one another, perturbation terms neglected in the approximate treatment can result in a mixing of the states, so that the combination or overtone will assume some of the character of the fundamental and vice versa. The perturbation can also result in the two energy levels becoming spread apart or split. The extent of this resonance, if allowed by symmetry, will depend on the closeness of the energy levels and the terms in the Fermi resonance interaction matrix. This effect has recently been invoked to explain anomalies in the C—H stretching region of the Raman spectrum of polyethylene (10).

References

1 T. Shimanouchi, in "Physical Chemistry: An Advanced Treatise," Vol. 4, H. Eyring, D. Henderson, and W. Jost, Eds., Academic Press, New York, (1971), Chapter 6, p. 233.

2 D. C. McKean and J. L. Duncan, *Spectrochim. Acta* **27A**, 1879 (1971).

3 E. B. Wilson, Jr., J. C. Decius, and P. C. Cross, "Molecular Vibrations," McGraw-Hill, New York, 1955.

4 J. H. Schachtschneider and R. G. Snyder, *Spectrochim. Acta* **19**, 117 (1963).

5 R. G. Snyder, *J. Mol. Spect.* **23**, 224 (1967).

6 G. Zerbi, *Appl. Spect. Rev.*, **2**(2), 193 (1969).

7 D. Steele, "Theory of Vibrational Spectroscopy," Saunders, Philadelphia, 1971.

8 R. G. Snyder and J. H. Schachtschneider, *Spectrochim. Acta* **19**, 85 (1963).

9 J. Barnes and B. Fanconi, *J. Phys. Chem. Ref. Data* **7**(4), 1309 (1978).

10 R. G. Snyder, S. L. Hsu, and S. Krimm, *Spectrochim. Acta* **34A**, 395 (1978).

7

SOLVING THE VIBRATIONAL PROBLEM USING COMPUTER METHODS

Then at the balance let's be mute,
We never can adjust it;
What's done we partly may compute
But know not what's resisted.
Robert Burns

In our development of the theory of vibrational spectroscopy presented in the preceding chapters we have not attempted to solve a secular matrix whose dimensions are greater than three by three. Although there are manual methods that can be applied to the determination of the eigenvalues and eigenvectors of larger secular determinants (1,2), the advent and widespread application of computer methods has made such laborious computations redundant. In this chapter we will consider the mathematical methods used to calculate a force field from a set of observed frequencies; discuss the problems associated with such calculations; outline computer routines that can be applied to the vibrational problem; and discuss the calculation of eigenvectors and the potential energy distribution. In the next chapter we will apply the theory developed thus far to the calculation of the normal modes of benzene.

Before describing the least-squares methods that can be used to calculate a set of force constants we will review the problems associated with obtaining sufficient data points. It has been mentioned a number of times that for most polyatomic molecules the number of force constants to be determined is greater than the number of observed frequencies. This situation also holds for the more useful and therefore more complex model force fields, particularly in the early stages of a refinement procedure where it can be difficult to decide which off-diagonal terms are significant and which, to a good approximation, can be set equal to zero. Even though the inclusion of data from isotopic molecules and from the corefinement of the observed frequencies of similar molecules would appear to provide more than enough data points, this is unfortunately not always the case. Much of the spectral information overlaps.

For example, consider the case of the B_{2u} symmetry species of benzene. The secular equation for this species written in terms of symmetry coordinates is

$$\begin{vmatrix} \bar{G}_{11}\bar{F}_{11}+\bar{G}_{12}\bar{F}_{12}-\lambda & \bar{G}_{11}\bar{F}_{12}+\bar{G}_{12}\bar{F}_{22} \\ \bar{G}_{12}\bar{F}_{11}+\bar{G}_{22}\bar{F}_{12} & \bar{G}_{12}\bar{F}_{12}+\bar{G}_{22}\bar{F}_{22}-\lambda \end{vmatrix}=0 \qquad (7.1)$$

so that there are three symmetry force constants and two observed frequencies. The inclusion of data from the completely deuterated molecule superficially appears to introduce two additional vibrational frequencies. However, the four observed vibrational frequencies are not independent, because the ratio of the product of the frequencies is related to the nuclear masses and symmetry of the molecule through the product rule. Nevertheless, there are still three independent pieces of frequency data. Unfortunately, the quadratic form of the secular equation (7.1) results in two possible sets of solutions. Duinker and Mills (3) plotted $\bar{F}_{11}$ and $\bar{F}_{22}$ against $\bar{F}_{12}$, as shown in Figure 7.1, since the two sets of curves should intersect at the same values of $\bar{F}_{12}$. The lack of a common intersection point is a measure of anharmonicity and uncertainties in the experimental data. If we use a least-squares procedure to calculate the force constants, either of these two solutions might be determined, depending on the starting point of the refinement. Such factors have to be taken into account in any normal coordinate analysis. With this in mind we will now consider methods for the evaluation of force constants.

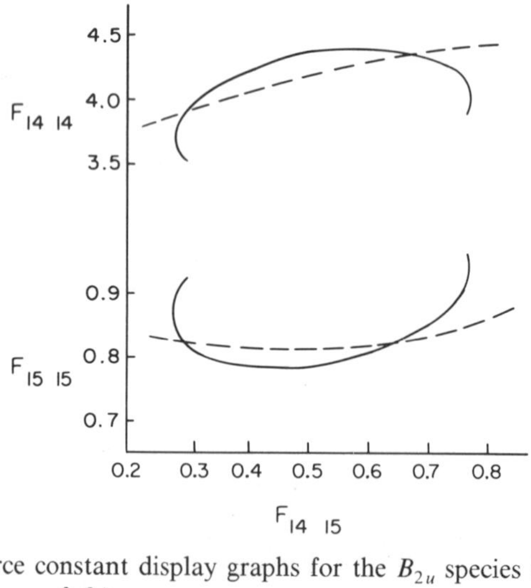

Figure 7.1. Force constant display graphs for the B_{2u} species of benzene. (reprinted by permission from ref. 3.)

7.1 Calculation of Force Constants

In order to calculate a force field it is necessary first to assume values of the force constants, using wherever possible data transferred from similar molecules (and on occasion a degree of chemical intuition). The frequencies calculated with these initial force constant values are then compared to the observed values and the force field is adjusted to minimize the differences between the two. Several methods have been proposed for the systematic adjustment of the force constants using perturbation and least-squares techniques (4–7). However, some of these methods are ineffective because they fail to guarantee convergence, or because their results converge to an unrealistic force field. Perhaps the simplest method of reaching a convergent set of force constants is based on a stepwise adjustment procedure. Let $\Delta\lambda_i$ be the difference between the observed and calculated frequencies for the ith normal mode:

$$\Delta\lambda_i = \lambda^{\text{obs}} - \lambda^{\text{calc}} \tag{7.2}$$

The weighted sum of the squared deviations, **S**, can be defined as

$$\mathbf{S} = \sum \Delta\lambda_i \mathbf{W} \Delta\lambda_i \tag{7.3}$$

where **W** is a matrix of weighting factors representing the confidence in a particular experimental frequency value. Naturally, the force constant refinement procedure has the aim of minimizing **S**. Each force constant is incremented in turn by a value δ. For example, the first force constant F_1 is adjusted to $(F_1 + \delta)$ and $(F_1 - \delta)$. The vibrational frequencies are calculated for each of these values and the weighted sum of the squared deviations **S** is also determined. If either of these new force constants gives a smaller value of **S**, this force constant is kept. This procedure is repeated for the next force constant, and so on. The refinement is continued until $2M$ adjustments to the force constants (where M is the number of independent force constants) have been made without obtaining a squared deviation that is smaller than the value of **S** in the preceding cycle. The value of δ can then be reduced and the force constants again refined.

For large polyatomic molecules this procedure can be extremely time consuming and expensive. An alternative method is based on the assumption that provided $\Delta\lambda_i$ is small, a small correction to the force constant matrix **F** of $\Delta\mathbf{F}$ required to minimize **S** will result in only insignificant changes in the eigenvectors $\mathbf{L}_r$, so that we can write in matrix form

$$\mathbf{G}(\mathbf{F} + \Delta\mathbf{F})\mathbf{L}_r = \mathbf{L}_r(\boldsymbol{\Lambda} + \Delta\boldsymbol{\Lambda}) \tag{7.4}$$

where $\mathbf{\Delta\Lambda}$ is the matrix of eigenvalue deviations $\Delta\lambda_i$. Subtracting

$$\mathbf{GFL}_r = \mathbf{L}_r\mathbf{\Lambda} \tag{7.5}$$

we obtain

$$\mathbf{G\Delta FL}_r = \mathbf{L\Delta\Lambda} \tag{7.6}$$

In fact, it has been pointed out (8) that when $\mathbf{\Delta\Lambda}$ and the changes in the eigenvectors are small there is no need to assume that $\mathbf{L}_r$ is unchanged. Substituting $\mathbf{G}=\mathbf{L}_r\mathbf{L}_r^T$ and premultiplying (7.6) by $\mathbf{L}_r^{-1}$, we obtain

$$\mathbf{\Delta\Lambda} = \mathbf{L}_r^T\mathbf{\Delta FL}_r \tag{7.7}$$

or in expanded form

$$\Delta\Lambda_i = \sum_{jk} L_{ij}\Delta F_{jk} L_{ki} \tag{7.8}$$

The Jacobian matrix $\mathbf{J}$ is defined by

$$\mathbf{\Delta\Lambda} = \mathbf{J\Delta F} \tag{7.9}$$

and the elements of $\mathbf{J}$ can be written

$$J_{k,jk} = \frac{\delta\lambda_i}{\delta F_{jk}} \tag{7.10}$$

which, allowing for the fact that off-diagonal terms are multiplied by one another twice, is given by

$$J_{i,jk} = (2-\delta_{jk})L_{ij}L_{ik} \tag{7.11}$$

The increment $\mathbf{\Delta F}$ that is to be added to $\mathbf{F}$ in order to give a better fit is calculated from the normal equation

$$\mathbf{J}^T\mathbf{WJ\Delta F} = \mathbf{J}^T\mathbf{W\Delta\Lambda} \tag{7.12}$$

The adjustment of force constants by equation (7.12) is repeated, using revised normal coordinates to set up a new $\mathbf{J}$ matrix, until the convergent set is obtained. The uncertainties of these values are given by

$$\boldsymbol{\sigma}^2(\mathbf{F}) = (\mathbf{J}^T\mathbf{WJ})^{-1}\frac{\mathbf{S}}{\mathbf{N}-\mathbf{M}} \tag{7.13}$$

from statistical theory, where **N** is the number of data points (observed frequencies) and **M** is the number of force constants.

It should be noted that because the elements of **F** are not always independent, a matrix $\mathbf{\Phi}$ of independent force constants is often defined as

$$\mathbf{F}=\mathbf{Z\Phi} \tag{7.14}$$

where $Z_{k,ij}$ gives the coefficient of the kth element of $\mathbf{\Phi}$ in F_{ij}. Equation (7.12) then becomes

$$(\mathbf{JZ})^T\mathbf{W}(\mathbf{JZ})\,\Delta\mathbf{\Phi}=(\mathbf{JZ})^T\mathbf{W}(\Delta\mathbf{\Lambda}) \tag{7.15}$$

The weight matrix **W** is usually taken to be $\mathbf{W}=\mathbf{\Lambda}_{\text{obs}}^{-1}$.

One problem when applying this method is that in certain cases divergence of the process can occur. In general, $\Delta\mathbf{\Lambda}$ is a nonlinear function of the force constants. If the nonlinearity is considerable, the $\mathbf{J}^T\mathbf{WJ}$ matrix may turn out to be nearly singular, leading to corrections $\Delta\mathbf{\Phi}$ that are too large. The weighted sum of the squared frequency deviations **S** may then increase from one step to the next, so that wild oscillation in the calculated values of $\Delta\mathbf{\Phi}$ occurs and the refinement diverges. Papousek et al. (13) proposed the application of the Levenberg modification (14) of the standard method of least square in order to overcome this difficulty. We have incorporated this method into our normal coordinate analysis program and (in general) it has been our experience that convergence to a reasonable solution occurs quickly. The method is a damped least-squares procedure that can be adapted to thc force constant refinement process in the following form:

$$\left[(\mathbf{JZ})^T\mathbf{W}(\mathbf{JZ})+\mathbf{bE}\right]\Delta\mathbf{\Phi}=(\mathbf{JZ})^T\mathbf{W}\Delta\mathbf{\Lambda} \tag{7.16}$$

where **b** is a damping factor and **E** is the unit matrix. Levenberg (14) presented a convergence proof demonstrating that **b** can be chosen so that the sums of the squares of the frequency deviations and the corrections are minimized in each step. Papousek et al. (13) have pointed out that appropriate values of **b** could be calculated automatically in each iteration, but in force constant calculations it is usually adequate to estimate a value empirically. The damping factor should be chosen so that it is sufficiently small not to slow down the convergence appreciably, but is adequate to remove any possible near singularity of the $(\mathbf{JZ})^T\mathbf{W}(\mathbf{JZ})$ matrix. In our laboratories we frequently use a value of $\mathbf{b}=0.01$.

In all least-squares refinement calculations it has to be kept firmly in mind that a good fast convergence does not guarantee a physically significant solution. This is particularly true when the force constants to be determined are to some extent dependent.

7.2 Calculation of Eigenvectors

The aim of most normal coordinate calculations is to obtain an understanding and description of the normal modes observed in the vibrational spectrum. This is usually accomplished in two ways. The first is to obtain the eigenvector matrix **L**, usually in cartesian coordinates (i.e., $\mathbf{L}_x$). The second is to calculate the potential energy distribution as a convenient method for band assignments. We will discuss the latter method in the next section. Previously, it was mentioned that only the ratios, and not the absolute displacements, of the atoms during a particular normal mode of vibration A_{ij} could be determined. We could obtain a description of the normal modes by arbitrarily letting the displacement of a particular atom equal unity, but went on to show that if we defined a normalized ratio

$$l_{ij} = \frac{A_{ij}}{\left[\Sigma_i (A_{ij})^2\right]^{1/2}} \tag{7.17}$$

the coefficients l_{ij} not only represent a description of the normal modes but also give the transformation from mass-adjusted cartesian coordinates to normal coordinates. It is customary to describe the modes of motion in each normal coordinate by drawing arrows representing atomic displacements on a diagram of the molecule being considered such that the length of the arrow is proportional to the appropriate l_{ij} value. The diagram then indicates the relative amplitudes of the atomic displacements during the normal mode of motion *j*.

When the vibrational problem is solved by computer methods, the **L** matrix is calculated during the diagonalization of the secular determinant. We have considered two forms of the secular equation, one in terms of cartesian displacement coordinates and the other in terms of internal coordinates. We have seen that the former equation can be written in such a way that the force constants can be defined in terms of internal coordinates. There are some advantages to the cartesian coordinate description, the one mentioned most frequently in our preceding discussions being the symmetric nature of the $\mathbf{D}^T\mathbf{F}\mathbf{D}$ matrix in the secular equation

$$|\mathbf{D}^T\mathbf{F}\mathbf{D} = \lambda\mathbf{E}|\mathbf{L}_X = \mathbf{0} \tag{7.18}$$

An additional advantage is that the eigenvectors $\mathbf{L}_X$ are described in terms of the cartesian displacements of the normal coordinates and the form of the vibration is easier to visualize by plotting than the equivalent eigenvectors expressed in internal coordinates, $\mathbf{L}_r$. As mentioned earlier, the eigenvector matrix is determined during the diagonalization process. Consider the equation

$$\mathbf{D}^T\mathbf{F}\mathbf{D} = \lambda\mathbf{E} \tag{7.19}$$

By a similarity transformation $\mathbf{D}^T\mathbf{F}\mathbf{D}$ is diagonalized:

$$\mathbf{L}_X^T\mathbf{D}^T\mathbf{F}\mathbf{D}\mathbf{L}_X = \mathbf{\Lambda} \tag{7.20}$$

where $\mathbf{\Lambda}$ is the diagonal eigenvalue matrix and by definition the matrix $\mathbf{L}_X$ has to be the eigenvector matrix. Of course, the elements of $\mathbf{L}_X$ required for the diagonalization are not determined *a priori*. In practice, numerical methods suitable for machine programming [e.g., the Jacobi (11) or Householder (12) methods] are utilized that simultaneously construct $\mathbf{\Lambda}$ and $\mathbf{L}_X$ by an iteration procedure.

Wherever possible it is advantageous to use molecular symmetry to reduce the dimensions of the secular equation to block diagonal form. The eigenvector matrix is then calculated in terms of symmetry coordinates. The reduced secular equation can be written

$$|\mathbf{U}\mathbf{D}^T\mathbf{F}\mathbf{D}\mathbf{U}^T - \lambda\mathbf{E}| = \mathbf{0} \tag{7.21}$$

where $\mathbf{U}$ is the matrix defining the symmetry coordinates in terms of cartesian displacements. For simplicity, if we define

$$\mathbf{A} = \mathbf{U}\mathbf{D}^T\mathbf{F}\mathbf{D}\mathbf{U}^T \tag{7.22}$$

then the similarity transformation required to diagonalize $\mathbf{A}$ is given by

$$\mathbf{S}_{LX}^T\mathbf{A}\mathbf{S}_{LX} = \mathbf{\Lambda} \tag{7.23}$$

where the matrix $\mathbf{S}_{LX}$ is the eigenvectors in terms of the external symmetry coordinates. The eigenvectors in terms of the cartesian displacement coordinates are obtained by performing the inverse transformation from external symmetry coordinates:

$$\mathbf{L}_X = \mathbf{U}^T\mathbf{S}_{LX}\mathbf{M}^{-1/2} \tag{7.24}$$

If the vibrational problem is originally set up in the Wilson **GF** matrix form, a similar transformation from eigenvectors in internal symmetry coordinates, $\mathbf{S}_R$, can be constructed from

$$\mathbf{S}_{LR} = \mathbf{U}_R\mathbf{B}\mathbf{L}_X \tag{7.25}$$

where $\mathbf{U}_R$ is the matrix defining the symmetry coordinates in terms of internal coordinates. However, to obtain **L** the inverse of **B** has to be formed, which involves some difficulty, since in general **B** is not square and therefore cannot be inverted. However, as shown previously, the product $\mathbf{M}^{-1}\mathbf{B}^T\mathbf{G}^{-1}$ serves as the inverse of **B**, since

$$\mathbf{B}(\mathbf{M}^{-1}\mathbf{B}^T\mathbf{G}^{-1}) = (\mathbf{M}^{-1}\mathbf{B}^T\mathbf{G}^{-1})\mathbf{B} = \mathbf{E} \tag{7.26}$$

Hence,

$$\mathbf{L}_X = \mathbf{M}^{-1}\mathbf{B}^T\mathbf{G}^{-1}\mathbf{U}^T\mathbf{S}_{LR} \tag{7.27}$$

7.3 The Potential Energy Distribution

In many publications a quantity known as the potential energy distribution (PED) is used to describe the modes. The PED is usually defined in one of two ways. The first method determines the contribution of each force constant f_{ij} to the normal frequencies of vibration, while the second method expresses the PED as the contribution of each internal displacement coordinate to the normal modes (15–17). The PED is expressed in terms of the percentage contribution of each force constant or displacement coordinate to the potential energy of each normal mode. For example, the contribution of the C=O stretching coordinate to the characteristic band near 1700 cm^{-1} in the spectrum of a ketone would be calculated as close to 100%. However, terms that are larger than 100% or negative in value occasionally arise, and it is perhaps better to view the PED elements describing a given mode simply as terms whose sum equals 100.

We will first consider the PED expressed as a contribution of each force constant to the normal modes. It will be recalled that the potential energy can be expressed in terms of normal coordinates as

$$2V = \sum_i \lambda_i Q_i^2 \tag{7.28}$$

Consequently, if we consider unit displacements of the normal coordinate Q_i, all other normal coordinates being at rest, then λ_i is a measure of the potential energy. According to equation (3.55), each normal frequency is given by

$$\lambda_i = \sum_{jk} L_{ji} L_{ki} f_{jk} \tag{7.29}$$

Hence,

$$\sum \frac{(L_{ji} L_{ki} f_{jk})}{\lambda_i} = 1 \tag{7.30}$$

so that the terms

$$\frac{2 L_{ji} L_{ki} f_{jk}}{\lambda_i} \tag{7.31}$$

and

$$\frac{L_{ji}^2 f_{jj}}{\lambda_i} \tag{7.32}$$

represent the fractional contributions of each of the off-diagonal and diagonal elements, respectively, to the potential energy of the normal mode whose vibrational frequency is λ_i. The factor of 2 in equation (7.31) arises from the fact that $f_{jk} = f_{kj}$. Note that the eigenvector matrix $\mathbf{L_{ij}}$ in these equations is in terms of the internal coordinates. Using equation (7.11) to write equation (7.30) in terms of the Jacobian matrix, we obtain

$$\text{PED} = \frac{JF}{\mathbf{\Lambda}} \tag{7.33}$$

When the secular equation is set up in terms of cartesian coordinates, the Jacobian matrix and hence PED can be calculated in the following manner. First, the force constant matrix $\mathbf{F}$ is considered as the sum of j matrices, each of which contains nonzero elements only in those row and column positions defining a particular force constant (15–17); that is,

$$\mathbf{F} = \sum_j Z_j K_j \tag{7.34}$$

where K_j is the numerical value of the force constant and the elements of Z_j are either 0 or 1, depending on whether or not a particular force constant is an element in Z_j. The secular equation can now be rearranged in the following manner:

$$|\mathbf{U}\mathbf{D}^T\mathbf{F}\mathbf{D}\mathbf{U}^T - \lambda\mathbf{E}| = \mathbf{0} \tag{7.35}$$

let

$$\mathbf{Z}_j^* = \mathbf{U}\mathbf{D}^T\mathbf{Z}_j\mathbf{D}\mathbf{U}^T \tag{7.36}$$

then

$$\left|\sum_j \mathbf{Z}_j^*\mathbf{K}_j - \lambda\mathbf{E}\right| = \mathbf{0} \tag{7.37}$$

The matrix $\sum \mathbf{Z}_j^*\mathbf{K}_j$ is, of course, diagonalized by the eigenvector matrix $\mathbf{S}_{LX}$

$$\mathbf{S}_{LX}^T\left(\sum_j \mathbf{Z}_j^*\mathbf{K}_j\right)\mathbf{S}_{LX} = \mathbf{\Lambda} \tag{7.38}$$

The Jacobian matrix **J** can then be formed from the relation [see equation (7.10)]

$$\mathbf{J}_{ij} = \frac{\delta\lambda_i}{\delta k_j} = \left(\mathbf{S}_{LX}^T \mathbf{Z}_j^* \mathbf{S}_{LX}\right)_{ii} \tag{7.39}$$

Consequently, for a given vibrational frequency λ_i the PED is then

$$(\mathrm{PED})_{ij} = \frac{\mathbf{J}_{ij}\mathbf{K}_j}{\boldsymbol{\lambda}_i} \tag{7.40}$$

The attraction of using this method to obtain the PED is that it allows us to determine the contribution to the PED of off-diagonal as well as diagonal terms of the force constant matrix. However, the method is in general time consuming (hence expensive!), particularly for large, unsymmetrical molecules. A second disadvantage becomes apparent upon calculating the PED of a molecule where several internal coordinates are given the same force constant value (e.g., the C—C stretching coordinate of benzene, where all the bonds are equivalent). Since the PED in this method is given as a function of the force constant rather than of the internal coordinates, the PED does not provide information concerning which specific internal coordinate or combination of coordinates is involved in the vibrational mode. When this information is required, the cartesian displacements for the mode in question also have to be plotted.

An alternative procedure for calculating the PED, which is faster and also yields the PED as a function of the internal coordinates, is to use equation (7.30), where the **L** matrix is expressed in terms of internal displacement coordinates. Usually only the diagonal force constants are considered, so the PED in internal coordinate displacements can be defined as

$$(\mathrm{PED})_{ij} = (L_{ji})^2 \bar{F}_{jj} \left(\frac{100}{\lambda_i}\right) \tag{7.41}$$

where $\bar{F}_{jj}$ is the jth symmetrized diagonal force constant, defined as

$$\bar{\mathbf{F}}_{jj} = \mathbf{UFU}^T \tag{7.42}$$

Of course, the disadvantage of this method is that it neglects vibrational interactions between internal coordinates.

7.4 Computer Programs for Solving the Vibrational Problem

It is beyond the scope of this book to discuss in detail the mathematical logic involved in the computer programs that are available for solving the secular equation. We will confine our discussion to a description of the forms of the secular problem that are most suitable for solving on the computer, and will briefly outline the steps in the program used in our laboratories.

There are two major points that should be addressed when setting up the secular equation in a form suitable for computerized calculations. The first is that it is not always necessary to eliminate redundant coordinates. However, the problems associated with redundancy cannot just be ignored. Internal symmetry coordinates should be set up so that one of them expresses the redundancy condition while the rest are constructed orthogonal to this symmetry coordinate. In addition, as a result of redundancies, certain force constants can become indeterminate, as discussed in Chapter 6.

The second point concerning computerized methods that should be considered is that the most efficient methods for solving the secular determinant employ matrix diagonalization methods that are applicable only to symmetric matrices. Consequently, the secular equation is not usually set up directly in the Wilson **GF** matrix form. There are two alternative forms of the secular problem that are commonly used. The first of these we mentioned in Section 3.5, where it was shown that the secular determinant can be written as

$$|\mathbf{D}^T\mathbf{F}\mathbf{D}-\lambda\mathbf{E}|=0 \tag{7.43}$$

where **F** is the force constant matrix expressed in terms of internal coordinates, while $\mathbf{D}=\mathbf{B}\mathbf{M}^{-1/2}$. In effect, this is a cartesian coordinate representation, so that external symmetry coordinates have to be used to simplify the problem by reduction to a block diagonal form. The eigenvectors of $\mathbf{D}^T\mathbf{F}\mathbf{D}$ are defined as the matrix **Y**, which is related to $\mathbf{L}_x$ by $\mathbf{L}=\mathbf{D}\mathbf{Y}$, provided that **Y** is normalized so that $\mathbf{Y}\mathbf{Y}^T=\mathbf{E}$ and its eigenvalues are the same as those of the matrix **GF**. It is this form of the secular equation that we are most familiar with, so we will outline the steps in the computer program used to obtain solutions before we discuss the other form of the secular problem in common usage.

The particular advantage of this form of the secular equation is that internal coordinate redundancies are unimportant in terms of the construction of the symmetry coordinates, since the symmetry coordinates are defined in terms of a linear combination of cartesian displacements. However, since the advantages of defining the force constants in terms of the internal coordinates is retained, the problems of zero and indeterminate force constants are not removed and have to be taken into account. A flowchart for the complete program used in our laboratories to solve this form of the secular equation is shown as an example in Figure 7.2 (9). It consists of a mainline routine, which controls the flow of the program and does the force constant refinement (if required), and several subroutines, which perform the calculations associated with various parts of the vibrational problem. The subroutine entitled Normal is shown in Figure 7.3. This is also divided into several subroutines. In the subroutine entitled Chain 1, shown in Figure 7.4, the **B** matrix elements are calculated from the input description of the atomic masses, cartesian coordi-

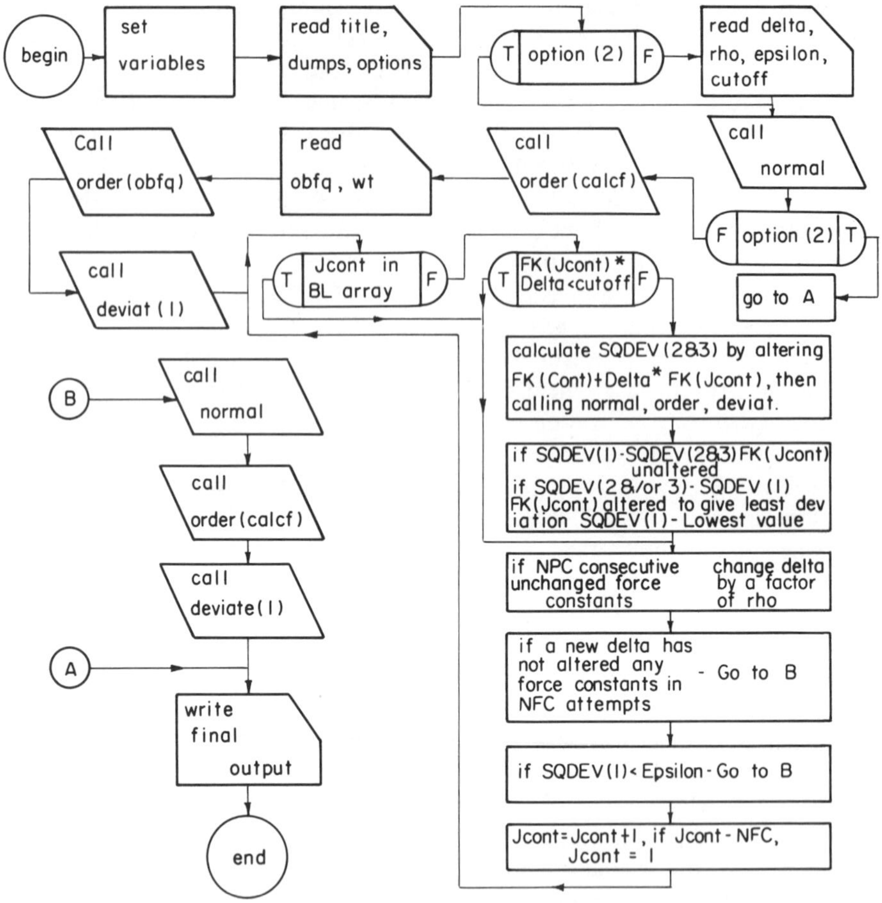

Figure 7.2. Flowchart of the main computer program.

nates, and definition of the internal coordinates, using the well-known relations of Wilson (see Chapter 3). For polymers, the **B** matrix has to be phased (see Chapters 12 and 13), and this is accomplished by also phasing the symmetry coordinates in this subroutine. Subroutine Chain 2, shown in Figure 7.5 creates the **F** matrix, which for polymers also has to be phased. Subroutine Chain 3, also shown in Figure 7.5, calculates the real phased symmetry matrix (for polymers), while subroutine Chain 4 solves the secular equation by the Jacobi diagonalization procedure, yielding the eigenvalues and eigenvectors. The eigenvectors obtained are expressed in terms of symmetry coordinates and are transformed by the program to cartesian coordinates. The subroutine entitled PED calculates the potential energy distribution according to the equations given in the preceding section. The flow diagram is outlined in Figure 7.6. Referring back to Figure 7.3, we see that two other subroutines, namely, the subroutines Order and Deviat, are called by the main program as part of the force constant refinement. Respectively, these subroutines arrange

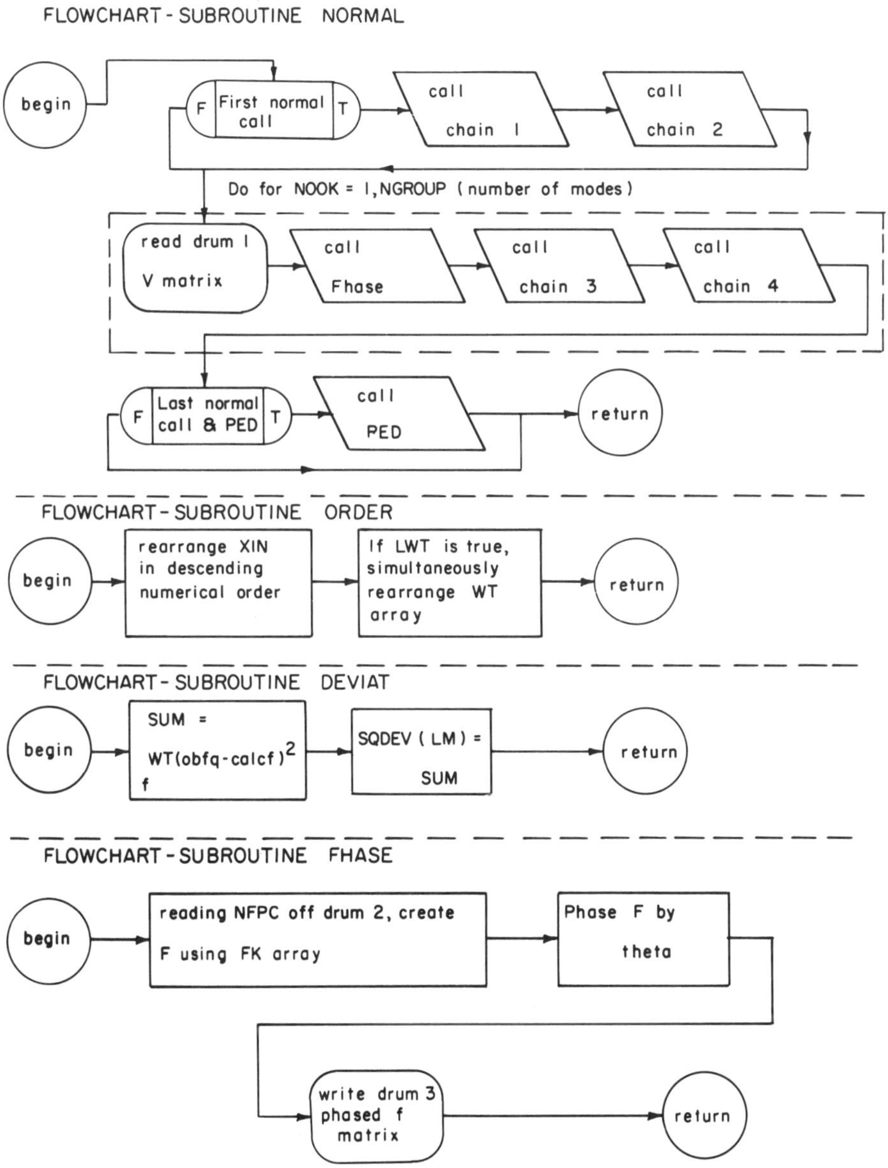

Figure 7.3. Flowchart of subroutines Normal Order, Deviat, and Fhase.

the calculated frequencies in descending numerical order and calculate the sum of the squared deviations between the observed and calculated frequencies.

The force constant refinement procedure used in this program is a variation on the simple one described in Section 7.1. As shown in the flowchart, each force constant is altered by an amount that is a certain fraction of the original force constant, rather than by the constant amount for each force constant that was described in Section 7.1. In addition, certain force constants are dropped

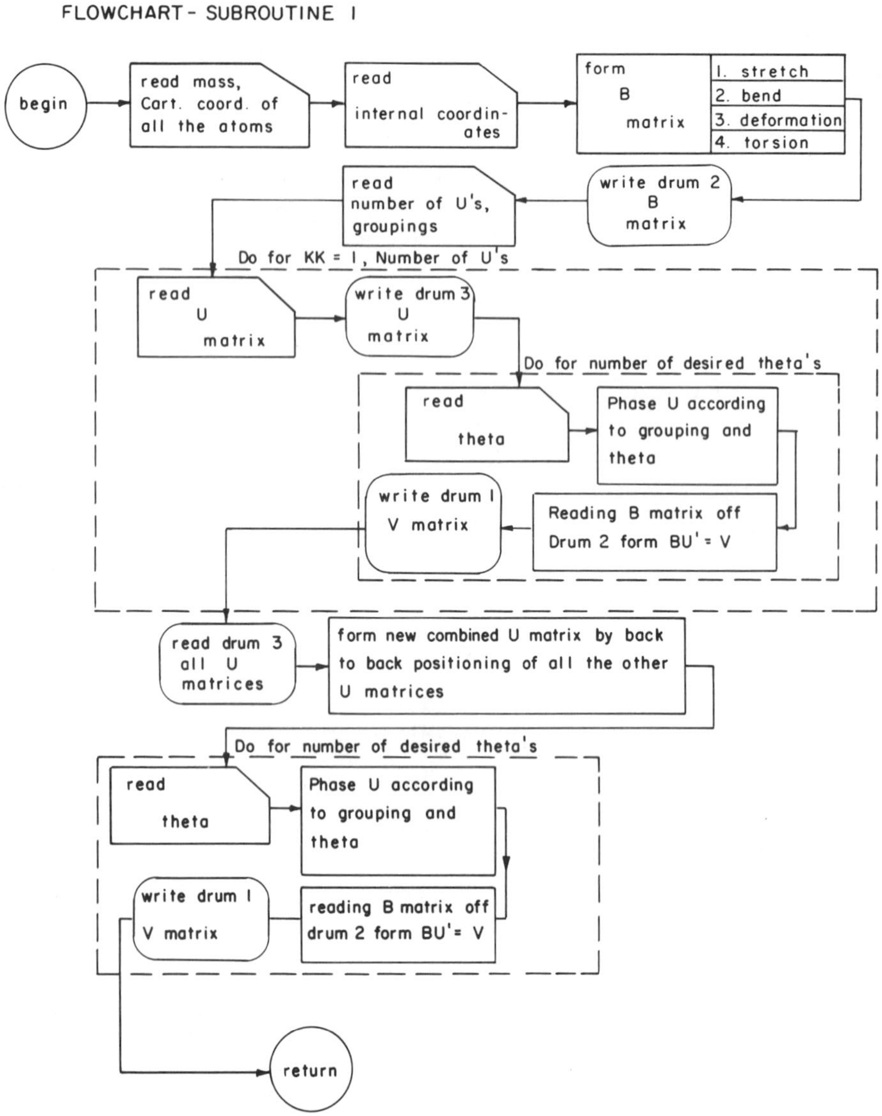

Figure 7.4. Flowchart of subroutine Chain 1.

from the refinement when the refinement on this force constant is smaller than a predetermined minimum (cutoff). The refinement can be set for a given number of iterations or continued until either of two criteria are met. If during an iteration the alterations (δ) in the force constants are discarded because they do not result in an improved frequency fit, the refinement is stopped. Alternatively, if the sum of the squared deviations becomes smaller than a predetermined value, the refinement is also stopped.

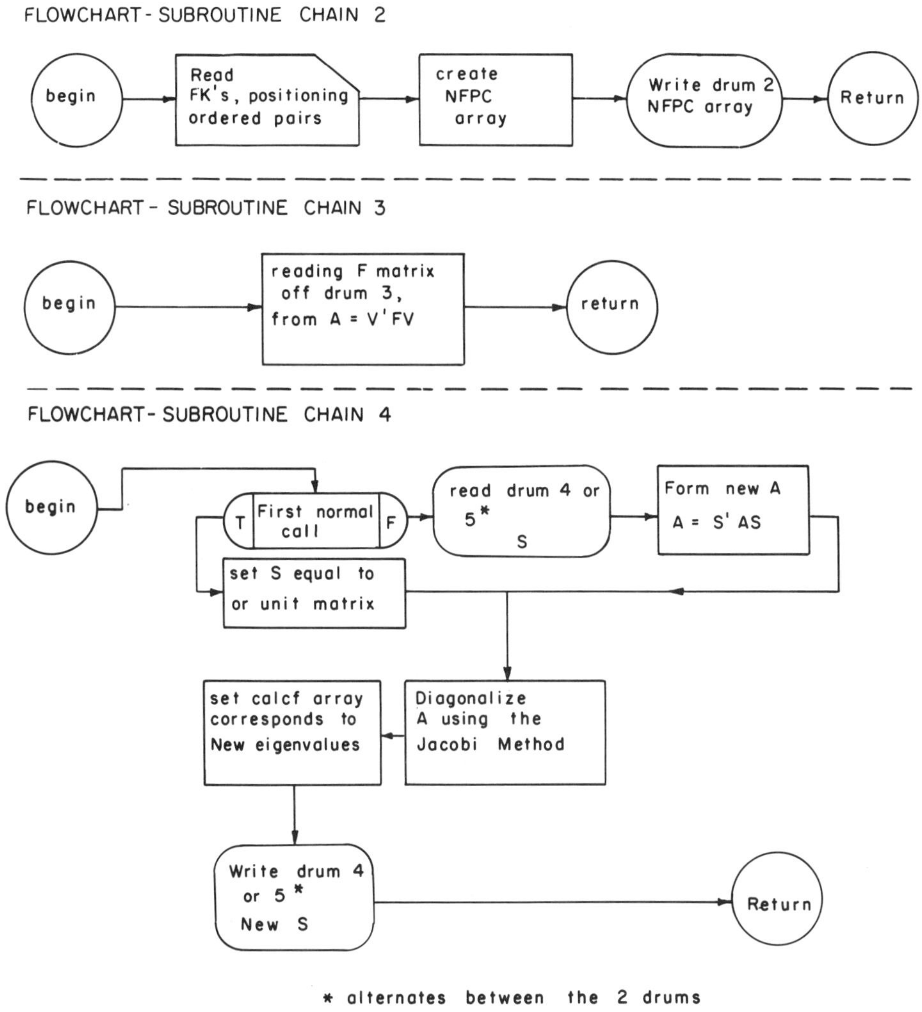

Figure 7.5. Flowchart of subroutines Chain 2, Chain 3, and Chain 4.

Finally, the second form of the secular problem commonly adapted for computerized solutions should be mentioned. In this form, the vibrational problem is set up in terms of internal valence coordinates. However, as mentioned previously, the **GF** matrix is not symmetric, and so cannot be easily diagonalized. In this method, solutions are obtained by diagonalizing two symmetric matrices as follows: let **A** and **Γ** be the eigenvectors and eigenvalues, respectively, of the **G** matrix.

$$\mathbf{GA} = \mathbf{A\Gamma} \tag{7.33}$$

Since **G** is symmetric, **A** and **Γ** can be determined by standard computer diagonalization methods. If **A** is normalized so that $\mathbf{AA}^T = \mathbf{E}$, then we can

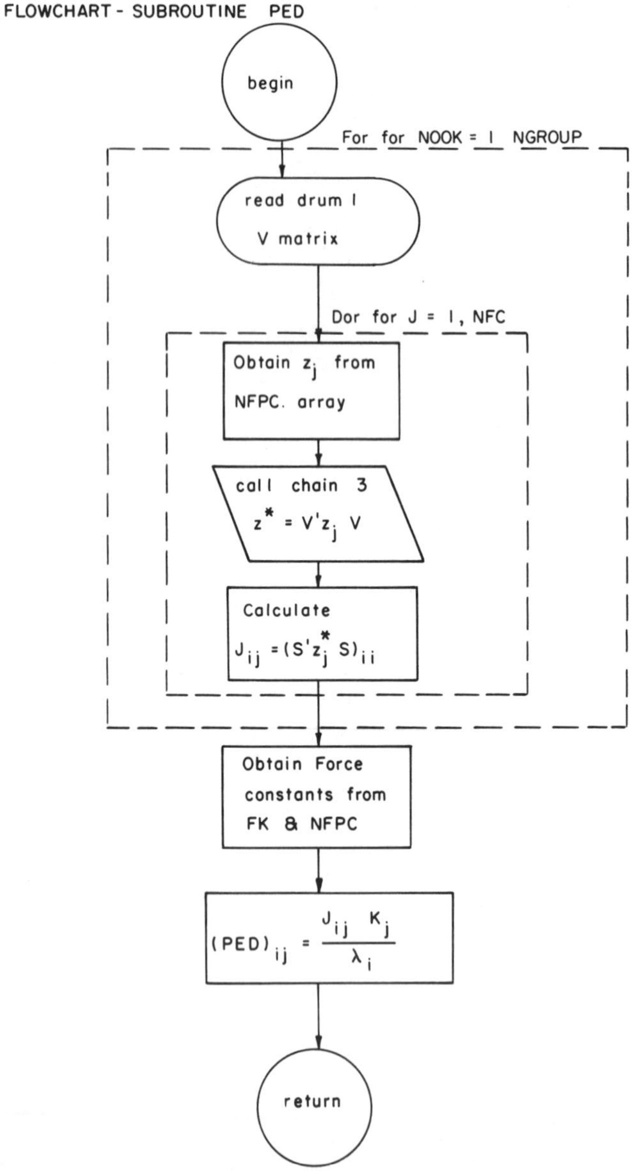

Figure 7.6. Flowchart of subroutine PED.

define a matrix

$$\mathbf{W} = \mathbf{A}\boldsymbol{\Gamma}^{1/2} \tag{7.45}$$

such that $\mathbf{W}\mathbf{W}^T = \mathbf{G}$. $\mathbf{W}$ is a square matrix. The transformation described in equation (7.45) can be applied to the **F** matrix to give

$$\mathbf{H}=\mathbf{W}^T\mathbf{F}\mathbf{W} \tag{7.46}$$

where **H** is symmetric. If **H** is diagonalized, that is, if we solve the secular equation

$$\mathbf{H}\mathbf{C}=\mathbf{C}\mathbf{\Lambda} \tag{7.47}$$

we obtain eigenvalues $\mathbf{\Lambda}$ that correspond to those of the **GF** matrix and eigenvectors **C** related to $\mathbf{L}_r$ by $\mathbf{L}_r=\mathbf{W}\mathbf{C}^T$. This result follows from writing the secular equation

$$\mathbf{W}^T\mathbf{F}\mathbf{W}\mathbf{C}=\mathbf{C}\mathbf{\Lambda}' \tag{7.48}$$

Then premultiplying by **W** yields

$$\mathbf{W}\mathbf{W}^T\mathbf{F}\mathbf{W}\mathbf{C}=\mathbf{W}\mathbf{C}\mathbf{\Lambda}' \tag{7.49}$$

and since $\mathbf{G}=\mathbf{W}\mathbf{W}^T$,

$$\mathbf{G}\mathbf{F}\mathbf{W}\mathbf{C}=\mathbf{W}\mathbf{C}\mathbf{\Lambda}' \tag{7.50}$$

Therefore $\mathbf{\Lambda}'=\mathbf{\Lambda}$ and $\mathbf{L}_r$ is related to **WC** by some simple normalization matrix. As with the first method it is not necessary to eliminate redundancies, which in this case will give zero roots in $\mathbf{\Gamma}$.

References

1 E. B. Wilson, Jr., J. C. Decius, and P. C. Cross, "Molecular Vibrations," McGraw-Hill, New York, 1955.

2 D. Steele, "Theory of Vibrational Spectroscopy," Saunders, Philadelphia, 1971.

3 J. C. Duinker and I. M. Mills, *Spectrochim. Acta* **24A**, 417 (1968).

4 T. Miyazawa, *J. Chem. Soc. Japan* **76**, 1132 (1955).

5 W. T. King, I. M. Mills, and B. L. Crawford, *J. Chem. Phys.* **27**, 455 (1957).

6 D. E. Mann, T. Shimanouchi, J. H. Meal, and J. Faro, *J. Chem. Phys.* **27**, 43 (1957).

7 T. Shimanouchi and I. Suzuki, *J. Chem. Phys.* **42**, 296 (1965).

8 D. A. Long, R. B. Gravenor, and M. Woodger, *Spectrochim. Acta* **19**, 937 (1963).

9 M. J. Hannon, Ph.D. Thesis, Case Western Reserve University, Cleveland, Ohio.

10 F. J. Boerio, Ph.D. Thesis, Case Western Reserve University, Cleveland, Ohio.

11 W. R. Givens, "Numerical Computation of the Characteristic Values of a Real Symmetric Matrix," Oak Ridge National Laboratory Report 1574 (1954).

12 P. A. Bussinger, *Communs. ACM* **42**, 1744 (1965).

13 D. Papousek, S. Toman, and J. Pliva, *J. Mol. Spect.* **15**, 502 (1965).

14 K. Levenberg, *Quart. Appl. Math.* **2**, 164 (1944).

15 M. J. Hannon, F. J. Boerio, and J. L. Koenig, *J. Chem. Phys.* **50**, 2829 (1969).

16 J. Overend and J. R. Scherer, *J. Chem. Phys.* **32**, 1289 (1960).

17 Y. Mikawa, J. W. Brasch, and R. J. Jakobsen, *J. Mol. Spect.* **24**, 314 (1967).

8

VIBRATIONAL ANALYSIS OF BENZENE

There's not the smallest orb which thou behold'st
But in this motion like an angel sings
William Shakespeare

As an example of the normal coordinate analysis of a small molecule we have decided to tackle benzene. The knowledgeable reader will recall that the vibrational analysis of this molecule was also considered in detail by Wilson, Decius, and Cross in their classic text, "Molecular Vibrations" (1). Although this earlier work considerably eases our task, the choice of benzene was made on fundamental grounds rather than on the principle of conservation of energy. As two of the present authors know to their cost (2–4), the calculation of the normal modes of this molecule introduces practically all of the problems inherent in defining coordinates and calculating force constants that a masochistic spectroscopist could hope to meet. Finally, since the publication of "Molecular Vibrations" there have been numerous attempts to determine a meaningful force field for this elusive molecule. Pulay et al., (5) have completed the most recent and possibly the most complete calculations. Fortunately, our task here is not to discuss the rival merits of various force fields, but to consider the nuts and bolts of defining cartesian, internal, and symmetry coordinates, together with the problems involved in determining a valence force field. With respect to defining and calculating force constants we will rely heavily on the work of Whiffen (6) and Duinker and Mills (7).

The central problem in any normal coordinate analysis is obtaining sufficient data with which to calculate a reasonable force field. Furthermore, the correct observed and calculated frequencies must be paired in the refinement, so that a good preliminary analysis of the vibrational spectrum is a necessity. We will therefore begin our analysis of benzene with a symmetry analysis of the molecule and a discussion of the infrared and Raman spectra.

8.1 Symmetry Analysis

The structure of benzene is illustrated in Figure 8.1, which also shows the symmetry elements possessed by this molecule. The structure is assumed to be planar, and valence theory holds that all six C—C bonds are equivalent. In fact, the vibrational spectrum provides very strong evidence for this structure. The symmetry of this molecule is that of the point group D_{6h}. The character table for this group is presented in Table 8.1.

The normal modes of benzene can be classified according to symmetry species by the methods of Chapter 4. If we choose cartesian coordinates as the basis for our reducible representations, then we can determine the reducible representation as in Table 8.2. We simply list the number of unshifted atoms for each symmetry operation and then multiply this number by the characteristic factor for that operation. Applying

$$n^{(\gamma)} = \frac{1}{g} \sum_{j=1}^{k} g_j x_j^{(\gamma)} x_j \tag{8.1}$$

and subtracting the rotations and translations, we obtain

$$\Gamma_{\text{VIB}} = 2A_{1g} + 2B_{2g} + E_{1g} + 4E_{2g} + A_{2u} + 2B_{1u} + 2B_{2u} + 3E_{1u} + 2E_{2u}$$

The infrared and Raman activity of these modes is summarized in Table 8.1. An immediate difficulty is already apparent. Only A_{2u} and E_{1u} modes are infrared active, while Raman-active modes are those with symmetry A_{1g}, E_{1g}, and E_{2g}. The limitations imposed by these selection rules have seriously hindered the vibrational analysis of this molecule. As we will show later, it is necessary to determine the frequencies of the inactive modes by a careful analysis of the weak overtones and combination Raman lines and infrared bands. This is a difficult task that initially resulted in an erroneous assignment

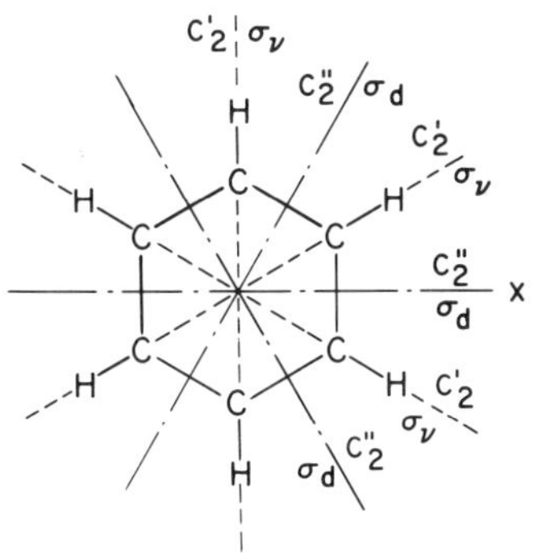

Figure 8.1. The symmetry elements of the benzene molecule. The z axis (a C_6 and S_6 symmetry axis) is perpendicular to the plane of the figure, which is the σ_h symmetry plane.

Table 8.1 Character Table for the Point Group D_{6h}

D_{6h}	E	$2C_6$	$2C_3$	C_2	$3C_3'$	$3C_2''$	i	$2S_3$	$2S_6$	σ_h	$3\sigma_d$	$3\sigma_v$	Activity: IR	Activity: Raman
A_{1g}	1	1	1	1	1	1	1	1	1	1	1	1		$\alpha_{xx}+\alpha_{yy}, \alpha_{zz}$
A_{2g}	1	1	1	1	−1	−1	1	1	1	1	−1	−1	R_z	
B_{1g}	1	−1	1	−1	1	−1	1	−1	1	−1	1	−1		
B_{2g}	1	−1	1	−1	−1	1	1	−1	1	−1	−1	1		
E_{1g}	2	1	−1	−2	0	0	2	1	−1	−2	0	0	(R_x, R_y)	$(\alpha_{xz}, \alpha_{yz})$
E_{2g}	2	−1	−1	2	0	0	2	−1	−1	2	0	0		$(\alpha_{xx}-a_{yy}, \alpha_{xy})$
A_{1u}	1	1	1	1	1	1	−1	−1	−1	−1	−1	−1		
A_{2u}	1	1	1	1	−1	−1	−1	−1	−1	−1	1	1	T_z	
B_{1u}	1	−1	1	−1	1	−1	−1	1	−1	1	−1	1		
B_{2u}	1	−1	1	−1	−1	1	−1	1	−1	1	1	−1		
E_{1u}	2	1	−1	−2	0	0	−2	−1	1	2	0	0	(T_x, T_y)	
E_{2u}	2	−1	−1	2	0	0	−2	1	1	−2	0	0		

Table 8.2 Reducible Representations of Benzene

D_{6h}	E	$2C_6$	$2C_3$	C_2	$3C_2'$	$3C_2''$	i	$2S_3$	$2S_6$	σ_h	$3\sigma_d$	$3\sigma_v$
Factor[a]	3	2	0	−1	−1	−1	−3	−2	0	1	1	1
No. of unshifted atoms	12	0	0	0	4	0	0	0	0	12	0	4
Γ_{Red}	36	0	0	0	−4	0	0	0	0	12	0	4

[a]From Section 4.9.

of a B_{2u} mode, thus consigning about 10 years worth of calculations to oblivion (such is the life of a spectroscopist).

It is useful in analyzing molecules of high symmetry to perform a symmetry analysis based on internal coordinates, even if the secular equation is to be solved in terms of cartesian displacements. In effect, by constructing internal symmetry coordinates we can get an approximate idea of the form of the normal modes in each symmetry species. This idea is of considerable use in making band assignments and in matching up observed frequencies with an initially calculated set. For symmetry species with only one mode (e.g., A_{2u}) there is no mixing with other coordinates. However, even if there are three or four modes with the symmetry properties of a particular species, it is still possible to make approximate assignments. For example, C—H stretching modes are readily identified in the spectrum and do not mix significantly with other coordinates.

The number of symmetry coordinates in each species calculated by using internal coordinates as a basis set is shown in Table 8.3. This table shows the number of internal coordinates of a given set (e.g., C—C stretching) that are left invariant (or reversed in sign) by a symmetry operation of a particular class. For simplicity of presentation, only those classes of operations that give nonzero results for some set are shown. There are two points to be noted. First, we have introduced a coordinate β that is a combination of C—C—H bends.

Table 8.3 Calculation of Symmetry Species for Benzene in Terms of Internal Coordinates

	E	$3C_2'$	$3C_2''$	σ_h	$3\sigma_d$	$3\sigma_v$	Species
s(CH)	6	2		6		2	$A_{1g}+E_{2g}+B_{1u}+E_{1u}$
T(CC)	6		2	6	2		$A_{1g}+E_{2g}+B_{2u}+E_{1u}$
ϕ(HCC)	12			12			$A_{1g}+A_{2g}+2E_{2g}+B_{1u}+B_{2u}+2E_{1u}$
Ω	6	2		6		2	$A_{1g}+E_{2g}+B_{1u}+E_{1u}$
β	6	−2		6		−2	$A_{2g}+E_{2g}+B_{2u}+E_{1u}$
μ(CH)	6	−2		−6		2	$B_{2g}+E_{1g}+A_{2u}+E_{2u}$
z(CCCC)	6		2	−6	−2		$B_{2g}+E_{1g}+A_{1u}+E_{2u}$

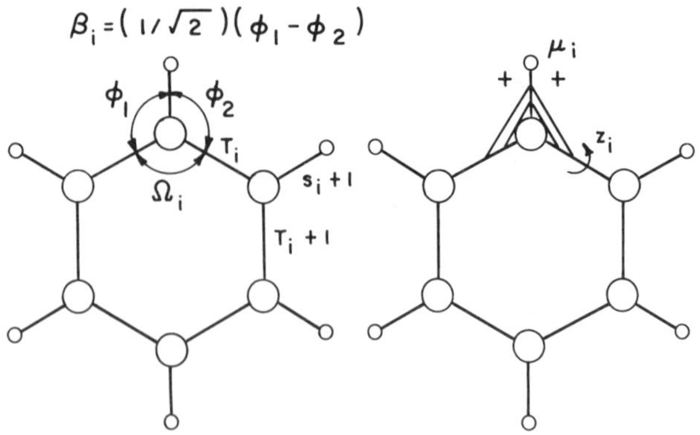

Figure 8.2. Definition of the internal coordinates of benzene.

Table 8.4 Planar Symmetry Coordinates for Benzene, Including E_{1u} Redundancy

Sym. Coord.		Coefficient for $j=$ 1	2	3	4	5	6	N	Int. Coord.
A_{1g}	S_1	+1	+1	+1	+1	+1	+1	$6^{-1/2}$	T
	S_2	+1	+1	+1	+1	+1	+1	$6^{-1/2}$	S
A_{2g}	S_3	+1	+1	+1	+1	+1	+1	$6^{-1/2}$	β
B_{1u}	S_{12}	−1	+1	−1	+1	−1	+1	$6^{-1/2}$	Ω
	S_{13}	−1	+1	−1	+1	−1	+1	$6^{-1/2}$	S
B_{2u}	S_{14}	−1	+1	−1	+1	−1	+1	$6^{-1/2}$	T
	S_{15}	−1	+1	−1	+1	−1	+1	$6^{-1/2}$	β
E_{2g}	S_{7a}	−2	+1	+1	−2	+1	+1	$12^{-1/2}$	S
	S_{7b}	0	−1	+1	0	−1	+1	2	S
	S_{8a}	−1	+2	−1	−1	+2	−1	$12^{-1/2}$	T
	S_{8b}	−1	0	+1	−1	0	+1	2	T
	S_{9a}	0	−1	+1	0	−1	+1	2	β
	S_{9b}	+2	−1	−1	+2	−1	−1	$12^{-1/2}$	β
	S_{6a}	−2	+1	+1	−2	+1	+1	$12^{-1/2}$	Ω
	S_{6b}	0	−1	+1	0	−1	+1	2	Ω
E_{1u}	S_{18a}	0	+1	+1	0	−1	−1	2	β
	S_{18b}	+2	+1	−1	−2	−1	+1	$12^{-1/2}$	β
	S'_{19a}	−1	0	+1	+1	0	−1	2	T
	S'_{19b}	+1	+2	+1	−1	−2	−1	$12^{-1/2}$	T
	S_{20a}	−2	−1	+1	+2	+1	−1	$12^{-1/2}$	S
	S_{20b}	0	+1	+1	0	−1	−1	2	S
	S'_{21a}	−2	−1	+1	+2	+1	−1	$12^{-1/2}$	Ω
	S'_{21b}	0	+1	+1	0	−1	−1	2	Ω

The symbols used to define various internal coordinates are illustrated in Figure 8.2. Second, a comparison with the results of the cartesian methods shows that the ring structure of this molecule introduces three redundancies among the C—C—H angle bends and three among the out-of-plane coordinates. As we pointed out in previous chapters, the problem of redundant coordinates can best be solved by an appropriate choice of symmetry coordinates. At this point we will simply list (in Table 8.4) as an illustration the internal symmetry coordinates derived by Duinker and Mills (7) for the in-plane modes. The numbering of these coordinates (S_1, S_2, etc.) corresponds to the numbering of the modes in the Wilson system (9). These coordinates can be determined by the methods described in Chapter 5, paying particular attention to the problem of redundancy. [In the treatment presented in Wilson et al. (1) the redundancies were not eliminated.] It is important to keep in mind that these coordinates, which are essentially pure C—C stretching, C—C—C bending and so on, are not a true description of the normal modes, since coordinates in the same symmetry species can mix. However, in certain cases, such as the C—H stretching mode ν_2 in species A_{1g}, this description is a very good approximation and aids in the assignment of modes in the observed vibrational spectrum. (We would expect ν_2 to be a strong polarized Raman line near 3000 cm^{-1}.) Kydd (10) has discussed the problems associated with redundant coordinates for the out-of-plane modes. We will discuss symmetry coordinates in more detail later when we consider the definition of external symmetry coordinates. Deriving external symmetry coordinates will allow us to illustrate the problem of constructing orthogonal coordinates in degenerate symmetry species.

8.2 The Vibrational Spectrum of Benzene and Its ^{13}C Isotopes

We will consider in detail the vibrational spectrum of benzene, paying particular attention to the ^{13}C isotopes, since we have firsthand experience of these materials (3). The vibrational spectrum of benzene and its deuterated derivatives has been considered by a number of authors, and this work is summarized in Varsanyi's book (8). We will not review or seek to condense this extensive body of work, but simply illustrate methods used to assign bands by working through the ^{13}C isotopes.

The Raman spectra of benzene, benzene-1-^{13}C, and benzene-(ul)-^{13}C are shown in Figure 8.3, 8.4, and 8.5, respectively. Benzene-(ul)-^{13}C is the manufacturer's description of what turned out to be a mixture of $^{13}C_6H_6$ and $^{13}C_5H_5{}^{12}CH$. (In our original paper a typographical error resulted in this molecule's being labeled "benzene-μl-^{13}C." We were initially alarmed but finally resigned to seeing this error perpetuated through the literature by workers using our data.) The spectra are shown between 300 cm^{-1} and 1700 cm^{-1} and the strongest lines have been cut off in order to show clearly the weaker modes. Of initial concern is the frequency of the ring "breathing"

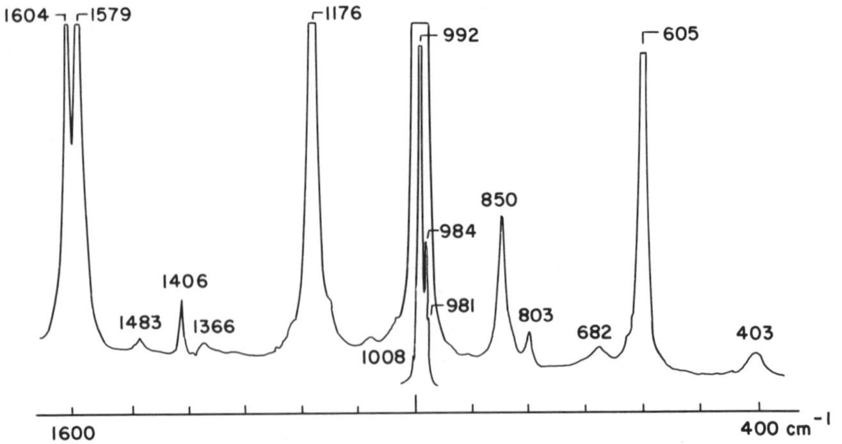

Figure 8.3. The Raman spectrum of benzene; $^{12}C_6H_6$. (Reprinted by permission from ref. 3.)

mode, observed at 992 cm^{-1} in $^{12}C_6H_6$. This vibrational mode involves large relative displacements of the carbon atoms and is therefore significantly affected by ^{13}C-substitution. Furthermore, this line is the strongest in the spectrum and can be detected at extremely small slit widths. The ensuing high resolution allows the qualitative determination of the ^{13}C isotopes present through the observed frequencies of this breathing mode.

Natural ^{13}C abundance is approximately 1.1% and therefore normal $^{12}C_6H_6$ samples contain approximately 7% mono-^{13}C-benzene. We assign the shoulder observed at 984 cm^{-1} in the Raman spectrum of $^{12}C_6H_6$ (Figure 8.3) to the

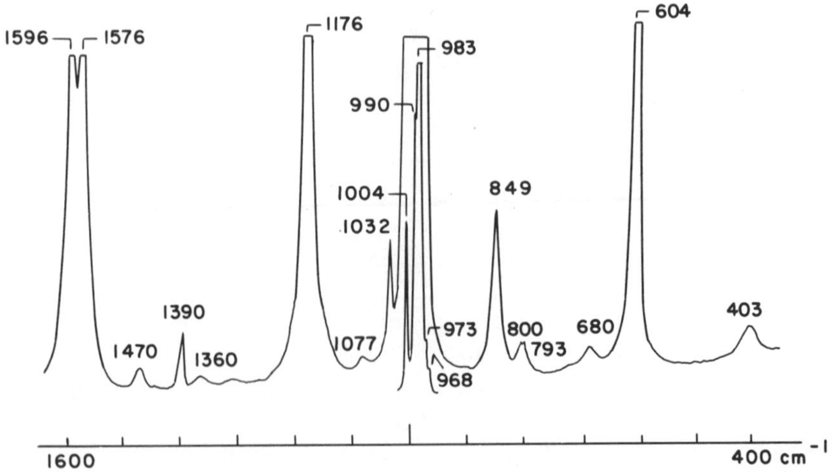

Figure 8.4. The Raman spectrum of benzene-1-^{13}C. (Reprinted by permission from ref. 3.)

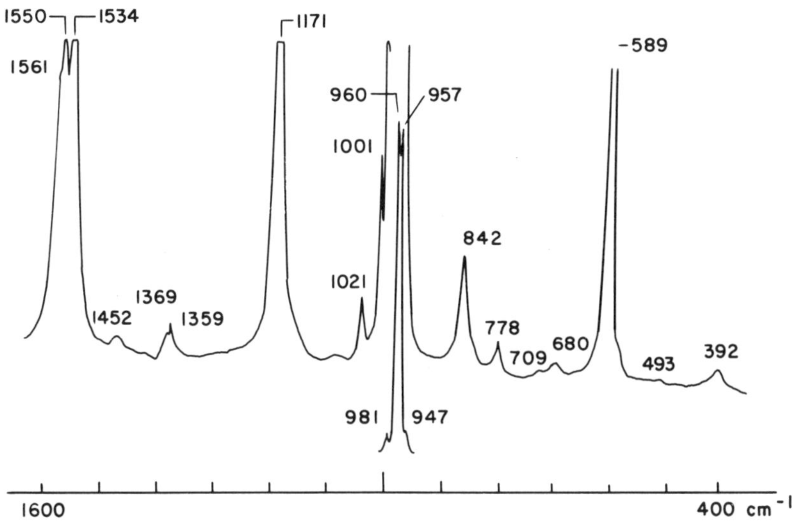

Figure 8.5. The Raman spectrum of benzene-(ul)-^{13}C. (Reprinted by permission from ref. 3.)

ring breathing mode of this isotope, since it corresponds to the very strong line observed at 983 cm^{-1} in the spectrum of benzene-1-^{13}C (Figure 8.4). Similarly, in this spectrum a shoulder near 990 cm^{-1} is observed that can be assigned to $^{12}C_6H_6$. The frequency shift upon monosubstitution corresponds to the value expected (5–10 cm^{-1}) for vibrations in which the ^{13}C atom participates strongly. Clearly, the predominant isotope present in this sample is benzene-1-^{13}C. The weak line at 973 cm^{-1} could possibly be due to a very small amount of disubstituted ^{13}C-benzene. The spectrum of benzene-(ul)-^{13}C is characterized by two very strong lines of almost equal intensity observed at 960 cm^{-1} and 957 cm^{-1}. These lines can be assigned to $^{13}C_5H_5{}^{12}CH$ and $^{13}C_6H_6$, respectively.

The frequency shifts observed between the lines assigned to $^{12}C_6H_6$, $^{13}CH^{12}C_5H_5$, and $^{13}C_6H_6$ isotopes correspond to expected values based on contemporary knowledge of the normal modes (6–8). A complete list of the observed frequencies will be given later. As already mentioned, we have numbered the modes of benzene as ν_1 through ν_{20}, according to the system described by Wilson (9).

Overtones and combination bands observed between 2000 cm^{-1} and 2900 cm^{-1} are particularly useful in assigning inactive fundamentals, and the Raman spectra of benzene and benzene-(ul)-^{13}C between 2700 cm^{-1} and 2200 cm^{-1} are compared in Figure 8.6. Doublets are observed for many of the lines in the latter sample, again because of the presence of both $^{13}C_5H_5{}^{12}CH$ and $^{13}C_6H_6$ isotopes, indicating that these overtones and combinations involve modes associated with stretches and bends of the ring carbon atoms. In the C—H stretching region the spectra of benzene and benzene-1-^{13}C were very

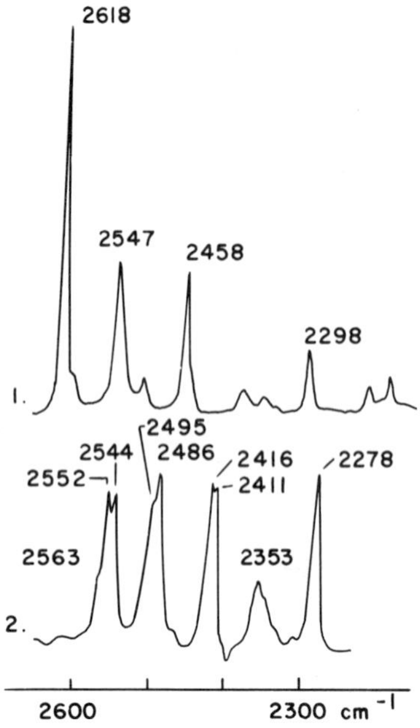

Figure 8.6. Spectrum 1, Raman spectrum of benzene ($^{12}C_6H_6$) between 2700 cm^{-1} and 2200 cm^{-1}. Spectrum 2, Raman spectrum of benzene-(ul)-^{13}C. (Reprinted by permission from ref. 3.)

similar, the strong $\nu_2(A_{1g})$ mode being observed at 3059 cm^{-1} and 3058 cm^{-1}, while the $\nu_7(E_{2g})$ line is observed as a shoulder at 3047 cm^{-1} and 3043 cm^{-1}, respectively. However, the ν_7 mode was not directly observable in the spectrum of benzene-(ul)-^{13}C. A strong line at 3034 cm^{-1} and a prominent shoulder at 3039 cm^{-1} are assigned to the ν_2 mode of $^{13}C_6H_6$ and $^{13}C_5H_5{}^{12}CH$, respectively. These lines are strongly polarized, whereas the $\nu_7(E_{2g})$ line is depolarized. Consequently, by placing a polarizer between the scattered light and monochromator, the ν_7 mode of $^{13}C_6H_6$ was observed at 3024 cm^{-1}, as shown in Figure 8.7.

The IR spectra of benzene, benzene-1-^{13}C, and benzene-(ul)-^{13}C are compared in Figures 8.8, 8.9, and 8.10, respectively. The scale expansion facility of the computerized Fourier transform system was utilized to superimpose spectra plotted on an absorbance scale of 0.0–0.3 units upon spectra plotted on a scale of 0.0–3.0 units. This procedure allows an accurate measurement of the frequencies of the weak combination bands, which are of considerable importance in the determination of the inactive fundamentals. The observed frequency shifts upon increasing ^{13}C-isotopic substitution are clearly discernible. There are some interesting observations that deserve comment before we proceed to a discussion of the assignments. Strictly, benzene-1-^{13}C belongs to the symmetry species C_{2v}. However, because of the small percentage change in mass, the substitution of a single ^{13}C atom for a ^{12}C atom results only in a perturbation of the vibrational system. The intensities are then nearly the same as the corresponding fundamentals of $^{12}C_6H_6$. Nevertheless, one or two of the

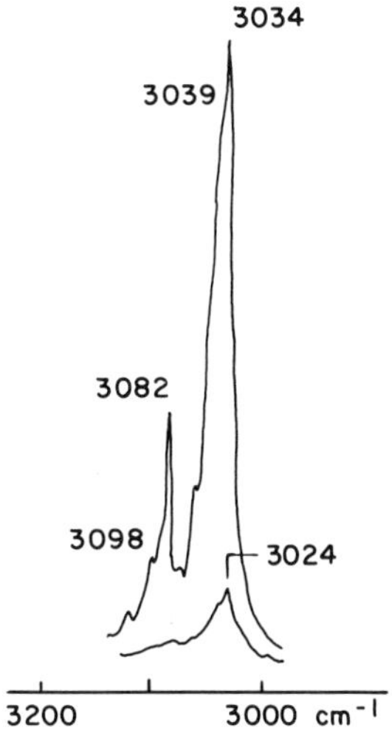

Figure 8.7. The polarized Raman spectrum of benzene-ul-^{13}C in the C—H stretching region. Top, spectrum taken with polarizer parallel; bottom, spectrum taken with polarizer perpendicular. (Reprinted by permission from ref. 3.)

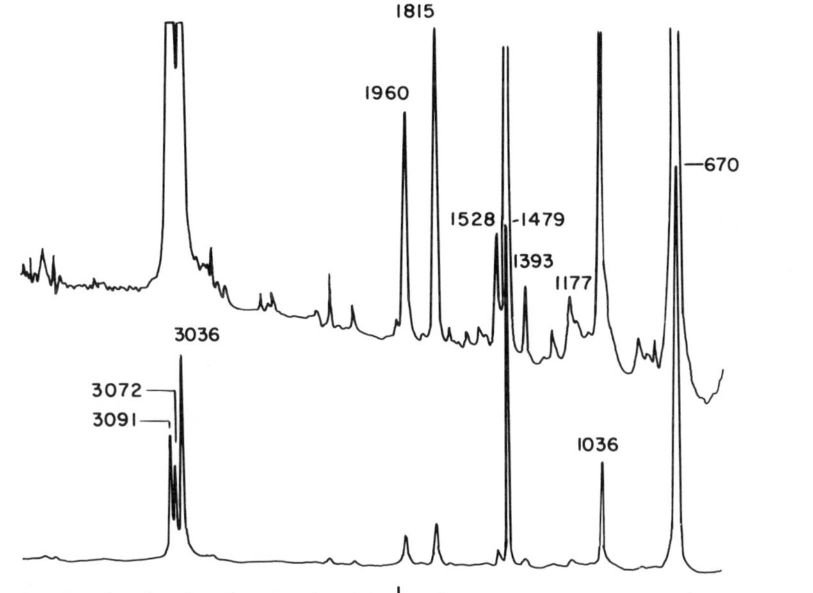

Figure 8.8. The infrared spectrum of benzene ($^{12}C_6H_6$). Top, absorbance scale of 0–0.3 units; bottom, 0–3.0 units. (Reprinted by permission from ref. 3.)

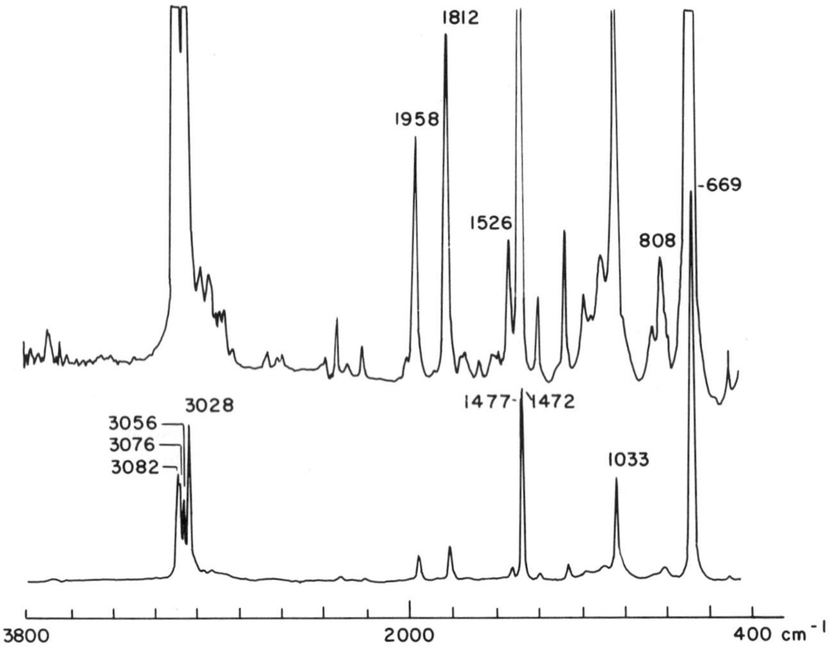

Figure 8.9. Infrared spectrum of benzene-l-^{13}C. Top, absorbance scale of 0–0.3 units; bottom, 0–3.0 units. (Reprinted by permission from ref. 3.)

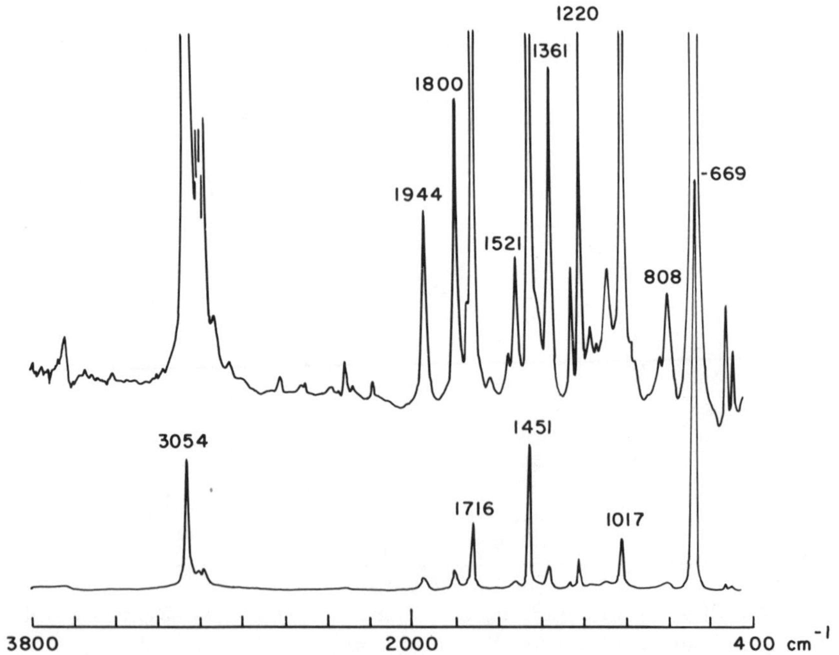

Figure 8.10. Infrared spectrum of benzene-ul-^{13}C. Top, absorbance scale of 0–0.3 units; bottom, 0–3.0 units. (Reprinted by permission from ref. 3.)

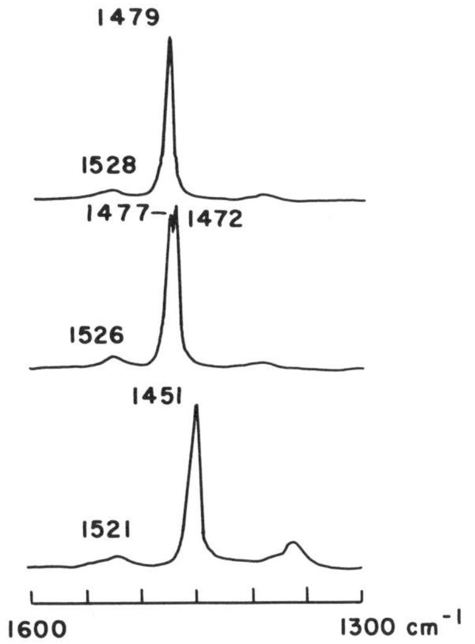

Figure 8.11. Comparison of the infrared spectra of (top to bottom) benzene ($^{12}C_6H_6$), benzene-l-^{13}C, and benzene-(ul)-^{13} between 1600 and 1300 cm^{-1}. (Reprinted by permission from ref. 3.)

inactive fundamentals have been detected with weak intensity, and we will discuss these individually later. In addition, the ν_{19} degenerate E_{1u} fundamental is split into two components at 1477 cm^{-1} and 1472 cm^{-1}, clearly shifted from the singularity at 1479 cm^{-1} in the spectrum of $^{12}C_6H_6$. This can be seen more clearly in Figure 8.11, which shows the 1600–1300 cm^{-1} region of the spectra of the three samples on an expanded scale. For this vibrational mode the selection rules for C_{2v} symmetry are then strictly obeyed, since an examination of the correlation table for D_{6h} and C_{2v} groups (1) demonstrates that a doubly degenerate E_{1u} mode of $^{12}C_6H_6$ is transformed to an A_1 and a B_2 mode upon monosubstitution. The C—C stretching vibration contributes significantly to this mode and it is therefore reasonable to expect that this would be the only E_{1u} mode in which the splitting would be sufficient to be observable. The complex of bands observed between 3000 cm^{-1} and 3100 cm^{-1} is also of interest. The three bands observed at 3091 cm^{-1}, 3072 cm^{-1}, and 3036 cm^{-1} in $^{12}C_6H_6$ are considered to be due to Fermi resonance between the fundamental ν_{20} and the combinations $\nu_8+\nu_{19}$ and $\nu_1+\nu_6+\nu_{19}$ (8). This is nicely confirmed through a comparison with the spectra of the ^{13}C isotopes. For benzene-l-^{13}C the higher frequency band is split into two components, at 3982 cm^{-1} and 3076 cm^{-1}, confirming the involvement of a combination of ν_{19}, which as we noted earlier is split upon monosubstitution of a ^{13}C atom. Furthermore, in the spectrum of benzene-(ul)-^{13}C the $\nu_8+\nu_{19}$ and $\nu_1+\nu_6+\nu_{19}$ combinations are shifted to a greater degree (approximately 50 cm^{-1}) than the ν_{20} fundamental, so that Fermi resonance no longer occurs. A singularity at 3054 cm^{-1} is then observed in this region of the spectrum (assigned to the ν_{20} fundamental), while the combination bands are observed as weak absorption between 3005 cm^{-1} and 2987 cm^{-1}.

A complete list of the observed frequencies of benzene-l-^{13}C and benzene-(ul)-^{13}C is given in Tables 8.5 and 8.6, respectively. The calculated values of

Table 8.5 Observed Frequencies and Assignments of Benzene-1-^{13}C(>60% Isotopic Purity[a]

Observed Wavenumber (cm^{-1}) Raman	Infrared	Assignment	Calc. Wavenumber (cm^{-1})
	3688 vw	$\nu_6+\nu_{20}$	3669
	3676 vw	($\nu_6+\nu_{13}$?)	
	3632 vw		
			3065
	3082 s	ν_{20}	3063
	3076 s	$\nu_8+\nu_{19}$	3058
	3056 s	$\nu_1+\nu_6+\nu_{19}$	3064
	3028 s		3059
3058 vs		ν_2	
3043 sh s		ν_7	
3010 vw			
	2962 vw		
	2952 sh vw		2954 2944
			2928
2946 w		$2\nu_{19}$	
2935 w		$\nu_3+\nu_8$	
	2925 vw		
	2920 vw		
2915	2917 vw		
	2870 vw		
	2850 vw		
	2816 vw		
	2806 vvw		
2765 vw		$\nu_8+\nu_9$	2762
2741 vw			
2708 ww			
		$2\nu_3$	2684
2684 vvw	2647 vw	$\nu_9+\nu_{19}$	2653 2684
2599 w	2599 vvw	$2\nu_{14}$	2600
	2580 vvw	$\nu_8+\nu_{12}$	2597
		$(\nu_1+\nu_6)+\nu_{12}$	2591
2532 w			
2444 w		$\nu_{14}+\nu_{15}$	2444
2375 vw	2376 ww	$\nu_3+\nu_{18}$	2375
	2323 w	$\nu_9+\nu_{15}$	2320
2289 w		$2\nu_{15}$	2288
	2278 vvw		
	2207 vw	$\nu_9+\nu_{18}$	2209
2195 vw		$\nu_6+\nu_8, \nu_3+\nu_{10}$	2190
2173 vw		$\nu_{15}+\nu_{18}$	
	1977 vw	$\nu_5+\nu_{12}$	1991

Table 8.5 Continued

Observed Wavenumber (cm^{-1})		Calc. Wavenumber	
Raman	Infrared	Assignment	(cm^{-1})
	1958 w	$\nu_5+\nu_{17}$	1951
	1812 w	$\nu_{10}+\nu_{17}$	1812
	1752 vvw		1748
	1746	$\nu_6+\nu_{15}$	
1691 vw		$2\nu_{10}$	1696
	1665 vvw	$\nu_4+\nu_{17}$	1661
	1608 vvw	$\nu_6+\nu_{12}$	1608
1596 s	1594 vvw	$\nu_1+\nu_6$	1587
1576 s	1575 vvw	ν_8	
	1526 w	$\nu_{10}+\nu_{11}$	1517
	1477 s		
1470 vw	1472 s	ν_{19}	
1400 sh vw			
1394 w		$2\nu_4$	1394
	1390 w	$\nu_5+\nu_{16}$	1390
1360 vw		$\nu_{16}+\nu_{17}$	1367
1300	1302 vw	ν_{14}	
	1178 sh vw		
1176 s	1174 vw	ν_9	
	1144 vvw	ν_{15}	
	1097 vw		
1077	1075 sh vvw	$\nu_{11}+\nu_{16}$	1072
1027 m	1033 m	ν_{18}	
	1009 sh w	$\nu_6+\nu_{16}$	1007
1004 m		ν_{12}	
	1000 sh vw		
990 sh s		$\nu_1[^{12}C_6H_6]$	
983 vs	984 vvw	ν_1	
973 sh w			
968 sh w		ν_{17}(?)	964
	848 vw		
848 m		ν_{10}	
	820 sh w }	Impurities	
	808 w }		
800 w		$2\nu_{16}$	806
793 w			
	778 vw		
	772 vw	$\nu_9-\nu_{16}$	773
	698 w	ν_4	
680 vw	676 sh s		
	669 vs	ν_{11}	
604 s		ν_6	
	501 w		
403 vw		ν_{16}	

[a]Abbreviations: vs, very strong; s, strong; m, medium; w, weak; vw, very weak; vvw, very, very weak.

Table 8.6 Observed Frequencies and Assignments of Benzene-(ul)-^{13}C (90% Isotopic Purity)[a]

Observed Wavenumbers (cm^{-1}) Raman	Infrared	Assignment	Calc. Wavenumber (cm^{-1})
	3640 vw	$\nu_6+\nu_{20}, (\nu_6+\nu_{13}?)$	3643
3119 w			
3098 w		$2(\nu_1+\nu_6), (\nu_1+\nu_6)+\nu_8; [^{13}C_5H_5CH]$	3098, 3097
3091 sh w		$2(\nu_1+\nu_6), (\nu_1+\nu_6)+\nu_8$	3092, 3088
m		$2\nu_8$	3084
3071 w			
3056 w		$\nu_{20}?[^{13}C_5H_5{}^{12}CH]$	
	3054 s	ν_{20}	
3039 sh vs		$\nu_2[^{13}C_5H_5{}^{12}CH]$	
3034 vs		ν_2	
3024 sh w		ν_7	
	3005 w	$\nu_8+\nu_{19}, \nu_1+\nu_6+\nu_{19}; [^{13}C_5H_4{}^{12}CH]$	3005, 3005
	2993 w	$\nu_1+\nu_6+\nu_{19}$	2996
	2987 w	$\nu_8+\nu_{19}$	2992
	2928 vw		
	2916 sh vw		
2904 sh w		$2\nu_{19}[^{13}C_5H_5{}^{12}CH]$	2914
2896 w		$2\nu_{19}$	2900
2889 w		$\nu_3+\nu_8$	2874
	2855 vw		
2724 vw		$\nu_{14}+\nu_{19}$	2727
2717 vw		$\nu_8+\nu_9$	2714
2700 vw			
2690 vw			
2667 vw		$2\nu_3$	2666
	2619 vw	$\nu_9+\nu_{19}$	2622
2552 w		$2\nu_{14}[^{13}C_5H_5{}^{12}CH]$	2552
2544 w		$2\nu_{14}$	2544
	2520 vw	$(\nu_1+\nu_6)+\nu_{12}$	2527
	2505 vw	$\nu_8+\nu_{12}$	2521
2495 sh w		$\nu_1+\nu_8[^{13}C_5H_5{}^{12}CH]$	2507
2486 w		$\nu_1+\nu_8$	2499
2416 w		$\nu_{14}+\nu_{15}[^{13}C_5H_5{}^{12}CH]$	2416
2411 w		$\nu_{14}+\nu_{15}$	2411
2353 vw			
2342 vw		$2\nu_9$	2342
	2313 vw	$\nu_9+\nu_{15}[^{13}C_5H_5{}^{12}CH]$	2312
	2308 sh vw	$\nu_9+\nu_{15}$	2311
2278 w		$2\nu_{15}$	2278
	2188 w		$2\nu_{15}$
	2188 vw	$\nu_9+\nu_{18}[^{13}C_5H_5{}^{12}CH]$	2194
	2183 vw	$\nu_9+\nu_{18}$	2189
	1959 sh w	$\nu_5+\nu_{12}$	1960
	1944 w	$\nu_5+\nu_{17}$	1939
	1800 w	$\nu_{10}+\nu_{17}$	1800
	1747 w		

Table 8.6 Continued

Observed Wavenumbers (cm^{-1})			Calc.
Raman	Infrared	Assignment	Wavenumber (cm^{-1})
	1716 m	(Acetone)	
	1640 vw	$\nu_4+\nu_{17}$	1638
1559 vvw		$\nu_9+\nu_{16}$	1564
1561 sh s		$\nu_8[^{13}C_5H_5{}^{12}CH]$	
1550 s		ν_8	
1534 s		$\nu_1+\nu_6$	
	1521 w	$\nu_{10}+\nu_{11}$	1511
1456 w	1457 sh m	$\nu_{19}[^{13}C_5H_5{}^{12}CH]$	
1452 w	1451 s	ν_{19}	
	1419 sh vw		
	1374 sh w	$\nu_5+\nu_{16}$	1373
1369 w		$\nu_{12}+\nu_{16}$	1370
1359 w		$2\nu_4$	1360
	1361 w		
1275 vw		$\nu_4+\nu_6$	1269
	1220 w		
1171 s	1170 vw	ν_9	
	1140 vvw	ν_{15}	1139
	1098 sh vw		
	1091 vw		
	1076 sh vw		
1021 w	1022 sh m	$\nu_{18}[^{13}C_5H_5{}^{12}CH]$	
	1017 m	ν_{18}	
1001 m			
981 m		ν_{12}	
	979 vvw	$\nu_6+\nu_{16}$	981
960 vs	960 vvw	$\nu_1[^{13}C_5H_5{}^{12}CH]$	
957 vs		ν_1	
947 m			
881 vvw			
	843 vw	ν_{10}	
842 m	820 sh vw 798 sh w	Impurities	
	798 sh w		
786 sh w		$2\nu_{16}[^{13}C_5H_5{}^{12}CH]$	
778 w		$2\nu_{16}$	784
	780 sh vvw	$\nu_9-\nu_{16}$	779
709 vvw			
680 w	676 sh s	ν_4(?)	682
	669 vvs	ν_{11}	
589 s		ν_6	
	530 w		
	501 w		
493 vvw		$\nu_{19}-\nu_{17}$	492
392 w		ν_{16}	

[a]Abbreviations as in Table 8.5.

the combinations and overtones given in the final column are based on the assignments discussed in the next section.

8.3 Assignment of Observed Frequencies

We will first outline the method used to assign the modes and will then discuss these assignments individually. The procedure of assignment used here differs somewhat from that of the original work of Brodersen and Langseth (11), principally because of the availability of high quality Raman data. First, a preliminary set of fundamental frequencies was chosen, based on the theoretically active modes. This is not completely straightforward, since modes ν_8 and ν_{20} are involved in Fermi resonance with combination bands. Next, the frequencies of as many of the inactive fundamentals as possible were determined from the first Raman overtone bands, using previously reported results (8) as a guide to assignment. It was determined that the use of Raman overtones and combinations gave more consistent results than the use of the IR overtones. For example, ν_{14} and ν_{15} were determined from the first overtone bands observed at 2544 cm^{-1} and 2278 cm^{-1}, respectively, in $^{13}C_6H_6$. Using these values, we calculate the combination $\nu_{14}+\nu_{15}$ to be 2411 cm^{-1}, precisely the observed value. It is well known that discrepancies in the position of overtone and combination bands are due to anharmonic terms in the potential energy function, and that anharmonic corrections are different for various isotopes, being less for the heavier species (1). The results of this study suggest that such discrepancies are also smaller for overtones and combinations that are totally symmetric, since the calculated values of many of these bands (see below) are in excellent agreement with observed values. This procedure allowed the determination of the frequencies of modes ν_3, ν_4, ν_{14}, and ν_{15}. Vibrational modes ν_{12} and ν_{16}, although forbidden, appear with weak intensity in the Raman spectrum. Finally, modes ν_5 and ν_{17} were determined from the most prominent infrared combination band that involved an observed fundamental, $\nu_5+\nu_{12}$ near 1340 cm^{-1} and $\nu_{10}+\nu_{17}$ near 1800 cm^{-1}. This leaves unassigned only the C—H stretching vibration ν_{13}. These fundamental frequencies are presented in Table 8.7. From this table we constructed a list of all allowed overtones (quantum number=2) and combinations (excluding C—H stretching vibrations), which we present in Tables 8.8, 8.9, and 8.10. These calculated values were used to assign the observed frequencies (Table 8.5 and 8.6). The relatively few ternary and difference combinations were assigned by reference to previous work (8, 11).

Tables 8.8, 8.9, and 8.10 can be used in concert with the observed frequencies in Tables 8.5 and 8.6 to refine further the values of the inactive frequencies. However, an excellent agreement with observed frequencies is already apparent and we consider that even a least-squares refinement would alter calculated values by only 1 or 2 cm^{-1} at most. Furthermore, in this study we are particularly concerned with the frequency shift of the ^{13}C isotopes from

Table 8.7 Assigned Fundamental Frequencies

Symmetry		Freq.			
D_{6h}	C_{2v}	No.	$^{12}C_6H_6$	$^{13}CH^{12}C_5H_5$	$^{13}C_6H_6$
A_{1g}	A_1	1	992	983	957
		2	3059	3058	3034
A_{2g}	B_1	3	1346	1342	1333
B_{2g}	A_2	4	703	697	680
		5	989	987	981
E_{1g}	A_2+B_2	10	849	848	842
E_{2g}	A_1+B_1	6	606	604	589
		7	3046	3043	3024
		8	1596	1586	1542
		9	1178	1176	1171
A_{2u}	B_1	11	670	669	669
B_{1u}	B_2	12	1008	1004	981
		13	[3062][a]		[3024][b]
B_{2u}	A_2	14	1309	1300	1272
		15	1149	1144	1139
E_{1u}	A_2+B_2	18	1036	1033	1017
				1477	
		19	1479	1472	1450
		20	3073	3065	3054
E_{2u}	A_1+B_1	16	404	403	392
		17	966	964	958

[a] From (8).
[b] Calculated using the Teller-Redlich product rule.

the observed values of $^{12}C_6H_6$. For many of the vibrational modes, particularly for benzene-1-^{13}C, this shift is small. Consistency in method of assignment for the three isotopes therefore has priority over the determination of a best fit to all observed overtones and combinations. Such a consistency in method of assignment should be equally important if other isotopes become available for future studies, and we have therefore tabulated our results in detail (3).

The A_{1g}, E_{1g}, and E_{2g} Raman-Active Fundamentals, ν_1, ν_2, ν_6, ν_7, ν_8, ν_9, and ν_{10}

The totally symmetric modes ν_1 and ν_2 are the strongest lines in the Raman spectrum and are easily assigned. In the Raman spectrum of benzene-(ul)-^{13}C the line at 960 cm^{-1} assigned to $^{13}C_5H_5{}^{12}CH$ is slightly stronger than the 957 cm^{-1} line of $^{13}C_6H_6$. However, for most other Raman lines and infrared bands of the spectrum (e.g., ν_{19}) assigned to these two isotopes this intensity relationship is reversed and we tentatively conclude that $^{13}C_6H_6$ is the major species

Table 8.8 Raman-Active Overtones (Quantum No. = 2)

Overtone	Symmetry Species	$^{12}C_6H_6$	$^{13}CH^{12}C_5H_5$	$^{13}C_6H_6$
$2\nu_8$	$A_{1g}+E_{2g}$	3192	3172	3084
			2954	
$2\nu_{19}$	$A_{1g}+E_{2g}$	2958	2944	2900
$2\nu_3$	A_{1g}	2692	2684	2666
$2\nu_{14}$	A_{1g}	2618	2600	2544
$2\nu_9$	$A_{1g}+E_{2g}$	2356	2352	2344
$2\nu_{15}$	A_{1g}	2298	2288	2278
$2\nu_{18}$	$A_{1g}+E_{2g}$	2072	2066	2034
$2\nu_{12}$	A_{1g}	2022	2016	1956
$2\nu_5$	A_{1g}	1978	1974	1962
$2\nu_{17}$	$A_{1g}+E_{2g}$	1932	1928	1916
$2\nu_1$	A_{1g}	1984	1966	1914
$2\nu_{10}$	$A_{1g}+E_{2g}$	1698	1696	1684
$2\nu_4$	A_{1g}	1406	1394	1359
$2\nu_{11}$	A_{1g}	1340	1338	1338
$2\nu_6$	$A_{1g}+E_{2g}$	1212	1208	1178
$2\nu_{16}$	$A_{1g}+E_{2g}$	808	806	784

present. The E_{2g} mode ν_7 was discussed earlier. Mode ν_8 is involved in Fermi resonance with $\nu_1+\nu_6$ to give the familiar doublet between 1500 cm^{-1} and 1600 cm^{-1}. We have assumed that the unperturbed frequency is equal to the mean value. Broderson and Langseth (11) assumed ν_8 to be 1 cm^{-1} less than this value, since the lower frequency component of the doublet is slightly more intense, which led them to the conclusion that it contained more of the fundamental than the high frequency component. However, in the spectrum of benzene-(ul)-^{13}C a shoulder at 1561 cm^{-1} is observed on the high frequency component at 1550 cm^{-1}. The frequency difference between the combination $\nu_1+\nu_6$ for $^{13}C_5H_5{}^{12}CH$ and $^{13}C_6H_6$ is only of the order of 3 cm^{-1}. We therefore suggest that the 11 cm^{-1} frequency difference between the shoulder and main line is derived predominantly from a shift in mode ν_8 for the two isotopes. This frequency difference is of the right order of magnitude for a mode that is predominantly a ring C—C stretching vibration and comparable to the shift of 10 cm^{-1} between $^{12}C_6H_6$ and $^{13}CH^{12}C_5H_5$.

The A_{2u} and E_{1u} IR-Active Fundamentals ν_{11}, ν_{18}, ν_{19}, and ν_{20}

The observed splitting of the degenerate fundamental ν_{19} observed in benzene-1-^{13}C and the absence of the Fermi resonance complex of bands between 3000 cm^{-1} and 3100 cm^{-1} for benzene-(ul)-^{13}C was discussed earlier. Brodersen and Langseth (11) computed a value of 3053 cm^{-1} for the ν_{20} of

Table 8.9 Raman-Active Binary Combinations

Combination	Symmetry Species	$^{12}C_6H_6$	$^{13}CH^{12}C_5H_5$	$^{13}C_6H_6$
$\nu_3+\nu_8$	E_{2g}	2942	2928	2875
$\nu_{14}+\nu_{19}$	E_{2g}	2788	2777 2772	2727
$\nu_8+\nu_9$	$A_{1g}+A_{2g}+E_{2g}$	2774	2762	2714
$\nu_{15}+\nu_{19}$	E_{2g}	2628	2612 2616	2589
$\nu_5+\nu_8$	E_{1g}	2585	2573	2523
$\nu_3+\nu_9$	E_{2g}	2524	2518	2505
$\nu_1+\nu_8$	E_{2g}	2588	2569	2499
$\nu_{18}+\nu_{19}$	$A_{1g}+A_{2g}+E_{2g}$	2515	2511 2505	2467
$\nu_{12}+\nu_{19}$	E_{2g}	2487	2481 2476	2430
$\nu_{14}+\nu_{15}$	A_{1g}	2458	2444	2411
$\nu_{17}+\nu_{19}$	$B_{1g}+B_{2g}+E_{1g}$	2455	2441 2436	2408
$\nu_8+\nu_{10}$	$B_{1g}+B_{2g}+E_{1g}$	2445	2434	2384
$\nu_{14}+\nu_{18}$	E_{2g}	2345	2333	2289
$\nu_{14}+\nu_{17}$	E_{1g}	2275	2264	2230
$\nu_4+\nu_8$	E_{1g}	2299	2283	2222
$\nu_3+\nu_{10}$	E_{1g}	2195	2190	2175
$\nu_{15}+\nu_{18}$	E_{2g}	2185	2177	2156
$\nu_5+\nu_9$	E_{1g}	2167	2163	2154
$\nu_6+\nu_8$	$A_{1g}+A_{2g}+E_{2g}$	2202	2190	2131
$\nu_1+\nu_9$	E_{2g}	2170	2159	2129
$\nu_{11}+\nu_{19}$	E_{1g}	2149	2146 2141	2119
$\nu_{15}+\nu_{17}$	E_{1g}	2115	2108	2098
$\nu_9+\nu_{10}$	$B_{1g}+B_{2g}+E_{1g}$	2027	2024	2014
$\nu_{12}+\nu_{18}$	E_{2g}	2044	2037	1997
$\nu_{17}+\nu_{18}$	$B_{1g}+B_{2g}+E_{1g}$	2002	1997	1975
$\nu_{12}+\nu_{17}$	E_{1g}	1974	1968	1938
$\nu_3+\nu_6$	E_{2g}	1952	1946	1922
$\nu_4+\nu_9$	E_{1g}	1881	1873	1852
$\nu_{16}+\nu_{19}$	$B_{1g}+B_{2g}+E_{1g}$	1883	1880 1875	1842
$\nu_5+\nu_{10}$	E_{2g}	1838	1835	1823
$\nu_1+\nu_{10}$	E_{1g}	1841	1831	1799
$\nu_6+\nu_9$	$A_{1g}+A_{2g}+E_{2g}$	1784	1780	1761
$\nu_{11}+\nu_{18}$	E_{1g}	1706	1702	1686
$\nu_{14}+\nu_{16}$	E_{1g}	1713	1703	1664
$\nu_4+\nu_5$	A_{1g}	1691	1684	1661
$\nu_{11}+\nu_{17}$	E_{2g}	1636	1633	1627
$\nu_5+\nu_6$	E_{1g}	1595	1591	1570

Table 8.9 Continued

Combination	Symmetry Species	$^{12}C_6H_6$	$^{13}CH^{12}C_5H_5$	$^{13}C_6H_6$
$\nu_1+\nu_6$	E_{2g}	1598	1587	1546
$\nu_{15}+\nu_{16}$	E_{1g}	1553	1547	1532
$\nu_4+\nu_{10}$	E_{2g}	1551	1544	1522
$\nu_6+\nu_{10}$	$B_{1g}+B_{2g}+E_{1g}$	1455	1452	1431
$\nu_{16}+\nu_{18}$	$B_{1g}+B_{2g}+E_{1g}$	1440	1436	1409
$\nu_{12}+\nu_{16}$	E_{1g}	1415	1411	1370
$\nu_{16}+\nu_{17}$	A_{1g}	1370	1367	1340
$\nu_4+\nu_6$	E_{1g}	1309	1301	1269
$\nu_{11}+\nu_{16}$	E_{2g}	1074	1072	1061

liquid $^{12}C_6H_6$ by assuming that the sum of the squares of the three unperturbed frequencies is equal to the sum of the squares of the three observed frequencies. Since ν_{20} is observed at 3054 cm^{-1} for $^{13}C_6H_6$, this assignment now seems unlikely. The lower frequency band of $^{12}C_6H_6$ at 3036 cm^{-1} is considerably more intense than that of the other two components and therefore possibly has a larger contribution from the ν_{20} fundamental. If we assume that the calculated frequencies of the combinations $\nu_8+\nu_{19}$ at 3075 cm^{-1} and $\nu_1+\nu_6+\nu_{19}$ at 3077 cm^{-1} are correct, then a value of 3073 cm-1 for ν_{20} would seem reasonable, and as we shall see later is confirmed by the application of the Teller-Redlich product rule. The assumption that the higher frequency bands involve a larger contribution from the combinations is supported by a consideration of the spectrum of benzene-1-^{13}C where the band originally at 3091 cm^{-1} in $^{12}C_6H_6$ is now observed as a doublet at 3083 cm^{-1} and 3077 cm^{-1}, probably because of the involvement of the degenerate fundamental ν_{19}. Since the lower frequency band is shifted 7 cm^{-1} to 3029 cm^{-1}, we assume a similar isotopic shift in the position of the ν_{20} fundamental, to 3066 cm^{-1}, for benzene-1-^{13}C.

The Inactive B_{1u} Fundamentals ν_{12} and ν_{13}

Mair and Hornig (12) observed the ν_{12} fundamental at 1010 cm^{-1} in the spectrum of benzene in the solid state. A line at 1008 cm^{-1} observed in the Raman spectrum of liquid $^{12}C_6H_6$ (Figure 8.3) can also be assigned to this mode. Even though this line has not been reported in previous Raman studies, a consideration of the spectrum of benzene-1-^{13}C confirms the assignment. This line is now shifted to 1004 cm^{-1} but is considerably more intense. The relative intensities of the ν_1 and ν_{12} bands in the spectrum of benzene-1-^{13}C is very similar to the comparable doublet in monosubstituted benzene samples, probably because of a breakdown in the D_{6h} selection rules, since this mode is one of the distinct carbon ring frequencies. We have discussed a similar breakdown in selection rules for the ν_{19} C—C ring stretching mode. The line at

Table 8.10 Infrared-Active Binary Combinations

Combination	Symmetry Species	$^{12}C_6H_6$	$^{13}CH^{12}C_5H_5$	$^{13}C_6H_6$
$\nu_8+\nu_{19}$	$B_{1u}+B_{2u}+E_{1u}$	3075	3063 3058	2992
$\nu_8+\nu_{14}$	E_{1u}	2905	2886	2814
$\nu_3+\nu_{19}$	E_{1u}	2825	2819 2814	2783
$\nu_8+\nu_{15}$	E_{1u}	2745	2730	2681
$\nu_9+\nu_{19}$	$B_{1u}+B_{2u}+E_{1u}$	2657	2653 2648	2622
$\nu_8+\nu_{18}$	$B_{1u}+B_{2u}+E_{1u}$	2632	2619	2559
$\nu_8+\nu_{12}$	E_{1u}	2612	2597	2521
$\nu_8+\nu_{17}$	$A_{1u}+A_{2u}+E_{2u}$	2562	2550	2500
$\nu_9+\nu_{14}$	E_{1u}	2487	2476	2444
$\nu_1+\nu_{19}$	E_{1u}	2471	2460 2455	2407
$\nu_3+\nu_{18}$	E_{1u}	2382	2375	2350
$\nu_9+\nu_{15}$	E_{1u}	2327	2320	2311
$\nu_{10}+\nu_{19}$	$A_{1u}+A_{2u}+E_{2u}$	2328	2325 2320	2292
$\nu_9+\nu_{18}$	$B_{1u}+B_{2u}+E_{1u}$	2214	2209	2189
$\nu_9+\nu_{12}$	E_{1u}	2185	2180	2152
$\nu_9+\nu_{17}$	$A_{1u}+A_{2u}+E_{2u}$	2144	2140	2130
$\nu_6+\nu_{19}$	$B_{1u}+B_{2u}+E_{1u}$	2085	2081 2076	2039
$\nu_3+\nu_{11}$	A_{2u}	2016	2011	2002
$\nu_1+\nu_{18}$	E_{1u}	2028	2016	1974
$\nu_5+\nu_{12}$	A_{2u}	1997	1991	1961
$\nu_5+\nu_{17}$	E_{1u}	1955	1951	1939
$\nu_8+\nu_{16}$	$A_{1u}+A_{2u}+E_{2u}$	2000	1989	1934
$\nu_6+\nu_{14}$	E_{1u}	1915	1904	1861
$\nu_{10}+\nu_{18}$	$A_{1u}+A_{2u}+E_{2u}$	1885	1881	1859
$\nu_{10}+\nu_{17}$	$B_{1u}+B_{2u}+E_{1u}$	1815	1812	1800
$\nu_6+\nu_{15}$	E_{1u}	1755	1748	1728
$\nu_4+\nu_{12}$	A_{2u}	1711	1701	1661
$\nu_4+\nu_{17}$	E_{1u}	1669	1661	1638
$\nu_1+\nu_{18}$	A_{2u}	1662	1652	1626
$\nu_6+\nu_{18}$	$B_{1u}+B_{2u}+E_{1u}$	1642	1637	1606
$\nu_6+\nu_{12}$	E_{1u}	1617	1612	1567
$\nu_9+\nu_{16}$	$A_{1u}+A_{2u}+E_{2u}$	1582	1579	1564
$\nu_6+\nu_{17}$	$A_{1u}+A_{2u}+E_{2u}$	1572	1568	1547
$\nu_{10}+\nu_{11}$	E_{1u}	1519	1517	1511
$\nu_5+\nu_{16}$	E_{1u}	1393	1390	1373
$\nu_{10}+\nu_{16}$	$B_{1u}+B_{2u}+E_{1u}$	1253	1251	1234
$\nu_4+\nu_{16}$	E_{1u}	1107	1100	1072
$\nu_6+\nu_{16}$	$A_{1u}+A_{2u}+E_{2u}$	1010	1007	981

981 cm^{-1} in the spectrum of benzene-(ul)$-{}^{13}C$ is assigned to ν_{12} of ${}^{13}C_6H_6$.

The frequency of normal mode ν_{13} remains undetermined by these results. Varsanyi (8) in a comprehensive review of the literature concluded that the frequency of this C—H stretching vibration was at 3062 cm^{-1}.

The Inactive B_{2u} Fundamentals ν_{14} and ν_{15}

It has been noted earlier that these fundamentals can be calculated with a remarkable degree of internal consistency from the overtones $2\nu_{14}$, $2\nu_{15}$, and the combination $\nu_{14}+\nu_{15}$, which appear in the Raman spectrum. Furthermore, ν_{15} appears as a very weak broad band in the infrared spectrum of benzene-ul-${}^{13}C$, probably because of a breakdown of selection rules.

The Inactive E_{2u} Fundamentals ν_{16} and ν_{17}

Vibrational mode ν_{16} appears as a weak, broad band in the Raman spectrum of benzene and its isotopes. The frequency of ν_{17} was determined primarily from the overtone $\nu_{17}+\nu_{10}$ near 1800 cm^{-1}. This band is prominent in the infrared spectrum and involves a combination with an observed fundamental. Consequently, any errors are predominantly due to anharmonicity. Close agreement between the observed and calculated value of $\nu_{19}-\nu_{17}$ is obtained using this frequency.

The Inactive A_{2g} Fundamental ν_3

The frequency of this mode was determined from the Raman-active overtone $2\nu_3$.

The Inactive B_{2g} Fundamentals ν_4 and ν_5

Fundamental ν_4 was also determined from a Raman-active overtone, $2\nu_4$. This value gives good agreement between the observed and the calculated frequency of $\nu_4+\nu_{17}$ and is further observed with weak intensity in the IR spectrum of benzene-1-${}^{13}C$ at 698 cm^{-1}, in excellent agreement with the calculated value of 697 cm^{-1}.

Fundamental ν_5 is involved in a number of prominent combination bands, most of which, however, involve inactive modes. The combination $\nu_5+\nu_{16}$ was chosen to determine the frequency, since ν_{16} is observed in the Raman spectrum and the calculated value of ν_5 for ${}^{12}C_6H_6$ of 989 cm^{-1} is in excellent agreement with the generally accepted frequency of 990 cm^{-1} given by Brodersen and Langseth (11).

The Teller-Redlich Product Rule

The Teller-Redlich product rule can be applied to the assigned fundamental frequencies of ${}^{12}C_6H_6$ and ${}^{13}C_6H_6$ within each of the symmetry species as a

Table 8.11 Application of Teller-Redlich Product Rule to the Fundamental Frequencies of $^{12}C_6H_6$ and $^{13}C_6H_6$

Symmetry Species	Product Rule Formulas	Theoretical Value	Calc. Value from Assigned Frequencies
A_{1g}	$\frac{\nu_1'\nu_2'}{\nu_1\nu_2}=\left(\frac{m_{12}}{m_{13}}\right)^{1/2}$	0.961	0.957
A_{2g}	$\frac{\nu_3'}{\nu_3}=\left[\left(\frac{m_{12}}{m_{13}}\right)\frac{I_2'}{I_2}\right]^{1/2}$	0.992	0.993
B_{2g}	$\frac{\nu_4'\nu_5'}{\nu_4\nu_5}=\left(\frac{m_{12}}{m_{13}}\right)^{1/2}$	0.961	0.962
E_{1g}	$\frac{\nu_{10}'}{\nu_{10}}=\left[\left(\frac{m_{12}}{m_{13}}\right)\frac{I_\perp'}{I_\perp}\right]^{1/2}$	0.992	0.992
E_{2g}	$\frac{\nu_6'\nu_7'\nu_8'\nu_9'}{\nu_6\nu_7\nu_8\nu_9}=\frac{m_{12}}{m_{13}}$	0.923	0.928
A_{2u}	$\frac{\nu_{11}'}{\nu_{11}}=\left[\left(\frac{m_{12}}{m_{13}}\right)\frac{M'}{M}\right]^{1/2}$	0.997	0.998
B_{1u}	$\frac{\nu_{12}'\nu_{13}'}{\nu_{12}\nu_{13}}=\left(\frac{m_{12}}{m_{13}}\right)^{1/2}$	0.961	(?)
B_{2u}	$\frac{\nu_{14}'\nu_{15}'}{\nu_{14}\nu_{15}}=\left(\frac{m_{12}}{m_{13}}\right)^{1/2}$	0.961	0.963
E_{1u}	$\frac{\nu_{18}'\nu_{19}'\nu_{20}'}{\nu_{18}\nu_{19}\nu_{20}}=\frac{m_{12}}{m_{13}}\left(\frac{M'}{M}\right)^{1/2}$	0.958	0.959
E_{2u}	$\frac{\nu_{16}'\nu_{17}'}{\nu_{16}\nu_{17}}=\left(\frac{m_{12}}{m_{13}}\right)^{1/2}$	0.961	0.962

check on the accuracy of the assignments. The calculated and observed values are compared in Table 8.11 and are in excellent agreement. The small deviations represent a sum of contributions from experimental error and anharmonicity of the frequencies involved. Such deviations are apparently very small.

A calculated value was not determined for symmetry group B_{1u} because of uncertainty in the value of ν_{13}. Instead, the literature value of 3062 cm^{-1} was taken for $^{12}C_6H_6$ and used to calculate ν_{13} for $^{13}C_6H_6$.

8.4 Coordinate Definitions

Using the band assignments discussed in the preceding section and an approximate knowledge of the normal modes from symmetry analysis, we can determine a force field for benzene by refining the calculated to the observed frequencies. Using this calculated force field, we can then obtain a more precise knowledge of the normal modes. The availability of numerous isotopes (initially various deuterated species and more recently ^{13}C isotopes), together with the deceptive simplicity and symmetry of this molecule, has tempted numerous spectroscopists into a vibrational analysis from which few have emerged unscathed. Most of the problems are associated with the calculation of the force field, which we will consider in the following section. However, the problems introduced by redundant coordinates can also result in some difficulties, which we will consider here.

The coordinate definitions necessary for a normal coordinate analysis are the cartesian coordinates, defining the position in space and mass of each atom; the internal coordinates, which allow the definition of a valence force field; and the symmetry coordinates. Internal or external symmetry coordinates have to be defined, depending on whether the computer program utilized is based on the Wilson **GF** form of the secular equations,

$$(\mathbf{GF}_r - \lambda\mathbf{E})\mathbf{L}_r = 0$$

or whether the problem is defined in the symmetric form:

$$(\mathbf{D}^T\mathbf{F}_r\mathbf{D} - \lambda\mathbf{E})\mathbf{L}_x = 0$$

However, both methods allow the force constant to be defined in terms of internal coordinates. The cartesian coordinates can be determined from a knowledge of the bond angles and bond length. In our calculations we assumed C—H and C—C bond lengths to be 1.084 and 1.397 Å, respectively, with all bond angles equal to 120°. The axes of the cartesian system were placed at the center of the ring with the z axis perpendicular to the ring and the y axis going through carbon atoms 1 and 7 in Figure 8.12.

The internal coordinates are defined in a relatively straightforward manner and the **B** matrix elements (relating the internal and cartesian displacement

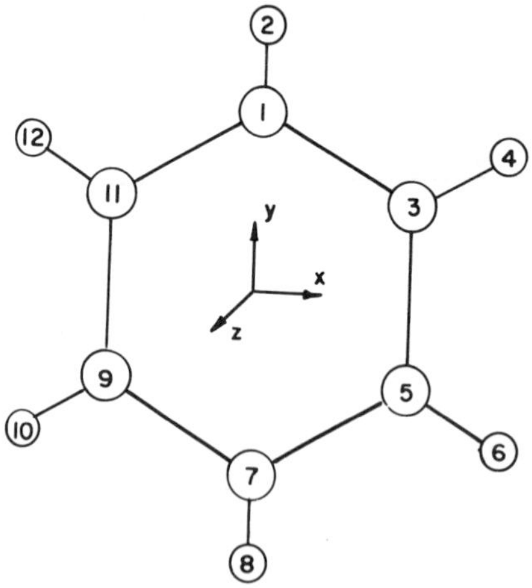

Figure 8.12. The definition of the cartesian axes and the numbering of atoms for the definition of the internal coordinates of benzene.

coordinates) are determined by the program. However, care must be taken to define the coordinates in a consistent manner. For example, in our normal coordinate analysis program the four atoms involved in the out-of-plane hydrogen bending coordinate, μH_2 in Figure 8.12, have to be defined in a given order, with the hydrogen atom first and the carbon atom last, for example, 2–11–3–1. This coordinate could also be defined 2–3–11–1, but this definition is for a displacement *below* the plane of the ring, in the $-z$ direction rather than the $+z$ direction. Either definition can be used, as long as all other hydrogen out-of-plane bending coordinates are defined in an equivalent manner, so that if we use 2–11–3–1, the other definitions are 4–1–5–3, 6–3–7–5, and so on. There is also a problem in defining the other out-of-plane coordinate, C—C torsions in the ring. Two different types of torsional coordinates have been used by various authors. Both have been explicitly defined by Whiffen (6) and discussed by Kydd (10). The first definition involves only four C atoms, while the second involves two H atoms as well. We chose to use the latter, summing two torsional coordinates τ_i and τ_i' round each bond so that

$$z_i = \frac{1}{\sqrt{2}}(\tau_i + \tau_i')$$

The torsional coordinates τ_i and τ_i' are defined so that the first and last atoms are trans. For example, the torsion of bond C_1—C_3 in Figure 8.12 is defined as the normalized sum of coordinates 2–1–3–5 and 11–1–3–4. Finally, defining two C—C—H in-plane bending coordinates, ϕ_i and ϕ_i', at each carbon atom is not convenient and can lead to all sorts of problems in defining interaction force constants. Consequently, a new bending coordinate, β_i, is defined:

$$\beta_i = \frac{1}{\sqrt{2}}(\phi_i - \phi_i')$$

This new coordinate is the change in the angle between the C—H bond and the bisector of the external C—C—C angle.

Although this coordinate definition removes the local bond angle redundancy at each carbon atom, it must be kept in mind that there are still redundancies among the bond angles as a result of the ring structure of this molecule. The redundancies can be removed by the appropriate definition of internal symmetry coordinates. As discussed in Chapter 5, a symmetry coordinate is first constructed so as to express the redundancy condition, then the remaining symmetry coordinates in the species are constructed orthogonal to it. However, if the purpose of the calculation is simply to reproduce the frequencies, then redundant symmetry coordinates can be defined. They simply result in additional zero roots.

We will not consider the details of the construction of internal symmetry coordinates. The method is discussed in Chapter 5, where the correct choice of internal coordinates for the generation of the symmetry coordinates of degenerate species was discussed, using the E_{1u} species of benzene as an example. In addition, the in-plane internal symmetry coordinates, as calculated by Duinker and Mills (7) and Crawford and Miller (13), were listed earlier in this chapter in Table 8.2. These authors eliminated the redundant coordinate in A_{1g} by inspection, but the four redundant E_{1u} coordinates S_{18}, S_{19}, S_{20}, and S_{21} present a more complicated situation. In this symmetry species the redundancy was removed by the orthogonal transformation

$$\begin{bmatrix} S_{19} \\ S_r \end{bmatrix} = (2)^{-1/2} \begin{bmatrix} +1 & -1 \\ +1 & +1 \end{bmatrix} \begin{bmatrix} S'_{19} \\ S'_{21} \end{bmatrix}$$

where S_r is the redundant coordinate, indentically equal to zero.

The problem of redundant internal coordinates is not encountered in the construction of external symmetry coordinates. However, extra coordinates are encountered in species that have the same symmetry properties as translations and rotations. These result in the calculation of zero frequencies. We have found this to be useful, in that errors in defining coordinates or typing up the computer cards are often revealed by the calculation of the wrong number of zeros. The external symmetry coordinates of the nondegenerate species are simply and easily calculated by the methods of Chapter 5. For the degenerate species it is essential to ensure that orthogonal symmetry coordinates are defined. For example, for species E_{2g} a symmetry coordinate can be constructed starting from y_1 (the y coordinate of atom 1 in Figure 8.12):

$$S_1^{(E_{2g})} = \frac{1}{4\sqrt{3}}\left(4y_1 - 4y_7 - \sqrt{3}\,x_3 - y_3 - \sqrt{3}\,x_5 + y_5 + \sqrt{3}\,x_9 + y_9 + \sqrt{3}\,x_{11} - y_{11}\right)$$

All degenerate sets of symmetry coordinates must be identically "oriented" (i.e., must have the same transformation coefficients). Consequently, x_1 cannot be used to generate the second symmetry coordinate. We have to use a coordinate or combination of coordinates that has the same symmetry properties as y_1. The coordinate combinations $y_3 + y_{11}$ or $x_3 - x_{11}$ are appropriate combinations. Using $y_3 + y_{11}$ we obtain

$$S_2'^{(E_{2g})} = \frac{1}{N}\left(-2y_1 + 2y_7 - \sqrt{3}\,x_3 + 5y_3 - \sqrt{3}\,x_5 \right.$$
$$\left. - 5y_5 + \sqrt{3}\,x_9 - 5y_9 + \sqrt{3}\,x_{11} + 5y_{11}\right)$$

This coordinate is independent of $S_1^{(E_{2g})}$ but not orthogonal to it. An orthogonal coordinate can be generated, as discussed in Chapter 5, by defining a new coordinate $S_2^{(E_{2g})}$ as

$$S_2^{(E_{2g})} = \frac{1}{N}\left[S_2'^{(E_{2g})} + kS_1^{(E_{2g})}\right]$$

where the number k is determined by the orthogonality condition.

8.5 A Force Field for Benzene

In this section we will first discuss the extent to which a general valence force field can be calculated and then present the results of our calculations. The diagonal and interaction force constants are defined in Table 8.12 according to the numbering of the internal coordinates shown in Figure 8.13. It is important to note that certain interactions are zero through symmetry, for example, $F_{s_1\beta_1}$. The vertical mirror through bond s_1 leaves this C—H stretch unaltered, but reverses the sign of the C—H bend, β_1. As we discussed in Chapter 5, for certain molecules, even if sufficient observed frequencies were available, it would not be possible to calculate all the valence force constants. Only certain combinations of them, the symmetry force constants can be calculated. Furthermore, for benzene even the symmetry force constants are underdetermined by the observed frequencies of C_6H_6 and C_6D_6. The inclusion of the observed

Table 8.12 Force Constants for Benzene

	In plane[a]											
	s_1	s_2	s_3	s_4	s_5	s_6	T_1	T_2	T_3	T_4	T_5	T_6
s_1	K_s	F_s^o	F_s^m	F_s^p	F_s^m	F_s^o	F_{Ts}^o	F_{Ts}^m	F_{Ts}^p	F_{Ts}^p	F_{Ts}^m	F_{Ts}^o
T_1							K_T	F_T^o	F_T^m	F_T^p	F_T^m	F_T^o
	Ω_1	Ω_2	Ω_3	Ω_4	Ω_5	Ω_6	β_1	β_2	β_3	β_4	β_5	β_6
s_1	$F_{s\Omega}$	$F_{S\Omega}^o$	$F_{s\Omega}^m$	$F_{s\Omega}^p$	$F_{s\Omega}^m$	$F_{s\Omega}^o$	0	$F_{s\beta}^o$	$F_{s\beta}^m$	0	$F_{s\beta}^{m\prime}$	$F_{s\beta}^{o\prime}$
T_1	$F_{T\Omega}^o$	$F_{T\Omega}^m$	$F_{T\Omega}^p$	$F_{T\Omega}^b$	$F_{T\Omega}^m$	$F_{T\Omega}^o$	$F_{T\beta}^o$	$F_{T\beta}^{o\prime}$	$F_{T\beta}^{m\prime}$	$F_{T\beta}^{p\prime}$	$F_{T\beta}^p$	$F_{T\beta}^m$
Ω_1	H_Ω	F_Ω^o	F_Ω^m	F_Ω^p	F_Ω^m	F_Ω^o	0	$F_{\Omega\phi}^{o\prime}$	$F_{\Omega\phi}^{m\prime}$	0	$F_{\Omega\phi}^m$	$F_{\Omega\phi}^o$
β_1							H_β	F_β^o	F_β^m	F_β^p	F_β^m	F_β^o
	Out of Plane											
	μ_1	μ_2	μ_3	μ_4	μ_5	μ_6	z_1	z_2	z_3	z_4	z_5	z_6
μ_1	H_μ	F_μ^o	F_μ^m	F_μ^p	F_μ^m	F_μ^o	$F_{z\mu}^o$	$F_{z\mu}^{pn}$	$F_{z\mu}^p$	$F_{z\mu}^{p\prime}$	$F_{z\mu}^{m\prime}$	$F_{z\mu}^{o\prime}$
z_1							H_z	F_z^o	F_z^m	F_z^p	F_z^m	F_z^o

[a] $F_s^{o\prime}\beta = -F_{s\beta}^o$; $F_{T\beta}^{o\prime} = -F_{T\beta}^o$, etc.
[b] $F_{z\mu}^{o\prime} = -F_{z\mu}^o$, etc.

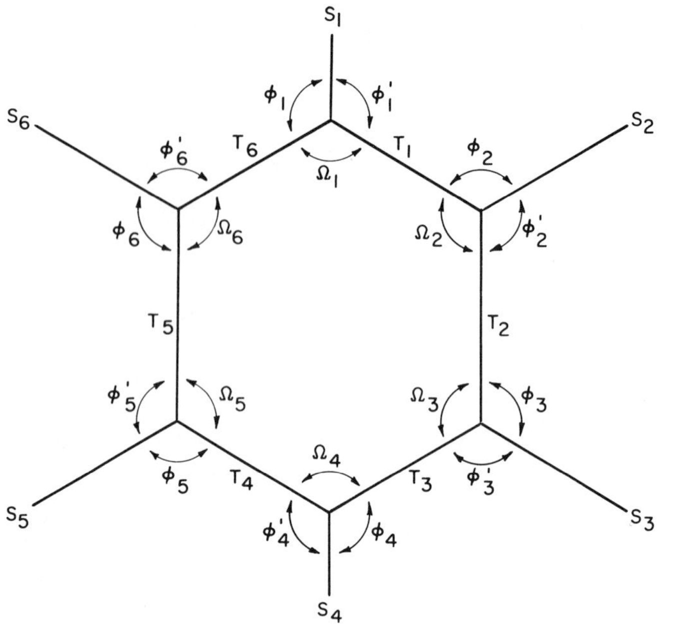

$$\beta_1 = 1/\sqrt{2}\,(\phi_1 - \phi_1')\ \text{etc.}$$

Figure 8.13. The numbering of the in-plane internal coordinates of benzene.

frequencies of $^{13}C_6H_6$ does not increase the data set, since for a molecule containing n nonequivalent sets of atoms only $n-1$ complete substitutions will give new equations (for benzene $n=2$). McKean and Duncan (14) have shown that theoretically an asymmetric isotopic substitution could give new equations through a breakdown of symmetry, and hence a recombination of valence force constants into new symmetry force constants. For benzene, however, this procedure has not been attempted, and it is uncertain whether the frequency shifts among the isotopic species would be sufficient to calculate the force constants with any precision. Consequently, the numerous normal coordinate calculations have relied on assuming certain force constants to be zero and on attempts to include other data, such as Coriolis constants (7). Even for those modes where the data are apparently sufficient for the determination of force constants there are still uncertainties. For example, in mode B_{2u} there are two normal modes and three symmetry force constants, $F_{14,14}$, $F_{15,15}$, and $F_{14,15}$. Duinker and Mills (7) showed that the data from C_6H_6 and C_6D_6 allowed the calculation of two possible solutions, as indicated by the force constant display graph shown previously in Chapter 7 (Figure 7.1). In situations of this type the spectroscopist has to rely on a degree of chemical intuition, which usually means choosing the solution with the smallest interaction force constant. In this case, however, Duinker and Mills (7) pointed out that in order for the

Table 8.13 Valence Force Field for Benzene

Force Constant	Coordinates Involved[a]	Φ^b	Dispersion (Uncertainty)
	In Plane		
K_T	T	7.0067	0.5122
K_s	s	5.0220	0.0949
H_Ω	Ω	1.4612	0.4385
H_ϕ	β	0.4956	0.0215
F_T^o	T_i, T_{i+1} (ortho)	0.6432	0.2615
F_T^m	T_i, T_{i+2} (meta)	−0.5908	0.2681
F_T^p	T_i, T_{i+3} (para)	0.6192	0.4793
$F_{T\phi}^o$	T_i, β_i	−0.2714	0.4725
$F'^o_{T\phi}$	T_i, β_{i+1}	0.2714	0.4725
$F_{T\phi}^m$	T_i, β_{i-1}	−0.0568	0.4545
$F'^m_{T\phi}$	T_i, β_{i+2}	0.0569	0.4545
$F_{T\phi}^p$	T_i, β_{i-2}	0.1019	0.4784
$F'^p_{T\phi}$	T_i, β_{i+3}	−0.1019	0.4783
$F_{T\Omega}$	T, Ω	0.3439	0.3792
F_{Ts}	T, s	−0.1150	0.2779
F_s^o	$s_i s_{i+1}$ (ortho)	0.0130	0.0478
F_Ω	$\Omega_i \Omega_{i+1}$ (ortho)	0.0847	0.2226
$F_{\Omega\phi}^o$	Ω_i, β_{i-1}	−0.1005	0.4599
$F'^o_{\Omega\phi}$	Ω_i, β_{i+1} (ortho)	0.1009	0.4599
F_ϕ^o	β_i, β_{i+1}	0.0127	0.0127
F_ϕ^m	β_i, β_{i+2} (meta)	−0.0007	0.0125
F_ϕ^p	β_i, β_{i+3} (para)	−0.0185	0.0221
	Out of place		
H_μ	μ	0.3237	0.0338
H_z	z	0.1190	0.0600
F_μ^o	μ_i, μ_{i+1} (ortho)	0.0184	0.0114
F_μ^m	μ_i, μ_{i+2} (meta)	−0.0235	0.0165
F_μ^p	μ_i, μ_{i+3} (para)	−0.0227	0.0175
F_z^o	z_i, z_{i+1} (ortho)	−0.0388	0.0343
$F_{z\mu}$	z_i, μ_i	0.0273	0.4551
$F'_{z\mu}$	z_i, μ_{i+1}	−0.0263	0.4551

[a]Coordinates are defined in Fig. 8.2. T_i, T_{i+1}, T_{i+2} represents a progression in a clockwise direction around the ring starting from the coordinate labeled T_i in Fig. 8.2.

[b]Units: stretch, mdyn/Å; bend, mdynÅ/rad^2, stretch bend interaction, mdyn/rad.

Table 8.14 Comparison of Observed and Calculated Frequencies for Benzene and Its Isotopes

Symmetry Species	ν_{obs} (cm^{-1})	ν_{calc} (cm^{-1})
	(a) $^{12}C_6H_6$	
A_{1g}	992	994
	3059	3063
A_{2g}	1346	1348
B_{2g}	703	704
	989	990
E_{1g}	849	849
E_{2g}	606	608
	1178	1178
	1596	1594
	3046	3043
A_{2u}	670	672
B_{1u}	1008	1012
	(3048)	3036
B_{2u}	1309	1309
	1149	1149
E_{1u}	1036	1038
	1479	1482
	3073	3055
E_{2u}	404	405
	966	968
	(b) $^{12}C_6D_6$	
A_{1g}	945	941
	2293	2288
A_{2g}	1055	1048
B_{2g}	599	597
	830	825
E_{1g}	664	660
E_{2g}	579	569
	869	860
	1553	1562
	2266	2273
A_{2u}	496	493
B_{1u}	963	961
	(2275)	2261
B_{2u}	1282	1284
	823	828
E_{1u}	812	806
	1330	1328
	2276	2280
E_{2u}	351	350
	789	792

Table 8.14 Continued

Symmetry Species	ν_{obs} (cm^{-1})	ν_{calc} (cm^{-1})
	(c) $^{13}C_6H_6$	
A_{1g}	957	959
	3034	3051
A_{2g}	1333	1338
B_{2g}	680	682
	981	982
E_{1g}	842	842
E_{2g}	589	587
	1171	1176
	1542	1533
	3024	3031
A_{2u}	669	670
B_{1u}	981	976
	(3026)	3025
B_{2u}	1272	1270
	1139	1137
E_{1u}	1017	1020
	1450	1451
	3054	3043
E_{2u}	392	393
	958	957

results to be consistent with those calculated for the E_{2g} species, the solution with the larger value of $F_{14,15}$ was chosen.

In our calculation we did not consider each symmetry species separately, but refined a set of valence force constants to give a close fit of all the observed frequencies of C_6H_6, C_6D_6, and $^{13}C_6H_6$ (4). Although the inclusion of the latter isotope does not increase the data set, it does increase the precision with which the force constants can be determined. The detailed examination, symmetry species by species, as performed by Duinker and Mills (7) and Whiffen (6) is a superior spectroscopic approach. However, our aim was eventually to develop a model force field that could be applied to the alkyl benzenes and hence to polymers such as polystyrene. In view of the difficulties (outlined earlier) involved in obtaining anything approaching a unique solution, we considered it sufficient to include predominantly those valence force constants considered important by previous authors. This force field is reproduced in Table 8.13 and the observed and calculated frequencies are compared in Table 8.14.

References

1 E. B. Wilson, Jr., J. Decius, and P. C. Cross, "Molecular Vibrations," McGraw-Hill, New York, 1955.

2 P. C. Painter and J. L. Koenig, *Spectrochim. Acta* **33A**, 1019 (1977).

3 P. C. Painter and J. L. Koenig, *Spectrochim. Acta* **33A**, 1003 (1977).

4 P. C. Painter and R. W. Snyder, *Spectrochim. Acta* **36A**, 337 (1980).

5 P. Pulay, G. Fogarasi, and J. E. Boggs, private communication (1979) of a paper to be published.

6 D. H. Wiffen, *Phil. Trans. Roy. Soc.* **A248**, 131 (1955).

7 J. C. Duinker and I. M. Mills, *Spectrochim. Acta* **24A**, 417 (1968).

8 G. Varsanyi, "Vibrational Spectra of Benzene Derivatives," Academic Press, New York, 1969.

9 E. B. Wilson, *Phys. Rev.* **45**, 706 (1934).

10 R. A. Kydd, *Spectrochim. Acta* **27A**, 2067 (1971).

11 S. Brodersen and A. Langseth, *Mat. Fys. Skr. Dan. Vid. Selsk.* **1**, 1 (1956).

12 R. D. Mair and D. F. Hornig, *J. Chem. Phys.* **17**, 1236, (1949).

13 B. L. Crawford, Jr., and F. A. Miller, *J. Chem. Phys.* **17** (3), 249 (1949).

14 D. C. McKean and J. L. Duncan, *Spectrochim. Acta* **27A**, 1879 (1971).

9

INFRARED AND RAMAN INTENSITIES

One never rises so high as when one doesn't know where one is going
Oliver Cromwell

The treatment of infrared and Raman intensities is, at least to the authors of this book, a difficult problem and one to be approached with some trepidation. In principle, a general theory could be developed from the exact solution of the Schrödinger equation; but since even the simplest molecules (let alone polyethylene) present practically insuperable problems, a number of semiempirical approaches have been attempted. A principal aim of most of this work has been to derive a relationship between intensities and parameters (e.g., bond dipole moments) that represents the electrical behavior of the molecule during vibrational motion. The rationale for this approach is based on an analogy with the calculation of normal mode frequencies. As we showed in Chapter 8, normal mode calculations are based on defining the potential energy in terms of a set of force constants in such a way that they can be transferred between similar molecules. Defining force constants in terms of internal coordinates has proved to be most useful, although a number of different force fields have been defined. Similarly, several parameters have been used in the study of intensities. Gussoni and Abbate (1) suggest that in describing infrared intensities the most widely employed parameters are:

1. The change in the total dipole moment with internal or symmetry coordinates (dipole moment derivatives).
2. The change in the total dipole moment with the cartesian displacements of each atom (polar tensors).
3. The change of individual bond moments with vibrational parameters plus a contribution from the equilibrium bond moments (bond charge parameters and electro-optical parameters).

Corresponding polarizability terms can be used to describe Raman intensities. However, because of the numerous difficulties in calculating intensities it is only very recently that an assault on the simplest polymer, polyethylene, has been attempted (2). Electro-optical parameters transferred from the *n*-paraffins were used in this work. The use of such electro-optical parameters, as expounded by Gribov (3) and recently reviewed by Gussoni (4) will be discussed here, but we will introduce the use of these parameters in an indirect fashion by initially considering the bond charge parameters described by van Straten and Smit (5). The relationship between bond charge parameters and electro-optical parameters is simple, and the theories are closely related. A preliminary discussion of bond charge parameters allows us to work through the details of calculating parameters for a simple molecule, sulfur dioxide.

No general discussion of intensities is complete without at least a brief discussion of the quantum mechanical basis of infrared absorption and Raman scattering. We will consider these arguments first, but it should be noted that many of the relevant quantum mechanical relationships presented in the following section are stated as if proved. It is beyond the scope of this text to present the relevant basic theoretical derivations. We are more concerned with the determination of essentially empirical electro-optical parameters that make possible the calculation of intensities in a manner similar to the calculation of vibrational frequencies from model force fields.

9.1 Quantum Mechanical Description of Infrared Absorption and Raman Scattering

Infrared absorption is readily described in quantum mechanical terms, but as we shall see later the theory of the Raman effect is more complicated. Essentially, we need to describe an interaction between a molecule and an electromagnetic wave, which can be regarded as an oscillating electric and magnetic field propagating through space. In order to interact with the electric field the molecule must have a charge distribution, which changes during the spectroscopic transition. Consequently, there must be an interaction of the instantaneous dipole moment with the electric field of the light wave. These transitions are known as electric dipole transitions. Using time-dependent quantum calculations it can be shown that transitions between states k and n, described by wave functions ψ_k and ψ_n, depend on quantities called Einstein transition probabilities. If we let E_k and E_n be the energies of states k and n

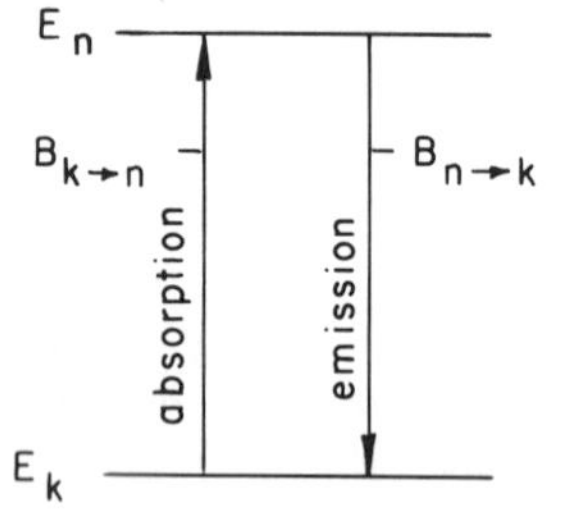

Figure 9.1. Schematic representation of induced emission and absorption [$\rho(\nu_{nk})\neq 0$].

and let N_k and N_n be the number of molecules with energy E_k and E_n, respectively, then the absorption process is related to the number of transitions $k \rightarrow n$ (per second) as

$$\frac{-dN_k}{dt} = N_k B_{k \rightarrow n} \rho(\nu_{nk}) \tag{9.1}$$

where $B_{k \rightarrow n}$ is the Einstein transition probability of induced absorption and $\rho(\nu_{nk})$ is the radiation density due to electromagnetic radiation of energy, $E_n - E_k = h(\nu_n - \nu_k)$. Emission can be described by a similar equation and the absorption and emission processes are illustrated schematically in Figure 9.1.

This treatment assumes that the particle has a permanent electric dipole moment and can therefore interact with the electromagnetic field of the incident light. Intuitively, one would expect that the probability of the transition occurring (i.e., the Einstein transition probability), and hence the intensity of the absorption band, should be related to the average value of the dipole moment. We will not derive this relationship but simply state the result of the quantum mechanical treatment, which relates the intensity of the infrared absorption band to the transition dipole moment as

$$A_{kn} = \frac{N_0 \pi g_i}{3c^2} \frac{\nu_i}{\omega_i} |(\mu)_{kn}|^2 \tag{9.2}$$

where N_0 is Avogadro's number, g_i is the degeneracy of the mode, c is the speed of light, ν_i is the wavenumber of the transition, and ω_i is the zero-order frequency resulting from anharmonic corrections to ν_i. The quantity $(\mu)_{kn}$ is the transition integral

$$(\mu)_{kn} = \int \psi_k \mu \psi_n \, dt \tag{9.3}$$

and μ is the dipole moment of the molecule, defined by

$$\mu = \sum_k e_k r_k \tag{9.4}$$

where e_k is the charge on the kth particle and r_k is the radius vector from the origin of the coordinate system of the particle k.

The absorption of infrared radiation is one-photon process. The scattering of radiation, however, is a typical two-photon process in which one photon is absorbed and one emitted at the same time. This process cannot be treated as two distinct one-photon processes; the annihilation of the incident photon and the creation of the scattered photon are nonseparable. In discussing the Raman effect we are concerned with the induced dipole moment M,* which is proportional to the field strength of the incident radiation, the proportionality constant α being called the polarizability of the molecule. Van Vleck (6) and others demonstrated that the intensity of the scattered light (averaged over all directions) is related to the polarizability by

$$I_{mn} = \frac{2^7\pi^5}{3^2c^4} I_0(\nu_0+\nu_{mn})^4 \sum_{\rho\sigma} |(\alpha_{\rho\sigma})_{mn}|^2 \tag{9.5}$$

where the subscripts m, n refer to a transition from state m to state n by a molecule that is perturbed by an electromagnetic wave while scattering light of frequency $(\nu_0+\nu_{mn})$. The subscripts ρ, α are independently x, y, z (i.e., α_{xx}, α_{xy}, etc.). According to the dispersion theory, $(\alpha_{\rho\sigma})_{mn}$ is given by

$$(\alpha_{\rho\sigma})_{mn} = \frac{1}{h} \sum_r \left[\frac{(M_\rho)_{rn}(M_\sigma)_{mr}}{\nu_{rm}-\nu_0} + \frac{(M_\rho)_{mr}(M_\sigma)_{rn}}{\nu_{rn}+\nu_0} \right] \tag{9.6}$$

where r is a vibronic state of the molecule; $(M_\rho)_{mr}$ represents the amplitude of the transition moment

$$(M_\rho)_{mr} = \int \psi_r m_\rho \psi_m \, dt \tag{9.7}$$

ψ_r and ψ_m are the wave functions describing states r and n and M_ρ is the ρth component of the electric moment operator. It is apparent from these equations that the Raman line corresponding to the transition $m \rightarrow n$ involves an intermediate level r. Such a transition is illustrated in Figure 9.2. Note that ν_{mn} can be either positive or negative. The states r are real states of the scattering molecule and the wave functions ψ_r make up the complete orthonormal set. However, this does not mean that during the scattering process the molecule "jumps" from state n to all possible states r; the summation over the states r is simply a result of using perturbation theory to derive an expression for the induced dipole moment.

Equation (9.6) can be derived in a more useful form by applying the zeroth-order Born-Oppenheimer approximation. This states that the electrons move so fast compared to the nuclei that the electron cloud immediately follows the movements of the nuclei. This allows us to give two labels to each state of the molecule, for example, ev, where e refers to the electronic state and v to the vibrational state. It is now assumed that the molecule passes from the state gi to gj where i and j are vibrational levels of the ground electronic state g. The vibronic-state function ψ_m can now be written

$$\psi_m = \theta_g(\varepsilon, Q)\phi_j^g(Q) \tag{9.8}$$

where $\theta_g(\varepsilon, Q)$ is the electronic wave function for state g and $\phi_j^g(Q)$ is the jth

*In Chapter 2 we used the symbol P for the induced dipole moment. We use M here to be consistent with the literature.

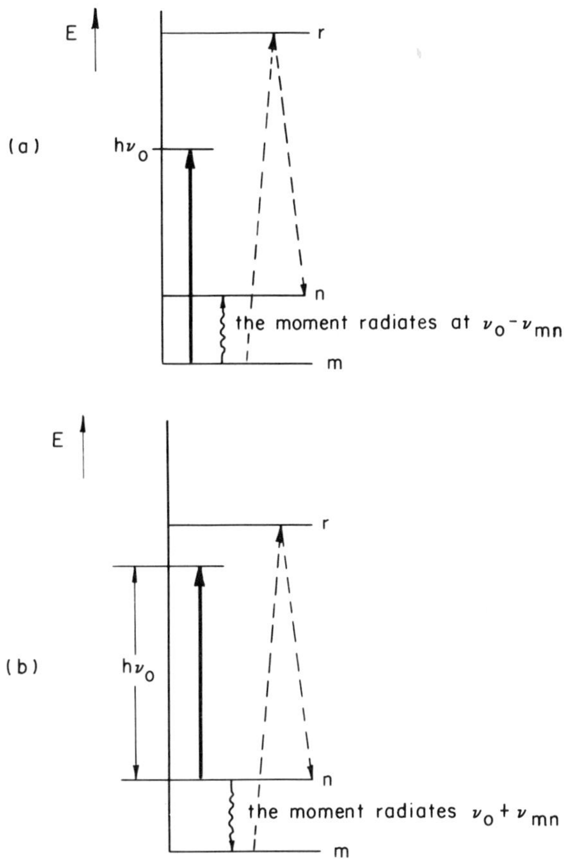

Figure 9.2. Schematic representation of spontaneous Stokes Raman scattering (*a*) and anti-Stokes Raman scattering (*b*).

vibrational state function of the ground electronic state. By obtaining the transition moments in terms of these functions and applying first-order perturbation theory we can obtain

$$M_{g\to e}(Q)=M^0_{g\to e}+\sum\lambda_{es}(Q)M^0_{g\to s} \tag{9.9}$$

where

$$\lambda_{es}(Q)=\sum_a \frac{h^0_{es}Q_a}{\Delta E^0_{es}} \tag{9.10}$$

and h^0_{es} is a perturbation energy per unit displacement of the ath normal mode due to the mixing of the ground state electronic wave functions of states e and s under vibrational perturbation. Upon substitution of these equations into

(9.6) the matrix element of the polarizability becomes

$$(\alpha_{\rho\sigma})_{gi\to gj}=A+B+C \tag{9.11}$$

where

$$A=\frac{1}{h}\sum_{en}\left(\frac{1}{\nu_{en\to gi}-\nu_0}+\frac{1}{\nu_{en\to gi}+\nu_0}\right)\left[(M_\rho)^0_{g\to e}(M_\sigma)^0_{g\to e}\right]$$

$$B=\frac{1}{h}\sum_{en}\left(\frac{1}{\nu_{en\to gi}-\nu_0}\right)\sum_{sa}\frac{h^0_{es}}{\Delta E^0_{es}}\left[(M_\rho)^0_{ge}(M_\sigma)^0_{gs}\int\psi_{gj}\psi_{en}\,d\tau\int\psi_{en}Q_a\psi_{gi}\,d\tau\right.$$

$$\left.+(M_\sigma)^0_{ge}(M_\rho)^0_{gs}\int\psi_{gi}\psi_{en}\,d\tau\int\psi'_{en}Q_a\psi_{gj}\,d\tau\right]$$

$$C=\frac{1}{h}\sum_{en}\left(\frac{1}{\nu_{en\to gj}+\nu_0}\right)\sum_{sa}\frac{h^a_{es}}{\Delta E^0_{es}}\left[(M_\rho)^0_{ge}(M_\sigma)^0_{gs}\int\psi_{gj}\psi_{en}\,d\tau\int\psi_{en}Q_a\psi_{gj}\,d\tau\right.$$

$$\left.+(M_\sigma)^0_{ge}(M_\rho)^0_{gs}\int\psi_{gj}\psi_{en}\,d\tau\int\psi_{en}Q_a\psi_{gi}\,d\tau\right]$$

These expressions are complex but demonstrate several important results. First, in the expressions for A and B the denominator $(\nu_{en\to gi}-\nu_0)$ will result in these terms becoming very large when ν_0 is near the absorption frequency $\nu_{gi\to en}$. This leads to the phenomenon of "resonance Raman scattering." Second, it can be shown that term A is responsible for Rayleigh scattering and terms B and C for Raman scattering. Furthermore, for $\nu_{ge}\gg\nu_0$ (i.e., the nonresonance case) then the sum rule developed by Van Vleck leads to

$$\sum\int\psi_{gi}\psi_{en}\int\psi_{en}\psi_{gj}=\delta_{ij} \tag{9.12}$$

$$\sum\int\psi_{gi}\psi_{en}\int\psi_{en}Q_a\psi_{gj}=\sum\int\psi_{gj}\psi_{en}\int\psi_{en}Q_a\psi_{gi}=\int\psi_{gi}Q_a\psi_{gj} \tag{9.13}$$

and

$$\int\psi_{gi}Q_a\psi_{gj}=\begin{cases}0 & \text{if } \nu_a^j\neq\nu_a^i\pm1\\ \left(\dfrac{\nu_a^i+1}{2\gamma_a}\right)^{1/2} & \text{if } \nu_a^j=\nu_a^i+1\\ \left(\dfrac{\nu_a^i}{2\gamma_a}\right)^{1/2} & \text{if } \nu_a^j=\nu_a^i-1\end{cases} \tag{9.14}$$

Substituting (9.14) into (9.11) and rearranging, we obtain a simplified expression for the polarizability:

$$(\alpha_{\rho\sigma})_{gi\to gj}=D+E \tag{9.15}$$

where

$$D=\frac{1}{h}\sum_{e}\left(\frac{2\nu_e}{\nu_e^2-\nu_0^2}\right)(M_\rho)_{ge}^0(M_\sigma)_{ge}^0\delta_{ij} \tag{9.16}$$

and

$$E=-\frac{2}{h^2}\sum_{\substack{es\\ s>e}}\sum_{a}\frac{(\nu_e\nu_s+\nu_0^2)h_{es}^a}{(\nu_e^2-\nu_0^2)(\nu_s^2-\nu_0^2)}$$
$$\times\left[(M_\rho)_{ge}^0(M_\sigma)_{gs}^0+(M_\rho)_{gs}^0(M_\sigma)_{ge}^0\right]\left[\frac{(n_a^i+1)h}{8\pi^2\mu V}\right]^{1/2} \tag{9.17}$$

The term D is responsible for Rayleigh scattering, while E is responsible for Raman scattering. These equations clearly demonstrate that the Raman effect depends on pairs of excited states. It is also required that transitions $g\to e$ and $g\to s$ be allowed by symmetry. For a totally symmetric ground state this means that the symmetry species Γ_s and Γ_e must correspond separately to Γ_ρ, the species of at least one of the translations. From this condition it is now possible to derive the selection rules for Raman scattering.

From equation (9.10) it can be seen that since $\Gamma_e=\Gamma_Q\Gamma_s$ for mixing to be possible (where Γ_Q is the symmetry species of vibration Q), then since the totally symmetry representation is

$$\Gamma_e\Gamma_e=\Gamma_e\Gamma_Q\Gamma_s \tag{9.18}$$

it is required that Γ_Q must correspond to any possible combination of Γ_e and Γ_s. These combinations are Γ_{xx}, Γ_{yy}, Γ_{zz}, Γ_{xy}, Γ_{xz}, Γ_{yz}, the well-known Raman selection rules.

The form of equation (9.17) does not make explicit the relationship between molecular polarizability and molecular vibration. This relationship is implicit in the dependence of electronic wave functions on the degree of vibrational excitement. However, we have presented what amounts to an early approximate form of vibronic theory. The relationship between Raman intensity and the change in molecular polarizability with vibration is more explicit in more recently derived relationships. The interested reader is referred to a review by Tang and Albrecht on Raman intensities (7).

9.2 The Application of Bond Charge Parameters to the Calculation of Infrared Intensities

The quantum mechanical description outlined in the preceding section is essential to an understanding of the processes that actually occur in molecules, but is not of much help to us in the practical calculation of intensities. One of the most useful analyses of infrared intensities accomplished so far has been based on "bond charge" parameters. We will present the definition of these bond charge parameters in this section and then apply these results to the analysis of sulfur dioxide. The treatment given here is essentially that of van Straten and Smit (5, 8).

In the preceding section it was shown that the intensity of an infrared absorption band A_{if} is given by

$$A_{if}=C\left[|(\mu_x)_{if}|^2+|(\mu_y)_{if}|^2+|(\mu_z)_{if}|^2\right] \tag{9.19}$$

where

$$C=\frac{N_0\pi g_i}{3c^2}\frac{\nu_i}{\omega_i}$$

and the $(\mu_x)_{if}$ are the transition integrals. N_0 is Avogadro's number in mole^{-1}, g_i is the degeneracy, c is the speed of light in msec^{-1}, ν_i is wavenumber of transition, and ω_i is the corresponding harmonic wavenumber. The transition integrals are

$$(\mu_x)_{if}=\int\psi_i\mu_x\psi_f\,d\tau \tag{9.20}$$

and the electric moment is given by

$$\mu_x=\sum_{\alpha}^{n}q_\alpha X_\alpha \tag{9.21}$$

$$\mu_y=\sum_{\alpha}^{n}q_\alpha Y_\alpha \tag{9.22}$$

$$\mu_z=\sum_{\alpha}^{n}q_\alpha Z_\alpha \tag{9.23}$$

In order to be consistent with the definitions of van Straten and Smit (8) we have defined q_α as the charge on the αth particle. This term should not be confused with a mass-adjusted cartesian coordinate. The terms X_α, Y_α, and Z_α are the space-fixed cartesian coordinates of the αth particle. These equations show the dependence of the transition integrals on the positions of the atoms in the space-fixed axis. However, the atoms move during the vibrational process and as the atoms move, the electron distribution or "effective charge" changes. In general, for small displacements the electric moment may be expanded as a power series in terms of the coordinates of the atoms. With normal coordinates as a basic set the result becomes

$$\mu_x = \mu_x^0 + \sum_{k=1}^{3N-6} \frac{\partial \mu_x}{\partial Q_k} Q_k + \text{higher terms} \tag{9.24}$$

Of course, similar equations apply to μ_y and μ_z but for simplicity and conciseness we will consider only μ_x. We now substitute the expression for μ_x in the transition integrals [equation (9.3)] to obtain

$$\int \psi_i^* \mu_x \psi_f \, d\tau = \mu_x^0 \int \psi_i^* \psi_f \, d\tau + \sum \frac{\partial \mu_x}{\partial Q_k} \int \psi_i Q_k \psi_f \, d\tau \tag{9.25}$$

The first term is equal to zero because of the orthogonality of the wave functions. The second term is nonzero only if for the ith normal mode $\eta_i = 1$, where η_i is the vibrational quantum number. So the infrared intensity is given by

$$A_i = C \left| \frac{\partial \boldsymbol{\mu}}{\partial Q_i} \right|^2 \tag{9.26}$$

where $\boldsymbol{\mu}$ is the total dipole moment. As pointed out by Gussoni (4), we need to express the quantity $\partial \boldsymbol{\mu} / \partial Q$ in terms of parameters that fulfill the following conditions:

1. They can be related to the maximum number of experimental data.
2. They can be used to predict the spectra of other molecules.
3. They can be related to *ab initio* values.

If we use the relationship between normal coordinates and internal coordinates

$$R_m = \sum_m (L_r)_{im} Q_i \tag{9.27}$$

we can obtain from equation (9.26):

$$A_i = C \left| \sum_m (L_r)_{im} \frac{\partial \boldsymbol{\mu}}{\partial R_m} \right|^2 \tag{9.28}$$

where $\partial \boldsymbol{\mu} / \partial R_m$ can be termed an internal molecular dipole moment derivative that relates a *molecular* electrical quantity, $\boldsymbol{\mu}$, to a *local* dynamical quantity R_m.

Other coordinate systems (e.g., symmetry or cartesian coordinates) can also be used to describe a local dynamical quantity. Infrared cartesian molecular parameters have been extensively studied by Morcillo et al. (9) and Person and Newton (10), and the interested reader should consult these papers or the review by Gussoni (4). Here we will continue our theoretical development in terms of internal coordinates.

Although we have obtained an expression for infrared intensities in terms of a local dynamical quantity, we are still using the molecular dipole moment $\boldsymbol{\mu}$, which is clearly a property of the individual molecule under consideration. If we are to obtain parameters that are transferable between molecules, it would be advantageous to represent the total dipole moment in terms of the dipole moment associated with individual bonds. If such a description is an accurate representation of the electrical properties of the molecule, we would then have transferable parameters that could be used in the calculation of the intensities of similar molecules, just as force constants are used in overlay calculations of vibrational frequencies. Eliashevich and Wolkenstein (11) were the first to assume that the molecular dipole moment $\boldsymbol{\mu}$ could be written as a vector sum of the individual bond moments $\boldsymbol{\mu}_k$. For an N-atom molecule having $N-1$ bonds

$$\boldsymbol{\mu}=\sum_{k}^{N-1}\boldsymbol{\mu}_k \tag{9.29}$$

and the bond moment of the kth bond is given as

$$\boldsymbol{\mu}_k=q_k r_k \mathbf{e}_k \tag{9.30}$$

where $\boldsymbol{\mu}_k$ is the effective moment of the kth bond, q_k the effective bond charge of the kth bond, r_k is the bond length, and $\mathbf{e}_k$ a unit bond direction vector, whose direction is along the bond from the negativity charged end to the positively charged end. For small changes in the internal coordinates, the corresponding change in the bond moments can be written as

$$\frac{\partial\boldsymbol{\mu}_k}{\partial R_j}=\left(\frac{\partial q_k}{\partial R_j}\right)r_k\mathbf{e}_k+\left(\frac{\partial r_k}{\partial R_j}\right)q_k\mathbf{e}_k+\left(\frac{\partial \mathbf{e}}{\partial R_j}\right)q_k r_k \tag{9.31}$$

The $\partial q_k/\partial R_j$ elements are the "bond charge reorganization" parameters. The second term $(\partial r_k/\partial R_j)$ represents the contribution from the change in bond length, and the third term $(\partial \mathbf{e}/\partial R_j)$ gives the bond moment change arising from a change in the bond direction. The bond charge parameters $(\partial q_k/\partial R_j)$ and q_k can only be determined from intensity data. The other quantities are known or can be determined from a knowledge of the dynamical behavior of the molecule.

Defining $\partial q_k/\partial R_j$ as the kjth element of the matrix $\partial \mathbf{q}/\partial \mathbf{R}$, we can write the following expressions for $\partial \boldsymbol{\mu}/\partial \mathbf{Q}$.

$$\left(\frac{\partial \boldsymbol{\mu}}{\partial \mathbf{Q}}\right)^T = \mathbf{L}_r^T\left[\left(\frac{\partial \mathbf{q}}{\partial \mathbf{R}}\right)^T \mathbf{re} + \left(\frac{\partial \mathbf{r}}{\partial \mathbf{R}}\right)^T \mathbf{qe} + \left(\frac{\partial \mathbf{e}}{\partial \mathbf{R}}\right)^T \hat{\mathbf{q}}\hat{\mathbf{r}}\right] \tag{9.32}$$

The $(\partial \boldsymbol{\mu}/\partial \mathbf{Q})^T$ is a rectangular $(3N-6)\times 3$ array:

$$\left(\frac{\partial \boldsymbol{\mu}}{\partial \mathbf{Q}}\right)^T = \begin{bmatrix} \dfrac{\partial \mu_x}{\partial Q_1} & \dfrac{\partial \mu_y}{\partial Q_1} & \dfrac{\partial \mu_z}{\partial Q_1} \\ \vdots & \vdots & \vdots \\ \dfrac{\partial \mu_x}{\partial Q_{3N-6}} & \dfrac{\partial \mu_x}{\partial Q_{3N-6}} & \dfrac{\partial \mu_z}{\partial Q_{3N-6}} \end{bmatrix} \tag{9.33}$$

It can be shown that for symmetrical molecules this matrix contains many zero elements. For example, in the case of C_{2v} and C_{3v} symmetry (defined in Chapter 4), there is only one nonzero element in each row.

The $(\partial \mathbf{q}/\partial \mathbf{R})^T$ term is a rectangular array $[(3N-6)\times(N-1)]$ of derivatives of bond charge with respect to the internal coordinates:

$$\left(\frac{\partial \mathbf{q}}{\partial \mathbf{R}}\right)^T = \begin{bmatrix} \dfrac{\partial q_1}{\partial R_1} & \dfrac{\partial q_2}{\partial R_2} & \cdots & \dfrac{\partial q_{N-1}}{\partial R_1} \\ \vdots & \vdots & & \vdots \\ \dfrac{\partial q_1}{\partial R_{3N-6}} & \dfrac{\partial q_2}{\partial R_{3N-6}} & \cdots & \dfrac{\partial q_{N-1}}{\partial R_{3N-6}} \end{bmatrix} \tag{9.34}$$

The $(\partial \mathbf{r}/\partial \mathbf{R})^T$ term is a rectangular $(3N-6)\times(N-1)$ array of the form

$$\left(\frac{\partial \mathbf{r}}{\partial \mathbf{R}}\right)^T = \begin{bmatrix} \dfrac{\partial r_1}{\partial R_1} & \dfrac{\partial r_2}{\partial R_1} & \cdots & \dfrac{\partial r_{N-1}}{\partial R_1} \\ \dfrac{\partial r_1}{\partial R_{3N-6}} & \dfrac{\partial r_2}{\partial R_{3N-6}} & \cdots & \dfrac{\partial r_{N-1}}{\partial R_{3N-6}} \end{bmatrix} \tag{9.35}$$

while $\mathbf{r}$ is an $(N-1)\times(N-1)$ diagonal matrix of the equilibrium bond lengths and $\mathbf{e}$ is an $(N-1)\times 3$ matrix of the type

$$\mathbf{e} = \begin{bmatrix} e_{1x} & e_{1y} & e_{1z} \\ e_{2x} & e_{2y} & e_{2z} \\ \vdots & \vdots & \vdots \\ e_{(N-1)x} & e_{(N-1)y} & e_{(N-1)z} \end{bmatrix} \tag{9.36}$$

The term **q** is an $(N-1)$ square diagonal matrix of the effective bond charges.

The $(\partial\mathbf{e}/\partial\mathbf{R})^T$ matrix is $(3N-6)\times(3N-3)$:

$$\left(\frac{\partial\mathbf{e}}{\partial\mathbf{R}}\right)^T = \begin{bmatrix} \frac{\partial e_{1x}}{\partial R_1} & \frac{\partial e_{1y}}{\partial R_1} & \frac{\partial e_{1z}}{\partial R_1} & \cdots & \frac{\partial e_{(N-1)z}}{\partial R_1} \\ \vdots & \vdots & \vdots & & \vdots \\ \frac{\partial e_{1x}}{\partial R_{3N-6}} & \frac{\partial e_{1y}}{\partial R_{3N-6}} & \frac{\partial e_{1x}}{\partial R_{3N-6}} & \cdots & \frac{\partial e_{(N-1)z}}{\partial R_{3N-6}} \end{bmatrix} \tag{9.37}$$

Finally, the term $\hat{\mathbf{q}}$ is a $(3N-3)\times(3N-3)$ diagonal matrix of "triplets" of the effective charges:

$$\hat{\mathbf{q}} = \begin{bmatrix} q_1 & & & & & \\ & q_1 & & & & 0 \\ & & q_1 & & & \\ 0 & & & q_2 & & \\ & & & & \ddots & \\ & & & & & q_{N-1} \end{bmatrix} \tag{9.38}$$

and $\hat{\mathbf{r}}$ is a $(3N-3)\times 3$ matrix of the bond lengths in the form

$$\hat{\mathbf{r}} = \begin{bmatrix} r_1 & 0 & 0 \\ 0 & r_1 & 0 \\ 0 & 0 & r_1 \\ r_2 & 0 & 0 \\ \vdots & \vdots & \vdots \\ 0 & 0 & r_{N-1} \end{bmatrix} \tag{9.39}$$

If we consider the kth bond of a molecule, the problem becomes the calculation of the variation of parameters defined above with unit change in internal coordinate, for example, $(\partial\mathbf{e}/\partial\mathbf{R})^T$. In order to calculate these quantities we need to determine the difference in the displacement vectors of the two atoms defining the bond, for which a transformation from the displacements in internal coordinates to a system of laboratory-fixed cartesian coordinates is required. We have the relation

$$\mathbf{R} = \mathbf{B}\mathbf{X} \tag{9.40}$$

but since **B** is a nonsquare matrix $[(3N-6)\times 3N]$, no simple inverse $(\mathbf{B}^{-1})$ exists. Therefore we must seek a solution **A** that satisfies the relationship

$$\mathbf{X} = \mathbf{A}\mathbf{R} \tag{9.41}$$

First, we will substitute $\mathbf{R}=\mathbf{BX}$, so

$$\mathbf{X}=\mathbf{ABX} \tag{9.42}$$

or it is required that

$$\mathbf{AB}=\mathbf{E} \tag{9.43}$$

Remember that

$$\mathbf{G}=\mathbf{BM}^{-1}\mathbf{B}^{T} \tag{9.44}$$

and multiplying both sides by $\mathbf{A}$ yields

$$\mathbf{AG}=\mathbf{ABM}^{-1}\mathbf{B}^{T} \tag{9.45}$$

and as a result

$$\mathbf{A}=\mathbf{M}^{-1}\mathbf{B}^{T}\mathbf{G}^{-1} \tag{9.46}$$

assuming that $\mathbf{G}^{-1}$ exists. A column of this matrix gives the cartesian displacements of the atoms under a unit displacement of R (i.e., $\partial x/\partial R$) and the desired transformation between the internal coordinates and the space-fixed cartesian coordinates is

$$\mathbf{X}=\mathbf{M}^{-1}\mathbf{B}^{T}\mathbf{G}^{-1}\mathbf{R} \tag{9.47}$$

Let us now compute $\partial \mathbf{e}/\partial \mathbf{R}$ and $\partial \mathbf{r}/\partial \mathbf{R}$. We begin by considering the kth bond between atoms ε_k and β_k. The deformation of this bond resulting from a unit change in the internal coordinate R_j will be equal to the difference in the displacement vectors of the initial atom (β) and the terminal (ε) atom:

$$\Delta r_{kj}=\Delta r_{\varepsilon j}-\Delta r_{\beta j} \tag{9.48}$$

The displacements of atoms ε and β can be in different planes, so that it is necessary to calculate Δr_{kj} values for each direction in the cartesian system. If we let $\mathbf{i}$, $\mathbf{j}$, and $\mathbf{k}$ be unit vectors parallel to the x, y, and z cartesian axes, respectively, then

$$\Delta\mathbf{r}_{kj}=(\Delta r_{kj})_x\mathbf{i}+(\Delta r_{kj})_y\mathbf{j}+(\Delta r_{kj})_z\mathbf{k} \tag{9.49}$$

By subtracting the displacement in the x direction of the initial atom β from the displacement in the x direction of the terminal atom ε we get the bond displacement component $(\Delta r_{kj})_x$, and the components $(\Delta r_{kj})_y$ and $(\Delta r_{kj})_z$ can be obtained in a similar fashion. The value of $\partial r_x/\partial R_j$ is equal to the scalar product $\Delta \mathbf{r}_{kj} \cdot \mathbf{e}_k$, so that in matrix notation we can write

$$\left(\frac{\partial \mathbf{r}}{\partial \mathbf{R}} \right)^T = \mathbf{A}^T \boldsymbol{\Delta}^T \hat{\mathbf{e}} \tag{9.50}$$

where the $k\alpha$th element of $\boldsymbol{\Delta}^T$ is equal to: 0 if α is not an atom of bond k; -1 if α is the initial atom of bond k; $+1$ if α is the terminal atom of bond k. So $\boldsymbol{\Delta}^T$ is a $3N \times (3N-3)$ matrix:

$$\boldsymbol{\Delta}^T = \begin{array}{c} \\ \vdots \\ x_\beta \\ y_\beta \\ z_\beta \\ \vdots \\ x_\varepsilon \\ y_\varepsilon \\ z_\varepsilon \\ \vdots \end{array} \begin{array}{c} \begin{array}{ccccccc} \cdots & k & & k & & k \cdots \end{array} \\ \left[\begin{array}{ccccccc} \cdots & -1 & & 0 & & 0 \cdots \\ & 0 & & -1 & & 0 \\ \cdots & 0 & \cdots & 0 & \cdots & -1 \\ & \vdots & & \vdots & & \vdots \\ \cdots & 1 & & 0 & & 0 \cdots \\ \cdots & 0 & & 1 & & 0 \cdots \\ \cdots & 0 & & 0 & & 1 \cdots \\ & \vdots & & \vdots & & \vdots \end{array} \right] \end{array} \tag{9.51}$$

and $\hat{\mathbf{e}}$ is a $(3N-1) \times (N-1)$ matrix:

$$\hat{\mathbf{e}} = \begin{bmatrix} e_{1x} & 0 & \cdots & 0 \\ e_{1y} & 0 & \cdots & 0 \\ e_{1z} & 0 & \cdots & 0 \\ 0 & e_{2x} & \cdots & 0 \\ 0 & e_{2y} & \cdots & 0 \\ 0 & e_{2z} & \cdots & 0 \\ \vdots & & & \\ 0 & 0 & \cdots & e_{(N-1)_x} \\ 0 & 0 & \cdots & e_{(N-1)_y} \\ 0 & 0 & \cdots & e_{(N-1)_z} \end{bmatrix} \tag{9.52}$$

If we now consider the $(\partial \mathbf{e}/\partial \mathbf{R})^T$ term, then the element $\partial e_{kx}/\partial R_j$ is equal to the scalar product $\Delta e_{kj} \cdot \mathbf{i}$ and it can be shown that

$$\Delta \mathbf{e}_{kj} = \left[\Delta \mathbf{r}_{kj} - (\Delta \mathbf{r}_{kj} \cdot \mathbf{e}_k) \mathbf{e}_k \right] \mathbf{r}_k^{-1} \tag{9.53}$$

So in the matrix notation introduced in equation (9.50)

$$\left(\frac{\partial \mathbf{e}}{\partial \mathbf{R}}\right)^T = (\mathbf{A}^T\mathbf{\Delta}^T - \mathbf{A}^T\mathbf{\Delta}^T\hat{\mathbf{e}}\hat{\mathbf{e}}^T)\hat{\mathbf{r}}^{-1} \tag{9.54}$$

where $\hat{\mathbf{r}}^{-1}$ is a $(3N-3)\times(3N-3)$ diagonal matrix containing the reciprocals of the equilibrium bond lengths. So substituting (9.50) and (9.54) in (9.32), one obtains

$$\left(\frac{\partial \boldsymbol{\mu}}{\partial \mathbf{Q}}\right)^T = \mathbf{L}_r^T\left[\left(\frac{\partial \mathbf{q}}{\partial \mathbf{R}}\right)^T \mathbf{re} + \mathbf{A}^T\mathbf{\Delta}^T\hat{\mathbf{e}}\mathbf{qe} + (\mathbf{A}^T\mathbf{\Delta}^T - \mathbf{A}^T\mathbf{\Delta}^T\hat{\mathbf{e}}\hat{\mathbf{e}}^T)\hat{\mathbf{q}}\right] \tag{9.55}$$

Theoretically, the observed intensities of the normal modes Q_i can be used to obtain values of $\partial\mu/\partial Q_i$. A set of simultaneous equations can then be solved to obtain values of the bond charge parameters $\partial q/\partial R$ and q. However, as in the case of the calculation of a force field from a set of observed frequencies, there are some practical difficulties. Just as the general valence force field is usually underdetermined by the frequencies, there are generally too many bond charge parameters relative to the number of available intensity data. A less common difficulty is that the intensity data supply only a knowledge of the absolute value of $\partial\mu/\partial Qi$, so that there is some indeterminacy in the values of calculated parameters. However, the sign of $\partial\mu/\partial Q_i$ can in some simple cases be determined from *ab initio* calculations. We will now consider as an example of the use of these equations the calculation of the infrared intensities of sulfur dioxide.

9.3 Example: Calculation of the Infrared Intensities of Sulfur Dioxide

The calculation of the intensities of a bent triatomic molecule has been performed by van Straten and Smit (8) using the equations presented earlier. The experimental data used by these authors are reproduced in Table 9.1. These experimental data were used to obtain the following matrix of dipole

Table 9.1 Observed Frequencies, Integrated Intensities, and Dipole Moment Derivatives for SO_2

	v_i (cm^{-1})	A_i (m/mol)	$\partial\mu/\partial Q$ (dyn/Å am $\mu^{1/2}$)
v_1	1151.4	25,000	0.775
v_2	517.7	25,200	0.777
v_3	1361.8	189,000	2.130

moment derivatives:

$$\frac{\partial \boldsymbol{\mu}}{\partial \mathbf{Q}} = \begin{array}{c} \\ Q_1 \\ Q_2 \\ Q_3 \end{array} \begin{array}{c} \begin{array}{ccc} \mu_x & \mu_y & \mu_z \end{array} \\ \begin{bmatrix} 0 & 0 & +0.775 \\ 0 & 0 & -0.777 \\ +2.130 & 0 & 0 \end{bmatrix} \end{array}$$

The signs of the various elements were determined from theoretical calculations. The dipole moment derivatives are given by

$$\left(\frac{\partial \boldsymbol{\mu}}{\partial \mathbf{Q}}\right)^T = \mathbf{L}^T\left[\left(\frac{\partial \mathbf{q}}{\partial \mathbf{R}}\right)^T \mathbf{re} + \mathbf{A}^T\boldsymbol{\Delta}^T\hat{\mathbf{e}}\mathbf{qe} + (\mathbf{A}^T\boldsymbol{\Delta}^T - \mathbf{A}^T\boldsymbol{\Delta}^T\hat{\mathbf{e}}\hat{\mathbf{e}}^T)\hat{\mathbf{q}}\right]$$

To illustrate the calculation, the first term of this equation, $\mathbf{L}^T(\partial \mathbf{q}/\partial \mathbf{R})^T\mathbf{re}$, will be evaluated; it involves the derivatives of bond charges with respect to the internal coordinate, $\partial \mathbf{q}/\partial \mathbf{R}$, which are, of course, the parameters we wish to determine.

The matrices involved in the first term have the form

$$\mathbf{L}^T = \begin{bmatrix} +0.19804 & +0.19804 & -0.08890 \\ -0.00463 & -0.00463 & +0.31416 \\ +0.23357 & -0.23357 & 0.00000 \end{bmatrix}$$

The $\mathbf{L}^T$ matrix was calculated by the standard methods of normal coordinate analysis using the internal symmetry coordinates and force field (mdyn/Å) shown in Figure 9.3, where

$$\begin{array}{llll} R_1 = \Delta r_1, & S_1 = \sqrt{2}\,(R_1 + R_2), & F_r = 10.42, & F_\alpha = 1.662 \\ R_2 = \Delta r_2, & S_2 = R_3, & F_r' = 00.12, & F_{r\alpha} = 0.525 \\ R_3 = r\,\Delta\alpha, & S_3 = \sqrt{2}\,(R_1 - R_2) & & \end{array}$$

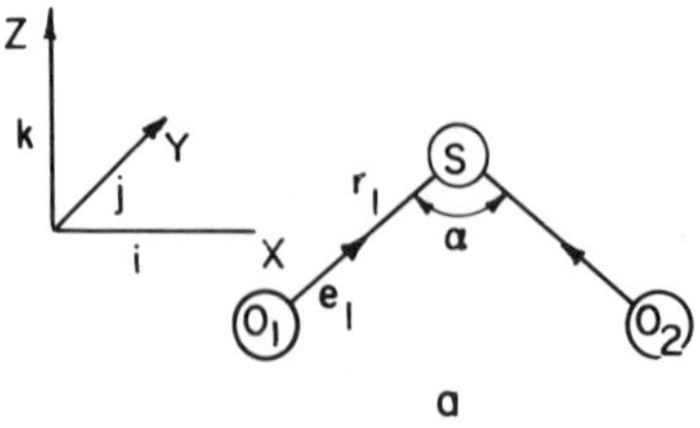

Figure 9.3. The definition of the cartesian coordinate axes, internal coordinates, and bond unit vectors of a molecular of SO_2. (Reprinted by permission from ref. 8.)

The form of the matrix $\partial\mathbf{q}/\partial\mathbf{R}$ is

$$\left(\frac{\partial \mathbf{q}}{\partial \mathbf{R}}\right)^T = \begin{array}{c} \\ R_1 \\ R_2 \\ R_3 \end{array} \begin{array}{c} \begin{array}{cc} q_{S-O_1} & q_{S-O_2} \end{array} \\ \begin{bmatrix} \left(\frac{\partial q}{\partial r}\right)_{SO} & \left(\frac{\partial q}{\partial r}\right)_{SO} \\ \left(\frac{\partial q}{\partial r}\right)'_{SO} & \left(\frac{\partial q}{\partial r}\right)_{SO} \\ \left(\frac{\partial q}{\partial \alpha}\right)_{SO} & \left(\frac{\partial q}{\partial \alpha}\right)_{SO} \end{bmatrix} \end{array}$$

where q_{S-O} is the effective charge on the SO bond; $(\delta q/\delta r)_{SO}$ is the bond charge reorganization resulting from a change in the "own" bond length; $(\delta q/\delta r)'_{SO}$ is the bond charge reorganization resulting from a change in the adjacent bond length; and $(\delta q/\delta\alpha)_{SO}$ is the bond charge reorganization resulting from a change in the valence angle.

The matrix $\mathbf{r}$ is given by

$$\mathbf{r} = \begin{bmatrix} r_{SO} & 0 \\ 0 & r_{SO} \end{bmatrix} = \begin{bmatrix} 1.4308 & \langle 0 \rangle \\ \langle 0 \rangle & 1.4308 \end{bmatrix}$$

where $r_{SO} = 1.4308$ Å is the S—O equilibrium band distance.

The unit $\mathbf{e}$ matrix is given by

$$\mathbf{e} = \begin{bmatrix} e_{1x} & e_{1y} & e_{1z} \\ e_{2x} & e_{1y} & e_{2z} \end{bmatrix} = \begin{bmatrix} 0.860 & 0 & 0.505 \\ -0.860 & 0 & 0.505 \end{bmatrix}$$

where

$$e_{1x} = \cos\left(90 - \frac{\alpha}{2}\right) = \cos 30.34 = 0.863 = -e_{2x}$$

and

$$\alpha = 119.32°$$

$$e_{1z} = \sin\left(90 - \frac{\alpha}{2}\right) = \sin 30.34 = 0.505 = e_{2z}$$

$$e_{1y} = e_{2y} = 0$$

So

$$\mathbf{L}^T\left(\frac{\partial \mathbf{q}}{\partial \mathbf{R}}\right)^T \mathbf{re}=\mathbf{L}^T\left(\frac{\partial \mathbf{q}}{\partial \mathbf{R}}\right)^T\begin{bmatrix} r_{\text{SO}} & 0 \\ 0 & r_{\text{SO}} \end{bmatrix}\begin{bmatrix} e_{1x} & 0 & e_{1z} \\ e_{2x} & 0 & e_{2z} \end{bmatrix}$$

$$=\mathbf{L}^T\left(\frac{\partial \mathbf{q}}{\partial \mathbf{R}}\right)^T\begin{bmatrix} r_{\text{SO}}e_{1x} & 0 & +r_{\text{SO}}e_{1z} \\ r_{\text{SO}}e_{2x} & 0 & +r_{\text{SO}}e_{2z} \end{bmatrix}$$

$$=\mathbf{L}^T\begin{bmatrix} \left(\frac{\partial \mathbf{q}}{\partial r}\right)_{\text{SO}} & \left(\frac{\partial \mathbf{q}}{\partial r}\right)'_{\text{SO}} \\ \left(\frac{\partial \mathbf{q}}{\partial r}\right)'_{\text{SO}} & \left(\frac{\partial \mathbf{q}}{\partial r}\right)_{\text{SO}} \\ \left(\frac{\partial \mathbf{q}}{\partial \alpha}\right)_{\text{SO}} & \left(\frac{\partial \mathbf{q}}{\partial \alpha}\right)_{\text{SO}} \end{bmatrix}\begin{bmatrix} r_{\text{SO}}e_{1x} & 0 & r_{\text{SO}}e_{1z} \\ r_{\text{SO}}e_{2x} & 0 & r_{\text{SO}}e_{2z} \end{bmatrix}$$

$$=\mathbf{L}^T\begin{bmatrix} \left(\frac{\partial \mathbf{q}}{\partial r}\right)_{\text{SO}} r_{\text{SO}}e_{1x}+\left(\frac{\partial \mathbf{q}}{\partial r}\right)'_{\text{SO}} r_{\text{SO}}e_{2x} & 0 & \left(\frac{\partial \mathbf{q}}{\partial r}\right)_{\text{SO}} r_{\text{SO}}e_{1z}+\left(\frac{\partial \mathbf{q}}{\partial r}\right)'_{\text{SO}} r_{\text{SO}}e_{2z} \\ \left(\frac{\partial \mathbf{q}}{\partial r}\right)'_{\text{SO}} r_{\text{SO}}e_{1x}+\left(\frac{\partial \mathbf{q}}{\partial r}\right)_{\text{SO}} r_{\text{SO}}e_{2x} & 0 & \left(\frac{\partial \mathbf{q}}{\partial r}\right)'_{\text{SO}} r_{\text{SO}}e_{1z}+\left(\frac{\partial \mathbf{q}}{\partial r}\right)_{\text{SO}} r_{\text{SO}}e_{2z} \\ \left(\frac{\partial \mathbf{q}}{\partial \alpha}\right)_{\text{SO}}(r_{\text{SO}}e_{1x}+r_{\text{SO}}e_{2x}) & 0 & \left(\frac{\partial \mathbf{q}}{\partial \alpha}\right)_{\text{SO}}(r_{\text{SO}}e_{1z}+r_{\text{SO}}e_{2z}) \end{bmatrix}$$

For normal coordinate Q_1 we only need to determine $(\partial\mu_2/\partial Q_1$; so, concentrating on the first term, we have

$$\left(\frac{\partial \mu_z}{\partial Q_1}\right)_{\text{first term}}=0.775$$

$$=0.198048\left(\frac{\partial q}{\partial r}\right)_{\text{SO}}(e_{1z}+e_{2z})r_{\text{SO}}$$

$$+0.19804\left(\frac{\partial q}{\partial r}\right)'_{\text{SO}}(e_{1z}+e_{2z})r_{\text{SO}}$$

$$-0.0889\left(\frac{\partial q}{\partial \alpha}\right)_{\text{SO}}(e_{1z}+e_{2z})r_{\text{SO}}$$

Substituting calculated values of **e** and **r** yields

$$\left(\frac{\partial \mu_z}{\partial Q_1}\right)_{\text{first term}}=0.775=0.286\left(\frac{\partial q}{\partial r}\right)_{\text{SO}}+0.286\left(\frac{\partial q}{\partial r}\right)'_{\text{SO}}-0.129\left(\frac{\partial q}{\partial \alpha}\right)_{\text{SO}}$$

For the term $\partial\mu_z/\partial Q_2$ we need only one term:

$$\left(\frac{\partial\mu_z}{\partial Q_2}\right)_{\text{first term}} = -0.0463\left(\frac{\partial q}{\partial r}\right)_{\text{SO}}(e_{1z}+e_{2z})r_{\text{SO}}+\left(\frac{\partial q}{\partial r}\right)'(e_{1z}+e_{2z})r_{\text{SO}}$$

$$+0.31416\left(\frac{\partial q}{\partial \alpha}\right)(e_{1z}+e_{2z})r_{\text{SO}}$$

Substitution yields

$$\left(\frac{\partial\mu_z}{\partial Q_2}\right)_{\text{first term}} = -0.007\left(\frac{\partial q}{\partial r}\right)_{\text{SO}} -0.007\left(\frac{\partial q}{\partial r}\right)'_{\text{SO}} +0.454\left(\frac{\partial q}{\partial \alpha}\right)_{\text{SO}}$$

For the term $\partial\mu_x/\partial Q_3$ we obtain

$$\left(\frac{\partial\mu_z}{\partial Q_3}\right)_{\text{first term}} =0.23357\left(\frac{\partial q}{\partial r}\right)_{\text{SO}}(e_{1x}-e_{2x})-0.23357\left(\frac{\partial q}{\partial r}\right)'_{\text{SO}}(e_{2x}-e_{1x})$$

Substitution yields

$$\left(\frac{\partial\mu_x}{\partial Q_3}\right)_{\text{first term}} =0.577\left(\frac{\partial q}{\partial r}\right)_{\text{SO}} -0.577\left(\frac{\partial q}{\partial r}\right)_{\text{SO}}$$

In order to conserve space, the results for the other two terms will be stated without proof, but the method of derivation should be obvious from the treatment above for the first term. The complete equations are as follows:

$$\left(\frac{\partial\mu_z}{\partial Q_1}\right):\quad +0.775= +0.310q_{\text{SO}}+0.286\left(\frac{\partial q}{\partial r}\right)_{\text{SO}} +0.286\left(\frac{\partial q}{\partial r}\right)'_{\text{SO}} -0.129\left(\frac{\partial q}{\partial \alpha}\right)_{\text{SO}}$$

$$\left(\frac{\partial\mu_z}{\partial Q_2}\right):\quad -0.777= -0.393q_{\text{SO}}-0.007\left(\frac{\partial q}{\partial r}\right)_{\text{SO}} -0.007\left(\frac{\partial q}{\partial r}\right)'_{\text{SO}} +0.454\left(\frac{\partial q}{\partial \alpha}\right)_{\text{SO}}$$

$$\left(\frac{\partial\mu_x}{\partial Q_3}\right):\quad +2.130=0.462q_{\text{SO}}+0.577\left(\frac{\partial q}{\partial r}\right)_{\text{SO}} -0.577\left(\frac{\partial q}{\partial r}\right)'_{\text{SO}} +0.000\left(\frac{\partial q}{\partial \alpha}\right)_{\text{SO}}$$

A fourth relationship is obtained by noting that

$$1.610=\boldsymbol{\mu}=qr_{\text{SO}}e_1+qr_{\text{SO}}e_2$$

$$=qr_{\text{SO}}(0.863+0.505)+qr_{\text{SO}}(-0.863+0.505)$$

$$1.610=q(1.4308)(1.010)$$

$$1.610=1.446q_{\text{SO}}$$

So

$$q_{SO}=1.114\,D/\text{Å}$$

Simultaneous solution of the $\partial\mu/\partial Q_i$ equations yields

$$\left(\frac{\partial q}{\partial r}\right)_{SO}=+1.986\,D/\text{Å}^2$$

$$\left(\frac{\partial q}{\partial r}\right)'_{SO}=-0.814\,D/\text{Å}^2$$

$$\left(\frac{\partial q}{\partial r}\right)_{SO}=-0.730\,D/\text{Å}^2$$

For this example, the bond charge parameters are completely determined.

9.4 The Application of Electro-Optical Parameters to the Calculation of Infrared Intensities

The bond charge parameter method discussed in Sections 9.2 and 9.3 is closely related to the electro-optical theory formulated by Gribov (3). In this theory the molecular dipole moment $\boldsymbol{\mu}$ is considered to be an additive function of the bond dipole moments:

$$\boldsymbol{\mu}=\sum_{k}^{K}\mu_k\mathbf{e}_k \tag{9.56}$$

where K is the number of bonds in the molecule, μ_k the dipole moment of the kth bond, and $\mathbf{e}_k$ the unit vector in the direction of the kth bond. Comparison with the bond charge model reveals that

$$\mu_k=r_kq_k \tag{9.57}$$

Using the electro-optical approach, we can write

$$\frac{\partial\boldsymbol{\mu}}{\partial R_m}=\sum_{k}^{K}\left(\frac{\partial\mu_k}{\partial R_m}\mathbf{e}_k+\mu_k\frac{\partial\mathbf{e}_k}{\partial R_m}\right) \tag{9.58}$$

so

$$\frac{\partial\boldsymbol{\mu}}{\partial Q_i}=\sum_{m}^{M}\frac{\partial\boldsymbol{\mu}}{\partial R_m}L_{mi}=\sum_{m}^{M}\left[\sum_{k}^{K}\left(\frac{\partial\mu_k}{\partial R_m}\right)\mathbf{e}_k+\mu_k\left(\frac{\partial\mathbf{e}_k}{\partial R_m}\right)\right]L_{mi} \tag{9.59}$$

There are two kinds of electro-optical parameters, or eop's: the equilibrium values of the bond dipole moments μ_k, and their variations $\partial\mu_k/\partial R_m$ with all of the vibrational parameters.

Use of a derivation similar to that given earlier for the bond charge model leads to the following matrix equation:

$$\frac{\partial \boldsymbol{\mu}}{\partial \mathbf{Q}_i} = \left\{ (\mathbf{e})^T \left(\frac{\partial \boldsymbol{\mu}}{\partial \mathbf{R}} \right) + (\boldsymbol{\mu}^0)^T \hat{\mathbf{r}}^{-1} [\boldsymbol{\Delta}\mathbf{A} - (\bar{\mathbf{e}}:\mathbf{0})] \right\} \mathbf{L}_i \tag{9.60}$$

where the matrices have the same form, with $\bar{\mathbf{e}}$ being a $K \times K$ diagonal matrix containing all the $\mathbf{e}_k^0$ and $:\mathbf{0}$ a $K \times (m-K)$ null matrix whose columns correspond to the nonstretching coordinates. The introduction of $|\bar{\mathbf{e}}:\mathbf{0}|$ ensures that the $\partial \mathbf{e}_k/\partial R_m$ vanishes when R_m coincides with the stretching of the kth bond. $(\mu^0)^T$ is a $3 \times (3N-3)$ matrix containing the bond dipole moments in the form

$$\begin{bmatrix} \mu_1 & 0 & 0 & \mu_2 & \cdots & 0 \\ 0 & \mu_1 & 0 & 0 & \cdots & 0 \\ 0 & 0 & \mu_1 & 0 & \cdots & \mu_{N-1} \end{bmatrix}$$

Using this equation, one can in principle calculate the infrared intensities. Unfortunately, the inverse process is complicated by the same difficulties encountered in the calculation of bond charge parameters, namely, the uncertainty in the sign of $\partial\mu/\partial Q$ (and of $\partial\alpha_{uv}/\partial Q$) and the large number of parameters relative to the experimental data.

In practice, the number of sign choices can be reduced in various ways. According to Gussoni (4), these are:

1. Intensities from isotopic derivatives can be used, together with the assumption that the electronic structure is invariant for isotopic substitution.
2. Physical constraints can be introduced.
3. We can compare $\partial\mu/\partial R$ or $\partial\alpha_{uv}/\partial R$ computed from quantum mechanical calculations with those derived from intensity data with various sign choices.
4. The intensity data of similar molecules can be used as constraints in an overlay calculation of electro-optical parameters.

We will consider an example of the use of infrared electro-optical parameters after a discussion of Raman intensities.

9.5 The Application of Electro-Optical Theory to the Calculation of Raman Intensities

Our discussion of Raman intensities is based on the treatment of Gussoni (4). In a manner analogous to the treatment given earlier we can first express the

Raman scattered intensities (for an isotropic material) as;

$$I_i = C(\nu_i)\left[45\left(\frac{\partial\bar{\alpha}}{\partial Q_i}\right)^2 + 13\left(\frac{\partial\bar{\gamma}}{\partial Q_i}\right)^2\right] \tag{9.61}$$

where $\partial\bar{\alpha}/\partial Q_i$ and $\partial\bar{\gamma}/\partial Q_i$ are the mean molecular polarizability and molecular anisotropy derivatives, respectively, and $C(\nu_i)$ is a function of the exciting and scattered frequency. If we denote the polarizability tensor $((\alpha))$ as α_{uv}, where

$$((\alpha)) = \begin{bmatrix} \mathbf{i} \\ \mathbf{j} \\ \mathbf{k} \end{bmatrix} \begin{bmatrix} \alpha_{xy} & \alpha_{xy} & \alpha_{xz} \\ \alpha_{yx} & \alpha_{yy} & \alpha_{yz} \\ \alpha_{zx} & \alpha_{zy} & \alpha_{zz} \end{bmatrix} [\mathbf{i} \quad \mathbf{j} \quad \mathbf{k}] \tag{9.62}$$

then

$$\frac{\partial\bar{\alpha}}{\partial Q_i} = \frac{1}{3}\sum_u \frac{\partial\alpha_{uv}}{\partial Q_i} \tag{9.63}$$

and

$$\frac{\partial\gamma}{\partial Q_i} = \frac{1}{2}\left[3\sum_{\substack{u,v\\=x,y,z}}\left(\frac{\partial\alpha_{uv}}{\partial Q_i}\right)^2 - 9\left(\frac{\partial\bar{\alpha}}{\partial Q_i}\right)^2\right] \tag{9.64}$$

A parametric method should then be capable of expressing $\partial\alpha_{uv}/\partial Q_i$ in terms of bond parameters. In a fashion similar to our treatment of infrared intensities we first assume that we can express the molecular polarizability $((\alpha))$ as a sum of k bond polarizabilities; that is,

$$((\alpha)) = \sum_k^k \alpha^k \tag{9.65}$$

We now let $\mathbf{e}^{kL}$ be the unit vector in the direction of the kth bond (longitudinal axis L) and $\mathbf{e}^{kT}$, $\mathbf{e}^{kT'}$ unit vectors perpendicular (transverse) to the bond direction, as illustrated in Figure 9.4. We can therefore express $((\alpha))$ as

$$((\alpha)) = \sum_k^K \alpha^k = \sum_k^K (\alpha^{kL}\mathbf{e}^{kL}\mathbf{e}^{kL} + \alpha^{kT}\mathbf{e}^{kT}\mathbf{e}^{kT} + \alpha^{kT'}\mathbf{e}^{kT'}\mathbf{e}^{kT'}) \tag{9.66}$$

The general element of $((\alpha))$ in the chosen cartesian system is given by

$$\alpha_{uv} = \mathbf{e}_u((\alpha))\mathbf{e}_v \tag{9.67}$$

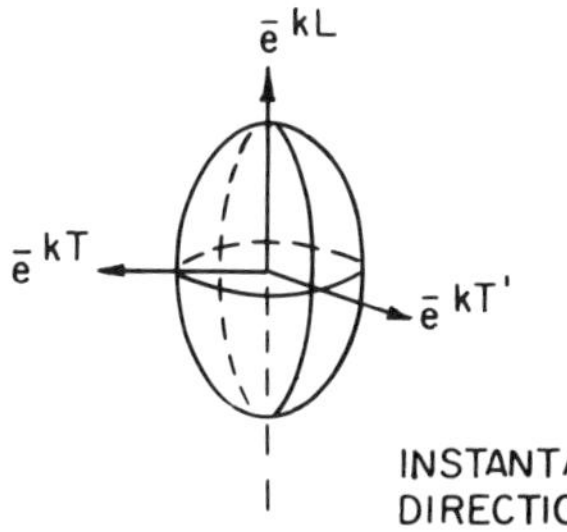

Figure 9.4. Plot of the polarizability tensor referred to its principal axes.

where $\mathbf{e}_u$ and $\mathbf{e}_v$ are unit vectors along the u and v axes, respectively. Therefore, from equation (9.61) we can obtain

$$\alpha_{uv} = \sum_k^K \left[\alpha^{kL} e_u^{kL} e_v^{kL} + \alpha^{kT} \left(e_u^{kT} e_v^{kT} + e_u^{kT'} e_v^{kT'} \right) \right] \tag{9.68}$$

where e_u^{kT} is the direction cosine of $\mathbf{e}^{kT}$ with respect to the u axis.

The matrix expression for $\partial\alpha_{uv}/\partial Q_i$ can now be written

$$\begin{aligned} \frac{\partial \alpha_{uv}}{\partial Q_i} &= \sum_m^M \frac{\partial \alpha_{uv}}{\partial R_m} L_{mi} \\ &= \sum_m^M \sum_{p=L,T,T'} \left\{ \sum_k^K \frac{\partial \alpha^{kp}}{\partial R_m} e_u^{kp0} e_v^{kp0} \right. \\ &\quad \left. + \sum_k^K \left[\alpha^{kp0} \left(\frac{\partial e_u^{kp}}{\partial R_m} e_v^{kp0} + e_u^{kp0} \frac{\partial e_v^{kp}}{\partial R_m} \right) \right] \right\} L_{mi} \end{aligned} \tag{9.69}$$

The first sum over k in this equation gives the valance term and contains the valence electro-optical parameters, $\partial\alpha^{kL}/\partial R_m$, $\partial\alpha^{kT}/\partial R_m$, and $\partial\alpha^{kT'}/\partial R_m$. The second sum over k is a deformation term and contains the equilibrium electro-optical parameters α^{kp0}. As with the corresponding expression for infrared intensity, the remaining terms are known or can be computed from a knowledge of the dynamical behavior of the molecule, with one important exception. The term $\partial\mathbf{e}^{kp}/\partial R_m$ with $p=L$ can be calculated as for infrared intensities [see equation (9.54)], but the terms with $p=T$ and T' are not known and cannot be completely determined by the dynamics of the system, but are affected by the rearrangement of the electron cloud. The terms $\partial\mathbf{e}^{kT}/\partial R_m$ and $\partial\mathbf{e}^{kT'}/\partial R_m$ have been expressed as bond orientation parameters and a long and cumbersome general equation for $\partial\alpha_{uv}/\partial Q_i$ has been derived (e.g., see (4)]). However, if we assume a cylindrical shape for the polarizability tensor, then the equilibrium bond "transverse" anisotropy vanishes, as does the bond orientation parameter. We can then obtain a simplified expression for Raman scattering intensities in terms of the electro-optical parameters $\partial\alpha^{kp}/\partial R_m$ and α^{kp0}. These two parameters are more useful if written in terms of the mean polarizability and the anisotropy, as defined in equations (9.58) and (9.59). The appropriate combinations as defined by Gussoni (4) are

1 Mean derived bond polarizability:

$$\left(\frac{\partial\bar{\alpha}}{\partial R}\right)_{km} \equiv \frac{\partial\bar{\alpha}^k}{\partial R_m} = \frac{\partial}{\partial R_m}\left(\frac{\alpha^{kL}+\alpha^{kT}+\alpha^{kT'}}{3}\right) \tag{9.70}$$

2 Derived "longitudinal" bond anisotropy:

$$\left(\frac{\partial\gamma}{\partial R}\right)_{km} \equiv \frac{\partial\gamma^k}{\partial R_m} = \frac{\partial}{\partial R_m}\left(\alpha^{kL} - \frac{\alpha^{kT}+\alpha^{kT'}}{2}\right) \tag{9.71}$$

3 Derived "transverse" bond anisotropy:

$$\left(\frac{\partial\gamma_t}{\partial R}\right)_{km} \equiv \frac{\partial\gamma_t^k}{\partial R_m} = \frac{\partial}{\partial R_m}\left(\frac{\alpha^{kT}-\alpha^{kT'}}{2}\right) \tag{9.72}$$

4 Equilibrium "longitudinal" bond anisotropy:

$$(\gamma^0)_k \equiv \gamma^{k0} = \alpha^{kL0} - \frac{\alpha^{kT0}+\alpha^{kT'0}}{2} \tag{9.73}$$

5 Equilibrium "transverse" bond anisotropy:

$$(\gamma_t^0)_k \equiv \gamma_t^{k0} = \frac{\alpha^{kT0}-\alpha^{kT'0}}{2} \tag{9.74}$$

Assuming a cylindrical shape for the polarizability tensor and using orthonormality relations among unit vectors (4), we have for equation (9.64)

$$\frac{\partial\boldsymbol{\alpha}_{uv}}{\partial\mathbf{Q}_i} = \left[\delta_{uv}\tilde{\underline{1}}\frac{\partial\boldsymbol{\alpha}}{\partial\mathbf{R}} + \left(\mathbf{e}_u^{L0}\mathbf{e}_v^{L0} - \frac{1}{3}\delta_{uv}\tilde{\underline{1}}\right)\frac{\partial\boldsymbol{\gamma}}{\partial\mathbf{R}} + \tilde{\gamma}^0\frac{\partial\mathbf{e}_u^L\mathbf{e}_v^L}{\partial\mathbf{R}}\right]\mathbf{L}_i \tag{9.75}$$

where the tilde is used to indicate the transpose of a matrix to avoid confusion with the superscript T used to indicate transverse; δ_{uv} is the Kronecker delta, and $\underline{\mathbf{1}}$ is a k vector of all unit entries.

9.6 Infrared and Raman Electro-Optical Parameters for the Alkanes

A detailed review of the application of electro-optical parameters to the calculation of the infrared and Raman intensities of the alkanes has been

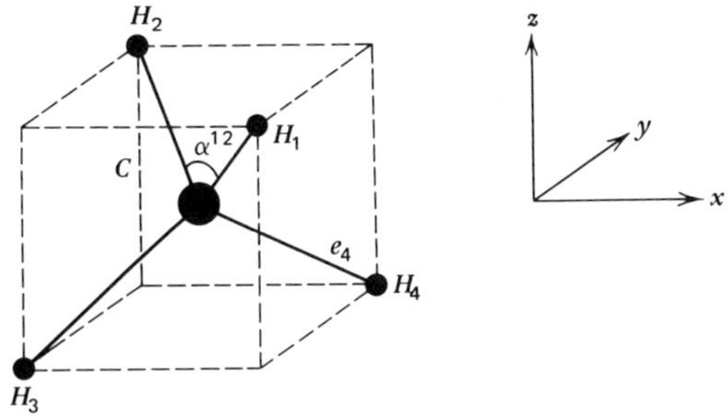

Figure 9.5. Definition of the internal coordinates, bond vectors, and cartesian axes for methane. (Reprinted by permission from ref. 17.)

presented by Gussoni et al. (4, 16, 17). Here we will simply outline the results; in Chapter 13 we will discuss the application of such parameters to a calculation of the intensities of polyethylene.

The simplest alkane is methane, which has two infrared-active species. Accordingly, there are four possible combinations of the signs for the values of $\partial\mu/\partial Q_i$ [(+ −), (+ +), (− +), (− −)]. The possibilities can be reduced to two by the use of isotopic derivatives. The coordinates used in these calculations are defined in Figure 9.5 and the parameters refined are given at the bottom of Tables 9.2 and 9.3, where the observed and calculated intensities are also reported. The choice of the most likely sign combinations was made on the principle of isotopic invariance. Upon performance of a least-squares refinement of the observed and calculated intensities, the most likely signs were determined to be either (+ −) or (− +). *A priori* calculations suggest (− +) as the best sign combination, but Gussoni (4) has pointed out that the results of applying such calculations to apolar molecules are not always reliable. A choice between these signs was made by corefining the data from methane with the intensity data from ethane and its deutero derivatives, so that the (− +) combination was finally selected. The data from these calculations were then used as a starting point for overlay calculations that also included propane. The electro-optical parameters reported for the alkanes by Gussoni (4) are reproduced in Table 9.4 and the internal coordinate definitions are shown in Figure 9.6. The equilibrium bond dipole moment for the C—H bond was assumed to be the same in CH_2 and CH_3 groups and the equilibrium bond dipole moment for a C—C bond was assumed to be zero.

The calculation of electro-optical parameters describing Raman intensities is not (at the time of writing this book) so advanced, principally because of a paucity of data. Only for methane have all the intensities and depolarization ratios been measured. Nevertheless, sufficient data have been amassed for some preliminary calculations that have allowed a good prediction of the intensities observed in the Raman spectrum of polyethylene. We will consider this further in Chapter 13. Here we conclude with the observation that the calculation of intensities is more difficult than the equivalent calculation of

Table 9.2 Least-Squares Refinement of Parameters P_1 and P_2 on the Intensity Data for Methane and Its Deutero Derivatives (13)[a]

Molecule	ν_{obs}	Species	A_{obs} [b]	A^{A}_{calc} [c]		A^{B}_{calc} [c]	
CH_4	3019	t_2	900	882	−	791	+
	1306	t_2	450	432	+	441	+
CD_4	2259	t_2	349	398	−	530	+
	996	t_2	259	267	+	236	+
CH_2D_2	3013	b_1 }	466	439		408	
	2976	a_1 }					
	2234	b_2 }	185	193		252	
	2202	a_1 }					
	1436	a_1	—	54		52	
	1234	b_2 }	287	301		284	
	1090	b_1 }					
	1033	a_1 }					
CH_3D	3021	e }	658	654		598	
	2945	a_1 }					
	2200	a_1	84	95		122	
	1471	e }	391	394		387	
	1300	a_1 }					
	1155	e }					
CD_3H	299	a_1	210	281		207	
	2200	e }	280	296		389	
	2141	a_1 }					
	1291	e	89	99		90	
	1036	e }	204	214		196	
	1003	a_1 }					
		χ_A		18.4		70.8	
		$P_1 = \dfrac{\partial \mu^k}{\partial r_k} - \dfrac{\partial \mu^k}{\partial r_j}$		−0.65		0.53	
		$P_2 = \mu^0 - \sqrt{2}\left(\dfrac{\partial \mu^k}{\partial \alpha_{kj}} - \dfrac{\partial \mu^k}{\partial \alpha_{jm}}\right)$		0.33		0.30	

[a] $\tilde{\nu}$ in cm^{-1} and A in 10^{10} $cm^{-1}S/atm$. χ_A is defined in ref. 13 as a parameter that measures the success of a given refinement.
[b] From Hiller and Straley (13).
[c] Quoted in ref. 4.

Table 9.3 Least-Squares Refinement of Parameters P_1 and P_2 on the (14 Intensity Data for Methane and Its Deutero Derivatives (4)

Molecule	$\tilde{\nu}$ (cm^{-1})	Species	Γ_{obs} [b] (cm^2/mol)	Γ^{A}_{calc} [c] (cm^2/mol)		Γ^{B}_{calc} [d] (cm^2/mol)	
CH_4	3019	t_2	2310	2309	−	1981	+
	1306	t_2	2554	2578	+	2699	+
CD_4	2259	t_2	1466	1394	−	1783	+
	996	t_2	1906	2090	+	1902	+
CH_3D	3021 2945	e a_1	1791	1735		1517	
	2200	a_1	269	340		422	
	1471 1300 1155	e a_1 e	2700	2401		2419	
CD_3H	2993	a_1	581	577		524	
	2200 2142	e a_1	875	1038		1314	
	1291	e	580	603		558	
	1036 1003	e a_1	1450	1628		1539	
		χ_Γ		133.5		235.8	
$P_1=\dfrac{\partial\mu^k}{\partial r_k}-\dfrac{\partial\mu^k}{\partial r_j}$				−0.67		+0.54	
$P_2=\mu^0-\sqrt{2}\left(\dfrac{\partial\mu^k}{\partial\alpha_{kj}}-\dfrac{\partial\mu^k}{\partial\alpha_{jm}}\right)$			+0.34	+0.31			

[a]All $\tilde{\nu}$ in cm^{-1} and Γ in cm^2/mol. χ_Γ is defined in ref. 14 as a parameter that measures the success of a given refinement.
[b]From Heickler (14).
[c]From Kondo and Saeki (15).
[d]Quoted in ref. 4.

Table 9.4 Electro-Optical Parameters Obtained from a Refinement[16] of Infrared Intensity Data[15] for Ethane, Propane, and Their Deutero Derivatives

CH_3		CH_2
$\frac{\partial\mu^k}{\partial r_k}$ -0.918 Debye/Å		$\frac{\partial m^k}{\partial d_k} = -0.863$ Debye/Å
$\frac{\partial\mu^k}{\partial r_j}$ -0.138 Debye/Å		$\frac{\partial m^k}{\partial d_j} = -0.063$ Debye/Å
$\frac{\partial\mu^k}{\partial\alpha_{jt}}$ $+0.004$ Debye/rad		$\frac{\partial m^k}{\partial\omega} = +0.022$ Debye/rad
$\frac{\partial\mu^k}{\partial\alpha_{kj}}$ -0.029 Debye/rad		$\frac{\partial m^k}{\partial\gamma_{kj}} = +0.022$ Debye/rad
$\frac{\partial\mu^k}{\partial\beta_{kj}} = -\frac{\partial\mu^k}{\partial\beta_{ji}} = +0.006$ Debye/rad		$\frac{\partial m^k}{\partial\gamma_{jl}} = -0.038$ Debye/rad
$\mu^0_{CH} = 0.245$ Debye		$m^0_{CH} = 0.245$ Debye
	CC	
$\frac{\partial\mu^k}{\partial R} = -0.422^b$ D/Å		$\frac{\partial m^k}{\partial R} = 0.233$ D/Å
$\frac{\partial M^k}{\partial R_k} = -0.017$ D/Å		$\frac{\partial M}{\partial d_k} = 0.011$ D/Å
$\frac{\partial M^k}{\partial\beta_{kj}} = -0.084$ D/Å		$\frac{\partial M}{\partial\omega} = 0.021$ D/Å
$\frac{\partial M^k}{\partial\gamma_{kj}} = 0.098$ D/Å		$\frac{\partial M^k}{\partial\gamma_{jl}} = -0.061$ D/Å
		$M^0_{C-C} = 0.0$ D

[a]The bond dipole moments of C—C are indicated by M, those of the C—H bonds (CH_2) by m, and those of the C—H bonds (CH_3) by μ.

vibrational frequencies, but the success so far achieved using electro-optical parameters suggests that they are transferable between similar molecules and that it is therefore a viable approach to intensity calculations.

References

1. M. Gussoni, and S. Abbate, *J. Chem. Phys.* **65**(9), 3439 (1976).
2. S. Abbate, M. Gussoni, G. Masetti, and G. Zerbi, *J. Chem. Phys.* **67**, 1519 (1977).
3. L. A. Gribov, "Intensity Theory for Infrared Spectra of Polyatomic Molecules," Consultants Bureau, New York, 1964.
4. M. Gussoni, in "Advances in Infrared and Raman Spectroscopy," Vol. 7, Heyden, London, 1980, Chapter 2, p. 61.
5. A. J. Van Straten and W. M. A. Smit, *J. Mol. Spect.* **65**, 202 (1977).

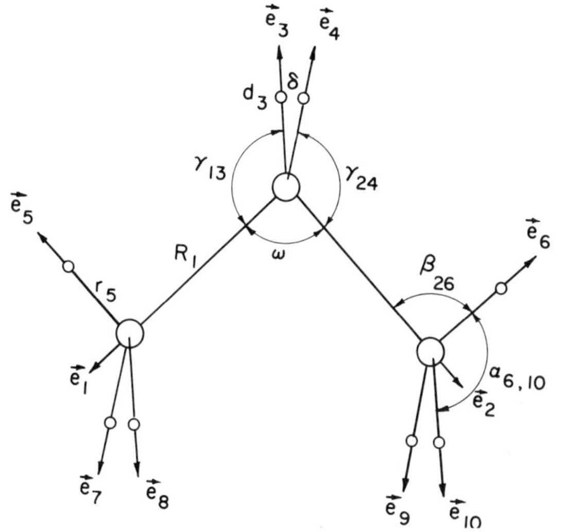

Figure 9.6. Internal coordinates and bond unit vectors of C_3H_8.

6. J. H. Van Vleck, *Proc. Natl. Acad. Sci.* **15**, 754 (1929).
7. J. Tang and A. C. Albrecht, in "Raman Spectroscopy," H. A. Szymanski, Ed., Plenum Press, New York, 1974, Vol. 2, p. 33.
8. A. J. Van Straten and W. M. A. Smit, *J. Mol. Spect.* **62**, 297 (1976).
9. J. F. Biarge, J. Herranz, and F. Morcillo, *An. Fis. Quim.* **A57**, 81 (1961).
10. W. B. Person and J. H. Newton, *J. Chem. Phys.* **61**, 1040 (1974).
11. M. Eliashevich and M. Wolkenstein, *Zh. Eksp. Teor. Fiz.* **9**, 101 (1945).
12. M. Gussoni, S. Abbate, and G. Zerbi, Nato paper presented at the Advanced Study Institute, Florence (1976).
13. R. E. Hiller and J. W. Straley, *J. Mol. Spect.* **5**, 24 (1960).
14. J. Heickler, *Spectrochim. Acta* **17**, 201 (1961).
15. S. Kondo and S. Saeki, *Spectrochim. Acta* **A29**, 735 (1973).
16. M. Gussoni, S. Abbate, and G. Zerbi, in "Vibrational Spectroscopy, Modern Trends," A. J. Barnes and W. J. Orville-Thomas, Eds., Elsevier, Amsterdam, 1977.
17. M. Gussoni, S. Abbate, and G. Zerbi, *J. Chem. Phys.* **71**(8), 3428 (1979).

10

THE LATTICE DYNAMICS OF CHAIN MOLECULES

Nothing in life shall sever
the chain that is round us now.
William Cory

At first glance, the enormous number of atoms present in a polymer chain would seem to make the calculation of its vibrational frequencies an impossible task. However, it will be shown in Chapter 11 that for a theoretically infinite polymer chain in a regular, well-defined conformation, the problem can be reduced by symmetry to the determination of the normal modes of the translational repeat unit. In effect, an ordered polymer chain can be considered to be a one-dimensional crystal or linear lattice. Physically, this is because the intermolecular forces between chains packed in a three-dimensional lattice are an order of magnitude less than the intramolecular interactions between repeat units in the same chain. Consequently, their effect on the vibrational spectrum can be neglected or treated as a perturbation. Before proceeding to an examination of the symmetry properties of polymer chains, it is useful to discuss the vibrations of simple model systems. This discussion will provide a basis for defining concepts and terms frequently used to describe the normal modes of polymers and will afford us some insight into the vibrations of a chain of identical units. We will start with the simplest model: an infinite one-dimensional chain of point masses. This model can be easily extrapolated to the simplest model of a polymer chain, a planar zigzag chain of point masses. It is convenient first to consider solutions in terms of cartesian displacement coordinates. We will introduce the Wilson **GF** method when we consider a chain of coupled oscillators.

10.1 The Infinite Monatomic Chain

The dynamics of a chain of particles is a classic problem in solid state physics and is discussed in a number of texts (1–4). Consider an infinite linear array of atoms of mass m separated by a distance d, as shown in Figure 10.1. Let the

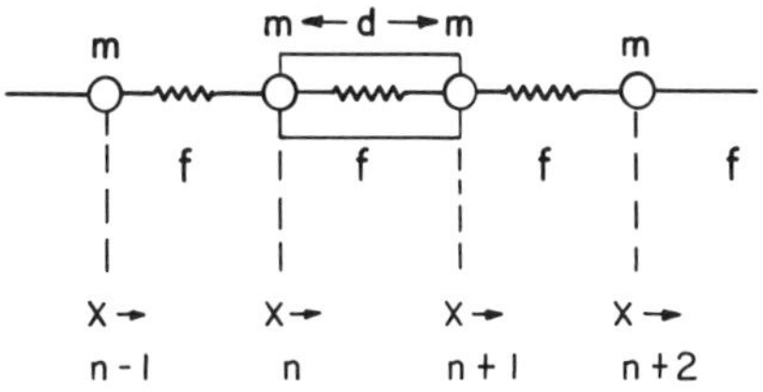

Figure 10.1. Schematic representation of a monatomic chain.

longitudinal displacement of the nth particle from the equilibrium position be x_n and the potential energy of the displacements of the atoms be determined by the force f acting between adjacent particles. In this treatment we will assume that only nearest neighbor forces are significant. Applying Newton's second law of motion, we obtain the force F_n acting on a given atom n:

$$F_n = m\ddot{x}_n = f(x_{n+1} - x_n) - f(x_n - x_{n-1}) \tag{10.1}$$

$$m\ddot{x}_n = fx_{n+1} - 2fx_n + fx_{n-1} \tag{10.2}$$

These equations can be more formally derived by expanding the potential energy, which is a function of the distance between adjacent particles, as a Taylor series in the displacements (see Chapter 2) and imposing the harmonic approximation. The displacements are again defined relative to the equilibrium position of the atoms. Since the atoms are spaced uniformly at a distance d from each other, the coordinate of the nth atom is related to the origin by a factor nd. The displacements are therefore propagated as a wave along the chain if the physical problem admits a solution of the type

$$x_n = Ae^{-2\pi i(\nu t - knd)} = Ae^{-i(\omega t - \phi n)} \tag{10.3}$$

where ν is the frequency, t represents time, ω (the circular frequency) $= 2\pi\nu$, k is the wave vector, ϕ (the phase angle) $= 2\pi kd$, and A is the amplitude. The quantity ϕ defines the difference in phase between the displacements of particle n and those of particle $n+1$. The magnitude of the wave vector k is defined here as equal to the reciprocal of the wavelength, $1/\lambda$*. (It should be noted that in many texts the definition $k = 2\pi/\lambda$ is used.) It is a vector quantity and its direction is the direction of propagation of the wave along the lattice.

Upon use of equation (10.3), equation (10.2) becomes (after division by x_n)

$$-4\pi^2\nu^2 m = f(e^{2\pi ikd} - 2 + e^{-2\pi ikd}) = -2f(1 - \cos 2\pi kd) \tag{10.4}$$

*It is conventional to use the symbol λ to specify wavelength as well as frequency, $\lambda = 4\pi^2c^2\nu^2$. Care has to be taken so as to avoid confusion.

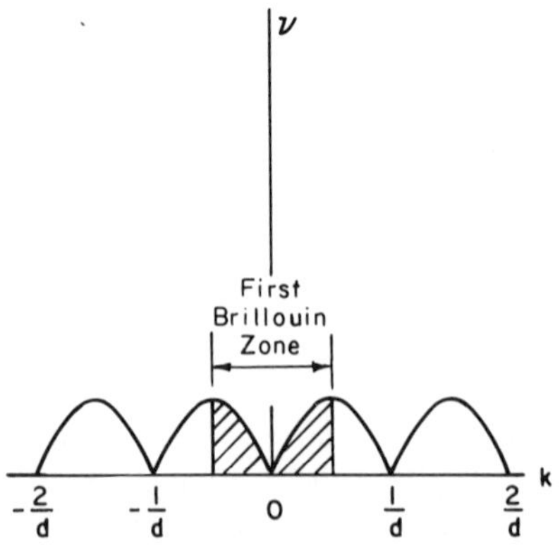

Figure 10.2. Plot of dispersion curve of the monatomic chain showing the first Brillouin zone.

and the positive root is given by

$$\nu=\frac{1}{\pi}\left(\frac{f}{2m}\right)^{1/2}(1-\cos 2\pi kd)^{1/2} \tag{10.5}$$

which can be written as

$$\nu=\frac{1}{\pi}\left(\frac{f}{m}\right)^{1/2}|\sin \pi kd|=\frac{1}{\pi}\left(\frac{f}{m}\right)^{1/2}\left|\sin\frac{\phi}{2}\right| \tag{10.6}$$

Figure 10.2 shows a plot of ν against k, demonstrating that ν is periodic in k. The period is equal to $1/d$ and a series of maxima occur at a frequency of

$$\nu_c=\frac{1}{\pi}\left(\frac{f}{m}\right)^{1/2} \tag{10.7}$$

the first maxima being at a point $k=1/2d$ from the origin.

On account of the periodic properties of the chain, it is sufficient to discuss the values of ν inside one period of k or ϕ. The most convenient choice is

$$-\pi\leqslant\phi\leqslant\pi, \qquad \frac{1}{2d}\leqslant k\leqslant\frac{1}{2d} \tag{10.8}$$

This range of values is referred to as the first *Brillouin zone* of the linear chain. Note that the symmetry of the zones implies that k can take positive and negative values of equal magnitude, so that waves associated with any zone can travel in the positive or negative x direction along the chain. The plot of ν against k or ϕ is the *dispersion curve*, and it is usual to show only the positive half, as in Figure 10.3. Also shown in Figure 10.3 is the form of the vibrations for various values of k. Naturally, the displacements are longitudinal (i.e., in the lattice direction), since we are considering only the one-dimensional case. However, in order to demonstrate the sinusoidal nature of the lattice vibrations, the displacements are displayed perpendicular to the axis of the chain. If

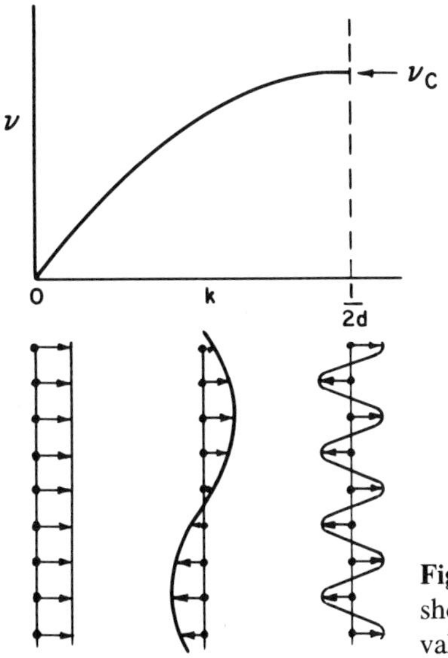

Figure 10.3. Plot of half of the first Brillouin zone showing the form of the normal modes for certain values of k.

$k=0$, the displacements of the atoms have the same amplitude and direction and are in phase. This vibration represents a translation of the lattice and has zero frequency. If $k=1/2d$, it can be shown that the displacements represent a standing wave in which all the particles have the same amplitude but alternate ones are out of phase by π radians. Traveling wave motion is exhibited at intermediate values of k and a typical example is also shown in Figure 10.3.

10.2 Vibrations of an Infinite Diatomic Chain

A lattice consisting of two or more types of atoms displays new features in the dispersion relationship that throw additional light onto the vibrations of polymer chains. Consider a chain consisting of two different atoms arranged alternately, as shown in Figure 10.4.

In the general case the atoms are linked by bonds of different types, characterized by force constants f_1 and f_2. If we again assume only nearest

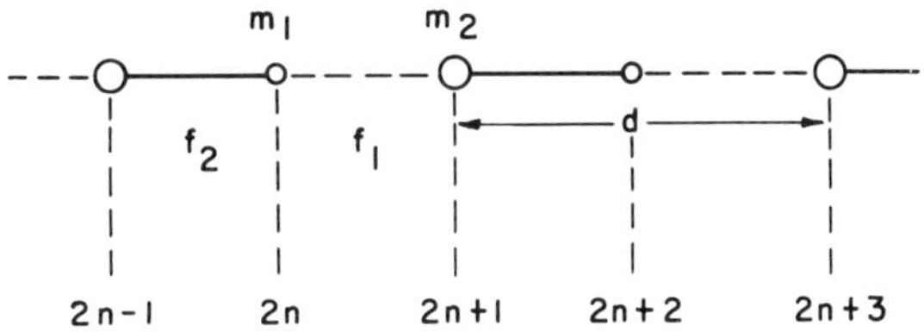

Figure 10.4. Schematic representation of a diatomic chain.

neighbor forces, we can write the following two equations of motion for the two different atoms, in the same way as the single equation for the monatomic chain described earlier.

$$F_{2n}=m_1\ddot{x}_{2n}=-f_2(x_{2n}-x_{2n-1})+f_1(x_{2n+1}-x_{2n}) \tag{10.9}$$

$$F_{2n+1}=m_2\ddot{x}_{2n+1}=-f_1(x_{2n+1}-x_{2n})+f_2(x_{2n+2}-x_{2n+1}) \tag{10.10}$$

Periodic solutions of the type

$$x_{2n}=A_1e^{-2\pi i(\nu t-nkd)} \tag{10.11}$$

$$x_{2n+1}=A_2e^{-2\pi i(\nu t-(n+1/2)kd)} \tag{10.12}$$

can again be used. Note that for the diatomic chain the distance d contains two particles and is the translational repeat unit or unit cell of the diatomic linear lattice. Substitution of equations (10.11) and (10.12) into (10.9) and (10.10) leads to the following pair of simultaneous equations, which will be recognized as the expanded secular equation:

$$\left[4\pi^2\nu^2m_1-(f_1+f_2)\right]A_1+\left(f_1e^{\pi ikd}+f_2e^{-\pi ikd}\right)A_2=0 \tag{10.13}$$

$$\left(f_2e^{\pi ikd}+f_1e^{-\pi ikd}\right)A_1+\left[4\pi^2\nu^2m_2-(f_2+f_1)\right]A_2=0 \tag{10.14}$$

which has nontrivial solutions only if the secular determinant is zero.

$$\begin{vmatrix} f_1+f_2-4\pi^2\nu^2m_1 & -\left(f_1e^{\pi ikd}+f_2e^{-\pi ikd}\right) \\ -\left(f_1e^{-\pi ikd}+f_2e^{\pi ikd}\right) & f_2+f_1-4\pi^2\nu^2m_2 \end{vmatrix}=0 \tag{10.15}$$

Expanding this determinant, we can determine two roots:

$$\nu^2=\frac{f_1+f_2}{8\pi^2}\left(\frac{1}{m_1}+\frac{1}{m_2}\right)$$

$$\pm\frac{1}{4\pi^2}\left[\left(\frac{f_1+f_2}{2}\right)^2\left(\frac{1}{m_1}+\frac{1}{m_2}\right)^2-\frac{4f_1f_2}{m_1m_2}\sin^2\pi kd\right]^{1/2} \tag{10.16}$$

It is instructive to consider two special cases, the first with $f_1=f_2$ and the second with $f_1\gg f_2$.

A model with $f_1=f_2=f$ implies equal spacing of the atoms and was first used by Born and von Karman (10) as a model of a simple ionic crystal. With

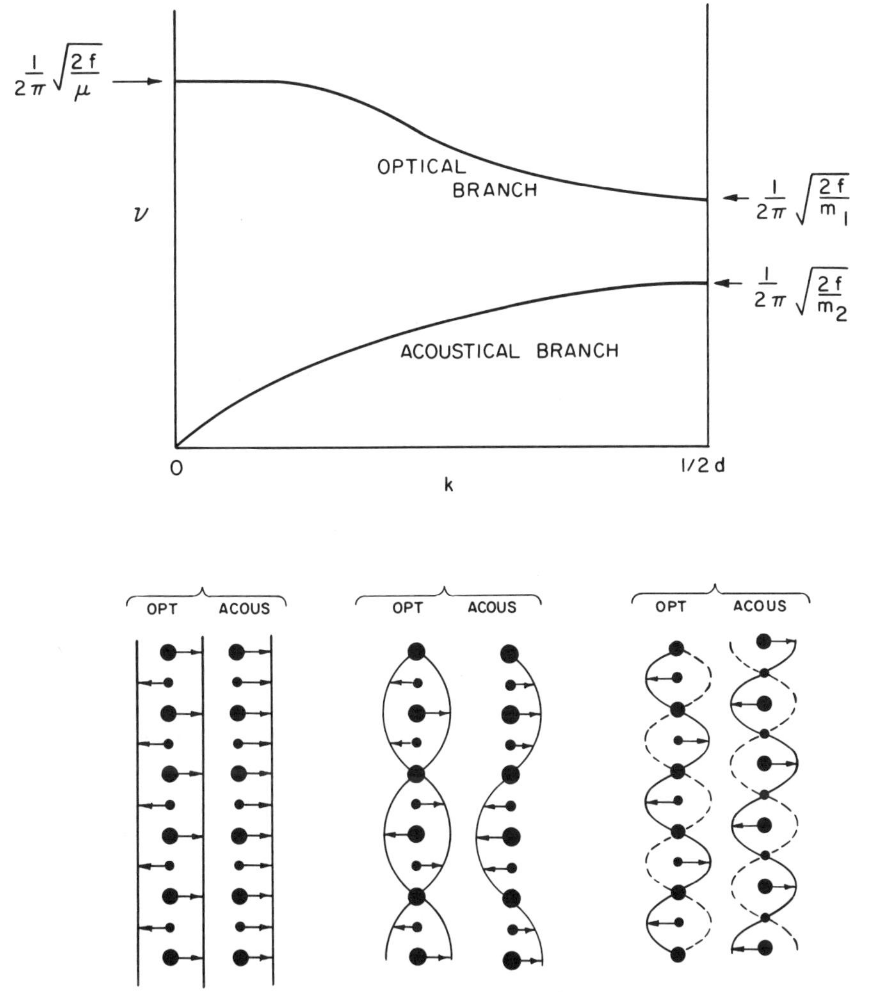

Figure 10.5. Dispersion curve of diatomic chain showing the form of the normal modes for certain values of k.

this assumption, equation (10.16) is reduced to

$$\nu^2 = \frac{f}{4\pi^2}\left\{\left(\frac{1}{m_1}+\frac{1}{m_2}\right) \pm \left[\left(\frac{1}{m_1}+\frac{1}{m_2}\right)^2 - \frac{4}{m_1 m_2}\sin^2 \pi k d\right]\right\}^{1/2} \qquad (10.17)$$

The two solutions of this equation result in two frequency branches of the dispersion relationship, as shown in Figure 10.5. With the plus sign we obtain the upper curve, known as the optical branch since it gives rise to fundamentals in the infrared and Raman spectra. The solution obtained using the negative sign gives the lower curve, called the acoustic branch because its frequencies fall in the region of sonic or ultrasonic waves. It can be seen that wave solutions do not exist for frequencies between $(1/2\pi)(2f/m_1)^{1/2}$ and $(1/2\pi)(2f/m_2)^{1/2}$, a region referred to as the **frequency gap**. Frequencies in this range require that $\sin \pi kd > 1$, which is only possible for complex k when the wave is attenuated and does not propagate freely.

The displacements observed at various points on the dispersion curve are also illustrated (rotated 90°) in Figure 10.5. At $k=0$ the acoustic branch has zero frequency and this mode corresponds to a translation of the lattice as a whole, without alteration of the distance between particles. The optical branch at $k=0$ is a normal vibration consisting of a simple stretching of the bond between two given atoms so that their center of mass remains fixed. Its frequency is equal to $(1/2\pi)(2f/\mu)^{1/2}$ where μ is the reduced mass, defined by

$$\frac{1}{\mu} = \left(\frac{1}{m_1} + \frac{1}{m_2}\right)$$

At $k=1/2d$ the acoustic vibrational mode involves the heavy atoms vibrating back and forth against each other with the light atoms remaining fixed. Conversely, in the optical mode the heavy atoms are fixed and the light atoms vibrate. Clearly, in the special case of $m_1=m_2$ there is no frequency gap and these two points on the frequency branches coincide. Typical waves observed for values of k intermediate between zero and $1/2d$ are also shown in Figure 10.5.

The significance of the dispersion curve derived for a diatomic lattice is that it indicates the existence of frequency banding for the normal modes of periodic structures. Only those frequencies located in the optical and acoustic branches can propagate along the chain. If we had a more complex chain containing N atoms per unit cell, there would be N bands (for vibrations in one dimension). A vibrational frequency located in a forbidden region or frequency gap, corresponding, for example, to a normal mode of a defect (a unit with a mass or structure different from that of the rest of the chain), would normally dampen very quickly with increasing distance from the defect site and would not propagate along the chain. The analysis of such imperfect systems is complex and will be considered in more detail in Chapter 15.

The case of a lattice with $f_1 \gg f_2$ is a simple model for a one-dimensional crystal with loose coupling between adjacent molecules. The force constant f_1 corresponds to the "stiffness" of the covalent bond linking pairs of atoms, while f_2 represents the much weaker intermolecular force. The two roots of equation (10.16) are then given approximately by

$$\nu_{\text{int}} = \frac{1}{2\pi}\left(\frac{f_1}{\mu}\right)^{1/2} \tag{10.18}$$

and

$$\nu_{\text{ext}} = \frac{1}{\pi}\left(\frac{f_2}{m_1+m_2}\right)^{1/2} |\sin \pi k d| \tag{10.19}$$

The first solution, ν_{int}, appears to be independent of k (i.e., its dispersion curve is a horizontal straight line) and is approximately equal to the vibrational frequency of an isolated diatomic molecule. There is in fact a slight dependence on k, which in this case has been neglected because f_2 is small but not zero. Vibrations of this type, involving nonrigid molecular motion, are known as **internal vibrations**. The frequencies of these modes differ from gas phase vibrations only to the extent that there is a coupling between different molecules in the lattice.

The second solution, ν_{ext}, corresponds to a vibration of the molecule as a whole, each molecule moving as a rigid body. Motions of this type are known as **external vibrations** or **lattice vibrations** and can be separated into two types; translational modes and rotational or librational modes. Translational modes, as the name suggests, involve a translation of the molecule as a whole. Rotational modes involve quasi-rotation of the molecules about their center of gravity. These modes are optically active.

10.3 Lattice Vibrations in Three Dimensions; Longitudinal and Transverse Waves

The treatment developed above for the one-dimensional case can be extended to vibrations in three dimensions, but in general the mathematics becomes algebraically more complicated. The simple diatomic chain can be fairly easily treated by defining a bending force constant f_α for displacements perpendicular to the chain. It can be shown that transverse motions are independent of longitudinal vibrations (for small displacements); in addition, transverse motions at right angles to one another are also mutually independent. If we define $\Delta\alpha_{2n}$ as the change in angle of the two bonds joining atom $2n$ to its nearest neighbors (see Figure 10.4), the force on atom $2n$ is given by

$$F_{2n} = f_\alpha(\Delta\alpha_{2n-1} + \Delta\alpha_{2n+1} - 2\,\Delta\alpha_{2n}) \tag{10.20}$$

The change in angle $\Delta\alpha_{2n}$ can be expressed in terms of the perpendicular displacements of atoms, σ:

$$\Delta\alpha_{2n} = \frac{2}{d}(\sigma_{2n} + \sigma_{2n-1} - 2\sigma_{2n}) \tag{10.21}$$

where σ can correspond to either the y or z cartesian displacement coordinate.

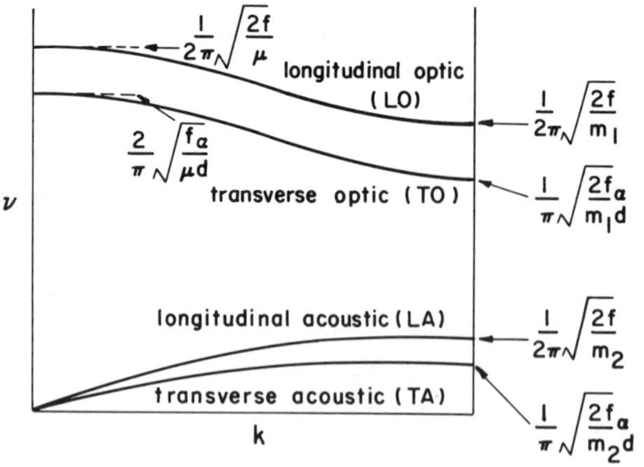

Figure 10.6. Dispersion curves for the longitudinal and transverse modes of a diatomic chain.

Then, as in the case of longitudinal vibrations, the equations of motion

$$F_{2n}=\frac{2f_\alpha}{d}(\sigma_{2n-2}-4\sigma_{2n-1}+6\sigma_{2n}-4\sigma_{2n+1}+\sigma_{2n+2})=m_1\ddot{\sigma}_{2n} \qquad (10.22)$$

$$F_{2n+1}=\frac{2f_\alpha}{d}(\sigma_{2n-1}-4\sigma_{2n}+6\sigma_{2n+1}-4\sigma_{2n+2}+\sigma_{2n+3})=m_2\ddot{\sigma}_{2n+1} \qquad (10.23)$$

can be used to derive the secular determinant, the solutions of which are

$$4\pi^2\nu^2=\frac{2f_\alpha}{m_1m_2d}\Big\{(m_1+m_2)(3+\cos 2\pi kd)\pm\big[(m_1+m_2)^2(3+\cos\pi kd)^2$$

$$-4m_1m_2(\cos^2 2\pi kd-2\cos 2\pi kd+1)\big]^{1/2}\Big\} \qquad (10.24)$$

In the case of the three-dimensional motions of a diatomic chain the two transverse directions are equivalent. Consequently, both of these branches of the dispersion curve are described by equation (10.24); that is the frequencies of the perpendicular vibrations are doubly degenerate. The dispersion curves of this simple diatomic system are illustrated in Figure 10.6.

The situation becomes more complex when we consider a three-dimensional crystal rather than a linear lattice. In general, the transverse vibrational frequencies of molecular crystals are not degenerate. There are three dispersion branches for each particle in the unit cell, as illustrated in Figure 10.7 for a given lattice direction of the wave vector k. If there are m molecules in the unit cell, each of which is composed of N atoms, there are $3mN$ branches. Three of these are acoustic modes, allowing up to $3mN-3$ optical branches (again depending on degeneracy). Since in many crystals the intramolecular forces are

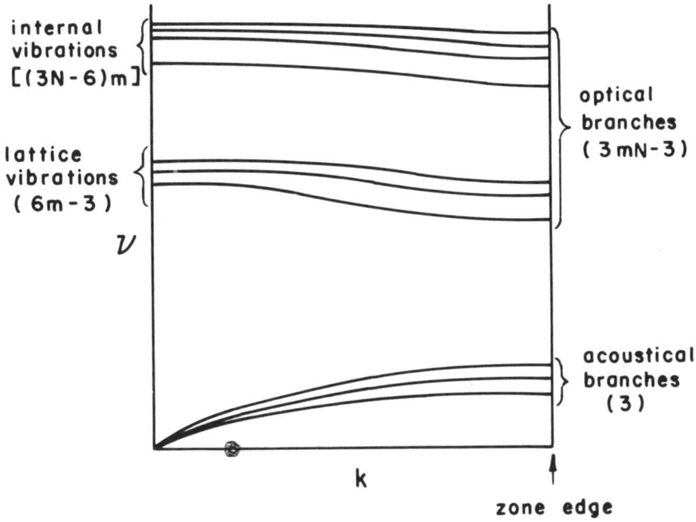

Figure 10.7. General form of the dispersion curves for a molecular crystal.

usually much stronger than the intermolecular interactions (equivalent to the $f_1 \gg f_2$ condition for the simple diatomic lattice), there will be $m(3N-6)$ fundamental modes that will correspond closely to the $3N-6$ vibrations of the molecules in the gas phase. These internal modes may have a slight dependence on k, depending on the extent of intermolecular forces.

The remaining $6m$ degrees of freedom are derived from the $3m$ translational and $3m$ rotational degrees of freedom of the m molecules in the unit cell. As mentioned earlier, three of these degrees of freedom are responsible for the acoustic modes of the crystal. This leaves $6m-3$ optical external modes, referred to as lattice vibrations. These modes are usually observed in the low frequency region of the infrared and Raman spectra.

In many cases, the vibrations of complex three-dimensional crystals cannot be described as simply longitudinal or transverse. The nature of the waves depends on the details of the intra- and intermolecular forces. However, the symmetry of the crystal can determine that in certain special directions the normal modes are strictly transverse or longitudinal, since the displacements in a given vibrational mode can be confined to a crystallographic plane. As an example, the dispersion relation of a diatomic lattice with a face-centered-cubic structure is illustrated in Figure 10.8 for wave vectors in the (100), (110), and (111) directions. For this crystal these planes correspond to the longitudinal and two transverse modes. Note that the transverse optical modes are degenerate. Such dispersion relationships are calculated as three separate linear lattice problems.

There are only three zero frequency modes in three-dimensional crystals (corresponding to the acoustic modes at $k=0$), as opposed to the six zero

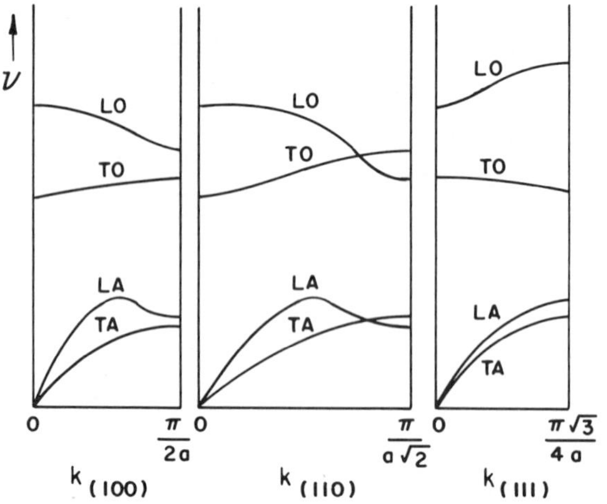

Figure 10.8. Dispersion curves (of an infinite three-dimensional diatomic lattice) for a typical face-centered-cubic crystal. Left to right, dispersion relations for phonons in (100), (110), and (111) directions. L and T denote longitudinal and transverse waves; O and A, optical and acoustic branches.

frequency modes of a finite molecule. This is a consequence of the periodic form of the solutions of the equation of motion for a three-dimensional lattice. A translation is consistent with a solution of this form, while a rotation is not. For a polymer chain, however, one zero frequency mode or degree of freedom, corresponding to a rotation about the chain axis, is allowed, providing the crystal is finite in directions perpendicular to the chain axis. Furthermore, polymer crystals can in most cases be treated as a single isolated chain because intermolecular forces are weak compared to those between atoms in the same chain. The atomic displacements of the four branches for which $\nu \rightarrow 0$ as $k \rightarrow 0$ are illustrated in Figure 10.9 for polyethylene.

In early theoretical studies of the vibrations of polymer chains, a simple one-dimensional chain was used as a model. The chain consisted of point masses arranged in a planar zigzag fashion, so that in-plane angle bending vibrational modes had to be considered. Originally, Kirkwood (6) calculated the in-plane normal vibrations using appropriate stretching and bending force constants. Pitzer (7) added out-of-plane force constants, and Krimm, Liang, and Sutherland (8) used this simple model to gain some insight into the normal modes of polyethylene. We will not discuss the details of these calculations, since they have been supplanted by an adaption of the Wilson **GF** method, as first proposed by Higgs (9). This adaptation will be considered in detail in Chapters 10 and 11.

10.4 Vibrations of a Finite Lattice

For an infinite lattice we have seen that the vibrational frequencies form continuous bands as a function of the wave vector k. In contrast, it will be

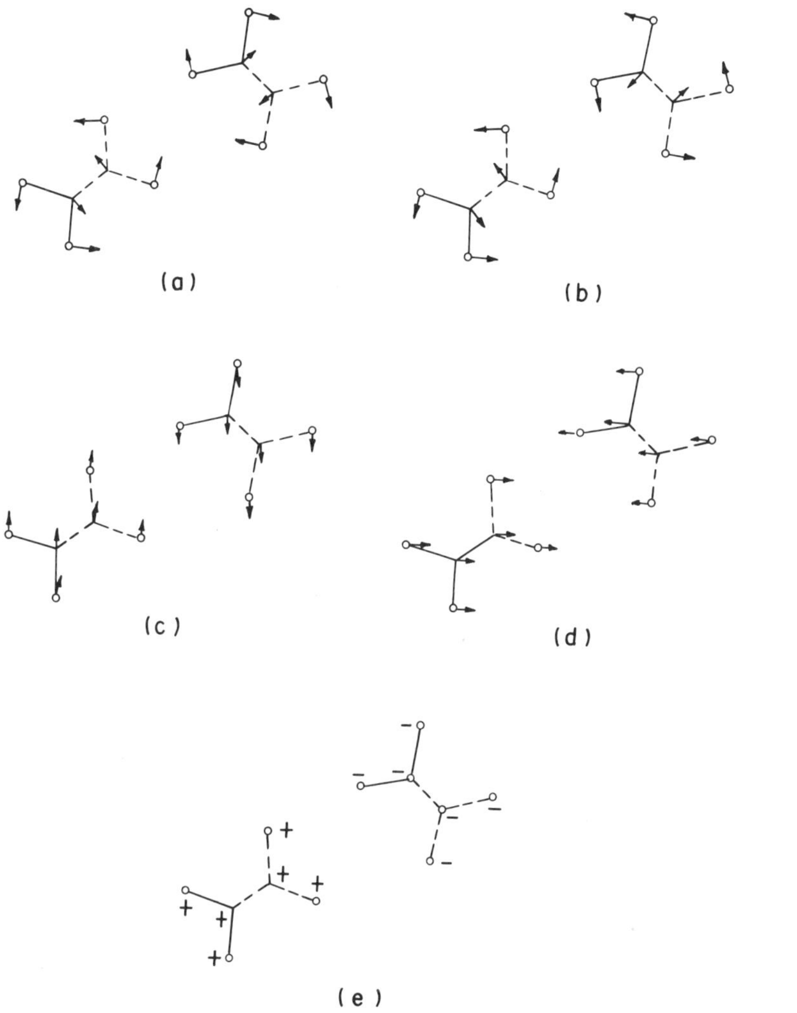

Figure 10.9. Lattice vibrations of orthorhombic polyethylene crystal. (*a*) A_g rotational mode; (*b*) B_{3g} rotational mode; (*c*) B_{1u} translational mode; (*d*) B_{2u} translational mode; (*e*) A_u translational mode. [Reprinted by permission from M. Tasumi and T. Shimanouchi, *J. Chem. Phys.* **43**(4), 1245 (1965).]

shown that the dispersion curves of a finite lattice are discontinuous, with only solutions corresponding to particular values of k being allowed. If we have a simple monatomic lattice of N point masses, then there are N homogeneous linear equations describing the vibrations of the system. In order to solve the vibrational problem of a finite chain, it is necessary to assume conditions at the boundary of the lattice. Zbinden (2) has discussed two solutions, corresponding to chains with fixed and free ends. The third boundary condition, which is usually applied to finite crystals, assumes that the initial and final particles in the chain are neighbors. This is the Born-von Karman (10) cyclic boundary

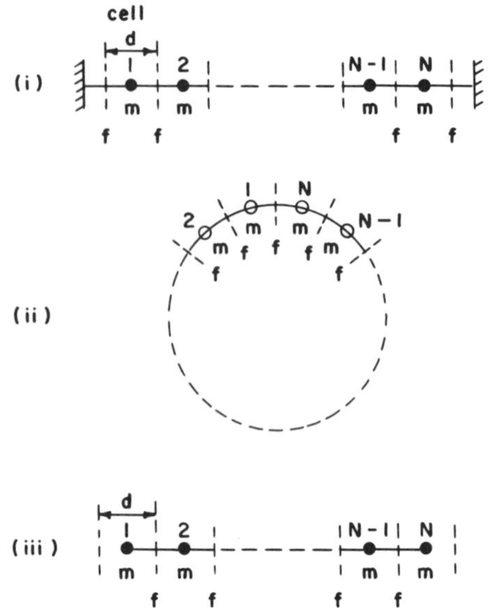

Figure 10.10. End conditions for a monatomic chain. (*i*) First and last atoms attached to a rigid wall; (*ii*) first and last atoms attached to each other (periodic boundary condition); (*iii*) first and last atoms as free ends.

condition. These three boundary conditions are illustrated in Figure 10.10. Since the boundary conditions only affect the equations of motion of the end particles, it can be intuitively perceived that as N becomes larger, the particular end condition assumed becomes less critical. We will now consider the mathematical requirements of the end conditions in more detail.

Consider a chain of N atoms, labeled $n=1,2,\ldots, N$, whose ends are fixed or attached to rigid walls, as in Figure 10.10(i). In effect, we impose the condition that atoms $n=0$ and $n=N+1$ have zero displacement, so that the boundary condition can be written

$$x_0=x_{N+1}=0 \tag{10.25}$$

If we rewrite equation (10.3) to include both halves of the Brillouin zone (i.e., to include both negative and positive values of k), the general solution becomes

$$\begin{aligned} x_n &= A^{(+)}e^{-2\pi i(\nu t-knd)}+A^{(-)}e^{-2\pi i(\nu t+knd)} \\ &= \left(A^{(+)}e^{2\pi iknd}+A^{(-)}e^{-2\pi iknd)}\right)e^{-2\pi i\nu t} \end{aligned} \tag{10.26}$$

where the two terms in these equations represent waves traveling in opposite directions along the chain. Substituting the first fixed-ends condition,

$$x_0=0$$

we obtain

$$A^{(+)}+A^{(-)}=0$$

and the general solution becomes

$$x_n = B\sin(2\pi knd)e^{2\pi i\nu t} \tag{10.27}$$

where B is a constant. Substituting the second condition, $x_{N+1}=0$, into this equation, we obtain

$$\sin[2\pi k(N+1)d]=0 \tag{10.28}$$

so that

$$2\pi k(N+1)d = 2\pi kL = s\pi \tag{10.29}$$

where $L=(N+1)d$ is the length of the chain and s is an integer. This can be rewritten in terms of the phase angle $\phi(=2\pi kd)$ to give

$$\phi = \frac{s\pi}{N+1} \tag{10.30}$$

The only values of s yielding solutions are $s=1,2,3,\ldots,N$. If $N=5$, for example, there will be five values of k allowed in the first Brillouin zone and five normal modes of vibration of the system.

The preceding derivation of the phase relationship between vibrations in adjacent unit cells based on the algebraic properties of the equations of the system can be more easily obtained from simple physical arguments. We will now apply these arguments to all three boundary conditions. The zero displacement at the wall implies that the distance between the walls is equal to an integral number of half-wavelengths in the length of the chain $L=(N+1)d$. Consequently, we can write this requirement as

$$2\pi kd(N+1)=s\pi$$

as in equation (10.29).

In the case of a chain of N atoms with **free** ends, it can be argued that the force acting at the boundaries of the end unit cells are zero (i.e., there are antinodes in the displacement function at these extremities), so that

$$2\pi kNd = s\pi \tag{10.31}$$

where s can take values of $0,1,2,\ldots,(N-1)$. The phase relationship can be written

$$\phi = \frac{s\pi}{N} \tag{10.32}$$

or if an immediate comparison to a chain with free ends is required, s can be redefined to take the values $1,2,\ldots, N$ and equation (10.32) becomes

$$\phi=\frac{(s-1)\pi}{N} \tag{10.33}$$

For a chain with cyclic boundary conditions [Figure 10.11(*iii*)] the closed system of N unit cells must contain an integral number of whole wavelengths; so

$$2\pi k N d=2\pi s \tag{10.34}$$

with $s=1,2,\ldots, N$ and the phase relationship is given by

$$\phi=\frac{2\pi s}{N} \tag{10.35}$$

This equation can be derived algebraically by imposing the condition $x_n=x_{n+N}$. This boundary conditions means that there is a traveling wave

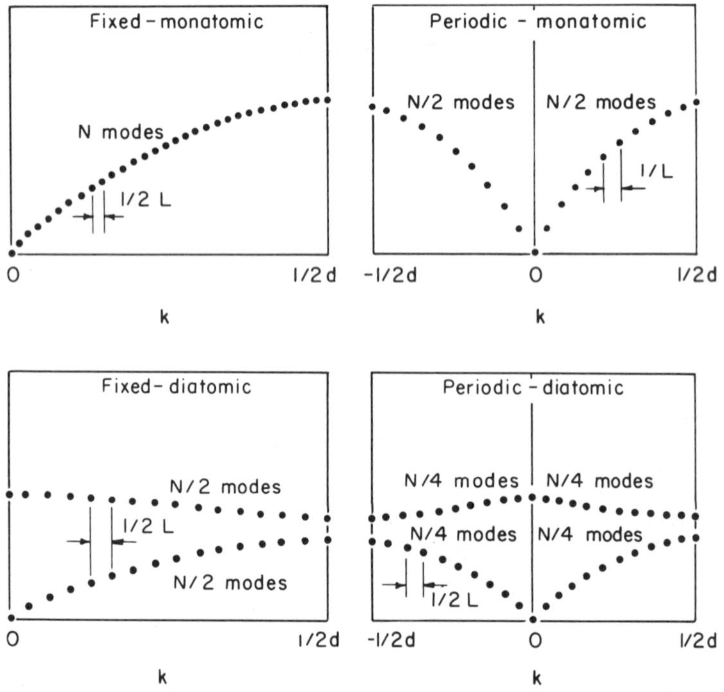

Figure 10.11. Dispersion relation for a finite linear lattice of $N=30$ atoms (longitudinal modes only). A comparison of fixed with periodic boundary conditions for the monatomic and diatomic chains.

Table 10.1 Dispersion Relationships for Finite One-Dimensional Linear Lattice

Boundary Condition	k		ϕ	Dispersion Relationship
Fixed Ends	$\dfrac{s}{2(N+1)d}$,	$s=1,2,\ldots,N$	$\dfrac{s\pi}{N+1}$	$\nu_s=\dfrac{1}{\pi}\left(\dfrac{f}{m}\right)^{1/2}\sin\dfrac{\phi}{2}$
				$=\dfrac{1}{\pi}\left(\dfrac{f}{m}\right)^{1/2}\sin\dfrac{s\pi}{N+1}$
Free Ends	$\dfrac{(s-1)}{2Nd}$,	$s=1,2,\ldots,N$	$\dfrac{(s-1)\pi}{N}$	$\nu_s=\dfrac{1}{\pi}\left(\dfrac{f}{m}\right)^{1/2}\sin\dfrac{(s-1)\pi}{N}$
Cyclic	$\dfrac{s}{Nd}$,	$s=1,2,\ldots,N$	$\dfrac{2\pi s}{N}$	$\nu_s=\dfrac{1}{\pi}\left(\dfrac{f}{m}\right)^{1/2}\sin\dfrac{2\pi s}{N}$

solution, as in the case of the infinite chain. It can be shown that the modes are doubly degenerate with the waves propagating in opposite directions for each degenerate pair. Consequently, it is sometimes chosen to specify the integer s by

$$s=0,\pm 1,\pm 2,\ldots,\pm\mu$$

where $\mu=N/2$ if N is even and $\mu=\pm(N-1)/2$ if N is odd.

To summarize, the dispersion relationship for a linear lattice is not a continuous curve, but is limited to discrete values by the boundary conditions imposed. Figure 10.11 illustrates the dispersion relationship for a lattice consisting of 30 atoms, comparing fixed and cyclic boundary conditions for monatomic and diatomic chains. (Note that for cyclic boundary conditions we have considered values of $s=\pm 1,\pm 2,\pm 3,\ldots,\pm 15$.) There is some confusion of definitions in the literature, so we have summarized the results derived here in Table 10.1.

10.5 The Distribution of Lattice Frequencies

In a polymer chain or crystal the number of repeat units is very large, so that the dispersion relation can be considered to be continuous. The frequency distribution of the normal modes, also known as the **density of vibrational states**, has often been used in the determination of the thermodynamic properties of crystals. For polymeric materials, the density of states has also been an important theoretical tool in spectroscopic studies of nonideal systems (e.g., chains with configurational or conformational defects). These systems will be discussed more fully in Chapter 15. In this section we will confine ourselves to defining the density of states.

We will define the frequency distribution function $g(\nu)$ in such a way that $g(\nu)\,d\nu$ is the number of normal modes in the range from ν to $\nu+d\nu$. A frequency distribution function defined in terms of squared frequencies can also be used, where $G(\nu^2)\,d\nu^2$ is the number of squared frequencies in the interval ν^2 to $\nu^2+d\nu^2$. The two functions are related by

$$g(\nu)=2\nu G(\nu^2) \tag{10.36}$$

If the dispersion relation has j branches, a density function $g_j(\nu)$ can be described for each branch. The normalization condition for a three-dimensional lattice consisting of n atoms in N unit cells is

$$\int_0^{\nu_c} g(\nu)\,d\nu=\sum_{j=1}^{3n}\int_0^{\nu_c} g_j(\nu)\,d\nu=3nN \tag{10.37}$$

where ν_c is the maximum frequency of the lattice. It should be noted that in many texts the distribution is normalized to unity and $g(\nu)\,d\nu$ is defined as the fraction of frequencies in the range from ν to $\nu+d\nu$.

For a simple monatomic chain of point masses it is convenient to calculate the distribution of lattice frequencies by first defining $g(k)$, the number of lattice frequencies in a unit range of the wavenumber k, or the density of states in k-space, since the vibrational states are uniformly distributed in k-space. From equation (10.29) we know that for a chain with fixed ends

$$k=\frac{s}{2L} \tag{10.38}$$

where $s=1,2,\ldots,N$. Consequently, there is one vibrational mode in each interval $1/2L$ in k-space. Hence,

$$g(k)=\begin{cases} 2L & \text{for} \quad k<\dfrac{1}{2d} \\ 0 & \text{for} \quad k>\dfrac{1}{2d} \end{cases} \tag{10.39}$$

Similarly, for a cyclic lattice we can use equation (10.32) to obtain an analogous result:

$$g(k)=\begin{cases} L & \text{for} \quad \dfrac{-1}{2d}<k<\dfrac{1}{2d} \\ 0 & \text{otherwise} \end{cases} \tag{10.40}$$

The number of frequencies in the interval ν to $\nu+d\nu$ can be written

$$g(\nu)\,d\nu=g(k)\frac{dk}{d\nu}d\nu=g(k)\left(\frac{d\nu}{dk}\right)^{-1}d\nu \tag{10.41}$$

The dispersion relation can be used to obtain $d\nu/dk$ directly. For positive k,

$$\nu = \frac{1}{\pi}\left(\frac{f}{m}\right)^{1/2} \sin \pi k d = \nu_c \sin \pi k d$$

which can be arranged to give

$$k = \frac{1}{\pi d} \sin^{-1}\left(\frac{\nu}{\nu_c}\right) \tag{10.42}$$

Hence,

$$\frac{dk}{d\nu} = \frac{1}{\pi d} \frac{1}{\left(\nu_c^2 - \nu^2\right)^{1/2}} \tag{10.43}$$

and

$$g(\nu) = \frac{2L}{\pi d} \frac{1}{\left(\nu_c^2 - \nu^2\right)^{1/2}} \tag{10.44}$$

for $k < 1/2d$.

This result is illustrated in Figure 10.12. It can be seen that this density of states tends to infinity near ν_c, where $d\nu/dk$ approaches zero (i.e., where the dispersion curve is horizontal, near the Brillouin zone boundary). Such points on the dispersion curve are known as **critical points** and the resulting features in the density of states are **singularities**. These features are extremely important

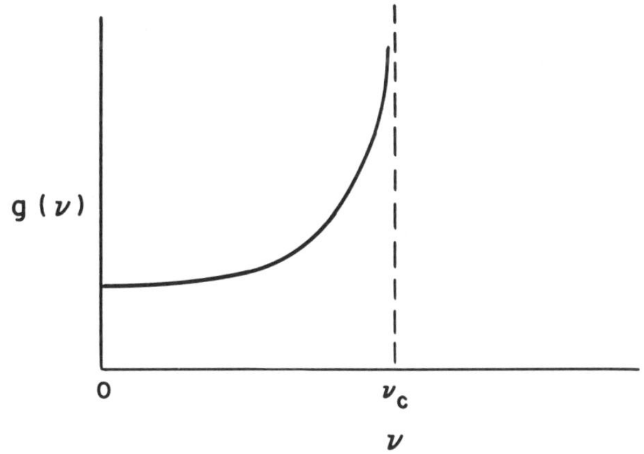

Figure 10.12. The frequency distribution or density of states for a monatomic chain.

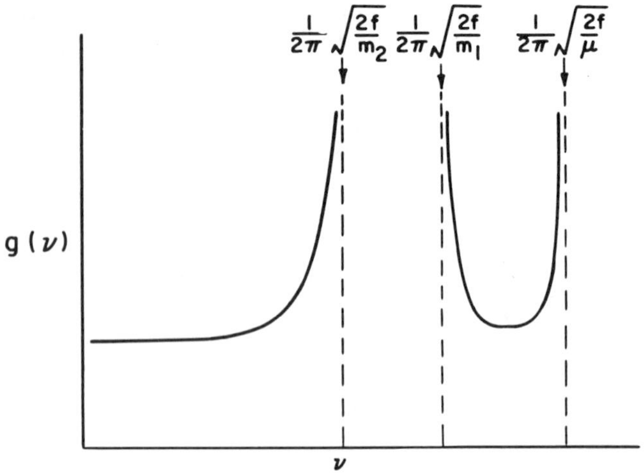

Figure 10.13. The frequency distribution or density of states for the diatomic chain.

in the interpretation of the spectra of disordered polymers and multiphonon effects (combinations and overtones) in the spectra of polymer crystals.

For a simple diatomic lattice the frequency distribution is split into two parts, corresponding to the acoustic and optical modes, as shown in Figures10.13. The optical branch has singularities at each limiting frequency determined by the zone boundaries.

There have been a number of calculations of the frequency distribution of three-dimensional lattices. These calculations are much more complex than for the simple one-dimensional lattice mentioned earlier. However, Maradudin et al. (11) have noted that there is one representation of the relation between $g_j(\nu)$ and $\nu_j(k)$ from which all others have been derived. Let $n(\nu)$ be the number of modes whose frequencies are less than or equal to ν. Then,

$$n(\nu)=\int_0^{\nu}\sum_{k,j}\delta(x-\nu_j(k))\,dx \tag{10.45}$$

where $\delta(t)$ is the Dirac delta function. It is not a function in the usual sense, but an example of a generalized function or distribution. The delta function $\delta(t)$ is zero everywhere except where $t=0$, or in equation (10.45), when $x-\nu_j(k)=0$. In effect, the integral over the sum of the δ function adds 1 to the right-hand side of the expression each time the integration variable crosses one of the allowed values of $\nu_j(k)$ and so "counts" the number of modes $n(\nu)$. The distribution function is related simply to $n(\nu)$ by

$$g(\nu)\,d\nu=[n(\nu+d\nu)-n(\nu)] \tag{10.46}$$

so that

$$g(\nu)=\frac{d}{d\nu}n(\nu) \tag{10.47}$$

Then

$$g(\nu)=\sum_{k,j}\delta(\nu-\nu_j(k)) \tag{10.48}$$

Relations between $g(\nu)$ and $\nu_j(k)$ are then found by substituting expressions for the delta function in equation (10.48).

This formalism has recently been introduced into polymer spectroscopy (12). However, most determinations of the density of states of polymeric materials use either graphical methods for simple polymers (13) or the negative eigenvalue theory for more complex macromolecular systems (14).

10.6 Phonons; the Quantization of Lattice Vibrations

In discussing the vibrations of chain molecules it has been sufficient to use classical mechanics, since in the harmonic approximation the classical vibrational frequencies correspond to those in the quantum mechanical treatment. Nevertheless, quantum mechanics must be used to give a correct description and specifically requires the energy to be quantized:

$$E_i=h\nu_i(n+\tfrac{1}{2}), \qquad n=0,1,2,\ldots \tag{10.49}$$

For chain molecules the frequency ν_i is a function of the wave vector k. Consequently, by analogy with the term "photon," used to describe a quantum of electromagnetic energy, the term "phonon" is used to describe a quantum of lattice vibrational energy, $h\nu_i$. Since phonons have effective momentum, represented by hk, their interactions with photons of electromagnetic radiation, or particles such as neutrons, are governed by the laws of conservation of momentum. This principle is used in inelastic neutron scattering where an experimental determination of the energy of the scattered neutrons as a function of direction can be used to determine the phonon dispersion in polymers.

10.7 The Coupled Oscillator Model

In the preceding sections we have demonstrated how the classical mechanical methods discussed in Chapter 2 can be applied to determining the normal modes of a simple linear lattice. It will be recalled, however, that even for the case of linear CO_2 it was necessary to introduce interaction force constants to account for the coupling between the carbon–oxygen bond stretching vibrations. Obviously, such mechanical coupling will also occur in complex polymeric materials. Consider the C—C—C angle bending vibration in polyethylene. Displacements involve the motion of carbon atoms in adjacent cells, and it is therefore a highly coupled mode. Coupling can occur through other mechanisms. For example, dipole-dipole interactions between the C=O group in polypeptides have been used to interpret the band splitting of the amide I mode. (See Chapter 14.) In this section we will consider a simple coupled oscillator model for the vibrations of a polymer chain. Even though this approach does not give the insight of a complete normal coordinate analysis, it has proved extremely useful in the interpretation of band splittings and progressions.

The coupled oscillator model gives a result of the same type as the direct method of solving the Newtonian equations of motion; that is, the frequency of vibration of the nth atom or nth oscillator in a chain is given by an expression of the following form:

$$\nu^2 = A + B\cos\phi \tag{10.50}$$

A simple coupled oscillator system is illustrated in Figure 10.14. The model essentially consists of a diatomic harmonic oscillator with a force constant or bond stiffness of f, coupled by an interaction force constant f'. This model corresponds to the diatomic chain problem, discussed earlier, with $f > f'$. It will be recalled that the secular determinant is of the form

$$\begin{vmatrix} f+f'-4\pi^2\nu^2 m_1 & -\left(fe^{\pi ikd}+f'e^{-\pi ikd}\right) \\ -\left(fe^{-\pi ikd}+f'd^{\pi ikd}\right) & f+f'-4\pi^2\nu^2 m_2 \end{vmatrix} = 0 \tag{10.51}$$

Note that in this equation the force constants are defined in terms of the cartesian displacement coordinates. The dispersion relation and displacements can be derived as detailed earlier and given in Zbinden (2). However, it is instructive to demonstrate again how the form of the secular equation can be simplified by defining the potential function in terms of internal coordinates and using the Wilson **GF** matrix method. If we now let f represent the force constant associated with the stretching of the bond linking the atoms, then the

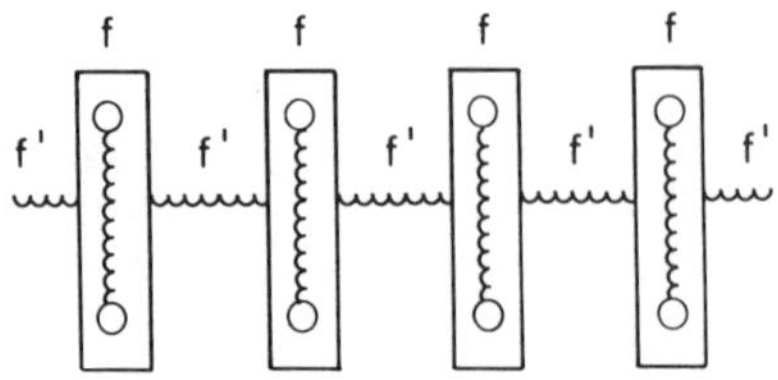

Figure 10.14. Schematic representation of a chain of coupled oscillators.

isolated oscillator frequency ν_0 is given by

$$4\pi^2\nu_0^2 = gf \tag{10.52}$$

In this simple system the g matrix is equal to $1/\mu$, where μ is the reduced mass, since for expository purposes there is no need to define any **B** matrix terms related to the geometry of the system. If we now consider the oscillators to be coupled by a force constant f', then the **F** and **G** matrices of the infinite chain of coupled oscillators are

$$\mathbf{F} = \begin{bmatrix} ff' & & & & \\ & f'ff' & & & \\ & & f'ff' & & \\ & & & f'ff' & \\ & & & & \ddots \end{bmatrix} \tag{10.53}$$

$$\mathbf{G} = \begin{bmatrix} g & & & & \\ & g & & & \\ & & g & & \\ & & & g & \\ & & & & \ddots \end{bmatrix} \tag{10.54}$$

It is convenient to write the **F** matrix in the form

$$\mathbf{F} = \mathbf{f} + \mathbf{f}'e^{-2\pi ikz} + \mathbf{f}'e^{2\pi ikz} \tag{10.55}$$

a result that is based on the translational symmetry of the infinite chain (see Chapter 11), which by convention is parallel to the z axis of a cartesian system. An expression of a similar form was obtained in the diatomic chain problem in cartesian displacement coordinates through the assumption of periodic solutions in time and space. The dispersion relation is then given by

$$\begin{aligned} 4\pi^2\nu^2 = \mathbf{GF} &= \mathbf{g}(\mathbf{f} + \mathbf{f}'e^{-2\pi \mathbf{ikz}} + \mathbf{f}'e^{2\pi \mathbf{ikz}}) \\ &= \mathbf{g}(\mathbf{f} + 2\mathbf{f}'\cos\phi) \end{aligned} \tag{10.56}$$

where $\phi = 2\pi kz$ is the phase shift between vibrations of adjacent oscillators. The dispersion relation for this system is shown in Figure 10.15. The frequency range for the normal modes, $g(f+2f')$ to $g(f-2f')$, is called the **lattice band**.

Equation (10.56) demonstrates that for the infinite chain ν is a continuous function of ϕ. In the case of the finite chain, however, ν is limited to discrete values. By imposing boundary conditions for a linear chain of N oscillators (as

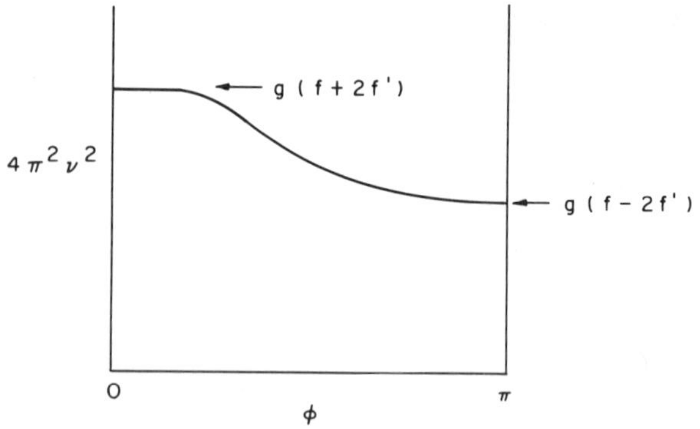

Figure 10.15. Dispersion relationship for a chain of coupled oscillators.

in Section 10.5) it can be shown that for a chain with fixed ends

$$\phi = \frac{s\pi}{N+1} \tag{10.57}$$

while for a chain with free ends

$$\phi = \frac{(s-1)\pi}{N} \tag{10.58}$$

where $s = 1, 2, \ldots, N$; that is, a chain of N oscillators has N discrete frequencies.

Similarly, the internal displacements of the nth oscillator r_n are given by

$$r_n = A_s(-1)^n \sin\left(\frac{sn\pi}{N+1}\right) \tag{10.59}$$

for a chain with fixed ends, while for a chain with free ends

$$r_n = A_s(-1)^r \sin\left[\frac{s(2n-1)\pi}{2N}\right] \tag{10.60}$$

where A_s is the normal coordinate associated with the normal vibration ν_s.

The expressions derived here are for a chain of **parallel** coupled oscillators. An example of a system where these equations could apply is the C=O stretching vibration (if such a polymer exists) of nylon-1, shown in Figure 10.16*a*. However, if we consider nylon-2, Figure 10.16*b*, then the chain of coupled oscillator consists of a set of antiparallel dipoles. The equations of motion of such a system are the same (apart from a change in the signs) as for the set of parallel dipoles. The dispersion relation is given by

$$4\pi^2\nu^2 = g(f - 2f'\cos\phi) \tag{10.61}$$

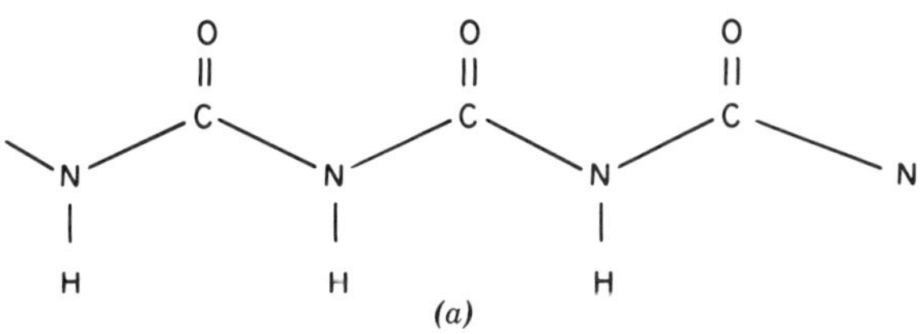

(a)

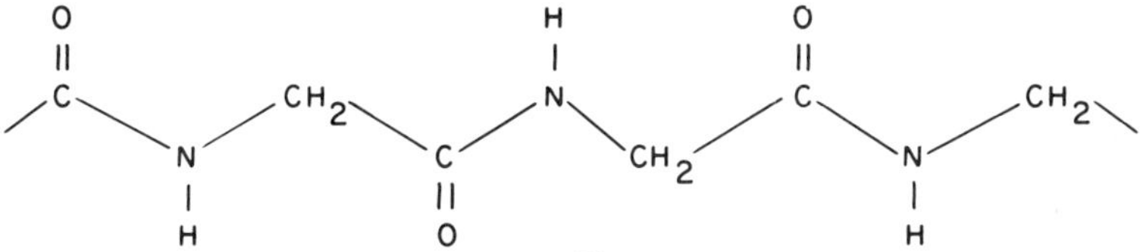

(b)

Figure 10.16. (*a*) Nylon-1, an example of a chain of parallel coupled oscillators. (*b*) Nylon-2 (polyglycine), an example of a chain of antiparallel coupled oscillators.

10.8 Progression Band Series and the Observed Infrared Intensity Distribution

Homologues of polymeric materials, particularly chain molecules of finite length, have played a central role in determining transferable force fields and "mapping" the polymer dispersion curve. The most elegant use of a series of oligomers is undoubtedly the studies of *n*-paraffins reported by Snyder (15) and Snyder and Schachtschneider (16), which culminated in the calculation of a valence force field of saturated hydrocarbons (17, 18). In this section we will show how the simple coupled oscillator theory can play a role in the interpretation of the characteristic features of the vibrational spectra of chain molecules. Most of the work reported in the literature has involved the infrared spectrum, and we will concentrate our discussion on this area. However, corresponding arguments also apply to the Raman spectra of these materials.

In order to develop our argument let us consider the case of a chain of five parallel oscillators with fixed ends. We have demonstrated that the five normal frequencies of this system are related by the expression

$$4\pi^2\nu^2 = gf + 2gf'\cos\phi \tag{10.62}$$

where

$$\phi = \frac{s\pi}{N+1}, \qquad s = 1, 2, \ldots, 5 \tag{10.63}$$

These five frequencies give five points, which will fall on the continuous dispersion curve of the theoretically infinite polymer chain, in which ϕ can take

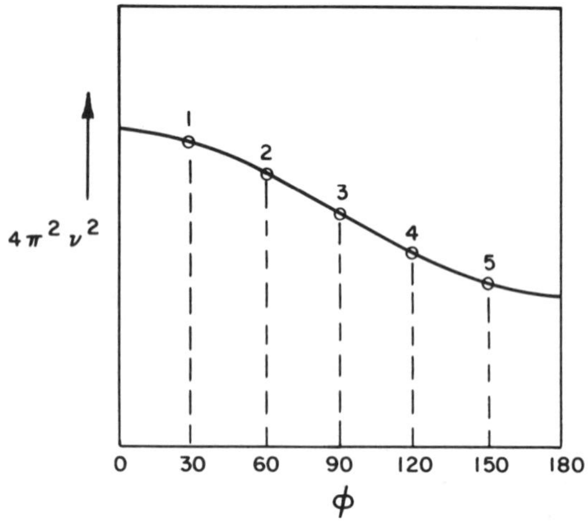

Figure 10.17. Dispersion relation for a chain of parallel coupled oscillators.

any value. This situation is illustrated in Figure 10.17. The form of the normal modes can be determined from

$$r_n = A_s(-1)^n \sin\left(\frac{sn\pi}{N+1}\right) \tag{10.64}$$

The displacements are illustrated in Figure 10.18.

The intensity of infrared absorption for each of these modes is proportional to the square of the change of the total dipole moment during the vibration.

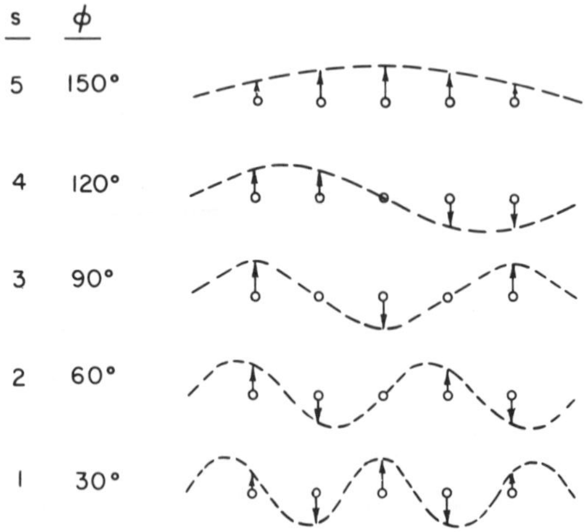

Figure 10.18. Form of the normal modes for a finite chain of five coupled parallel oscillators.

The dipole moment change for a specific oscillator depends on the displacements r_n given by equation (10.64), and the total dipole moment change is proportional to the vector sum of the r_n values. Thus by inspecting Figure 10.18 it is immediately apparent that the intensities of the bands corresponding to $s=2$ and $s=4$ will be zero, since the vector sum of the displacements is zero. The following arguments can be used to obtain quantitative expressions for the relative intensities of bands in a chain of coupled oscillators (2). If we let ΔM_s be the *relative* change in the dipole moment during a particular normal vibration, then this can be put equal to the sum of the displacements r_n.

$$\Delta M_s = \sum_{n=1}^{N} r_n \tag{10.65}$$

for a chain of antiparallel dipoles

$$\Delta M_s = \sum_{n=1}^{N} (-1)^n r_n \tag{10.66}$$

Substituting equation (10.64) and removing the constant factor A_s (since we are interested only in predicting relative values of the intensities), we obtain for parallel dipoles

$$\Delta M_s = \sum_{n=1}^{N} (-1)^n \sin\left(\frac{sn\pi}{N+1}\right) = \begin{cases} 0 & \text{if } s+N \text{ is odd} \\ -\tan\left[\dfrac{s\pi}{2(N+1)}\right] & \text{if } s+N \text{ is even} \end{cases} \tag{10.67a}$$

Similarly, for antiparallel dipoles,

$$\Delta M_s = \sum_{n=1}^{N} \sin\left(\frac{sn\pi}{N+1}\right) = \begin{cases} 0 & \text{if } s \text{ is even} \\ \cot\left(\dfrac{s\pi}{N+1}\right) & \text{if } s \text{ is odd} \end{cases} \tag{10.67b}$$

Corresponding expressions can be determined for a chain with free ends. Zbinden (2) has argued that the absorption is proportional not only to ΔM_s^2 but also to the square of the frequency ν_s^2, or the circular frequency $\omega_s^2 (= 4\pi^2\nu^2)$, thus treating the absorption as a forced vibration of the system. However, since the width of an observed absorption band (equal to the width of the lattice band in the dispersion curve and proportional to the size of the interaction force constant) is in most cases realtively narrow, ω_s^2 is approximately constant. Writing the **relative** band intensity I_s as

$$I_s = \omega_s^2 \Delta M_s^2 \tag{10.68}$$

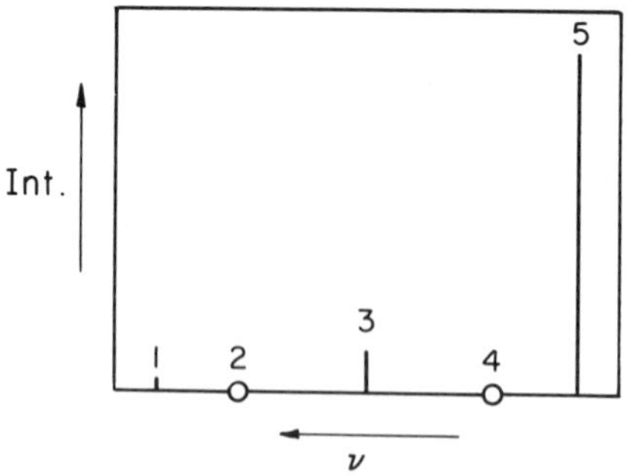

Figure 10.19. Relative intensities of the normal modes of a chain of five coupled parallel oscillators.

we obtain for parallel dipoles

$$I_s=0 \qquad \text{if } s+N \text{ is odd} \tag{10.69}$$

$$I_s=\omega_s^2\tan^2\left[\frac{s\pi}{2(N+1)}\right] \qquad \text{if } s+N \text{ is even} \tag{10.70}$$

and for antiparallel dipoles

$$I_s=0 \qquad \text{if } s \text{ is even} \tag{10.71}$$

$$I_s=\omega_s^2\cot^2\left[\frac{s\pi}{2(N+1)}\right] \qquad \text{if } s \text{ is odd.} \tag{10.72}$$

The expected infrared absorption intensity distribution in our model of five coupled antiparallel oscillators, as determined from these equations, is shown in Figure 10.19. It can be seen that the lowest frequency band, corresponding to $s=5$, is by far the strongest in the spectrum. In addition, as the chain length increases, $\phi=s\pi/[2(N+1)]$ will tend to zero and the frequency of this band asymptotically approaches the frequency of the theoretical infinite perfect chain. This model works very well in the interpretation of the band progressions of the CH_2 rocking modes of *n*-paraffins (16). The infrared spectrum of *n*-$C_{24}H_{50}$ is shown in Figure 10.20. The relative intensities of the series fit the predictions outlined earlier for a set of parallel dipoles; that is, the lowest frequency band corresponding in this case to $s=1$, is the most intense.

It has been shown that for the *n*-paraffins a better fit of the observed frequencies can be obtained by using the fixed-end model as opposed to the free-end model. The frequencies of the methylene rocking mode for the *n*-paraffins C_3H_8 through *n*-$C_{30}H_{62}$, as measured by Snyder and Schachtschneider (16), are displayed as an array in Figure 10.21. Note that the most intense $s=1$ modes quickly approach the asymptotic value characteristic of polyethylene. These observed frequencies ν_k are plotted against ϕ/π in Figure 10.22. Although the data for the longer molecules tend to a common curve, data for the shorter chains are displaced. This is due to an interaction between the methylene rocking modes and methyl end groups rocking modes.

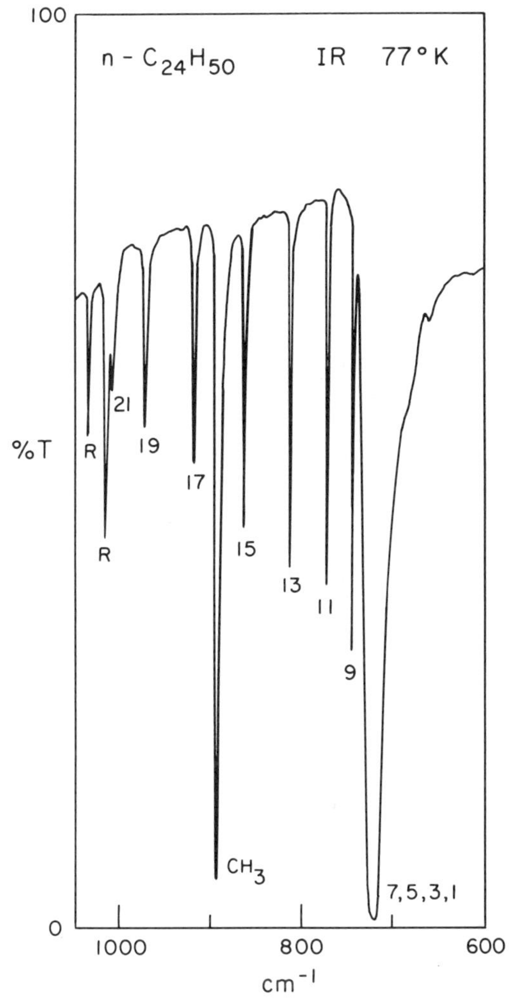

Figure 10.20. Observed infrared spectrum of $n-C_{24}H_{50}$ between 1000 and 600 cm^{-1}.

In addition, it should be noted that the upper portion of this dispersion curve corresponds to the CH_2 twisting mode, so that the rocking and twisting modes are mixed, except in the limit of the infinite chain where the selection rules are strict. It is apparent that in order to interpret fully such nuances of coupling, a complete normal coordinate analysis is required. Nevertheless, the simple coupled oscillator model gives substantial insight and greatly aids the interpretation and classification of the observed frequencies of a series of homologues. In the limit of very short and very long chains this approximation is less useful; in the former case because of the possibility of extensive coupling between the modes of the repeat unit and the end groups, and in the latter case because of the impossibility of resolving the bands, which makes detection and assignment of modes extremely difficult.

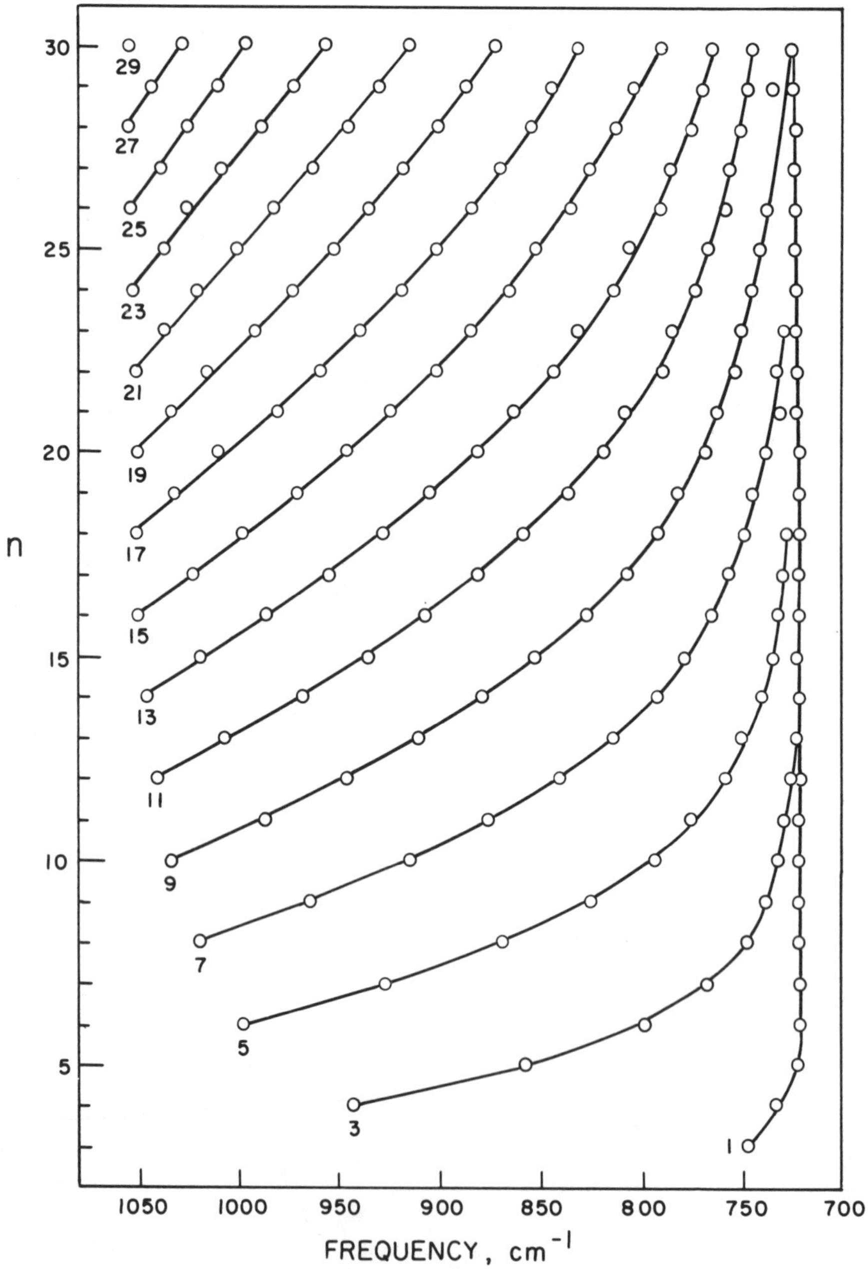

Figure 10.21. Methylene rocking and twisting mode array for *n*-paraffins C_3H_8 through $n-C_{30}H_{62}$ (*s* values at left) (Reprinted by permission from ref. 16.)

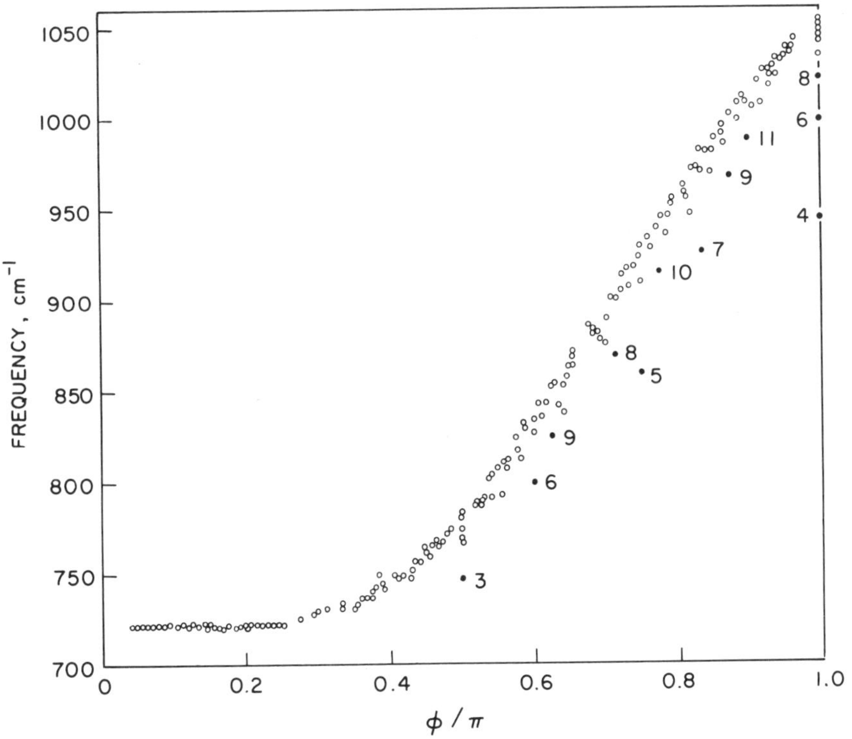

Figure 10.22. Frequency–phase curve for methylene rocking and twisting modes, C_3H_8 through n–$C_{30}H_{62}$ ($3 = C_3H_8$, etc.). (Reprinted by permission from ref. 16.)

References

1 G. Turrell, "Infrared and Raman Spectra of Crystals," Academic, New York, 1972.

2 R. Zbinden, "Infrared Spectroscopy of High Polymers," Academic, New York, 1964.

3 P. M. A. Sherwood, "Vibrational Spectroscopy of Solids," Cambridge University Press, Cambridge, Massachusetts, 1972.

4 M. J. P. Musgrave, "Crystal Acoustics," Holden-Day, San Francisco, 1970.

5 J. E. Lynch, Jr., Ph.D. Thesis, University of Michigan, Ann Arbor, 1968.

6 J. G. Kirkwood, *J. Chem. Phys.* **7**, 506 (1939)

7 S. Pitzer, *J. Chem. Phys.* **8**, 711 (1940).

8 S. Krimm, C. Y. Liang, and G. B. B. M. Sutherland, *J. Chem. Phys.* **25**, 549 (1956).

9 P. W. Higgs, *Proc. Roy. Soc.* (*London*) **A220**, 470 (1953).

10 M. Born and Th. von Karman, *Phys. Z.* **13**, 297 (1912).

11 A. A. Maradudin, E. W. Montroll, and G. H. Weiss, "Theory of Lattice Dynamics in the Harmonic Approximation," Academic, New York, 1963.

12 V. N. Kozyrenko, I. V. Kumpanerko, and I. D. Mikhailov, *J. Polymer Sci. Polymer Phys. Ed.* **15**, 1721 (1977).

13 L. Piseri and G. Zerbi, *J. Chem. Phys.* **48**, 3561 (1968).
14 G. Zerbi, *Appl. Spect. Revs.* **2**, 193 (1969).
15 R. G. Snyder, *J. Mol. Spect.* **4**, 411 (1960).
16 R. G. Snyder and J. H. Schachtschneider, *Spectrochim. Acta* **19**, 85 (1963).
17 J. H. Schachtschneider and R. G. Snyder, *Spectrochim. Acta* **19**, 117 (1963).
18 R. G. Snyder and J. H. Schachtschneider, *Spectrochim. Acta* **21**, 169 (1965).

11

SYMMETRY ANALYSIS OF POLYMER CHAINS AND CRYSTALS

Symmetry is tedious, and tedium is the very basis of mourning. Despair yawns.

Victor Hugo

We have seen that the symmetry operations that define the structure of nonpolymeric free molecules form a mathematical group, and all such molecules can be assigned to one of a limited number of point groups on this basis. The methods of group theory can then be used to classify and determine the activity of the normal modes of vibration according to the symmetry species of the point group. A corresponding treatment can be applied to an ordered polymer chain, considered as a one-dimensional crystal, or indeed to polymer crystals. This symmetry analysis depends on the assumption of translational symmetry of the unit cell, and modes are classified according to line group or space group symmetry. This type of analysis is particularly useful in polymer vibrational spectroscopy because for most macromolecules an oriented sample can be obtained. This analysis in turn allows a symmetry classification of the experimentally observed infrared bands and Raman lines according to dichroism and polarization measurements, respectively. Furthermore, in terms of normal coordinate calculations the application of symmetry arguments is a key factor in reducing the dimensions of the problem to a consideration of the normal modes of the repeat unit. It is the purpose of this chapter to examine in some depth the application of symmetry to the vibrational analysis of polymeric molecules. In order to do this, however, we first have to consider certain symmetry operations associated with chain molecules and three-dimensional crystals that were not defined in our earlier discussion of point groups, together with some further aspects of group theory.

11.1 Screw Axes and Glide Planes

In addition to the symmetry operations defined in Chapter 4, we have to define two new operations to account for the translational symmetry of ordered chain

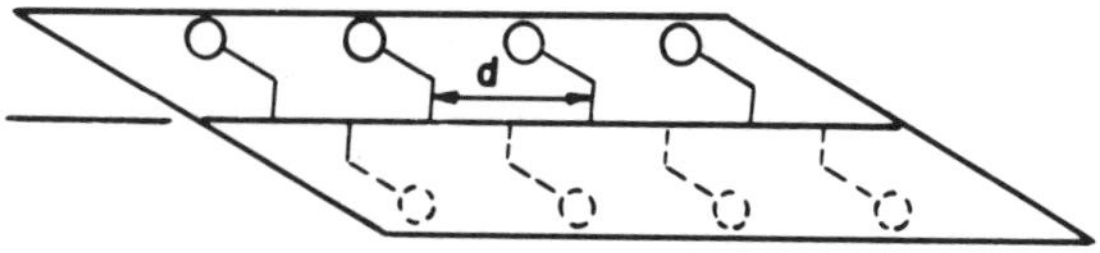

Figure 11.1. Illustration of a screw (rotation-translation) symmetry operation, 2_1.

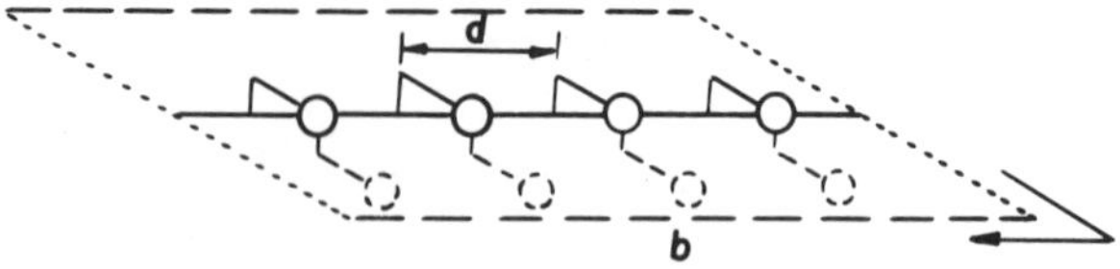

Figure 11.2. Illustration of a glide (reflection-translation) operation.

molecules and crystals. These operations are a rotation followed by a translation, or a screw axis; and a reflection followed by a translation, or a glide plane.

The screw rotation is a clockwise rotation about an axis by an angle $\phi = 2\pi/n$, where n can have values of 1, 2, 3, 4, or 6, followed by a translation in the direction of the axis. This operation is illustrated in Figure 11.1 and is given the symbol $\overline{C}_p$.

The glide reflection is illustrated in Figure 11.2 and consists of a reflection in a plane followed by translation along a direction parallel to this plane. For isolated polymer chains only translations along the axis of the polymer chain, by convention the z axis, are allowed, and the symbols $\bar{\sigma}_v'$, $\bar{\sigma}_v''$ are used to denote glide reflections at the yz and the xz plane, respectively, followed by translation along the z axis. These symmetry operations apply only to what we will define as infinite or cyclic groups.

11.2 Isomorphic and Cyclic Groups

In discussing the symmetry properties of ordered polymer chains we will find the concepts of cyclic groups and isomorphic groups extremely useful. The general properties can be illustrated by a multiplication table (Table 11.1) for a

Table 11.1 General Example of the Multiplication Table of a Cyclic Group

	1	i	-1	$-i$
1	1	i	-1	$-i$
i	i	-1	$-i$	1
-1	-1	$-i$	1	i
$-i$	$-i$	1	i	-1

Table 11.2 Multiplication Table for the Cyclic Group C_4

	E	C_4	$C_4^2(C_2)$	C_4^3
E	E	C_4	C_2	C_4^3
C_4	C_4	C_2	C_4^3	E
$C_4^2(C_2)$	C_2	C_4^3	E	C_4
C_4^3	C_4^3	E	C_4	C_2

group of order four consisting of the elements 1, i, -1, $-i$. Any element of this group multiplied by itself is another element of the group. The **period** of any of the elements of a finite group is defined as the power to which it must be raised in order to generate the identity element. For example, the period of 1 is 1, since this is the identity element. The period of both i and $-i$, however, is 4. These two elements both have the property that their period is equal to the order of the group, and the four powers of either of these elements also generate the whole group. A group containing any such elements is said to be cyclic. Note that the elements 1 and -1 are not capable of generating the whole group.

Cyclic groups must also be Abelian (every element is a class in itself). Consequently, the number of irreducible representations is equal to the order of the group and the representations all have one dimension. The representations of any cyclic group can be determined by a simple formula which we will discuss later when we derive the character table for a one-dimensional translation group.

The symmetry group of rotation by $2\pi/4$ (C_4) is also a cyclic group. Compare the multiplication table of this group (shown in Table 11.2) to the simple example illustrated in Table 11.1. If we let E correspond to 1, C_4 to i, C_2 to -1, and C_4^3 to $-i$, then these two groups have the same multiplication table. If two groups have the same multiplication table, they are said to be **isomorphic** and constitute only one abstract group. We will use the converse of this (i.e., if two groups are isomorphic, they have the same abstract multiplication table and the same character table) in our symmetry analysis of polymer chains and crystals.

11.3 The Irreducible Representations of a Lattice Translation Group

If we are to use symmetry in our analysis of polymer molecular vibrations we must have available a character table for molecules with translational symmetry. These tables are simply constructed.

Consider a linear infinite chain of identical atoms, such as that used as a simple model of a polymer chain in Chapter 3. This chain has translational symmetry in that the chain coincides with itself after a translation of nt, where

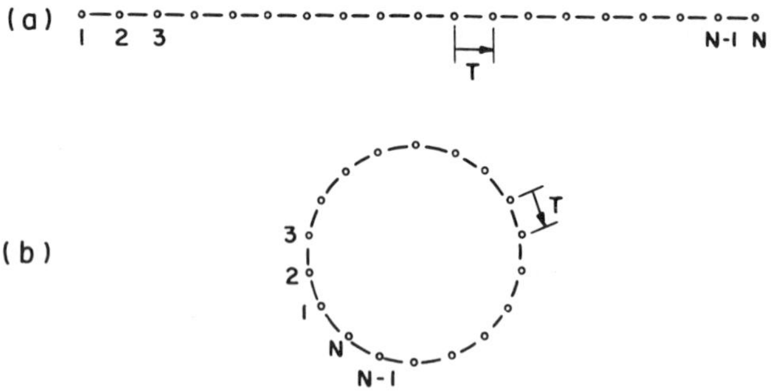

Figure 11.3. (*a*) One-dimensional lattice of N repeat units. (*b*) Cyclic boundary conditions for a chain of N repeat units (T is the lattice vector).

t is a unit vector along the chain axis and n is any integer. These symmetry operations,

$$T = nt \tag{11.1}$$

for a group of infinite order, since $-\infty \leqslant n \leqslant +\infty$. However, real polymer chains are finite and a translation is no longer a "perfect" symmetry operation because units at one edge of the chain are no longer replaced. This difficulty is circumvented by assuming Born's cyclic boundary conditions. If the chain has N units (with N very large), then a translation T_N is considered equivalent to the unit element of the translation group T_0, or $Nt = E$. This condition is illustrated in Figure 11.3. The translational symmetry operations now form a cyclic group of order N, which must therefore also be Abelian (see Section 11.2). Consequently, this group will have N one-dimensional irreducible representations.

Similarly, for a three-dimensional lattice we can write any translation T_{n_1,n_2,n_3} as

$$T_{n_1,n_2,n_3} = n_1t_1 + n_2t_2 + n_3t_3 \tag{11.2}$$

and by imposing the cyclic boundary conditions

$$N_1t_1 + N_2t_2 + N_3t_3 = E \tag{11.3}$$

can derive an Abelian group of order $N_1N_2N_3$, again equal to the number of irreducible representations. As we mentioned earlier, all of the irreducible representation matrices are one-dimensional or simple numbers. These characters, which can be written

$$\Gamma^{(\gamma)} = \chi_{100}^{(\gamma)}, \chi_{200}^{(\gamma)}, \ldots, \chi_{N_100}^{(\gamma)}; \ldots : \chi_{N_1,N_2,N_3} \tag{11.4}$$

obey the same rules of multiplication as the group operations (outlined in the preceding section):

$$\chi^{(\gamma)}_{n_1 n_2 n_3}\chi^{(\gamma)}_{l_1 l_2 l_3}=\chi_{n_1+l_1,\,n_2+l_2,\,n_3+l_3} \qquad (11.5)$$

In addition, the character of the first symmetry class or identity is equal to unity

$$\chi^{(\gamma)}_{000}=\chi^{(\gamma)}_{N_1 00}=\chi^{(\gamma)}_{00N_3}=\cdots=\chi^{(\gamma)}_{N_1 N_2 N_3}=1 \qquad (11.6)$$

These two conditions determine the characters of the irreducible representations (1,2), given by

$$\chi^{(k)}_T=\exp(2\pi i k T) \qquad (11.7)$$

where T is now the three-dimensional translation unit vector T_{n_1, n_2, n_3} defined in equation (11.3) and k is the wave vector, used in Chapter 10 to describe periodic solutions of the vibration of a chain. Under the Born cyclic boundary conditions the wave vector can be written in the form

$$k=\frac{p_1}{N_1}b_1+\frac{p_2}{N_2}b_2+\frac{p_3}{N_3}b_3 \qquad (11.8)$$

where p_1, p_2, and p_3 are integers running from 0 to N_1-1, N_2-1, N_3-1, respectively, and b_1, b_2, b_3 are the reciprocal lattice vectors, defined by

$$b_i t_j=\delta_{ij} \qquad (11.9)$$

Consequently, equation 11.7 represents a set of $(N_1 N_2 N_3)^2$ numbers that are the entries in the character table for this cyclic group. Naturally, the character table is simplified if we confine our attention to the one-dimensional lattice where the translation group is of order N_1. There are then N_1^2 characters,

Table 11.3 Character Table for the One-Dimensional Translation Group

	$T_0=E$	T_1	$T_2,\ldots,T_{N-1}$	
$\Gamma^{(0)}$	1	1	$1,\ldots,1$	
$\Gamma^{(1)}$	1	$\exp\left(2\pi i\frac{1}{N}\right)$	$\exp\left(2\pi i\frac{2}{N}\right),\ldots,\exp\left[2\pi i\frac{(N-1)}{N}\right]$	
$\Gamma^{(2)}$	1	$\exp\left(2\pi i\frac{2}{N}\right)$	$\exp\left(2\pi I\frac{4}{N}\right),\ldots,\exp\left[2\pi i\frac{2(N-1)}{N}\right]$	
$\Gamma^{(N-1)}$	1	$\exp\left(2\pi i\frac{N-1}{N}\right)$	$\exp\left[2\pi i\frac{2(N-1)}{N}\right],\ldots,\exp\left[2\pi i\frac{(N-1)^2}{N}\right]$	

given by

$$\chi_n^k = \exp\left[2\pi i\left(\frac{pbnt}{N}\right)\right] = \exp\left(\frac{2\pi ipn}{N}\right) \tag{11.10}$$

The character table for this group is given in Table 11.3. The selection rules for this group will be discussed in the following section.

11.4 Classification of the Normal Modes and Selection Rules of the Lattice Translation Group

In this section we turn our attention to the classification of the normal modes according to symmetry species and the selection rules of each of these species. In other words, if we have m atoms in each of the $N_1N_2N_3$ unit cells, how are the $3mN_1N_2N_3$ normal vibrations distributed among the irreducible representations? For now we will assume that the only symmetry elements present are lattice translations.

We demonstrated in Chapter 4 that the number of normal modes $n^{(\gamma)}$ in each symmetry species can be determined by calculating how many times the irreducible representations appear in a reducible representation that is constructed using the cartesian displacement coordinates or internal coordinates as a basis. The characters of the reducible representation can be easily found by considering the atoms or coordinates that remain unshifted by each symmetry operation. For a lattice translation symmetry operation only the identity operation $E(T_0)$ has this property. All other operations $T_1, \ldots, T_{N-1}$ are translations and shift the position of the m atoms in each unit cell. The trace χ_j of E is equal to $3mN_1N_2N_3$, since this is the order of the unit matrix. The trace of all the other operations $T_1, \ldots, T_{N-1}$ is zero. The number of normal modes in each symmetry species can now be calculated as in Chapter 4, using

$$n^{(\gamma)} = \frac{1}{g}\sum_j g_j \chi_j^{(\gamma)} \chi_j \tag{11.11}$$

where it will be recalled that χ_j is the character of the **reducible** representation, while $\chi_j^{(\gamma)}$ is the character of the irreducible representation, which for the lattice translation group was discussed in Section 11.3 [see equation (11.10)].

Since

$$\frac{1}{g} = \frac{1}{N_1N_sN_3} \tag{11.12}$$

$$\chi_j^{(\gamma)} = \chi_E^{(\gamma)} = 1 \tag{11.13}$$

$$\chi_j = \chi_E = 3mN_1N_2N_3 \tag{11.14}$$

then

$$n^{(\gamma)} = \frac{3mN_1N_2N_3}{N_1N_2N_3} = 3m \tag{11.15}$$

Naturally, for the one-dimensional case the answer is the same and the number of normal modes in each symmetry species $\Gamma^{(0)}, \Gamma^{(1)}, \ldots, \Gamma^{(N-1)}$ is $3m$.

In order to determine which of these symmetry species are active in the infrared and Raman we need to know to which species the components of the dipole moment vector and polarizability tensor belong. The dipole moment vector transforms in the same way as the components of the translation vector. It can be shown formally (1) that all three translations belong to the totally symmetric species $\Gamma^{(0)}$, and this is the only active infrared mode. In fact, any vector or tensor is totally symmetric with respect to the translation group. Consequently, the only representation that has infrared and Raman activity is Γ^0. This corresponds to values of k, the wavenumber vector, equal to zero.

11.5 Factor Groups, Space Groups, and Line Groups

So far, we have considered only the translational symmetry of three-dimensional crystals or one-dimensional chains. Many polymer crystals or individual chains have additional symmetry elements relating the position of atoms within the unit cell, such as rotations, reflections, screw axes, or glide planes. These will also play a part in determining spectral activity. The group of symmetry operations that collectively describe the total symmetry of a crystal is known as the space group, while in an analogous fashion those that describe a particular structure of a chain molecule form a line group. We will first discuss space groups.

Consider a lattice consisting of translational repeat units each of which can be described by H different symmetry elements such as rotations and reflections. The symmetry operations of the space group are then combinations of pure lattice translations with these H other symmetry elements, so that a point in one unit cell is carried into an equivalent point in another unit cell. However, we saw earlier that only modes for which the translations are equivalent to the identity operation (i.e., $k=0$) are Raman and infrared active, allowing a maximum of $3m$ normal modes of vibration (including translations and rotations), where m is the number of atoms in the translational repeat unit. We can therefore classify and determine the activity of these $3m$ modes according to the group of H symmetry operations. This group is called the **factor group** and any factor group is always isomorphous with one of the 32 point groups and has the same selection rules as these point groups. We can therefore use the appropriate point group character table and the methods of Chapter 4 to classify the modes of a translational repeat unit. These will be the only active modes, since by definition factor group modes are symmetric with respect to translation.

Table 11.4 Line Groups Characterized by Their Factor Groups

Factor Group Order	Symmetry Elements of Factor Group	Isomorphous Point Group
	Point Symmetry Operations Only	
1	E	C_1
2	E, i	C_i
2	E, σ_h	C_s
2	$E, \sigma_v'; E, \sigma_v''$	C_s
2	$E, C_2'; E, C_2''$	C_2
p	$E, (p-1)C_p$	C_p
$2p$	$E, (p-1)C_p, p\sigma_v$	C_{pv}
4	$E, \sigma_h, \sigma_v', C_2''; E, \sigma_h, \sigma_v'', C_2'$	C_{2v}
4	$E, \sigma_h', C_2', i; E, \sigma_v'', C_2'', i$	C_{2h}
4	E, σ_h, C_2, i	C_{2h}
4	E, C_2, C_2', C_2''	$D_2 \equiv V$
8	$E, \sigma_h, \sigma_v', \sigma_v'', C_2, C_2', C_2'', i$	$D_{2h} \equiv V_h$
	Point Symmetry Operations, Screw Rotations, and Glide Reflections	
2	$E, \bar{\sigma}_v'; E, \bar{\sigma}_v''$	C_s
p	$E, (p-1)\bar{C}_p$	C_p
4	$E, \bar{\sigma}_v', \sigma_v'', \bar{C}_2; E, \sigma_v', \bar{\sigma}_v'', \bar{C}_2$	C_{2v}
4	$E, \bar{\sigma}_v', \bar{\sigma}_v'', C_2$	C_{2v}
4	$E, \sigma_h, \bar{\sigma}_v', C_2''; E, \sigma_h, \bar{\sigma}_v'', C_2'$	C_{2v}
4	$E, \bar{\sigma}_v', C_2', i; E, \bar{\sigma}_v'', C_2'', i$	C_{2h}
4	$E, \sigma_h, \bar{C}_2, i$	C_{2h}
4	$E, \bar{C}_2, C_2', C_2''$	$D_2 \equiv V$
8	$E, \sigma_h, \sigma_v', \bar{\sigma}_v'', \bar{C}_2, C_2', C_2'', i; E, \sigma_h, \bar{\sigma}_v', \sigma_v'', \bar{C}_2, C_2', C_2'', i$	$D_{2h} \equiv V_h$
8	$E, \sigma_h, \bar{\sigma}_v', \bar{\sigma}_v'', C_2, C_2', C_2'', i$	$D_{2h} \equiv V_h$

For many polymers, interactions between chains in a crystal lattice are negligibly small and a satisfactory analysis of the spectrum can usually be based on the line group. Since the line group is in effect a one-dimensional space group, the arguments presented earlier can be used to show that we have to consider only the factor group elements of the line group. A list of line groups is given in Table 11.4, together with the corresponding isomorphous point groups. It is convenient at this point to illustrate factor group analysis using polyethylene as an example.

Example 1: Factor Group Analysis of a Single Chain of Polyethylene

A planar zigzag chain of polyethylene is shown in Figure 11.4. The one-dimensional lattice translation repeat unit contains two CH_2 units. The factor group of the line group is of order 8 and contains the following symmetry

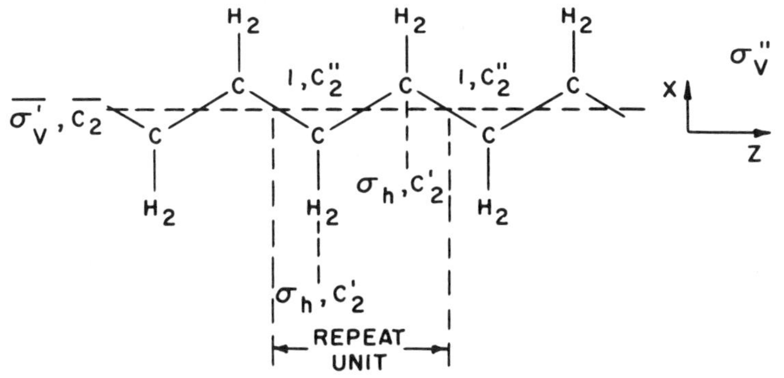

Figure 11.4. Polyethylene chain, line group V_h.

elements:

1 E
2 Mirror plane in plane of chain (xz)—σ_v''
3 Mirror plane through CH_2 group (xy)—σ_h
4 Glide plane along z, reflection at yz plane—$\bar{\sigma}_v'$
5 Center of inversion at midpoint of C—C bond $-i$
6 Two-fold screw axis along a—$\bar{C}_2$
7 Two-fold rotation axis at middle of C—C bond and perpendicular to xz plane—C_2''
8 Two-fold axis along x and bisecting CH_2 group—C_2'

By looking at Table 11.4 we can see that the factor group of polyethylene is isomorphous with the point group D_{2h} $(\equiv V_n)$. The character table of this point

Table 11.5 Character Table for the Point Group

D_{2h}	E	$C_2(z)$	$C_2(y)$	$C_2(x)$	i	$\sigma(xy)$	$\sigma(zx)$	$\sigma(yz)$	
A_g	1	1	1	1	1	1	1	1	$\alpha_{xx}, \alpha_{yy}, \alpha_{zz}$
B_{1g}	1	1	-1	-1	1	1	-1	-1	$R_z \alpha_{xy}$
B_{2g}	1	-1	1	-1	1	-1	1	-1	$R_y \alpha_{zx}$
B_{3g}	1	-1	-1	1	1	-1	-1	1	$R_x \alpha_{yz}$
A_u	1	1	1	1	-1	-1	-1	-1	
B_{1u}	1	1	-1	-1	-1	-1	1	1	T_z
B_{2u}	1	-1	1	-1	-1	1	-1	1	T_y
B_{3u}	1	-1	-1	1	-1	1	1	-1	T_x
m_j	6	2	0	0	0	6	0	2	
χ_j	18	-2	0	0	0	6	0	2	

group is shown in Table 11.5. Comparing the factor group and line group, we have

Operators in factor group	$E\sigma_v''\sigma_h\sigma_v' iC_2\bar{C}_2''C_2'$
Operators in D_{2h} point group	$E\sigma_v''\sigma_h\sigma_v' iC_2C_2''C_2'$

We can find all factor group representations compiled in tables if the factor group is itself a point group. If the factor group contains glide planes, we simply substitute the isomorphic point group by replacing the glide plane by a mirror plane in the same orientation. For screw axes, however, we can run into problems because the order of a screw axis can be larger than those tabulated in character tables and it is possible for the helix to have a noninteger number of residues per turn. An example of this type of helix is the 15/7 helix of poly(tetrafluoroethylene) (PTFE). We will consider PTFE as an example of a general helix in the following section.

Continuing with polyethylene, we determine the characters of the reducible representation by multiplying the number of unchanged atoms in a unit cell (u) by the contribution to the character per unshifted atom:

$$x(R)=u(2\cos\psi\pm 1)$$

	E	$\bar{C}_2$	C_2''	C_2'	i	σ_h	σ_v''	σ_v'
Number of unshifted atoms	6	0	0	2	0	6	2	0
Contribution	3	−1	−1	−1	−3	1	1	1
Γ_{red}	18	0	0	−2	0	6	2	0

Note that if the factor group operation moves an atom to a translationally equivalent atom in another unit cell, the atom is counted as unshifted. Using the character table for D_{2h}, the normal modes are then calculated as if the system were a point group:

$$\Gamma_{(A_g)}=\frac{18(1)-2(1)+6(1)+2(1)}{8}=3$$

$$\Gamma_{(B_{1g})}=\frac{18(1)-2(-1)+6(1)+2(-1)}{8}=3$$

and so on. In the same manner the contribution of the other irreducible factor group representations to the reducible representation can be made.

$$\Gamma_{\text{red}}=3A_g+3B_{1g}+2B_{2g}+1B_{3g}+A_u+2B_{1u}+3B_{2u}+3B_{3u}$$

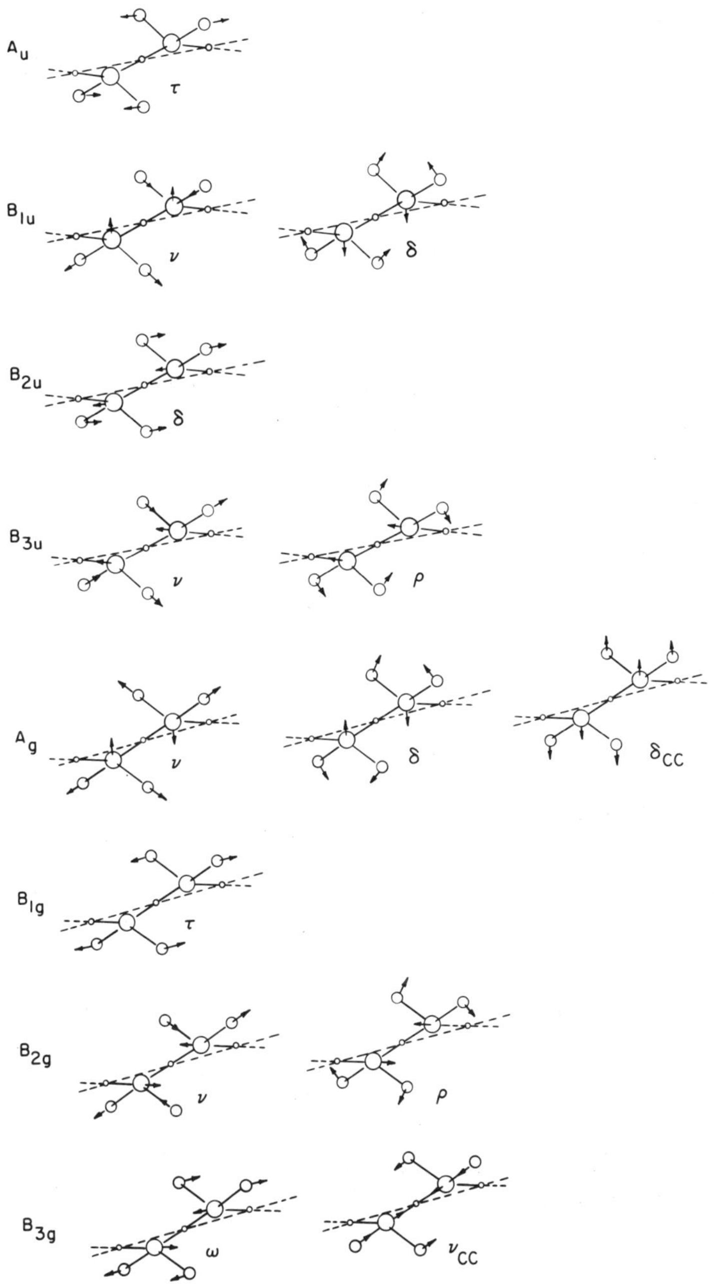

Figure 11.5. Approximate form of the normal modes of polyethylene.

Of the 18 ($=3N$) vibrations, 4 are nongenuine normal vibrations that are due to translation and rotation. Three of the nongenuine vibrations are due to pure translation of the chain (T_x, T_y, T_z); the remaining nongenuine mode is rotation around the chain axis (R_z). Rotations around the x and y axes are impossible, since we deal essentially with an infinitely long chain (because of the Born boundary conditions applied to a finite segment).

By inspection of the character table, it can be seen that the nongenuine modes are

$$\Gamma_{\text{rot-trans}} = B_{1g} + B_{1u} + B_{2u} + B_{3u}$$

By removing the translations and rotation we obtain the reducible representation for vibrations alone:

$$\Gamma_{\text{vib}} = 3A_g + 2B_{1g} + 2B_{2g} + 1B_{3g} + A_u + B_{1u} + 2B_{2u} + 2B_{3u}$$

The assignment of observed infrared bands and Raman lines to the appropriate symmetry species is greatly aided by dichroic infrared or anisotropic Raman measurements of oriented samples. We will discuss such measurements later. In addition, we can also obtain a rough idea of the form of the normal modes in each irreducible representation by applying symmetry arguments to an internal coordinate representation, rather than to the cartesian coordinate representation used earlier. In effect we generate internal symmetry coordinates. We applied this approach to the water molecule in Chapter 3 and noted that the true normal coordinates are linear combinations of these symmetry coordinates. Nevertheless, to a first approximation this mixing is sometimes neglected in order to obtain a rough idea of the form of the normal modes in each symmetry species. Such approximate modes derived for polyethylene (3) are illustrated in Figure 11.5.

11.6 Normal Modes of a General Helical Chain

A polymer such as polyethylene having an extended planar structure is relatively easy to analyze by the line group method. The translational repeat unit is readily identified as corresponding to two CH_2 units. However, many polymers have more complicated helical structures and in this section we will extend the line group method to such conformations. Consider the general infinite helix illustrated in Figure 11.6. We will discuss the case of n chemical units and m turns of the helix in the translational repeat unit, an n_m helix. For example, isotactic polystyrene has exactly three chemical repeat units in one turn of the helix (i.e., it is a 3_1 helix). Each complete turn is therefore the translational repeat unit. For PTFE, which has a 15_7 helical conformation, the translational repeat unit consists of 15 chemical repeat units in 7 turns.

A helix of identical repeating units can be generated by an operator $H_\ell(\ell\psi, \ell z_1)$, which transforms the coordinates of one monomer (x, y, z) into

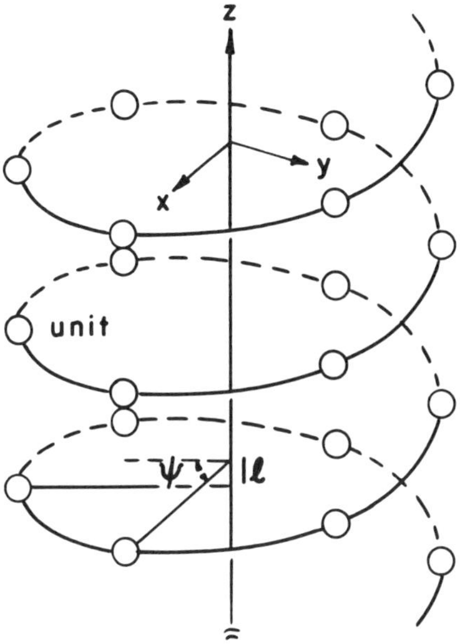

Figure 11.6. Definition of coordinate axes of a helix.

the coordinates of a monomer ℓ neighbors away by rotation through an angle $\ell\psi$ and translation along the helix axis by a distance ℓz_1. The angle ψ is equal to $2\pi m/n$.

$$H_\ell(\psi)=\begin{bmatrix}\cos\ell\psi & -\sin\ell\psi & 0\\ \sin\ell\psi & \cos\ell\psi & 0\\ 0 & 0 & 1\end{bmatrix}\begin{bmatrix}x\\ y\\ z\end{bmatrix}+\begin{bmatrix}0\\ 0\\ \ell z_1\end{bmatrix} \tag{11.16}$$

This symmetry operation is a screw rotation discussed in Section 11.5. All such operations together with the pure translation constitute the line group of the polymer, assuming for now that there are no additional symmetry elements in the chemical repeat unit. The factor group of this line group is denoted by the symbol $C(2\pi m/n)$ and is isomorphous with the point group C_n. The character tables can be derived by standard methods and are presented in Tables 11.6 and 11.7 for odd and even values of n, respectively. The symmetry operation C^1 is a rotation of $\psi=2\pi m/n$ about the z axis followed by a translation along the axis of $1/n$ of the unit cell length. The symmetry operation C^k indicates C^1 performed k times in succession. If each chemical unit contains p atoms, then there are $n\times 3p$ normal modes in the translational repeat unit. Applying the same arguments as presented in Section 11.5, we can easily show that there are $3p$ normal modes in each symmetry species. The assignment of normal modes to particular symmetry species is based on the phase difference between vibrations in adjacent chemical units. We will now consider how this phase angle is related to the geometry of the helix.

Let θ be the phase angle between vibrations in adjacent chemical units. The phase angle θ is separate and distinct and should not be confused with the angle ψ defined earlier as a symmetry operator determined by the geometry of

Table 11.6 Character Table, Numbers of Normal Modes, and Selection Rules for the Helical Polymer Molecule under the Group $C(2m\pi/n)$[a]

	E	C^1	C^2	$\cdots$	C^{n-1}	N	Infrared	Raman
A	1	1	1	$\cdots$	1	$3p(T_\pi, R_\pi)$	A	A
E_1	1	ε	ε^2	$\cdots$	ε^{n-1}	$3p(T_\sigma)$	A	A
	1	ε^{-1}	ε^{-2}	$\cdots$	$\varepsilon^{-(n-1)}$	$3p(T_\sigma)$	A	A
E_2	1	ε^2	ε^4	$\cdots$	$\varepsilon^{2(n-1)}$	$3p$	F	A
	1	ε^{-2}	ε^{-4}	$\cdots$	$\varepsilon^{-2(n-1)}$	$3p$	F	A
E_3	1	ε^3	ε^δ	$\cdots$	$\varepsilon^{3(n-1)}$	$3p$	F	F
	1	ε^{-3}	$\varepsilon^{-\delta}$	$\cdots$	$\varepsilon^{-3(n-1)}$	$3P$	F	F
$\vdots$	$\vdots$	$\vdots$	$\vdots$	$\cdots$	$\vdots$	$\vdots$	$\vdots$	$\vdots$
$E_{(n-1)/2}$	1	$\varepsilon^{(n-1)/2}$	$\varepsilon^{[(n-1)/2]2}$	$\cdots$	$\varepsilon^{[(n-1)/2](n-1)}$	$3p$	F	F
	1	$\varepsilon^{-(n-1)/2}$	$\varepsilon^{-[(n-1)/2]2}$	$\cdots$	$\varepsilon^{-[(n-1)/2](n-1)}$	$3p$	F	F

[a]In this table n is odd; A means active and F means forbidden; $\varepsilon = \exp[i(2\pi m/n)]$.

Table 11.7 n = even[a]

	E^1		C^1	C^2	$\cdots$	C^{n-1}	N	Infrared	Raman
A	1		1	1	$\cdots$	1	$3p(T_\pi, R_\pi)$	A	A
B	1		-1	1	$\cdots$	-1	$3p$	F	A
E_1	{	1	ε	ε^2	$\cdots$	$\varepsilon^{(n-1)}$	$3pT_\sigma$ $3p$ $3p$	A	A
		1	ε^{-1}	ε^{-2}	$\cdots$	$\varepsilon^{-(n-1)}$	$3p(T_\sigma)$	A	A
E_2	{	1	ε^2	ε^4	$\cdots$	$\varepsilon^{2(n-1)}$		F	A
		1	ε^{-2}	ε^{-4}	$\cdots$	$\varepsilon^{-2(n-1)}$		F	A
E_3	{	1	ε^3	ε^σ	$\cdots$	$\varepsilon^{3(n-1)}$	$3p$	F	F
		1	ε^{-3}	$\varepsilon^{-\sigma}$	$\cdots$	$\varepsilon^{-3(n-1)}$	$3p$	F	F
	$\vdots$		$\vdots$	$\vdots$	$\cdots$	$\vdots$	$\vdots$	$\vdots$	$\vdots$
$E_{(n/2)-1}$	{	1	$\varepsilon^{[(n/2)-1]}$	$\varepsilon^{[(n/2)-1]2}$	$\cdots$	$\varepsilon^{[(n/2)-1](n-1)}$	$3p$	F	F
		1	$\varepsilon^{-[(n/2)-1]}$	$\varepsilon^{-[(n/2)-1]2}$	$\cdots$	$\varepsilon^{-[(n/2)-1](n-1)}$	$3p$	F	F

[a] $\varepsilon = \exp[i(2\pi m/n)]$.

the chain. To satisfy the $k=0$ condition, vibrations in the first chemical unit of a translational repeat unit must be in phase with the corresponding modes in the first chemical unit of the adjacent translational unit. For example, isotactic polystyrene has three chemical units in one turn of a helix. For the phase condition specified to hold we could obviously have $\theta=0°$, which would mean that the vibrations in adjacent chemical repeat units are all in phase. However, when $\theta=120°$ the condition also holds, since the phase angle between the first chemical unit in each translational repeat unit would then be $3\times 120°=2\pi$, that is, they would also be in phase.

Generalizing this condition, we can require that $n\theta/m$ must be 2π times a whole number:

$$\frac{n\theta}{m}=2\pi r \tag{11.17}$$

or

$$\theta=\left(\frac{2\pi m}{n}\right)r=\psi r \tag{11.18}$$

with $r=0,1,2,\ldots,n-1$ (for $n>r-1$ we get nothing new, since $r=n$ is equivalent to $r=0$, and so on). This condition is more usually defined with $r=0,\pm 1,\pm 2,\ldots,\pm(n-1)/2$ for n odd and $r=0,\pm 1,\pm 2,\cdots$ for n even, so that θ now takes values $-\pi\leqslant\theta\leqslant\pi$.

Consequently, the frequency of a normal mode ν_i in an isolated chemical repeat unit gives rise to a band of frequencies $\nu_i(\theta)$ in a helix, where θ is the phase difference between the motions in adjacent units. This is the phonon dispersion curve, which resembles that derived in Chapter 10 for a simple lattice. However, θ is not continuous but has discrete values that are given by equation (11.18).

For each frequency $\nu_i(\theta)$ there is corresponding frequency $\nu_i(-\theta)$ that is the complex conjugate of $\nu_i(\theta)$ and is therefore equal to it, since frequencies are real quantities. With the exception of $\theta=0$ and $\theta=\pi$ these frequencies are then doubly degenerate and belong to one of the irreducible representations $\Gamma^{(\theta)}$ of the factor groups listed in Tables 11.6 and 11.7 [$\Gamma^{(0)}\equiv A$, $\Gamma^{(1)}\equiv E_1$, etc].

It should be realized that for many polymers we can construct two different dispersion curves, the first a plot of frequency as a function of k, the wave vector describing the phase relationship between units in adjacent translational repeat units; the second a plot of frequency as a function of the phase relationship θ, relating vibrations in adjacent chemical units. If the phase relationship between vibrations in adjacent translational repeat units is used, the selection rules limit Raman and infrared activity to $k=0$ modes, as discussed earlier. The selection rules using the phase relationship between chemical units are also easily derived (4).

For example, rotations about the z axis will not change T_z. This component transforms according to the totally symmetric representation A and is infrared

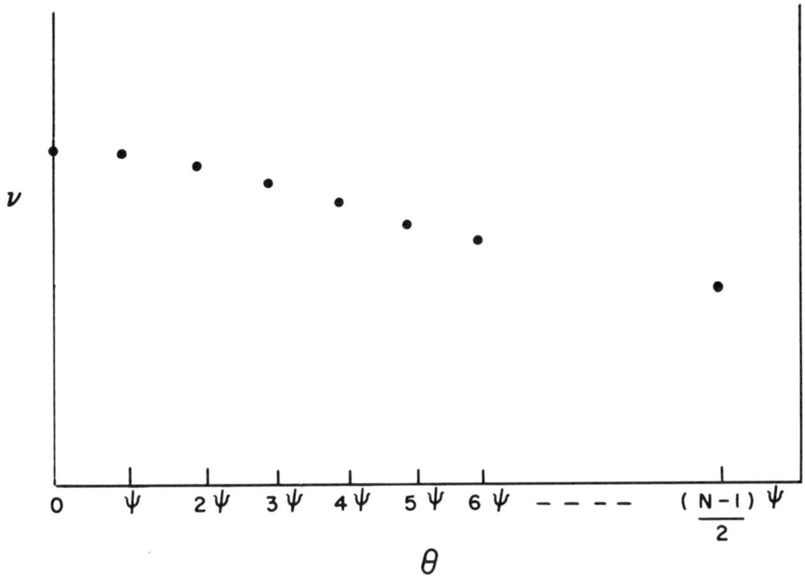

Figure 11.7. Dispersion relation for a general helical chain.

active. It can be shown that infrared-active modes belong to symmetry species A and E_1, while Raman-active modes belong to species A, E_1, and E_2. For symmetry species A, E_1, and E_2 the vibrations of motions in adjacent chemical repeat units are phase shifted 0, ψ, and 2ψ, respectively. A dispersion plot of frequency against phase shift in adjacent **chemical** units rather than translational repeat units is customary in polymer spectroscopy. Such a dispersion curve is illustrated schematically in Figure 11.7.

Using the information we have presented to this point, we see that a string of beads on a 15/7 helix would belong to a line group that is isomorphous with the point group C_{15}. The character table for this group is shown in Table 11.8. If we consider each bead to contain three atoms (so we can compare the helix to that of PTFE later), the characters of the reducible representations can be determined by multiplying the number of unchanged atoms in the unit cell by the contribution to the character per unshifted atom:

	E	$2C_{15}^{1}$	$2C_{15}^{2}$	$2C_{15}^{3}$	$2C_{15}^{4}$	$2C_{15}^{5}$	$2C_{15}^{6}$	$2C_{15}^{7}$
No. of unchanged atoms	45	0	0	0	0	0	0	0
Contribution	3	$(2\cos\psi+1)$	$(2\cos 2\psi+1)$	. . . . . .	. . . . . .	. . . . . .	. . . . . .	. . . . . .
Γ_{red}	135	0	0	0	0	0	0	0

Table 11.8 Character Table for the Point Group C_{15}[a]

C_{15}	E	$2C_{15}^1$	$2C_{15}^2$	$2C_{15}^3$	$2C_{15}^4$	$2C_{15}^5$	$2C_{15}^6$	$2C_{15}^7$	
A	1	1	1	1	1	1	1	1	$z;(x^2+y^2,z^2)$
E_1	1	ε	ε^2	ε^3	ε^4	ε^5	ε^6	ε^7	$(x,y);(yz,xz)$
	1	ε^{-1}	ε^{-2}	ε^{-3}	ε^{-4}	ε^{-5}	ε^{-6}	ε^{-7}	
E_2	1	ε^2	ε^4	ε^6	ε^8	ε^{10}	ε^{12}	ε^{14}	(x^2y^2,xy)
	1	ε^{-2}	ε^{-4}	ε^{-6}	ε^{-8}	ε^{-10}	ε^{-12}	ε^{-14}	
E_3	1	ε^3	ε^6	ε^9	ε^{12}	ε^{15}	ε^{18}	ε^{21}	
	1	ε^{-3}	ε^{-6}	ε^{-9}	ε^{-12}	ε^{-15}	ε^{-18}	ε^{-21}	
$\vdots$									
E_7	1	ε^{+7}	ε^{+14}	ε^{21}	ε^{28}	ε^{35}	ε^{42}	ε^{49}	
	1	ε^{-7}	ε^{-14}	ε^{-21}	ε^{-28}	ε^{-35}	ε^{-42}	ε^{-49}	

[a] $\varepsilon=\exp[i(2\pi m/n)]$

Using the character table for C_{15}, we calculate the number of normal modes as we have done before:

$$\Gamma_{\text{red}}=9A+9E_1+9E_2+9(E_3+E_4+E_5+E_6+E_7)$$

By inspection of the character table we have produced, it can be seen that the nongenuine modes due to rotation and translation are

$$\Gamma_{\text{rot-trans}}=2A_1+E_1$$

By removing the translation and rotation we obtain the reducible representation for vibrations alone:

$$\Gamma_{\text{vib}}=7A+8E_1+9E_2+9(E_3+E_4+E_5+E_6+E_7)$$

Using the selection rules we discussed earlier, we would predict the following optical activity:

Raman active	$7A+8E_1+9E_2$
IR active	$7A+8E_1$
Inactive	$9(E_3+E_4+E_5+E_6+E_7)$

So, using line group methods, we have analyzed an isolated general helix. We can see that the classification of normal modes according to symmetry for a helical polymer is generally straightforward. However, it can become more complex if there is symmetry in the chemical repeat unit. We will consider PTFE as an example of such a polymer.

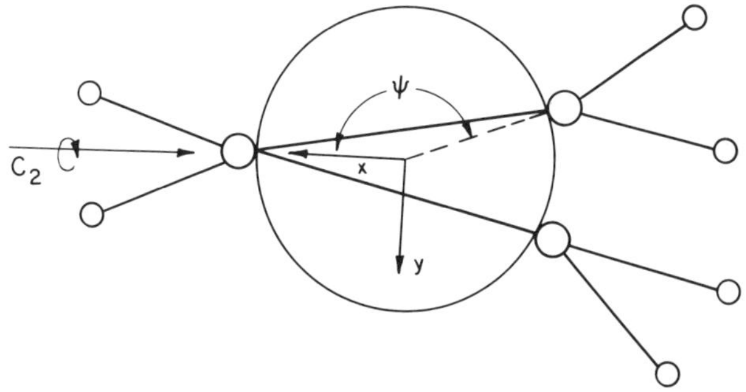

Figure 11.8. The helix of poly(tetrafluoroethylene).

Example 2: Classification of the Normal Modes of Poly(tetrafluoroethylene)

PTFE is a helical polymer with 15 CF_2 groups in 7 turns of the helix:

$$\psi = 2\pi \times \frac{7}{15} = \frac{14\pi}{15}$$

and its chain structure is illustrated in Figure 11.8. The symmetry elements of the PTFE helix are:

1 E
2 Two-fold rotation axes perpendicular to the chain axis, bisecting the CF_2 groups—$15C_2$
3 Screw axis along the z axis—$C^n(\theta)$:

$$\theta = \pm n\psi, \qquad n = 1, 2, 3, \ldots, 7$$

The factor group of the line group is of order 30 and is isomorphous with the point group D_{15}. The character table for D_{15} has been derived in the same manner as we did for C_{15} and is shown in Table 11.9. The characters of the reducible representation are found in the same manner as for C_{15}:

	E	$2C_{15}^1$	$2C_{15}^2$	$2C_{15}^3$		$2C_{15}^7$	$15C_2$
No. of unchanged atoms	45	0	0	0	$\cdots$	0	1
Factor	3	$(2\cos\psi + 1)$	$(2\cos 2\psi + 1)$	0	$\cdots$		-1
Γ_{red}	135	0	0	0	$\cdots$	0	-1

The number of normal modes contained in Γ_{red} are calculated by using the

Table 11.9 Character Table for the Point Group D_{15}

D_{15}	E	$2C_{15}^1$	$2C_{15}^2$	$2C_{15}^3$	$2C_{15}^4$	$2C_{15}^5$	$2C_{15}^6$	$2C_{15}^7$	$15C_2$	
A_1	1	1	1	1	1	1	1	1	1	x^2+y^2, z^2
A_2	1	1	1	1	1	1	1	1	-1	z
E_1	2	$2\cos(\theta)$	$2\cos(2\theta)$	$2\cos(3\theta)$	$2\cos(4\theta)$	$2\cos(5\theta)$	$2\cos(6\theta)$	$2\cos(7\theta)$	0	$(x, y)(xz, yz)$
E_2	2	$2\cos(2\theta)$	$2\cos(4\theta)$	$2\cos(6\theta)$	$2\cos(8\theta)$	$2\cos(10\theta)$	$2\cos(12\theta)$	$2\cos(14\theta)$	0	(x^2-y^2, xy)
E_3	2	$2\cos(3\theta)$	$2\cos(6\theta)$	$2\cos(9\theta)$	$2\cos(12\theta)$	$2\cos(15\theta)$	$2\cos(18\theta)$	$2\cos(21\theta)$	0	
⋮									0	
									0	
									0	
									0	
E_7	2	$2\cos(7\theta)$	$2\cos(14\theta)$	$2\cos(21\theta)$	$2\cos(28\theta)$	$2\cos(35\theta)$	$2\cos(42\theta)$	$2\cos(49\theta)$	0	

character table for D_{15}:

$$\Gamma_{red}=4A_1+5A_2+9E_1+9E_2+9(E_3+E_4+E_5+E_6+E_7)$$

By inspection of the character table we have produced, it can be seen that the nongenuine modes that are now due to rotation and translation are

$$\Gamma_{rot\text{-}trans}=2A_2+E_1$$

When these nongenuine modes are removed we have the reducible representation for normal vibrations alone:

$$\Gamma_{vib}=4A_1+3A_2+8E_1+9E_2+9(E_3+E_4+E_5+E_6+E_7)$$

Using the selection rules, we see the following optical activity:

Raman active	$4A_1+8E_1+9E_2$
IR active	$3A_2+8E_1$
Inactive	$9(E_3+E_4+E_5+E_6+E_7)$

It can be seen by this example that adding symmetry to the chemical unit changes the optical activity of the A modes *only*. These further selection rules arise because although z^2 and x^2+y^2 still transform by the totally symmetric representation, z now transforms by the representation in which the operator for C_2 has a trace of -1.

11.7 A Comparison of Line Group and Space Group Analysis

According to the line group treatment presented in the preceding section, the individual polyethylene chain has 18 factor group normal vibrations (including 3 translations and 1 rotation). A similar analysis can be performed on the space group of the orthorhombic crystal, illustrated in Figure 11.9. The factor group of the space group is of order 8 and its elements are $E, \bar{C}_2(a), \bar{C}_2(b), \bar{C}_2(c), \bar{\sigma}_v(bc), \bar{\sigma}_v(ac)$, and $\bar{\sigma}_h(ab)$. This factor group is isomorphous with the point group D_{2h} and therefore has the same character table as the isolated polyethylene chain. This correspondence between the factor group of the line group and the factor group of the space group is accidental and is not generally true for other polymers. The unit cell contains 4 carbons and 8 hydrogens, giving 36 fundamental vibrations.

There are only $36-3$ genuine normal vibrations, since we have to subtract 3 pure translations. (No free rotation of the crystal as a whole is possible because we have assumed an infinite lattice.) Using the same procedure as for an isolated chain, we can show that

$$\Gamma=6A_g+6B_{1g}+3B_{2g}+3B_{3g}+3A_u+3B_{1g}+6B_{2g}+3B_{3g}$$

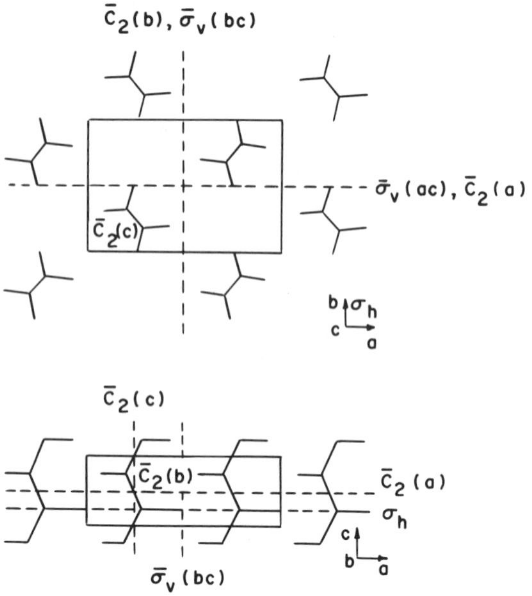

Figure 11.9. The orthorhombic unit cell of polyethylene. The polymer chain is parallel to the *c* axis of the crystal.

which is double the total number calculated for the isolated chain. Each vibration of the single chain corresponds to two vibrations of the space group, since there are two chains per unit cell. If the intermolecular interaction is small, these two modes would have the same frequency; however, doublet splitting is observed for some modes but not others. Although symmetry analysis tells us nothing about the size of the intermolecular force constants that determine if the splitting is sufficient to be observable, we shall see that it still allows us to predict which modes can appear as doublets and which cannot.

In discussing the spectra of crystals of small molecules, the concept of site symmetry is useful. A given site in a crystal can be characterized by the group of symmetry operations that leave it invariant. Obviously, this site group must be a subgroup of the crystal and characterizes the symmetry of the crystalline field that surrounds the site in question. However, the geometric structure of an individual molecule that occupies that site must also conform to the symmetry of the site, so that the site group is also a subgroup of the molecular point group. The site group can therefore be used to establish the correlation of the point group to the factor group of the space group. This correlation allows us to calculate the normal modes of individual chains and then establish their activity in the crystal through a correlation table. Correlation tables can be constructed in a formal fashion (5), but to illustrate their use in polymer spectroscopy it is sufficient to construct one by inspection. We will first discuss

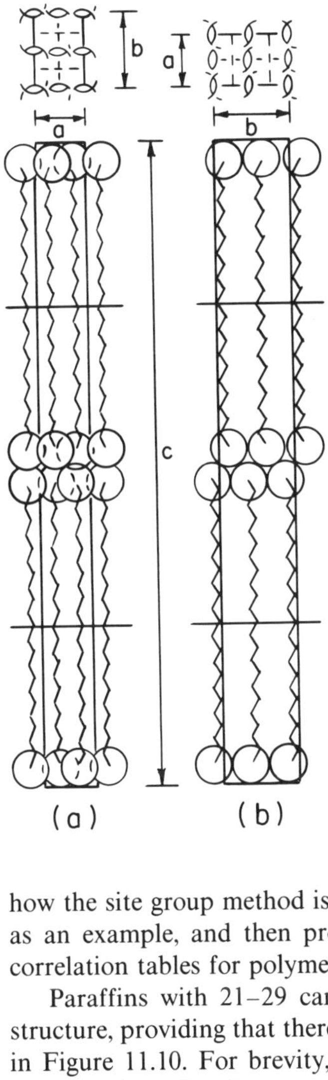

Figure 11.10. Structure of the unit cell of the paraffin *n*-tricosane: (*a*) *ac* projection; (*b*) *bc* projection. [Reprinted by permission from A. E. Smith, *J. Chem. Phys.* **21**, 229 (1953).]

how the site group method is applied to finite molecules, using the *n*-paraffins as an example, and then proceed to adapt the method to the derivation of correlation tables for polymers.

Paraffins with 21–29 carbon atoms form crystals with an orthorhombic structure, providing that there is an odd number of carbon atoms, as illustrated in Figure 11.10. For brevity, we will not consider in detail paraffins with an even number of carbon atoms. The symmetry operations of the factor group are E, $\bar{C}_2(z)$, $\bar{C}_2(y)$, $C_2(x)$, i, $\bar{\sigma}(yz)$, $\sigma(xz)$, and $\sigma(xy)$; the factor group is isomorphous with the point group D_{2h}. The operations E and $\sigma(xy)$ leave a given molecule fixed, so that the site group of a single chain is C_s.

The symmetry operations of a single molecule, illustrated in Figure 11.11, are E, C_2, $\sigma(xz)$, and $\sigma(yz)$, which form the C_{2v} point group.

The molecular axes x, y, z are parallel to the x, y, z crystallographic axes, respectively. To determine the activity of the normal vibrations of the crystal, we need first to determine the effect of the reduced symmetry at the site group; that is, we need to map the irreducible representations of the point group onto the site group. This can be done by comparing the characters of the irreducible

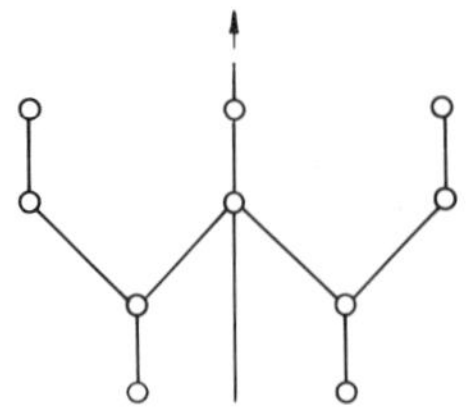

Figure 11.11. Symmetry properties of a finite hydrocarbon chain, n odd (C_{2v} symmetry).

Table 11.10 Character Table for the Point Group C_{2v}

C_{2v}	E	C_2	$\sigma(xz)$	$\sigma(xy)$
A_1	1	1	1	1
A_2	1	1	−1	−1
B_1	1	−1	1	−1
B_2	1	−1	−1	1

Table 11.11 Character Table for the Point Group C_S

C_s	E	$\sigma(xy)$
A'	1	1
A''	1	−1

representations. Consider the character tables of the point groups C_{2v} and C_s, shown as Tables 11.10 and 11.11. Only symmetry operations I and $\sigma(xy)$ are common to both tables. Comparing the characters for these symmetry operations, we can see that symmetry species A_1 and B_2 of C_{2v} correspond to A' of C_s, while A_2 and B_1 correspond to A''. The site group C_s can then be mapped onto the factor group of the space group, isomorphous with D_{2h}, in the same way. The correlation table can be written as in Table 11.12

Selection rules for the infrared and Raman fundamentals can be determined from published character tables. Modes appearing in symmetry species A_2 of the isolated molecule are infrared inactive. However, the correlation diagram demonstrates that the lower symmetry imposed by the site can, in principle, allow infrared activity of these vibrations. The correlation diagram also indicates that each normal mode of the isolated molecule has four components in the crystal. The activity of each of these modes naturally depends on the selection rules of point group D_{2h}.

We can consider the vibrations of a crystalline polymer with respect to those of the individual chain in a similar manner, using polyethylene as an example. There are four equivalent sites in the space group of polyethylene,

Table 11.12 **Correlation table for paraffins with an odd number of carbon atoms (orthorhombic crystal structure).**

Point Group (Molecule)	Site Group	Factor Group Unit Cell
C_{2v} $(n - odd)$	C_s	D_{2b}

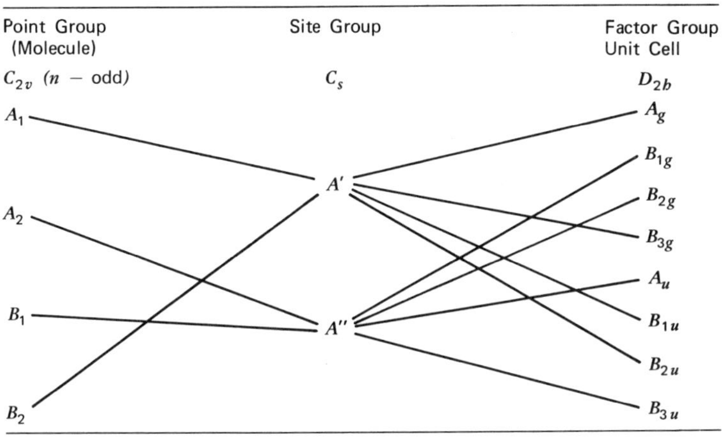

each occupied by a single CH_2 unit, as can be seen in Figure 11.9. However, unlike the situation with crystals of small molecules, there is strong coupling between CH_2 units covalently linked as part of the same chain, whereas between chains there are only weak forces. A more meaningful analysis can therefore be made on the basis of a chain segment containing two CH_2 units and having effective symmetry C_{2h}. Defining the molecular and crystal axes as in Figure 11.9 we can construct the correlation table shown as Table 11.13.

This analysis predicts that all of the single-chain infrared-active modes that transform in the same way as T_x and T_y (i.e., are polarized perpendicular to the

Table 11.13 **Correlation table for polyethylene.**

Single Infinite Chain	Chain Segment	Unit Cell
D_{2b}	C_{2b}	D_{2b}
A_g		A_g
B_{1g}	A_g	B_{1g}
B_{2g}		B_{2g}
B_{3g}	B_g	B_{3g}
A_u		A_u
$(T_z)\ B_{1u}$	A_u	$B_{1u}\ (T_z)$
$(T_y)\ B_{2u}$		$B_2\ (T_y)$
$(T_x)\ B_{3u}$	B_u	$B_{3u}\ (T_x)$

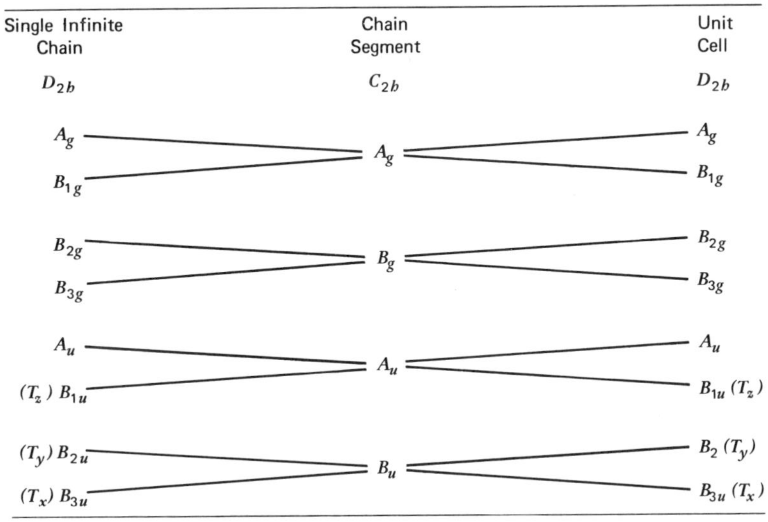

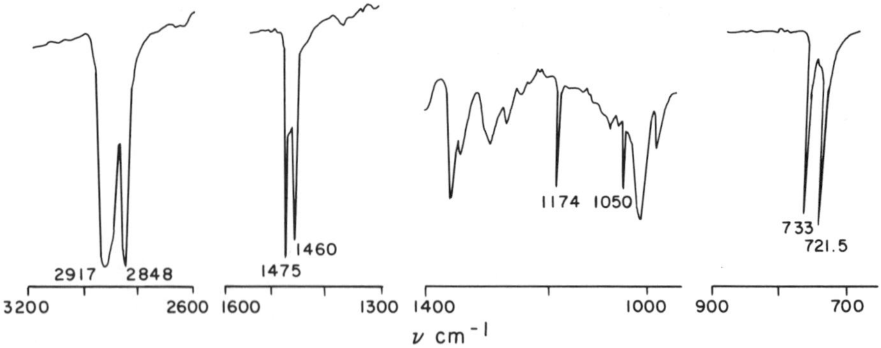

Figure 11.12. The infrared spectrum of polyethylene at 77°K. [Reprinted by permission from P. J. Hendra, H. P. Jobic, E. P. Marsden, and D. Bloor, *Spectrochim. Acta* **33A**, 445 (1977).]

chain axis) should be split into two components, B_{2u} and B_{3u}, in the crystal. (These crystal modes are polarized parallel to the *b* and *a* axes of the unit cell, respectively.) However, parallel chain vibrations in symmetry species B_{1u} correlate to A_u and B_{1u} modes in the crystal, the former being infrared inactive, so that we would not predict a splitting of these modes. Finally, the inactive A_u species of the isolated chain could possibly become infrared active, while two translatory lattice vibrations should be active in the far-infrared spectrum.

The infrared spectrum of polyethylene is shown in Figure 11.12. The doublets at 721.5, 733 cm^{-1}, and at 1460, 1475 cm^{-1}, are due to correlation splitting of the methylene rocking and bending modes, respectively of the single chain. The fundamentals at 1050 cm^{-1} and 1174 cm^{-1} are singlets, as predicted for B_{1u} modes.

References

1 R. Zbinden, "Infrared Spectroscopy of High Polymers," Academic Press, New York, 1964.

2 G. Turrell, "Infrared and Raman Spectra of Crystals," Academic Press, New York, 1972.

3 S. Krimm, C. Y. Liang, and G. B. B. Sutherland, *J. Chem. Phys.* **25**, 549 (1956).

4 P. W. Higgs, *Proc. Roy. Soc. (London)* **A220**, 470 (1953).

5 W. G. Fately, F. R. Dollish, N. T. McDevitt, and F. F. Bertley, "Infrared and Raman Selection Rules for Molecular and Lattice Vibrations: The Correlation Method," Wiley-Interscience, New York, 1972.

12

NORMAL COORDINATE ANALYSIS OF ISOLATED POLYMER CHAINS

Hereafter, when they come to model Heaven
And calculate the stars, how they will wield
Thy mighty frame, how build, unbuild, contrive
To save appearances, how gird the sphere
With centric and eccentric scribbled o'er
Cycle and epicycle, orb in orb.

John Milton

We have treated the general problem of the vibrations of a linear chain in Chapter 10. For infinite chains the vibrational frequencies of the translational repeat unit were expressed as a continuous dispersion curve. In Chapter 11 it was demonstrated that for helical polymers we could further simplify the problem by considering the vibrations of the chemical repeat unit, expressed as a function of the phase shift between corresponding modes in adjacent units. The selection rules for an infinite perfect chain permit only those modes corresponding to specific values of this phase angle to be Raman or infrared active. Higgs (1) first proposed that this simplification of the vibrational problem could be used in conjunction with the Wilson **GF** matrix method, appropriately extended to polymers by expressing the potential and kinetic energy as a function of the phase angle. In this chapter we will consider this theoretical development in more detail. Naturally, this treatment is only an approximation, since real polymers never have an infinite length and are imperfectly ordered. In addition, interchain interactions in polymer crystals have to be considered. Nevertheless, the normal coordinate analysis of theoretically isolated perfect chains is a powerful and useful aid in understanding the vibrational spectra of actual polymers and applying this knowledge to structural studies. In part this is because the effect of a small concentration of conformational defects or interchain interactions can to a first approximation be neglected or treated as a perturbation. We shall consider polymer crystals and disordered polymers in detail in later chapters.

12.1 The Derivation of the Secular Equation Describing an Infinite, Ordered Polymer Chain

In order to calculate the normal modes of polymer chains we need to express the vibrational problem in terms of a repeat unit, usually the chemical repeat unit, and a phase angle θ* relating the vibrations in adjacent units. Some of the basic concepts have already been developed in Chapter 11, where we considered the symmetry properties of ordered chains. Higgs (1) was the first to use the symmetry properties of helical chains in applying the Wilson **GF** method of polymers. However, the equations can be set up in a number of ways. We will consider essentially the original treatment of Higgs (1) with modifications discussed by Piseri and Zerbi (2).

The set of all operations H^s (where s is any integer) that transform one unit into the sth unit along the chain constitutes an infinite group isomorphous to the infinite cyclic group C_∞ (1). This set of operations can be written as

$$H^s(\psi)=\begin{vmatrix} \cos s\psi & -\sin s\psi & 0 \\ \sin s\psi & \cos s\psi & 0 \\ 0 & 0 & 1 \end{vmatrix}\begin{vmatrix} x_i \\ y_i \\ z_i \end{vmatrix}+s\begin{vmatrix} 0 \\ 0 \\ 1 \end{vmatrix} \tag{12.1}$$

The irreducible representations of this groups are one-dimensional and the characters are given by

$$\chi(\theta, H^s)=e^{is\theta} \tag{12.2}$$

as discussed in the preceding chapter. (Note that ψ refers to the rotation angle defining the relative positions of adjacent units in the polymer chain, while θ represents the phase angle between vibrations in these units.) Accordingly, every normal mode of a polymer in an ordered helical conformation belongs to one of the irreducible representations of the cyclic group. If a normal mode in a given unit cell vibrates with an amplitude A, then the sth unit down the chain must vibrate in the same manner with an amplitude $Ae^{-is\theta}$ or $Ae^{is\theta}$. For each frequency $\nu_i(\theta)$ there is a corresponding complex conjugate $\nu_i(-\theta)$. In other words, with the exception of $\nu_i(0)$ and $\nu_i(\pi)$, the frequencies are doubly degenerate. This periodic relationship will naturally determine the form of the secular equation, whether we express it in terms of internal coordinates using the Wilson **GF** matrix approach or in terms of cartesian coordinates. Higgs (1) originally derived the phase relationship by a reduction of the infinite **F** and **G** matrices made possible by a conversion to internal symmetry coordinates. An alternative derivation, used by Piseri and Zerbi (2), is to assume a periodic

*We will use θ to designate the phase difference between vibrations in a polymer, as opposed to the symbol ϕ used in Chapter 10 to designate the phase difference between vibrations in the units of a model finite chain.

solution, an approach we discussed earlier in our treatment of lattice vibrations. We consider it useful to examine both since there is a degree of variation in the nomenclature used in the literature that, at least to us, appears confusing and plagued by typographical errors.

In order to derive the secular equation we need to obtain expressions for the potential and kinetic energy. First, the potential energy is expanded in a Taylor series about the equilibrium position (as in Chapter 2) and the expansion truncated to quadratic terms within the harmonic approximation.

$$V = V_0 + \sum_{n,i} F_i^n R_i^n + \tfrac{1}{2} \sum_{n,r} \sum_{i,k} F_{ik}^{nr} R_i^n R_k^r \qquad (12.3)$$

where R_i^n represents the ith internal coordinate of the nth chemical repeat and

$$F_i^n = \left(\frac{\delta V}{\delta R_i^n} \right)_0 \qquad (12.4)$$

$$F_{ik}^{nr} = \left(\frac{\delta^2 V}{\delta R_i^n \, \delta R_k^r} \right)_0 \qquad (12.5)$$

At the equilibrium configuration F_i^n is zero, leaving

$$2V = \sum_{n,r} \sum_{i,k} F_{ik}^{nr} R^n R^r \qquad (12.6)$$

Because of the helical symmetry and periodicity of the chain we can write

$$F_{ik}^{nr} = F_{ik}^{s} \qquad (12.7)$$

where $s = |n - r|$ and represents interactions between the ith coordinate in one unit and the kth coordinate s units away. Imposing this condition on equation (12.6), we obtain

$$2V = \sum_{n,i,k} F_{ik}^0 R_i^n + \sum_{\substack{n,s \\ i,k}} \left(F_{ik}^s R_i^n R_k^{n+s} + F_{ki}^s R_i^n R_k^{n-s} \right) \qquad (12.8)$$

The kinetic energy can be written in an analogous form by using the kinetic energy matrix **G** and the momenta P_i^n conjugated to the coordinates R_i^n:

$$2T = \sum_{n,i,k} G_{i1}^0 P_i^n P_k^n + \sum_{n,s,i,k} \left(G_{ik} P_i^n P_k^{n+s} + G_{ki} P_i^n P_k^{n-s} \right) \qquad (12.9)$$

Piseri and Zerbi (2) then applied the approach discussed in our treatment of lattice vibrations. The expressions for the kinetic and potential energy are substituted into the Lagrangian equation of motion, in this case yielding an infinite number of second-order differential equations in R_i^{n+s} (or X_i^{n+s} if we choose a cartesian coordinate basis set). The solutions to such differential equations assume a periodic form:

$$R_i^{n+s} = A_i \exp[-i(\lambda t + s\theta)] \tag{12.10}$$

where θ is the phase difference between amplitudes of vibration of equivalent atoms on successive residues and λ is the circular frequency. Substitution of these equations into the set of differential equations obtained by using the Lagrangian equation of motion yields $3N$ simultaneous linear equations in the unknown amplitudes of vibration A_i. The nontrivial solutions are given by

$$|\mathbf{G}(\theta)\mathbf{F}(\theta) - \boldsymbol{\lambda}(\theta)\mathbf{E}| = 0 \tag{12.11}$$

When the secular equation is expressed in internal coordinates, $\mathbf{G}(\boldsymbol{\theta})$ and $\mathbf{F}(\boldsymbol{\theta})$ are given by

$$\mathbf{G}(\theta) = \mathbf{G}^0 + \sum_s (\mathbf{G}^s e^{is\theta} + \mathbf{G}^s e^{-is\theta}) \tag{12.12}$$

$$\mathbf{F}(\theta) = \mathbf{F}^0 + \sum_s (\mathbf{F}^s e^{is\theta} + \mathbf{F}_e^{s-is\theta}) \tag{12.13}$$

In some treatments the $\mathbf{G}(\boldsymbol{\theta})$ and $\mathbf{F}(\boldsymbol{\theta})$ matrices are defined in an equivalent form in terms of an integer n designating the unit cell:

$$\mathbf{G}(\theta) = \sum_{n=-\infty}^{\infty} \mathbf{G}_n e^{in\theta} \tag{12.14}$$

$$\mathbf{F}(\theta) = \sum_{n=-\infty}^{\infty} \mathbf{F}_n e^{in\theta} \tag{12.15}$$

in contrast to equations (12.12) and (12.13), which depend on the distance of interaction s, which is equal to the number of units between the n and n' unit cells (i.e., $s = |n - n'|$).

Equations (12.14), (12.15) correspond to those derived originally by Higgs (1) using a transformation to symmetry coordinates:

$$S_i(\theta) = (2\pi)^{-1/2} \sum_s R_i^{n+s} e^{is\theta} \tag{12.16}$$

This equation is simply derived by considering the equation for the generation of symmetry coordinates given in Chapter 5, rewritten in terms of the symmetry of a helix as

$$S_i(\theta) = N \sum \chi(\theta, H^s) H^s R_i^n \tag{12.17}$$

The character $X(\theta, H^s)$ is equal to $e^{is\theta}$, as discussed earlier, while the coordinate generated by the symmetry operation H^s upon the internal coordinate R_i^n is R_i^{n+s}. A Fourier transformation of equation (12.16) gives

$$R_i^{n+s}=(2\pi)^{-1/2}\int_{-\pi}^{\pi} S_i(\theta)e^{-is\theta}\,d\theta \tag{12.18}$$

where the summation in equation (12.16) has been replaced by an integration over the first Brillouin zone. Substituting into the expressions for the kinetic and potential energy, we obtain, as the infinite order of equations (12.8) and (12.9),

$$2V=\int_{-\pi}^{+\pi}\bar{\mathbf{S}}(\theta)\mathbf{F}(\theta)\mathbf{S}(\theta)\,d\theta \tag{12.19}$$

$$2T=\int_{-\pi}^{+\pi}\bar{\dot{\mathbf{S}}}(\theta)\mathbf{G}^{-1}(\theta)\dot{\mathbf{S}}(\theta)\,d\theta \tag{12.20}$$

where $\mathbf{S}$ is the Hermitian conjugate of $\bar{\mathbf{S}}$, and $\mathbf{F}(\boldsymbol{\theta})$ and $\mathbf{G}(\boldsymbol{\theta})$ are as defined earlier. Note that in equation (12.20) we define T in terms of $\mathbf{G}^{-1}$ as opposed to the dependence on $\mathbf{G}$ given in equation (12.9). This is because we are expressing the kinetic energy in terms of internal coordinates instead of their momentum conjugates (i.e., we can write $2T=\mathbf{P}^T\mathbf{G}\mathbf{P}$ or $2T=\mathbf{R}^T\mathbf{G}^{-1}\mathbf{R}$; see Chapter 3). The secular equation of order $3N$, as given in equation (12.11), is then derived directly by substituting the finite-order kinetic and potential energy expressions given by equations (12.19) and (12.20) in the Lagrange equation.

If there is local symmetry within the repeat unit, symmetry coordinates (U) can be introduced in the usual way. It should be kept in mind that such "local" symmetry coordinates have to be "phased" ($\mathbf{U}(\theta)$) to account for the helical or translational symmetry of the chain. For example, if we set up the secular equation using a cartesian coordinate basis (but expressing the force constant in terms of internal coordinates, as described in Chapter 3), we obtain

$$|\mathbf{U}(\theta)\mathbf{D}^T(\theta)\mathbf{F}(\theta)\mathbf{D}(\theta)\mathbf{U}^T(\theta)-\boldsymbol{\lambda}(\theta)\mathbf{E}|=0 \tag{12.21}$$

where $\mathbf{U}(\boldsymbol{\theta})$ is the phase-dependent external symmetry coordinate transformation matrix.

Equations (12.11) or (12.21) yields $3N$ solutions of the form

$$\lambda(\theta)=4\pi^2c^2\nu^2(\theta) \tag{12.22}$$

As in the case of the simple monatomic and diatomic chains discussed in Chapter 10, we have obtained a dispersion relationship, the difference being only in the level of complexity. For a polymer molecule having $3N$ atoms in every chemical repeat unit, we calculate $3N$ branches. The dispersion relation has a period of 2π [i.e., $\nu(\theta)=\nu(\theta+2\pi)$], and the first Brillouin zone is defined by $-\pi \leqslant \theta \leqslant \pi$. In addition, it can be shown that $\nu(\theta)=\nu(-\theta)$, so that it is possible to limit calculations to half of the Brillouin zone, $0 \leqslant \theta \leqslant \pi$. An examination of the selection rules, discussed in Chapter 11, demonstrates that the infrared- and Raman-active fundamentals of a helical polymer can be determined from three equations of the infinite set, namely, those corresponding to $\theta=0$, ψ, and 2ψ. However, it is often of value to plot the dispersion curve by obtaining, for example, solutions every $5°$. Unfortunately, there can be difficulties or ambiguities in determining the dispersion curves because they can cross or repel, and often a knowledge of the eigenvectors is required in order to trace the curves accurately. It should also be noted at this stage that two of the dispersion curves always reach zero for $\theta=0$. These branches correspond to the acoustic branches discussed in Chapter 10 for much simpler models.

Although the vibrational problem has been reduced to the dimensions of the chemical repeat unit, there remain two aspects of the problem that have to be considered. The first is trivial from a conceptual point of view and involves defining the phase-angle-dependent **B** matrix in a manner that allows a simple and straightforward definition of coordinates. We will discuss the form used by Small et al. (3). The second problem is revealed by an examination of equations (12.19) and (12.20), which demonstrate that in the general case this form of the secular equation is defined in terms of complex numbers.

We will first consider the problem of setting up the **B** matrix for a given chemical repeat unit. The internal displacement coordinates $\mathbf{R}^n$ of the nth chemical repeat unit are related to the cartesian displacement coordinates by

$$\mathbf{R}^n = \sum_{s=-m}^{m} \mathbf{B}^{n,n+s} \mathbf{X}^{n+s} \tag{12.23}$$

where $\mathbf{X}^{n+s}$ is the cartesian displacement coordinate of the atoms s units from n, and $\mathbf{B}^{n,n+s}$ is the usual **B** matrix, which now transforms the cartesian displacements of the $(n+s)$th unit into the internal coordinates of the nth unit. In other words, it is necessary to consider atoms outside the chemical repeat unit in order to define all the internal coordinates of the repeat unit. For example, in order to define a torsional internal coordinate for polyethylene (chemical repeat unit CH_2) we have to use four carbon atoms, τ_{1234}:

$$
\begin{array}{ccccc}
 & CH_2 & & CH_2 & \\
\diagdown & \diagup & \diagdown & \diagup & \diagdown \\
CH_2 & & CH_2 & & \\
-1 & 0 & 1 & & 2
\end{array}
$$

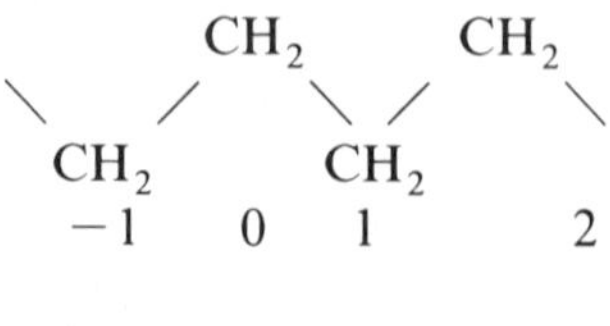

Therefore, we have to include the contributions to the **B** matrix from atoms one and two units removed [i.e., $s=2$ in equation (12.23). In general, the maximum value of s for which the corresponding element of the **B** matrix is nonzero will depend on the geometry of the helix and the extent of kinetic and potential interactions we wish to consider. But in most cases this is limited to adjacent or more rarely next nearest neighbor contributions.

In order to simplify the calculation of the **B** matrix, it is convenient to relate the coordinates of all atoms in the rth unit cell to an arbitrarily defined initial unit cell, $n=0$, by

$$\mathbf{X}^r = \mathbf{H}^r \mathbf{X}^0 \tag{12.24}$$

where $\mathbf{H}^r$ is as defined in equation (12.1) and the cartesian coordinates of each unit are all related by a rotation through an angle ψ followed by a translation z_1. However, in the following treatment only the rotational part of $\mathbf{H}^r$ has to be considered because the internal coordinate definitions are invariant to simple translation.

Then for $n=0$, equation (12.23) becomes

$$\mathbf{R} = \sum_{s=-m}^{m} \mathbf{B}^{0,s} \mathbf{X}^s \tag{12.25}$$

The matrices $\mathbf{B}^{0,s}$ are calculated by fixing the x and y axes of the helix. The cartesian displacements $n+s$ units away are related to the coordinates $\mathbf{X}^{0+s}$ by the relation

$$\mathbf{X}^{n+s} = \mathbf{H}^n \mathbf{X}^s \tag{12.26}$$

Substituting this expression into equation (12.23), we obtain

$$\mathbf{R}^n = \sum_s \mathbf{B}^{n,n+s} \mathbf{H}^n \mathbf{X}^s \tag{12.27}$$

Because the definition of the internal coordinates of the chemical repeat unit is independent of the value of n, we can equate equations (12.25) and (12.27):

$$\sum_s \mathbf{B}^{0,s} \mathbf{X}^s = \sum_s \mathbf{B}^{n,n+s} \mathbf{H}^n \mathbf{X}^s \tag{12.28}$$

By equating the coefficients of $\mathbf{X}^s$ we obtain the relationship

$$\mathbf{B}^{0,s} = \mathbf{B}^{n,n+s} \mathbf{H}^n \tag{12.29}$$

which can be rewritten as

$$\mathbf{B}^{n,n+s} = \mathbf{B}^{0,s} (\mathbf{H}^n)^{-1} = \mathbf{B}^{0,s} (\mathbf{H}^n)^T \tag{12.30}$$

If there are no redundant internal coordinates (apart from those due to translational and rotational degrees of freedom) there will be $3N$ internal coordinates in a chemical repeat unit containing N atoms. Consequently, the matrices $\mathbf{B}^{n,n+s}$ will be of dimension $3N \times 3N$. In addition,

$$(\mathbf{B}^{n,n+s})^T = \mathbf{H}^n (\mathbf{B}^{0,s})^T \tag{12.31}$$

This relationship is important because it allows us to obtain the values of $\mathbf{B}^{n,n+s}$ from $\mathbf{B}^{0,s}$ in a simple manner. Once the $\mathbf{B}^{0,s}$ matrix is calculated for the appropriate values of s, the infinite-order $\mathbf{G}^m$ matrix can be calculated as

$$\mathbf{G}^m = \sum_{n+s} \mathbf{B}^{0,n+s} \mathbf{M}_1^{-1} (\mathbf{B}^{m,n+s})^T \tag{12.32}$$

where $\mathbf{G}^m$ is the contribution to the kinetic energy due to interactions of internal coordinates separated by m units.

In practice, the calculation of the $\mathbf{B}$ matrix is a relatively straightforward process and is handled by computer using this type of approach. Essentially, the input has to consist of a set of cartesian coordinates for all the atoms necessary to define the internal coordinates of the chemical repeat plus a set of symmetry coordinates accounting for the helical symmetry of the chain and any other symmetry elements present.

Finally, we wish to consider the problem that in the general case the symmetry coordinates and the secular equation for a polymer will be complex. The exceptions are for values of $\theta = 0$ or π. Fortunately, it is possible to transform a complex matrix to an equivalent real matrix by means of a similarity transformation (2). If $\mathbf{H}$ is a complex matrix with a real part $\mathbf{A}$ and an imaginary part $\mathbf{B}$, it can be written

$$\mathbf{C} = \begin{bmatrix} \mathbf{H} & \mathbf{0} \\ \mathbf{0} & \mathbf{H}^* \end{bmatrix} = \begin{bmatrix} \mathbf{A} + \mathbf{iB} & \mathbf{0} \\ \mathbf{0} & \mathbf{A} - \mathbf{iB} \end{bmatrix} \tag{12.33}$$

The complete matrix $\mathbf{C}$ can be transformed to its real equivalent $\mathbf{Y}$ by the similarity transformation.

$$\mathbf{Y} = \mathbf{PCP}^{-1} = \begin{bmatrix} \mathbf{A} & -\mathbf{B} \\ \mathbf{B} & \mathbf{A} \end{bmatrix} \tag{12.34}$$

where

$$\mathbf{P} = \frac{1}{\sqrt{2}} \begin{bmatrix} \mathbf{E} & \mathbf{E} \\ -\mathbf{iE} & \mathbf{E} \end{bmatrix} \tag{12.35}$$

and $\mathbf{E}$ is the identity matrix of order $3N$. For the general example

$$\mathbf{B}(\boldsymbol{\theta}) = \sum_{n-\infty}^{\infty} \mathbf{B}^n \exp(\mathbf{i}n\boldsymbol{\theta}) = \sum_n (\mathbf{B}^n \cos n\boldsymbol{\theta} + \mathbf{iB}^n \sin n\boldsymbol{\theta}) \tag{12.36}$$

where $\cos n\boldsymbol{\theta}$ and $\sin n\boldsymbol{\theta}$ are column matrices we can write

$$\mathbf{B}(\boldsymbol{\theta})=\begin{bmatrix}\mathbf{B} & \mathbf{0}\\ \mathbf{0} & \mathbf{B}\end{bmatrix}$$

$$=\begin{bmatrix}\sum_n \mathbf{B}^n\cos n\boldsymbol{\theta}+\mathbf{iB}^n\sin n\boldsymbol{\theta} & \mathbf{0}\\ \mathbf{0} & \sum_n \mathbf{B}^n\cos n\boldsymbol{\theta}-\mathbf{iB}^n\sin n\boldsymbol{\theta}\end{bmatrix} \tag{12.37}$$

and the real matrix equivalent becomes

$$\mathbf{B}(\theta)=\begin{bmatrix}\sum_n(\mathbf{B}^n\cos n\theta) & -\sum_n(\mathbf{B}^n\sin n\theta)\\ \sum_n(\mathbf{B}^n\sin n\theta) & \sum_n(\mathbf{B}^n\cos n\theta)\end{bmatrix}$$

$$=\begin{bmatrix}\mathbf{B}^{\cos} & -\mathbf{B}^{\sin}\\ \mathbf{B}^{\sin} & \mathbf{B}^{\cos}\end{bmatrix} \tag{12.38}$$

where

$$\mathbf{B}^{\cos}=\sum_n \mathbf{B}^n\cos\theta n \tag{12.39}$$

$$\mathbf{B}^{\sin}=\sum_n \mathbf{B}^n\sin\theta n \tag{12.40}$$

Each element of equation (12.31) is a $3N\times 3N$ array. The $\mathbf{F}(\theta)$ and $\mathbf{G}(\theta)$ matrices are obtained in an equivalent form:

$$\mathbf{G}(\theta)=\begin{bmatrix}\mathbf{G}^{\cos} & -\mathbf{G}^{\sin}\\ \mathbf{G}^{\sin} & \mathbf{G}^{\cos}\end{bmatrix}\mathbf{F}(\theta)=\begin{bmatrix}-\mathbf{F}^{\cos} & -\mathbf{F}^{\sin}\\ \mathbf{F}^{\sin} & \mathbf{F}^{\cos}\end{bmatrix} \tag{12.41}$$

By the conversion to real matrices the size of the secular equation has been doubled. The latent roots of this new secular equation can be shown to consist of all the roots of the original equation plus their complex conjugates.

12.2 Example: The One-Dimensional Diatomic Chain

Even though the final result obtained in the preceding section, the dependence of the terms of the secular equation on the phase angle θ relating vibrations in successive units, is perhaps intuitively grasped, the specific application of certain equations is probably not clear. Consequently, before proceeding to consider the application of this method to a real polymer, polyethylene (see Chapter 13), we will briefly reconsider a simple example, the one-dimensional

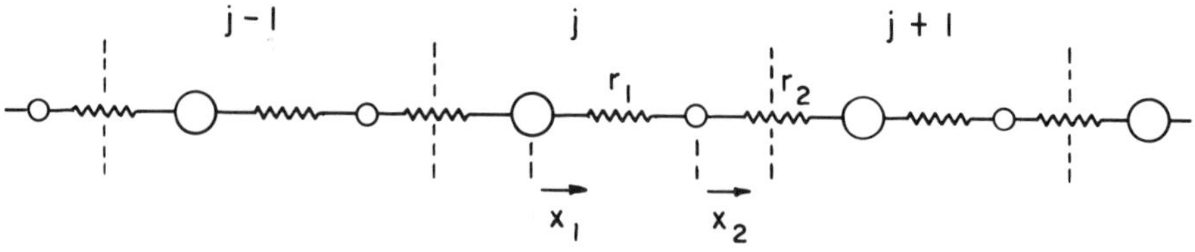

Figure 12.1. Internal coordinates of a linear diatomic chain.

vibrations of a diatomic chain. This problem was tackled in Chapter 10, but the vibrational problem was set up and solved in terms of cartesian displacement coordinates. We will now consider a description in terms of internal coordinates and set up the secular matrix in terms of the Wilson **GF** form as modified by Higgs (1) for polymer chains.

The diatomic chain is illustrated in Figure 12.1. The translational repeat unit contains two atoms, labeled 1 and 2, with masses m_1 and m_2, respectively. The unit cells are labeled $j-1$, j, $j+1,\ldots$. Accordingly, the cartesian and internal displacement coordinates are labeled x_1^j, x_2^j, r_1^j, r_2^j, for the jth unit cell. If we consider only the bond stretching force constants and assume that all the bonds in the lattice are equivalent, equation (12.12) becomes

$$\mathbf{F}(\theta)=\sum_{n=-\infty}^{\infty}\mathbf{F}^n\exp(in\theta)=\mathbf{F}^0 \tag{12.42}$$

where

$$\mathbf{F}^0=\begin{bmatrix}\mathbf{f} & \mathbf{0}\\ \mathbf{0} & \mathbf{f}\end{bmatrix} \tag{12.43}$$

and **f** is the bond stretching force constant. In order to set up the $\mathbf{B}(\theta)$ matrix we first need to construct the **B** matrix for the repeat unit, which can include contributions from adjacent units, depending on the internal coordinates being defined. For this simple example the transformation matrix between the internal coordinates r_j and the cartesian displacement coordinates is

	x_1^{j-1}	x_2^{j-1}	x_1^j	x_2^j	x_1^{j+1}	x_2^{j+1}
r_1^j	0	0	-1	1	0	0
r_2^j	0	0	0	-1	1	0

The matrices $\mathbf{B}^n$ then take the form

$$\mathbf{B}^{-1}=\begin{bmatrix}0 & 0\\ 0 & 0\end{bmatrix},\qquad \mathbf{B}^0=\begin{bmatrix}-1 & 1\\ 0 & -1\end{bmatrix},\qquad \mathbf{B}^1=\begin{bmatrix}0 & 0\\ 1 & 0\end{bmatrix} \tag{12.44}$$

We can then obtain $\mathbf{B}(\theta)$ in the complex form from equation (12.29) or in real form from equation (12.31). For this simple example we can use equation

12.29:

$$\mathbf{B}(\theta) = \sum_{n=-\infty}^{+\infty} \mathbf{B}^{n} \exp(in\theta)$$

$$= \begin{bmatrix} 0 & 0 \\ 0 & 0 \end{bmatrix} + \begin{bmatrix} -1 & 1 \\ 0 & -1 \end{bmatrix} + \begin{bmatrix} 0 & 0 \\ e^{i\theta} & 0 \end{bmatrix}$$

$$= \begin{bmatrix} -1 & 1 \\ e^{i\theta} & -1 \end{bmatrix} \tag{12.45}$$

The secular equation

$$|\mathbf{D}^{T}(\theta)\mathbf{F}(\theta)\mathbf{D}(\theta) - \boldsymbol{\lambda}(\theta)\mathbf{E}| = 0 \tag{12.46}$$

where $\mathbf{D}(\theta) = \mathbf{B}(\theta)\mathbf{M}^{-1/2}$ then becomes

$$\begin{bmatrix} \dfrac{2f}{m_1} & -\dfrac{f}{(m_1 m_2)^{1/2}}(1+e^{-i\theta}) \\ \dfrac{f}{(m_1 m_2)^{1/2}}(1+e^{i\theta}) & \dfrac{2f}{m_2} \end{bmatrix} - \begin{bmatrix} \lambda(\theta) & 0 \\ 0 & \lambda(\theta) \end{bmatrix} = 0 \tag{12.47}$$

The solution is the dispersion relation given in Chapter 10:

$$\lambda(\theta) = f\left\{ \frac{m_1 + m_2}{m_1 m_2} \pm \left[\left(\frac{m_1 + m_2}{m_1 m_2} \right)^2 - \frac{2}{m_1 m_2}(1 - \cos\theta) \right]^{1/2} \right\} \tag{12.48}$$

12.3 Force Fields for Polymer Molecules

In Section 12.2 the secular equation was set up in a general form to include interactions between coordinates separated by s units. In principle, there is no limitation on s, so that the definition of force constant interaction terms could become extraordinarily complex. However, for organic polymers the observed frequencies and dispersion curves can be reproduced by considering only nearest neighbor interactions (4). In fact, Schachtschneider and Snyder (5) determined that the magnitude of the interaction force constants decreases in a gratifying manner on going from nearest neighbor to next nearest neighbor interactions in n-paraffin chains. Nevertheless, even though the next nearest neighbor interaction terms were not insignificant, in subsequent calculations they were neglected (6). This observation brings us to a general discussion of the limitations of force fields applicable to polymer molecules. Because of familiarity and prejudice we will confine this discussion to valence force fields, although the general principles apply to other potential functions.

We noted in Chapter 6 that except for the simplest of molecules, a general valence force field cannot be determined. The vibrational problem is underdetermined by the observed data. Consequently, our enthusiasm for the results of any normal coordinate calculation has to be tempered by our reservations concerning the force field. This is particularly true of polymers, where not only do we make assumptions concerning which interaction terms within the repeat unit are significant, but also we invariably confine the definition of interunit interactions to nearest neighbor terms. Even so, the observed frequencies obtained from spectroscopic studies of polymers are usually insufficient to determine the force constants considered significant on the basis of previous experience or on the basis of something approaching chemical intuition. This difficulty has been circumnavigated by the use of the so-called overlay technique (5). It is assumed that the force constants describing chemically and structurally similar molecules are the same; in other words, transferability is assumed. The least-squares fitting of observed and calculated frequencies, the essence of normal coordinate analysis, is performed for a large number of molecules. This technique has proved particularly successful for the linear and branched hydrocarbons, as described in the seminal papers of Snyder and Schachtschneider (5, 6). In fact, various modifications of this force field have provided the basis for the normal vibrational analysis of many hydrocarbon polymers.

In calculating the normal modes of polymers the most critical area would therefore seem to be the degree of transferability of the calculated force field. It has been our experience that even a force field of proven utility, such as that developed for paraffins and polyethylene (5, 7), does not always produce a good fit between certain observed and calculated frequencies for the chain vibrations of other hydrocarbon polymers. Discrepancies can be due to differences in chain geometry or can arise because the nature of substituents (in the case of vinyl polymers) produces an unaccounted-for perturbation. A further refinement of one or two force constants can often improve the fit of the recalcitrant frequencies. However, the subsequent interpretation of the results has to be made with care, particularly in the case of structural work. Vibrational spectroscopists are often seriously tempted to draw structural conclusions on the basis of a fit between observed and calculated frequencies. If the force field has been inadequately defined, any such conclusions are essentially unreliable. We will reexamine in more detail the specific effects of changes in geometry in Chapter 13.

Taking into account the reservations discussed earlier, Zerbi (8) has listed the criteria that, in principle, should be used to judge the reliability of normal coordinate calculations of polymers:

1 The force field to be used should have been calculated with the most recent techniques and should have been shown to be transferable to many model

molecules. The statistical dispersion $\sigma\{\bar{\phi}_i\}$ of each force constant must be clearly shown. The quantity $\sigma\{\bar{\phi}_i\}$ does not have an absolute meaning [see (9)], but gives a rough estimate of the stability of the least-squares calculation.

2 Only zero-order calculations should be carried out on the polymer, and no adjustment of the force constants should be made in order to improve the fit. Conclusions of practical importance from the calculation must be derived after pointing out the degree of approximation accepted with the use of a given force field.

3 Pairing between calculated and experimental frequencies should be supported by experimental evidence. Dichroism helps to assign the symmetry species of each fundamental.

4 No attempt should be made to match several experimental and observed frequencies that all occur in a narrow frequency range. A slight modification of the interaction force constants may switch the order of the species of the calculated frequencies.

5 For structural works the fit is more meaningful for frequencies below 1200 cm^{-1}, where the most structure-sensitive modes occur and where large overlapping is less likely.

6 Since computers are available, how sensitive a calculated frequency is to changes of geometry should be tested.

7 The uncertainty of the normal modes should be evaluated by means of recently developed techniques (9, 10).

8 The description of the normal modes may slightly or drastically change, according to the force field adopted. Hence, in proposing a vibrational assignment, one must always specify the type of force field used and the approximations accepted.

At this point it is useful to discuss briefly the physical meaning and value of force constant calculations. On occasion, the calculation of force constants is justified on the grounds that it affords a physical insight into bond order, delocalization of electrons, and the forces between nonbonded atoms. However, as pointed out (once more) by Zerbi (8), the basic assumption of harmonic motion together with the other uncertainties and arbitrary compromises introduced into the calculations neutralizes the physical meaning of the force constants. They are reduced to useful parameters for predicting the spectra of molecules. This is not to say that the calculation of a model force field is a pointless exercise. If the criteria just listed are carefully observed, we can determine the form of the normal modes of a molecule with some confidence. A good assignment of the observed vibrational spectrum is not only a fundamental piece of knowledge in its own right, but can aid in understanding, for example, the chemical changes that occur upon reaction, such as the oxidation of a polymer. In addition, a good transferable force field can allow the plotting of phonon dispersion curves and the calculation of the density of states. This in turn provides a base for the interpretation of the results of experimental techniques such as neutron scattering. Finally, normal coordinate calculations are capable of providing insight into the conformational states of a molecule, as demonstrated by Snyder (7) in his study of polyethylene and the *n*-paraffins.

References

1 P. W. Higgs, *Proc. Roy. Soc. (London)* **A220**, 472 (1953).

2 L. Piseri and G. Zerbi, *J. Chem. Phys.* **48**, 3561 (1968).

3 E. W. Small, B. Fanconi, and W. L. Peticolas, *J. Chem. Phys.* **52**, 4369 (1970).

4 G. Zerbi, in "Vibrational Spectroscopy—Modern Trends," A. J. Barnes and W. J. Orville-Thomas, Eds., Elsevier, New York, 1977, Chapter 24, p. 379.

5 J. H. Schachtschneider and R. G. Snyder, *Spectrochim. Acta* **19**, 117 (1963).

6 R. G. Snyder and J. H. Schachtschneider, *Spectrochim. Acta* **21**, 169 (1965).

7 R. G. Snyder, *J. Chem. Phys.* **47**, 1316 (1967).

8 G. Zerbi, *Appl. Spect. Revs.* **2**, 193 (1968).

9 J. L. Duncan, *Spectrochim. Acta* **20**, 1197 (1964).

10 J. Scherer, *Spectrochim. Acta* **22**, 1179 (1966).

13

THE NORMAL VIBRATIONAL ANALYSIS OF POLYETHYLENE

He fixed thee mid this dance
Of plastic circumstance.
Robert Browning

Polyethylene has a special place in polymer science. As the simplest polymer in terms of chemical structure it has been the preferred subject of many studies aimed at elucidating the general physical laws that govern the behavior of all polymers. This macromolecule also has its charm for many vibrational spectroscopists, particularly because an extremely wide range of oligomers, the *n*-paraffins, are available for force field calculations. Consequently, polyethylene has been exhaustively analyzed. A classified bibliography for the years 1949 through 1976 has been presented by Tasumi (1), and we will not attempt a detailed review of the literature. Our aim in this chapter is far more limited: to present a discussion of the development of a valence force field for saturated hydrocarbons and to consider the details of the analysis of a single chain of polyethylene. This will serve as a general illustration of the methods discussed in Chapter 12.

In our opinion, the classic work on the application of normal coordinate calculations to polymer chains is the analysis of the paraffins and polyethylene by Schachtschneider and Snyder (2–4) and Snyder (5,6). This body of work is, in effect, a series of successively modified force fields. The original force field (3) was based on band assignments for the *n*-paraffins in the planar zigzag conformation. This force field was subsequently extended to include other types of paraffins and one or two nonplanar conformations (4). Snyder (5,6) then noted that the earlier force fields needed to be modified further, in part because of an incorrect assignment of the polyethylene B_{2g} wagging mode and also because these force fields gave significantly less accurate frequencies when applied to nonplanar conformations of the *n*-paraffins.

More recently, Barnes and Fanconi (7) have reviewed the earlier work and recalculated a force field, in this case combining the valence force constants in the form of local symmetry force constants. We will discuss these various force fields in more detail later. First we will turn our attention to the essential preliminary step of analyzing and assigning the observed vibrational spectra of the paraffins.

13.1 Preliminary Analysis of the Vibrational Spectra of *n*-Paraffins

We have noted in preceding chapters that the essence of normal coordinate analysis is the calculation of a suitable force field. This is obtained by refining an initial calculated set of frequencies to a set of experimentally observed frequencies. Naturally, the correct observed and calculated frequencies have to be matched in the refinement. This can be accomplished in part by using group frequencies as a guide to band assignments. Dealing with a series of oligomers entails the added complication of assigning the band series that can be observed for many modes. Snyder (5,6) and Snyder and Schachtschneider (2) used a simple coupled oscillator model in their original analysis. We have considered this type of model in Chapter 10, paying particular attention to the expected intensity distribution of the progression band series. Here we will explore in more detail the application of this simple approach in conjunction with a symmetry analysis.

Hydrocarbon chains with even or odd numbers of carbon atoms can be assigned to C_{2v} or C_{2h} point groups, respectively, assuming that the chain is in the extended zigzag conformation. Figure 13.1 illustrates the symmetry properties of each type of chain. An *n*-paraffin with *n* carbon atoms has $3n+2$ atoms and $9n[=3(3n+2)-6]$ degrees of freedom. A symmetry analysis can be

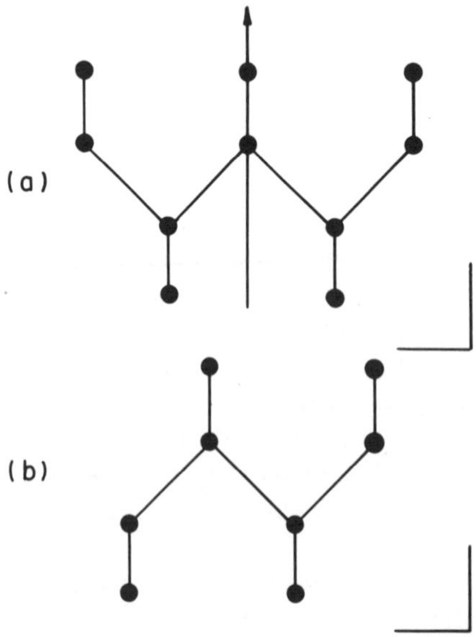

Figure 13.1. The symmetry properties of finite hydrocarbon chains: (*a*) *n* odd (C_{2v} symmetry); (*b*) *n* even (C_{2h} symmetry).

Table 13.1 Character Table of C_{2v} Point Group and Symmetry Analysis for the *n*-Paraffins[a]

C_{2v}	E	C_2	$\sigma(zx)$	$\sigma(yz)$		$n_{tot}^{(\gamma)}$	$n_{vib}^{(\gamma)}$
A_1	1	1	1	1	$T_z, \alpha_{xx}, \alpha_{yy}, \alpha_{zz}$	$\frac{5}{2}(n+1)$	$\frac{1}{2}(5n+3)$
A_2	1	1	-1	-1	R_z, α_{xy}	$2n$	$2n-1$
B_1	1	-1	1	-1	T_x, R_y, α_{zx}	$2(n+1)$	$2n$
B_2	1	-1	-1	1	T_y, R_x, α_{yz}	$\frac{1}{2}(5n+3)$	$\frac{1}{2}(5n-1)$
χ_j	$9n+6$	-1	3	$n+2$			

[a]Number of carbon atoms, n, odd.

performed in the usual manner, and the results are presented in Tables 13.1 and 13.2, together with the character table of the appropriate point group. It can be seen that the selection rules for chains with an odd number of carbon atoms differ markedly from those for chains with an even number of carbon atoms.

With the internal coordinates as a basis set, appropriate symmetry coordinates for the *n*-paraffins can be derived (see Chapter 11), and these are tabulated according to symmetry species in Table 13.3. These symmetry coordinates (methylene rocking, wagging, twisting, etc.) are assumed to provide a description of the normal modes, in other words, they correspond to the normal coordinates. The actual form of the vibrations, which can only be determined by normal coordinate analysis, could well turn out to be a linear combination of the symmetry coordinates in each symmetry species, but it is useful in this type of initial analysis to assume that there is no mixing. Furthermore, all interactions with the methyl end groups are neglected.

In the coupled oscillator approach, each of the coordinates is replaced by an oscillator having the vibrational frequency of the corresponding mode of the infinite chain. It was shown in Chapter 10 that if we assume only nearest neighbor interactions, the dispersion relationship is given by

$$4\pi^2c^2\nu_s^2=g(f+2f'\cos\phi) \tag{13.1}$$

Table 13.2 Character Table for Point Group C_{2h} and Symmetry Analysis of *n*-Paraffins[a]

C_{2h}	E	C_2	i	σ_v		$n_{tot}^{(\gamma)}$	$n_{vib}^{(\gamma)}$
A_g	1	1	1	1	$R_x, \alpha_{xx}, \alpha_{yy}$	$\frac{1}{2}(5n+4)$	$\frac{1}{2}(5n+2)$
B_g	1	-1	1	-1	$R_y, R_z, \alpha_{zz}, \alpha_{yz}$	$2n+1$	$2n-1$
A_u	1	1	-1	-1	$T_x, \alpha_{xy}, \alpha_{zx}$	$2n+1$	$2n$
B_u	1	-1	-1	1	T_y, T_z	$\frac{1}{2}(5n+4)$	$\frac{5}{2}n$
χ_j	$9n+6$	0	0	$n+2$			

[a]Number of carbon atoms n is even.

Table 13.3 Symmetry Coordinates for the *n*-Paraffins

Group	Mode	No. of Coordinates	C_{2v} (n odd) $A_1[R(p), IR(z)]$	$A_2[R(d)]$	$B_1[R(d), IR(x)]$	$B_2[R(d), IR(y)]$	C_{2h} (n even) $A_g[R(p)]$	$B_g[R(d)]$	$A_u[IR(x)]$	$B_u[IR(yz)]$
Methyl	Stretching	6	2	1	1	2	2	1	1	2
	Bending	6	2	1	1	2	2	1	1	2
	Rocking + wagging	4	1	1	1	1	1	1	1	1
	Torsion	2	0	1	1	0	0	1	1	0
Methylene	Stretching (ν)	$2(n-2)$	$\frac{1}{2}(n-1)$	$\frac{1}{2}(n-3)$	$\frac{1}{2}(n-1)$	$\frac{1}{2}(n-3)$	$\frac{1}{2}(n-2)$	$\frac{1}{2}(n-2)$	$\frac{1}{2}(n-2)$	$\frac{1}{2}(n-2)$
	Bending (δ)	$n-2$	$\frac{1}{2}(n-1)$	0	0	$\frac{1}{2}(n-3)$	$\frac{1}{2}(n-2)$	0	0	$\frac{1}{2}(n-2)$
	Rocking (ρ)	$n-2$	0	$\frac{1}{2}(n-3)$	$\frac{1}{2}(n-1)$	0	0	$\frac{1}{2}(n-2)$	$\frac{1}{2}(n-2)$	0
	Wagging (ω)	$n-2$	$\frac{1}{2}(n-3)$	0	0	$\frac{1}{2}(n-1)$	$\frac{1}{2}(n-2)$	0	0	0
	Twisting (τ)	$n-2$	0	$\frac{1}{2}(n-1)$	$\frac{1}{2}(n-3)$	0	0	$\frac{1}{2}(n-2)$	$\frac{1}{2}(n-2)$	0
Skelton	Stretching	$n-1$	$\frac{1}{2}(n-1)$	0	0	$\frac{1}{2}(n-1)$	$\frac{1}{2}n$	0	0	$\frac{1}{2}(n-2)$
	Bending	$n-2$	$\frac{1}{2}(n-1)$	0	0	$\frac{1}{2}(n-3)$	$\frac{1}{2}(n-2)$	0	0	$\frac{1}{2}(n-2)$
	Torsion	$n-3$	0	$\frac{1}{2}(n-3)$	$\frac{1}{2}(n-3)$	0	0	$\frac{1}{2}(n-4)$	$\frac{1}{2}(n-2)$	0
	Total	9n	$\frac{1}{2}(5n+3)$	$2n-1$	$2n$	$\frac{1}{2}(5n-1)$	$\frac{1}{2}(5n+2)$	$2n-1$	$2n$	$\frac{5}{2}n$

where ϕ is the phase shift between adjacent oscillators for the sth chain mode and is given by

$$\phi=\frac{\pi s}{N+1}, \qquad s=1,2\ldots N \tag{13.2}$$

Consequently, in order to assign a given observed infrared band (or Raman line) the value of s as well as the coordinate involved has to be specified. However, even if the coordinates are mixed, the frequencies are still a function of the phase angle. In the limit of an infinite chain the selection rules limit activity to modes with $\phi=0$ or $\phi=\pi$, in-phase or out-of-phase motions, respectively, of the two methylene units in the translational repeat unit. The correlation between the modes of the finite and infinite chains as described by Snyder and Schachtschneider (2) is given in Table 13.4. For each type of finite chain (n odd or n even) the symmetry of the modes is further distinguished by the parity of s. Note that the in-phase and out-of-phase combinations of the infinite chain (e.g., methylene rocking mode, $\phi=0$; methylene twisting mode, $\phi=\pi$) are correlated to each other through the point groups of the finite chains (via their common subgroup C_s). Consequently, even if our original assumption that the normal modes of the infinite chain correspond to our local symmetry coordinate definitions (rocking, twisting, etc.) is correct, mixing of certain pairs of these coordinates can be expected in the finite chains.

Table 13.4 Relation Between Symmetry Species and *k*; Phase Convention for Limiting Modes; Symmetry Species of Limiting Modes in Polyethylene

n-Paraffins			Polyethylene			
No. of Carbon Atoms	Symmetry Species[a] s odd	Symmetry Species[a] s even	ν Description of Modes	ϕ (rad)	ν (cm^{-1})	Symmetry Species
Odd	B_2	A_2	Methylene rocking-	0	722	B_{2u}
Even	A_u	B_g	twisting	π	1063	A_u
Odd	A_2	B_2	Methylene twisting-	0	1295[c]	B_{3g}
Even	B_g	A_u	rocking	π	1174	B_{1g}
Odd	B_1	A_1	Methylene wagging	0	1170	B_{1u}
Even	A_g	B_u		π	1415[c]	B_{2g}
Odd	A_1	B_1	C—C stretching	0	1133[c]	A_g
Even	A_g	B_u		π	1065[c]	B_{2g}
[b]			Methylene bending	0	1440[c]	A_g
				π	1475	B_{3u}

[a]Species A_1, B_1, B_2, A_u, and B_u are infrared active; species A_1, B_1, A_2, B_2, A_g, and B_g are Raman active.
[b]No detailed assignments made.
[c]From Snyder (5).

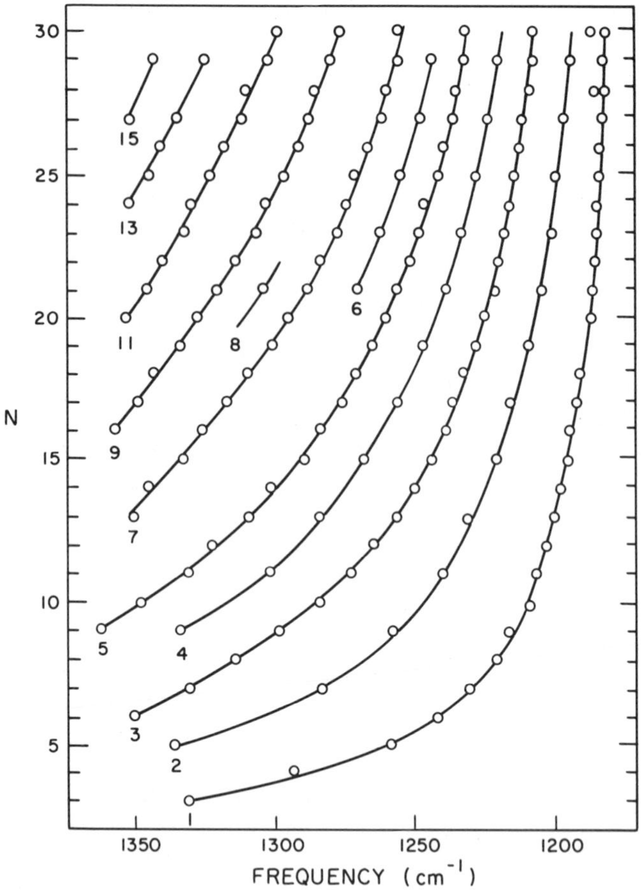

Figure 13.2. The methylene wagging mode array for *n*-paraffins C_3H_8 through *n*-$C_{30}H_{62}$. [Reprinted by permission from R. G. Snyder, *J. Mol. Spect.* **4**, 411 (1960).]

The application of the character tables given in Tables 13.3 and 13.4 to the analysis of the observed spectra of the paraffins can be illustrated by using the methylene wagging in the paraffin chain mode as an example. When the number of carbon atoms n is odd, the wagging vibrations can be assigned to symmetry species A_1 when s is even and to symmetry species B_2 when s is odd (point group C_{2v}). By referring to Table 13.3 it can be seen that there should be $\frac{1}{2}(n-3)$ wagging modes of species A_1 with polarization perpendicular to the chain axis. Similarly, we would expect $\frac{1}{2}(n-1)$ bands of species B_2 polarized parallel to the chain axis in the same region of the spectrum. Finite chains with an even number of carbon atoms can be analyzed in a corresponding manner. In this case, however, half the modes are predicted to belong to symmetry species A_g, which is infrared inactive.

This analysis is confirmed by a comparison of observed and predicted results (2). The methylene wagging bands are displayed as an array in Figure 13.2. The various branches are identified by their values of *s*. The experimental points were obtained from the observed band series. Snyder and Schachtschneider (2) noted that for odd-numbered *n*-paraffins more intense bands (*s* odd) alternate with less intense bands (*s* even), while for even-numbered paraffins only odd *s* bands are observed.

Using equation (13.2) we can obtain the dispersion curve shown in Figure 13.3. All points fall very close to a common curve, even for the shortest chains, indicating relatively little interaction between these vibrations and those of the

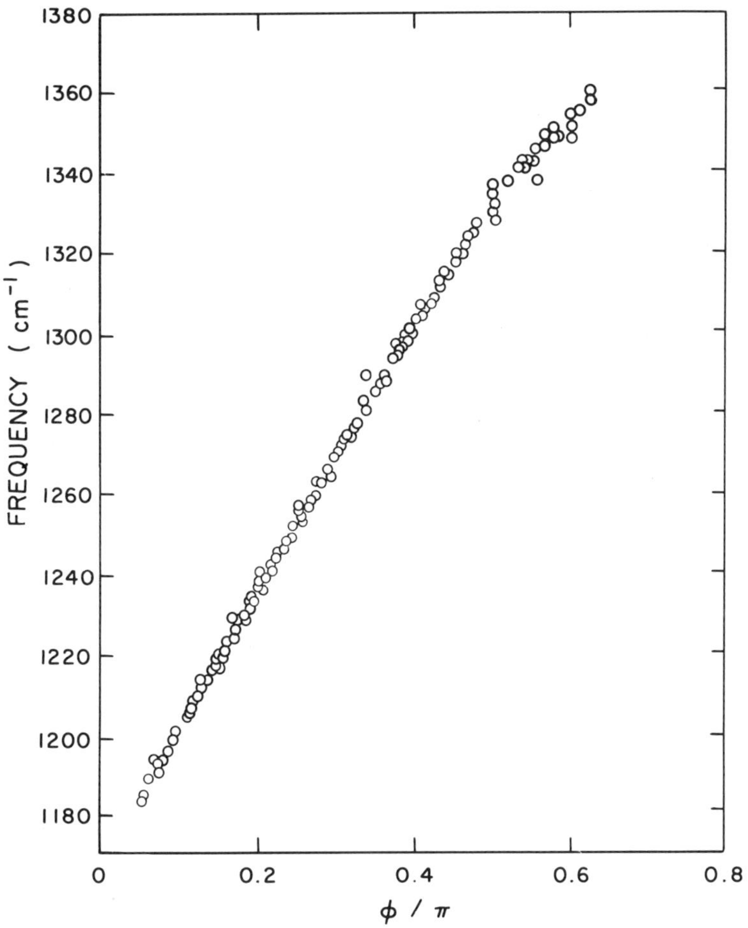

Figure 13.3. Frequency-phase curve for methylene wagging modes of C_3H_8 thru n-$C_{30}H_{62}$ (Reprinted by permission from ref. 2.)

methyl end group, and providing further support for the band assignments. The low frequency limit of this dispersion curve is near 1170 cm^{-1} and corresponds to a weak band in the infrared spectrum of polyethylene. However, the high frequency limit, corresponding to the $\phi=\pi$ mode of the infinite chain, is much more difficult to ascertain. An extrapolation to about 1420 cm^{-1} can be made, and on the basis of this and normal coordinate calculations the Raman line at 1415 cm^{-1} was originally assigned to this mode. As mentioned previously, the out-of-phase methylene wagging mode has subsequently been reassigned to 1382 cm^{-1}. The origin of the Raman line at 1415 cm^{-1} will be discussed later. This illustrates the dangers inherent in not only confirming assignments by normal coordinate analysis, but also in extrapolating dispersion curves to regions of the plot where there are few experimental data.

As a final example of band assignments using the coupled oscillator model, we will briefly discuss the methylene rocking-twisting modes. An example of the observed band progression series was presented in Chapter 10, where an array of the rocking-twisting modes of the finite chains was presented in Figure 10.21. Even though $(n-1)/2$ and $(n-2)/2$ infrared bands are predicted for odd- and even-numbered n-paraffins, respectively, $n/2$ bands are observed for even-numbered molecules because of the appearance of the out-of-plane methyl rocking mode. It is not possible to assign a particular band to this mode, since it mixes with all the methylene rocking modes of the series. Assignments of infrared bands to specific values of s were made simply by counting bands in order of increasing frequency, keeping in mind that only odd values of s are allowed. The dispersion curve was also shown previously in Figure 10.22. Displacements of points for the shorter molecules from the common curve are due to interactions between the methyl and methylene rocking modes. The lower frequency limit of this series corresponds to the strong characteristic 720 cm^{-1} rocking mode of polyethylene. The high frequency limit of the series corresponds to a CH_2 twisting mode. Except in this limit of the infinite chain, rocking and twisting coordinates are mixed. Nevertheless, the coupled oscillator model still allows an analysis, since even for mixed modes the vibrational frequency is a function of phase angle.

13.2 Intrachain Force Field for Paraffins

It may at first seem like putting the cart before the horse to discuss the force field before describing in detail the definition of coordinates involved in setting up the normal coordinate analysis of polyethylene. As we discussed in Chapter 12, however, the most useful and reliable force fields are not those obtained by directly refining the observed and calculated frequencies of the polymer itself, but those obtained from an overlay vibrational analysis of appropriate oligomers. We considered the normal coordinate analysis of low molecular weight materials in the first part of this book. At this point we will not consider the details of defining the various coordinates involved in solving the problem for the paraffins, but instead will concentrate most of our attention on the resulting force fields, which we will apply to an analysis of polyethylene in succeeding sections.

A series of force fields are available for paraffins. The first of these was obtained by corefining the frequencies of the *n*-paraffins (3) but also utilized an incorrect assignment for the B_{2g} methylene wagging mode of polyethylene (5). A revised force field was subsequently determined by Snyder (5), who pointed out that despite this single misassignment, all other 185 observed frequencies can be matched extremely well with either force field. In part this

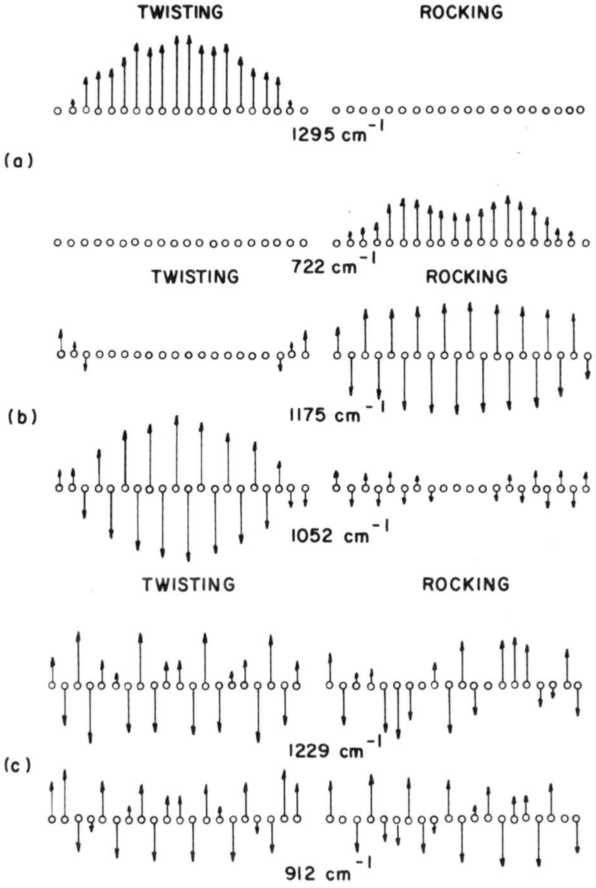

Figure 13.4. Form of normal modes for n-$C_{20}H_{42}$. The length of the arrows are proportional to the elements of $\mathbf{L}_r$. (*a*) In-phase ($\phi = 0$) twisting and rocking modes. (*b*) Out-of-phase ($\phi = \pi$) twisting and rocking modes. (*c*) Mixed twisting-rocking modes ($\phi = \pi/2$) [Reprinted by permission from J. H. Schachtschneider and R. G. Snyder, *J. Polymer Sci.* **7C**, 99 (1963).]

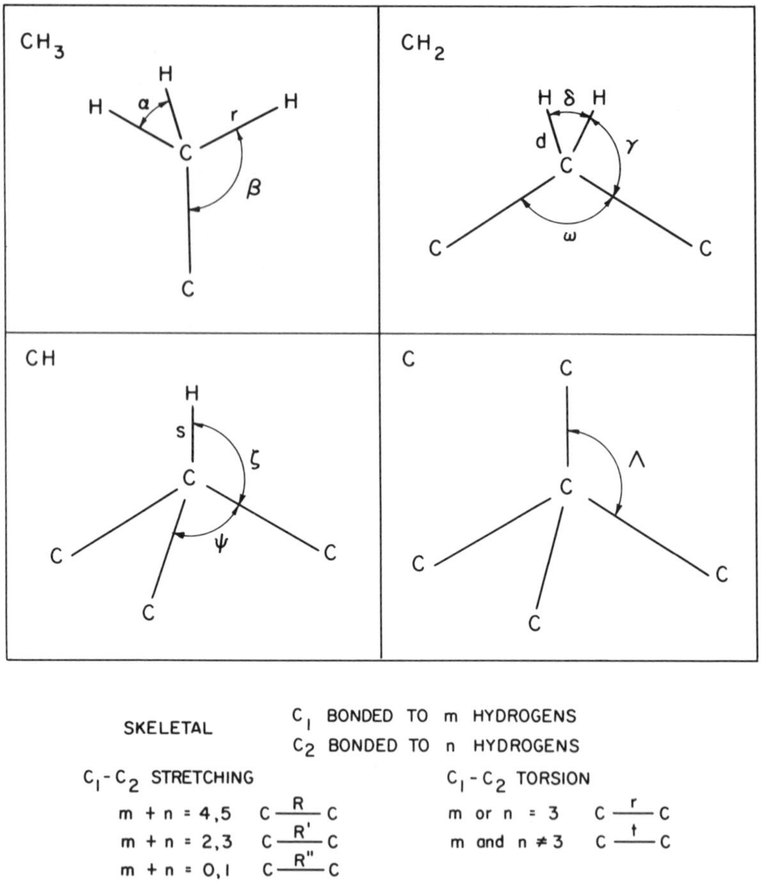

Figure 13.5. Defintion of internal coordinates of paraffins. (Reprinted by permission from ref. 4.)

is because the wagging fundamental of polyethylene is relatively isolated from other well-established wagging modes on the dispersion curve. We will not examine this force field further, but will concentrate our attention on subsequent calculations. However, one result of this analysis that is of interest is the calculated form of the **L** matrix for the rocking and twisting modes, since it illustrates some points made in Section 13.1. The elements of the **L** matrix for n-$C_{20}H_{42}$ are shown schematically in Figure 13.4. For $\phi=0$ (in-phase vibrations), modes near 722 cm^{-1} and 1295 cm^{-1} are almost pure rocking and twisting vibrations, respectively. Similarly, for $\phi=\pi$ (out-of-phase vibrations), modes calculated near 1175 cm^{-1} and 1052 cm^{-1} are essentially pure rocking and twisting, although a small degree of mixing is apparent. For intermediate values of ϕ, however, these two types of coordinates are thoroughly mixed. It also has been observed that contributions from the end groups are involved in a number of these modes.

The first force field we will discuss is the one derived for saturated hydrocarbons by Snyder and Schachtschneider (4). Our rationale for considering this first is that it is more general than the original calculations for *n*-paraffins in that it includes force constants for various branched structures. The molecules used in the refinement are those listed in Table 13.5. The internal coordinate definitions are illustrated in Figure 13.5. Interaction force constants are defined in detail in Figure 13.6. The force constant values are listed in Table 13.6. In calculating these force constants it was assumed that all valence angles are tetrahedral and that only staggered configurations occur about C—C bonds. The consequences of nontetrahedral bond geometry will be discussed later.

The force field listed in Table 13.6 defines 11 diagonal force constants. Methyl and methylene diagonal force constants (e.g., C—C—H bend) are separately defined, but interaction terms for these groups (e.g., C—C stretch, C—C—H bend interactions) are assumed to correspond. Even though next nearest neighbor interaction terms were not found to be negligible in the original calculations for the *n*-paraffins (3), only nearest neighbor interaction terms have been considered in subsequent work (4–6). It must be kept in mind, however, that the aim of these already complex calculations is the determination not of a general valence force field, but rather of an "effective" force field capable of accurately reproducing observed frequencies and indicating the form of the normal modes. In view of the various approximations (the assumed structural parameters and absence of corrections for anharmonicity, etc.), this is probably the best that can be expected. Nevertheless, in our (prejudiced) view the effort is not pointless in that one of the criteria for judging the effectiveness of any force field is its transferability, which in the case of saturated paraffins has made possible the analysis of the spectra of a number of complex hydrocarbons.

The refinement of observed and calculated frequencies for a complex series of molecules such as the saturated hydrocarbons is seldom a "pushbutton" process, where a list of observed frequencies and initial force constants are fed into a computer, which subsequently churns away and eventually spits out a force field. The process is subjective. It is essential that the observed and calculated frequencies be properly matched, which requires not only a good preliminary band assignment but also a good initial force field. In this respect, the original work on hydrocarbons (2–6) has been invaluable in providing a jumping-off point for the analysis of related materials. Snyder and Schachtschneider (3) obtained their original force field by essentially adding interaction force constants until no significant improvement in the fit between observed and calculated frequencies was obtained. This force field was modified (4) by first calculating the frequencies of the four branched hydrocarbons of highest symmetry from those listed in Table 13.5 (isobutane, neopentane, adamantane, and 2, 2, 3, 3-tetramethylbutane). The use of symmetry and the reasonably simple nature of the spectra of these molecules resulted in confident band assignments. A force constant refinement of the frequencies of these four molecules together with those of extended C_3H_8, n-C_4H_{10}, and n-C_5H_{12} was

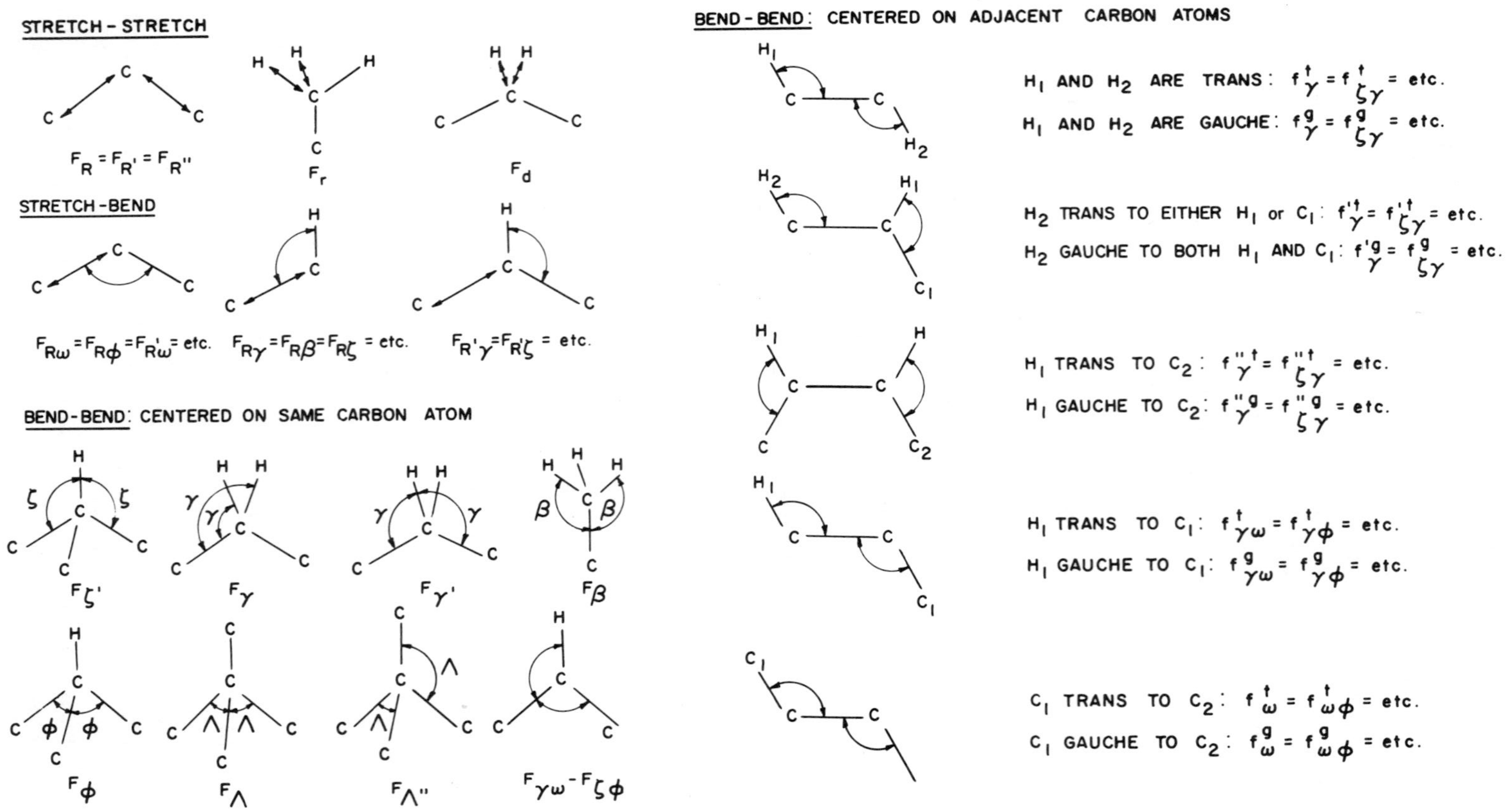

Figure 13.6. Defintion of interaction force constants. (Reprinted by permission from ref. 4.)

Table 13.5 List of Paraffin Molecules Used in Force Constant Calculation

Compound	Molecular Symmetry	No. Fundametals[a]	No. of Observed Frequencies Used to Determine the Force Constants	Error in Calculated Frequencies (%)
Propane	C_{2v}	27	23	0.70
trans-n-Butane	C_{2h}	36	26	0.45
trans-n-Pentane	C_{2v}	45	25	0.51
gauche-n-Butane	C_2	36	10	0.59
gauche-n-Pentane	C_1	45	8	1.30
Isobutane	C_{3v}	24	19	0.87
Neopentane	T_d	19	11	1.53
2-Methylbutane	C_s	45	—	—
2-Methylbutane	C_1	45	19	0.50
2,2-Dimethylbutane	C_s	54	16	1.44
2,3-Dimethylbutane	C_2	54	—	—
2,3-Dimethylbutane	C_{2h}	54	10	0.73
2,2,3-Trimethylbutane	C_s	63	16	1.46
2,2,3,3-Tetramethylbutane	D_{3d}	48	17	1.83
Adamantane	T_d	31	10	1.02
Cyclohexane	D_{3d}	32	24	0.88
κ-Methylcyclohexane	C_s	63	24	1.04
1,1-Dimethylcyclohexane	C_s	72	23	1.10
trans-Decalin	C_{2h}	90	27	1.04
			308	0.95[b]

[a]Degenerate fundamentals are counted only once.
[b]Average.

then performed. These three molecules were included so that the resulting force field could also be applied to the extended *n*-paraffins, a reasonable bias in that the force constants cannot be uniquely determined. This force field was then successively applied to more complex molecules, so that in the final analysis 308 observed frequencies from 17 molecules were used to refine 35 independent force constants.

Before proceeding to a discussion of other force fields we consider it useful to point out that in our experience of applying this and similar force fields it is easy to make errors in defining force constants, particularly bend-bend interactions in nonplanar structures. For example, f''_γ illustrated in Figure 13.6 is defined for a hydrogen atom trans to a hydrogen on an adjacent carbon **or trans to a next nearest neighbor carbon atom**. In order to clarify the internal coordinates involved in such force constant definitions, we have reproduced

Table 13.6 Valence Force Constants for Paraffins[a]

Force[ab] Constant	Φ_i	$\sigma(\Phi_i)^c$	Force[a, b] Constant	Φ_i	$\sigma(\Phi_i)^c$
K_r	4.699	0.008	$F_{R\gamma}$	0.328	0.015
K_d	4.554	0.010	$F'_{R\gamma}$	0.079	0.015
K_s	4.588	0.036	$F_{R\omega}$	0.417	0.019
K_R	4.387	0.035	F_β	−0.012	0.003
$K_{R'}$	4.337	0.045	F_γ	−0.021	0.004
$K_{R''}$	4.534	0.225	F'_γ	0.012	0.004
H_α	0.540	0.002	F_m	−0.041	0.023
H_β	0.645	0.004	$F_{\gamma\omega}$	−0.031	0.017
H_δ	0.550	0.005	f^t_γ	0.127	0.006
H_γ	0.656	0.004	f^g_γ	−0.005	0.005
H_ζ	0.657	0.008	f'^t_γ	0.002	0.004
H_ω	1.130	0.039	f'^g_γ	0.009	0.004
H_φ	1.084	0.038	f''^t_γ	−0.014	0.008
H_Λ	1.086	0.027	f''^g_γ	−0.025	0.007
H_τ	(0.024)[d]	—	$f^t_{\gamma\omega}$	0.049	0.011
H_r	(0.024)[d]	—	$f^g_{\gamma\omega}$	−0.052	0.010
F_Γ	0.043	0.006	f^t_ω	−0.011	0.029
F_d	0.006	0.010	f^g_ω	0.011	0.016
F_R	0.101	0.017			

[a]From Snyder (4). Stretch constants are in units of mdyn/Å; stretch-bend interaction constants in units of mdyn/rad. Bending constants are in units of mdyn-Å/(rad)^2.
[b]In this calculation we have set $F'_\zeta = F'_\gamma$, $F_\Lambda = F_\varphi$, and $F''_s = 0$.
[c]$\sigma(\Phi_i)$ is the standard error in Φ_i estimated from the standard error in frequency parameters and the variance-covariance matrix.
[d]Value taken to fit the torsional frequency of ethane (280 cm^{-1}).

the appropriate definitions for the trans-gauche structure of poly(ethyl ethylene) in Figure 13.7 and Table 13.7A and 13.7B. Force constant values shown in these tables will be discussed in more detail in Chapter 16. Note that the main-chain and side-chain internal coordinates were defined separately in this work (8).

The force field listed in Table 13.6 has been subsequently modified by including nonplanar conformations of the liquid *n*-paraffins (6). The aim of this work was to obtain an understanding of the contribution of nonplanar structures to the spectrum of polyethylene. Consequently, we will defer a more detailed discussion of this force field to Chapter 15, where we will discuss the vibrational spectra of disordered polymers. We conclude this section by discussing the local symmetry force field determined by Barnes and Fanconi (7).

In Chapter 11 we described a symmetry analysis of polyethylene based on a cartesian coordinate representation. A corresponding analysis in which the

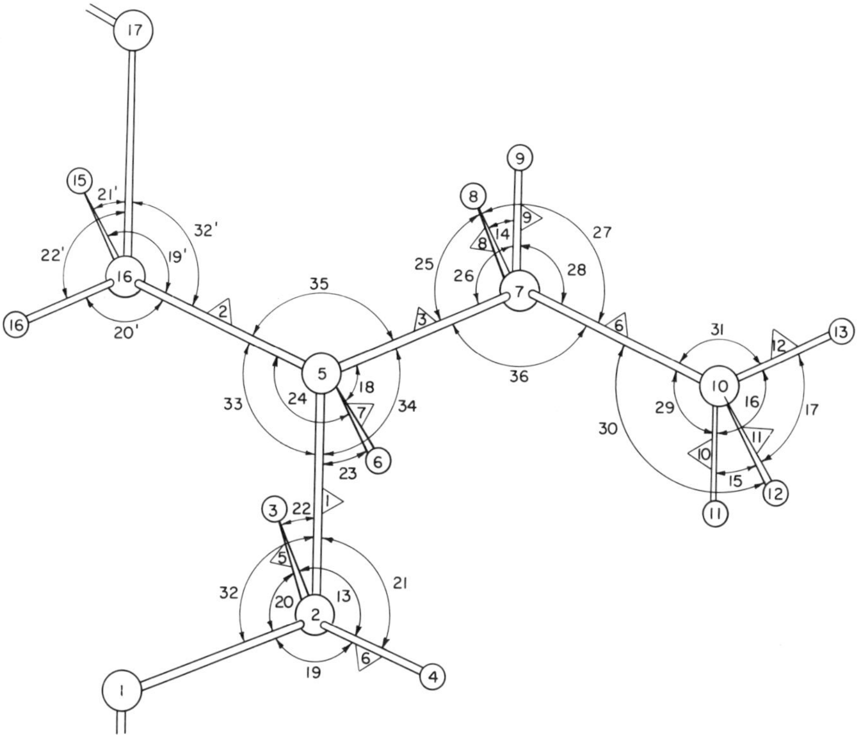

Figure 13.7. Definition of internal coordinates of poly(ethyl ethylene (excluding the torsional coordinates). (Reprinted by permission from ref. 8.)

Table 13.7A Definitions and Values of Elements of the *F* Matrix in Terms of Internal Coordinates[a]

Diagonal Elements					
Force Constant Symbol[b]	Internal Coordinates	Force Constant	Force Constant Symbol[b]	Internal Coordinates	Force Constant
K_R^I(s)	4	4.329	H_γ(s)	25,26,27,28	0.666
K_S(m)	1,2	4.337	H_δ(m)	13	0.55
K_S(s)	3	4.337	H_δ(s)	14	0.55
K_d(m)	5,6	4.554	H_ω(m)	32	1.13
K_d(s)	8,9	4.546	H_ω(m′)	33,34,35	1.084
K_s	7	4.588	$H_\omega(s)$	36	0.901
K_r	10,11,12	4.699	H_ζ(m)	18,23,24	0.657
H_α	15,16,17	0.54	H_τ(m)	37,38,39	0.024
H_β	29,30,31	0.637	H_τ(s)	40	0.024
H_γ(m)	19,20,21,22	0.66			

[a] From Holland-Moritz and Sausen (8).

[b] Parenthetical letters s, and m, mean side chain and main chain. H_ω(m) and H_ω(m′) refer to the angle deformation connected with the secondary and tertiary carbon atom of the main chain, respectively.

Table 13.7B Definitions and Values of the *F* Matrix in Terms of Internal Coordinates[a]

Off-Diagonal Elements

Force Constant Symbol	Internal Coordinates					Force Constant
F_r	10–11	10–12	11–12			0.029
$F_d(s)$	8–9					0.016
$F_d(m)$	5–6					0.006
$F_R(m)$	1–2	2′–1	1–3	2–3		0.101
$F_R(s)$	3–4					0.064
$F_{R\beta}(m)$	1–21	1–22	1–23	2–24	3–18	0.328
	2–19′	2–20′				
$F_{R\beta}(s)$	3–25	3–26	4–27	4–28	4–29	0.261
	4–30	4–31				
$F'_{R\gamma}(m)$	1–19	1–20	2–21′	2–22′	1–18	0.079
	1–24	3–24	3–23	2–18	2–23	
$F'_{R\gamma}(s)$	3–28	3–27	4–25	4–26		−0.004
	1–32	1–33	1–34	2′–32	2–35	
$F_{R\omega}(m)$	3–35	3–34	2–33	1–35	2–34	0.417
	3–33					
$F_{R\omega}(s)$	3–36	4–36				0.351
$F_\beta(s)$	29–30	29–31	30–31			−0.017
$F_\gamma(m)$	21–22	19–20	25–26	27–28		−0.021
$F'_\gamma(m)$	19–21	20–22	23–24	23–18	24–18	0.041
$F_\gamma(s)$	25–27	26–28				0.023
$F_{\gamma\omega}(m)$	32–19	32–20	32–21	33–23	33–24	−0.031
	34–18	34–23	35–18	35–24	32–22	
$F_{\gamma\omega}(s)$	36–25	36–26	36–27	36–28		−0.124
$f^t_\beta(m)$	22–23	19′–24	18–25			0.127
$f^t_\beta(s)$	28–29	27–30				0.088
$f^g_\beta(m)$	21–23	24–20′	18–26			−0.005
$f^g_\beta(s)$	28–30	27–29	27–31	28–31		−0.043
	21–24	20–23	22–18	22–24	23–25	
$f'^t_{\beta\gamma}(m)$	24–21′	19′–23	19′–18	20′–18	18–27	0.002
	23–26	25–24				
$f'^t_{\beta\gamma}(s)$	31–25	30–25	29–26	31–26		−0.002
$f'^g_{\beta\gamma}(m)$	32–19	21–18	20′–23	24–22′	18–28	0.009
	26–24					
$f'^g_{\beta\gamma}(s)$	29–25	30–26				0.002
$f''^t_\gamma(m)$	19–24	28–23	22′–18			−0.014
$f''^g_\gamma(m)$	20–24	20–18	21′–23	22′–23	19–18	−0.025
	24–28	21′–18	23–27	24–27		
$f^t_{\beta\omega}(m)$	21–33	20′–35				0.049
$f^t_{\beta\omega}(s)$	26–34	31–36				0.072

Table 13.7B Continued

Off-Diagonal Elements

Force Constant Symbol	Internal Coordinates					Force Constant
$f^g_{\beta\omega}$(m)	21–34	22–34	22–33	24–34′	20′–33	−0.052
	19′–33	19′–35	18–36	23–32		
$f^g_{\beta\omega}$(s)	25–35	25–34	26–35	29–36	30–36	−0.058
f^t_ω	32–34	35–36	33–32′			−0.011
f^g_ω	32–33	32–35	34–36			0.011
F_ω	33–34	33–35	34–35			−0.041

[a]The units of the force constants are stretch (mdyn/Å), angle deformation and torsion (mdyn-Å/rad), and angle stretch (mdyn Å).

reduced representation is based on the set of internal coordinates does not correspond to the reduction of the cartesian representation because of a bond angle redundancy. Since the bond angles describing a CH_2 group, four C—C—H bends (γ), one H—C—H bend (δ), and one C—C—C bend (ω), are a kinematically complete set, we cannot simply omit one of these coordinates. However, the redundant coordinate can be eliminated by using local symmetry coordinates, methylene rocking, twisting, wagging, and bending. These and

Table 13.8 Definition of Localized Symmetry Coordinates

Symmetry Coordinates	Internal Coordinates[a]							
	CH bond stretch	CH bond stretch	C_-CC_+ angle bend	HCH′ angle bend	C_-CH angle bend	C_-CH' angle bend	C_+CH angle bend	C_+CH' angle bend
Symmetric CH_2 stretch	$1/\sqrt{2}$	$1/\sqrt{2}$						
Antisymmetric CH_2 stretch	$1/\sqrt{2}$	$-1/\sqrt{2}$						
Skeletal angle bend			a	$-b$	$-c$	$-c$	$-c$	$-c$
CH_2 bending				d	$-e$	$-e$	$-e$	$-e$
CH_2 wagging					f	f	$-f$	$-f$
CH_2 rocking					f	$-f$	f	$-f$
CH_2 twisting					f	$-f$	$-f$	f

[a]Values are $a = 0.909633$, $b = 0.180325$, $c = 0.187116$, $d = 0.900871$, $e = 0.217044$, $f = 0.5$.

Table 13.9 General Force Field Parameters

No.	Type[a]	Chemical Repeat Unit	Value[b]	σ
1	1,1	0	4.517	0.011
2	1,3	0	0.460	0.066
3	1,4	0	0.143	0.086
4	1,5	0	−0.096	0.017
6	2,2	0	4.289	0.023
7	2,7	0	−0.660	0.043
8	3,3	0	4.187	0.014
9	3,4	0	0.116	0.033
10	3,5	0	0.129	0.057
11	3,6	0	0.286	0.003
13	4,4	0	1.005	0.024
14	4,5	0	−0.010	0.11
16	5,5	0	0.541	0.011
18	6,6	0	0.582	0.0028
20	7,7	0	0.8366	0.021
21	8,8	0	0.624	0.0045
22	9,9	0	0.085	0.0043
23	1,4	1	0.0785	0.0097
26	2,7	1	0.266	0.011
27	3,3	1	0.065	0.011
29	4,3	1	0.104	0.012
30	4,4	1	0.123	0.016
31	4,5	1	0.030	0.097
32	4,6	1	−0.101	0.0039
34	5,5	1	0.004	0.0008
35	5,6	1	−0.012	0.029
37	6,3	1	−0.004	0.027
38	6,6	1	−0.037	0.0013
40	7,7	1	−0.014	0.0095
41	7,8	1	0.018	0.0069
42	8,8	1	−0.066	0.0013
43	9,9	1	−0.0074	0.0088
45	4,4	2	0.005	0.016
46	4,5	2	−0.014	0.0026
47	4,6	2	0.048	0.02
49	5,5	2	0.002	0.03
51	7.7	2	0.0185	0.003
52	9,9	2	0.0007	0.0054

[a]Localized symmetry internal coordinates:
1. Symmetric C—H stretch.
2. Asymmetric C—H stretch.
3. Skeletal (CC) stretch.
4. Skeletal (CCC) angle bend.
5. Methylene scissors.
6. Methylene wagging.
7. Methylene rocking.
8. Methylene twisting.
9. Skeletal torsion.

[b]The units are mdyn/Å (10^2 N/m) for stretch-stretch terms; mdyn/Å (10^{-18} N/m) for bend-bend terms; and mdyn (10^{-8} N) for stretch-bend terms.

other local symmetry coordinates, as defined by Barnes and Fanconi (7), are presented in Table 13.8. (These local symmetry coordinates are based on nontetrahedal geometry.) By removal of redundant coordinates the correlation among valence force constants is minimized. Note that if the problem is set up in terms of cartesian coordinates with the force constants defined in terms of internal coordinates, such redundancies are removed but the correlation between force constants may not be, depending on the symmetry of the molecule under consideration. The force field parameters obtained by refining data from the *n*-alkanes (705 frequencies) are listed in Table 13.9. By using symmetry, force constants were separated into two sets in the refinement, one involving methylene rocking, methylene twisting, skeletal torsions, and asymmetric C—H stretching, while the other involved methylene wagging, methylene bending, C—H symmetric stretching, skeletal stretching, and angle deformation. Interactions in the crystal were also considered in this work, but we will defer a discussion of the results to Chapter 14.

13.3 Details of the Analysis of Polyethylene

In this section we will present in detail the definition of a set of coordinates that allow the normal modes of polyethylene to be calculated by standard

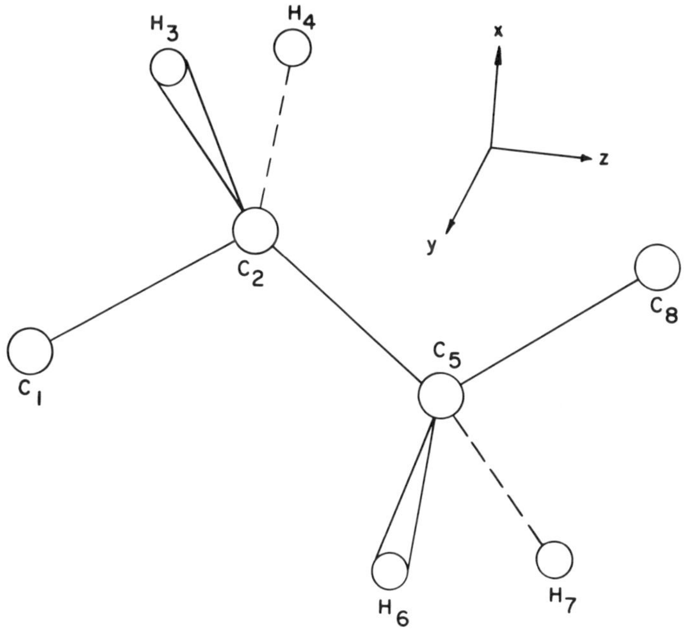

Figure 13.8. Numbering of atoms in polyethylene repeat unit for internal coordinate definitions.

Table 13.10 Internal Coordinate Definitions for Polyethylene

Number	Type	Atoms Involved		
1	CC stretch	C_2	C_5	
2	CH stretch	C_2	H_3	
3	CH stretch	C_2	H_4	
4	CCH bend	C_1	C_2	H_3
5	CCH bend	C_1	C_2	H_4
6	CCH bend	C_5	C_2	H_3
7	CCH bend	C_5	C_2	H_4
8	CCC bend	C_1	C_2	C_5
9	HCH bend	H_3	C_2	H_4
10	Torsion			

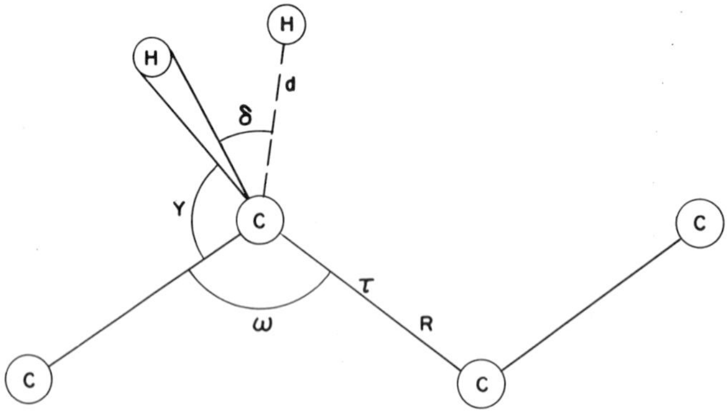

R = C-C STRETCH

d = C-H STRETCH

ω = ∠CCC BEND

Y = ∠CCH BEND

δ = ∠HCH BEND

$$\tau = \text{C-C TORSION} = \frac{1}{\sqrt{3}} \sum_{i,j}^{3} \tau_{imnj}$$

WHERE ATOMS i AND j ARE ALWAYS TRANS.

Figure 13.9. Internal coordinates of polyethylene.

computer programs. Since Schachtschneider and Snyder (3) have discussed the internal coordinate representation of the problem, we will present the alternative cartesian coordinate description, based on the secular equation written in the following form (see Chapter 12):

$$|\mathbf{U}(\theta)\mathbf{D}^T(\theta)\mathbf{F}(\theta)\mathbf{D}(\theta)\mathbf{U}^T(\theta)-\boldsymbol{\lambda}(\theta)\mathbf{E}|=\mathbf{0} \tag{13.3}$$

where $\mathbf{D}(\theta)=\mathbf{B}(\theta)\mathbf{M}^{-1/2}$ and $\mathbf{U}(\theta)$ is a set of external symmetry coordinates. In order to define the force constant matrix we first have to define a set of internal coordinates. The chemical repeat unit consists of a single CH_2 group, but in order to define a complete set of internal coordinates and to calculate $\mathbf{B}(\theta)$ the cartesian coordinates of all the atoms shown in Figure 13.8 have to be specified. We will define all angle bending coordinates around atom 2, giving a single redundancy condition and an extra internal coordinate. The internal coordinates are defined in Table 13.10 and illustrated in Figure 13.9. The cartesian coordinates were calculated using simple trigonometric principles and the following structural parameters:

C—C bond distance = 1.543 Å

C—H bond distance = 1.09 Å

All angles tetrahedral = 109° 28′

Helix rotation angle $\psi = 180° =$ planar zigzag

The coordinates are listed in Table 13.11 with these internal and cartesian coordinate definitions as input. The force constant definitions are those of Snyder (6), discussed briefly in the preceding section and listed in Table 13.12 so as to describe the coordinates involved.

It was shown in Chapter 11 that the normal modes of polyethylene can be classified according to symmetry as

$$\Gamma \text{vib}=3A_{1g}+1A_{1u}+3B_{1g}+2B_{1u}+3B_{2u}+1B_{3g}+3B_{3u}$$

Table 13.11 Cartesian Coordinates for the Atoms of Polyethylene Repeat Unit

Atom	x	y	z
1	−0.889165	0	−1.257364
2	0	0	0
3	0.631076	0.892402	0
4	0.631076	−0.892402	0
5	−0.889165	0	1.257364
6	0	0	2.514728

Table 13.12 Complete Force Field for Polyethylene

Force Constant[a]	Coordinates Involved	Value
K_R	C—C	4.532
K_d	C—H	4.538
H_γ	∠HCC	0.633
H_ω	∠CCC	1.032
H_δ	∠HCH	0.533
H_τ	—C—C—	0.024
F_d	C—H, C—H	0.019
F_R	C—C, C—C	0.083
$F_{R\gamma}$	C—C, ∠HCC	0.174
$F_{R_\gamma'}$	C—C, ∠HCC	−0.097
$F_{R\omega}$	C—C, ∠CCC	0.303
F_γ	∠HCC, ∠HCC	−0.019
$F_{\gamma'}$	∠HCC, ∠HCC	0.021
$F_{\gamma\omega}$	∠HCC, ∠CCC	−0.022
$f_\gamma t$	∠HCC, ∠HCC	0.073
$f_\gamma g$	∠HCC, ∠HCC	−0.058
$f'_\gamma t$	∠HCC, ∠HCC	−0.009
$f'_\gamma g$	∠HCC, ∠HCC	−0.004
$f''_\gamma t$	∠HCC, ∠HCC	0.010
$f''_\gamma g$	∠HCC, ∠HCC	0.012
$f_{\gamma\omega} g$	∠HCC, ∠CCC	−0.064
$f_\omega t$	∠CCC, ∠CCC	0.097

[a]Force constant units: stretch, mdyn/Å; stretch-bend, mdyn/rad; bend, mdyn-Å/rad^2.

The symmetry coordinates under each of these species can be derived by first considering the repeat unit to give "local symmetry coordinates" and then phasing these coordinates to give the coordinates under the line group symmetry of the polymer. The local symmetry coordinates can be derived using the standard equation:

$$S^{(\gamma)} = N \sum_{\rho} \chi_\rho^{(\gamma)} (\rho \chi_1) \tag{13.4}$$

where $S^{(\gamma)}$ is the symmetry coordinate for a particular irreducible representation γ, ρ the operation of a local symmetry element, $\chi_\rho^{(\gamma)}$ the character for this symmetry operation, ρ under the irreducible representation γ, and N a normalization factor, as described in Chapter 5. We can list the transformation $(\rho\chi_1)$ in a table representing the covering operations for the local symmetry of the molecule, which for polyethylene is C_{2v}. This transformation table is shown in

Table 13.13 Transformation Table for the Atoms of the Polyethylene Repeat Unit

	E	σ_v	σ_h	C_2
x_{C_2}	x_{C_2}	x_{C_2}	x_{C_2}	x_{C_2}
y_{C_2}	y_{C_2}	y_{C_2}	$-y_{C_2}$	$-y_{C_2}$
z_{C_2}	z_{C_2}	$-z_{C_2}$	z_{C_2}	$-z_{C_2}$
x_{H_3}	x_{H_3}	x_{H_3}	x_{H_4}	x_{H_4}
y_{H_3}	y_{H_3}	y_{H_3}	$-y_{H_4}$	$-y_{H_4}$
z_{H_3}	z_{H_3}	$-z_{H_3}$	z_{H_4}	$-z_{H_4}$
x_{H_4}	x_{H_4}	x_{H_4}	x_{H_3}	x_{H_3}
y_{H_4}	y_{H_4}	y_{H_4}	$-y_{H_3}$	$-y_{H_3}$
z_{H_4}	z_{H_4}	$-z_{H_4}$	z_{H_3}	$-z_{H_3}$

Table 13.13. The cartesian coordinates correspond to the atoms shown in Figure 13.8. Multiplying by the characters $\chi_\rho^{(\gamma)}$ in each symmetry species, we obtain the following local symmetry coordinates.

A_{1g} Species

$$S_1=(x_{C_2}+x_{C_2}+x_{C_2}+x_{C_2})=x_{C_2}$$
$$S_2=\frac{1}{\sqrt{2}}(x_{H_3}+x_{H_4})$$
$$S_3=\frac{1}{\sqrt{2}}(y_{H_3}-y_{H_4})$$

A_{1u} Species

$$S_1=\frac{1}{\sqrt{2}}(z_{H_3}-z_{H_4})$$

B_{1g} Species

$$S_1=y_{C_2}$$
$$S_2=\frac{1}{\sqrt{2}}(x_{H_3}-x_{H_4})$$
$$S_3=\frac{1}{\sqrt{2}}(h_{H_3}+y_{H_4})$$

B_{1u} Species

$$S_1=z_{C_2}$$
$$S_2=\frac{1}{\sqrt{2}}(z_{H_3}+z_{H_4})$$

These local symmetry coordinates should not be confused with those defined in the preceding section in terms of internal coordinates. Note that one symmetry coordinate is obtained for each normal coordinate previously predicted by the symmetry analysis. These include the translations and rotations, which represent vibrations having zero frequencies.

Upon the local symmetry coordinates we now impose the polymer chain symmetry through a phasing operation on the cartesian coordinates of atoms **for which we have defined equivalent atoms in adjacent chemical repeat units.** As shown in Chapter 12, this phasing operation has the form

$$\mathbf{X}(\theta)=\sum_{n}\mathbf{X}_{n}e^{in\theta} \tag{13.5}$$

which can be rewritten

$$\mathbf{X}(\theta)=\sum_{n=-\infty}^{+\infty}\mathbf{X}_{n}(\cos n\theta+i\sin n\theta) \tag{13.6}$$

For polyethylene only those frequencies are active for which $\theta=0$ and π. Therefore

$$\mathbf{X}(\theta)=\sum_{n}\mathbf{X}_{n}\cos n\theta \tag{13.7}$$

since $\sin n\theta=0$ for all values of n if $\theta=0$ or π.

In principle, **n** spans over all the chemical repeat units from $-\infty$ to 0 to $+\infty$. In practice, however, **n** spans the number of repeat units required to define the internal coordinates, since all others provide zero elements to the **B** matrix. In all other instances the symmetry element will be multiplied by a zero in the infinite **B** matrix. $\mathbf{X}_n$ represents that coordinate (or combination of coordinates) into which a particular **X** symmetry coordinate transforms under the nth screw operation.

Applying the transformation to the local symmetry coordinates, we proceed as follows for the phase angle $\boldsymbol{\theta}=0°$. For the A_{1g} species we start with

$$S_1=x_{C_2}$$

Applying equation (13.7), we obtain

$$\begin{aligned}S_1(0)=&-\underset{\substack{\uparrow\\ x_n}}{x_{C_1}}(\cos-\underset{\substack{\uparrow\\ n}}{1}\times\underset{\substack{\uparrow\\ \theta}}{0})+x_{C_2}(\cos 0\times 0)\\ &-x_{C_5}(\cos 1\times 0)+x_{C_8}(\cos 2\times 0)\end{aligned}$$

Therefore,

$$S_1(0)=-x_{C_1}+x_{C_2}-x_{C_5}+x_{C_8}$$

Similarly,

$$S_2=\frac{1}{\sqrt{2}}(x_{H_3}+x_{H_4}),\qquad S_2(0)=\frac{1}{\sqrt{2}}(x_{H_3}+x_{H_4}-x_{H_6}-x_{H_7})$$

$$S_3=\frac{1}{\sqrt{2}}(y_{H_3}-y_{H_4}),\qquad S_3(0)=\frac{1}{\sqrt{2}}(y_{H_3}-y_{H_4}+y_{H_6}-y_{H_7})$$

Likewise, for the A_{1u} species, starting with

$$S_1=\frac{1}{\sqrt{2}}(z_{H_3}-z_{H_4})$$

yields

$$S_1(0)=\frac{1}{\sqrt{2}}(z_{H_3}-z_{H_4}-z_{H_6}+z_{H_7})$$

We now have the following symmetry coordinates.

B_{1g} Species

$$S_1=y_{C_2},\qquad S_1(0)=-y_{C_1}+y_{C_2}-y_{C_5}+y_{C_8}$$

$$S_2=\frac{1}{\sqrt{2}}(x_{H_3}-x_{H_4}),\qquad S_2(0)=\frac{1}{\sqrt{2}}(x_{H_3}-x_{H_4}+x_{H_6}-x_{H_7})$$

$$S_3=\frac{1}{\sqrt{2}}(y_{H_3}+y_{H_4}),\qquad S_3(0)=\frac{1}{\sqrt{2}}(y_{H_3}+y_{H_4}-y_{H_6}-y_{H_7})$$

B_{1u} Species

$$S_1=z_{C_2},\qquad S_1(0)=z_{C_1}+z_{C_2}+z_{C_5}+z_{C_8}$$

$$S_2=\frac{1}{\sqrt{2}}(z_{H_3}+z_{H_4}),\qquad S_2(0)=\frac{1}{\sqrt{2}}(z_{H_3}+z_{H_4}+z_{H_6}+z_{H_7})$$

The approach is the same for the $\theta=\pi$ modes. The representations transform as shown in the following correlation table:

$\theta=0$		$\theta=\pi$
A_{1g}	$\rightarrow$	B_{3u}
A_{1u}	$\rightarrow$	B_{3g}
B_{1g}	$\rightarrow$	B_{2u}
B_{1u}	$\rightarrow$	B_{2g}

Table 13.14 Calculated Frequencies for Polyethylene

Species	Calculated Frequency	
	This Work	Snyder
A_{1g}	2864	2864
	1439	1439
	1133	1134
B_{1g}	2928	2928
	1170	1170
	0	0
B_{2g}	1385	1384
	1060	1060
B_{3g}	1301	1301
A_{1u}	1063	1063
B_{1u}	1176	1175
	0	0
B_{2u}	2910	2910
	716	717
	0	0
B_{3u}	2850	2850
	1472	1472
	0	0

Once again applying the transformation

$$\mathbf{X}(\theta)=\sum_n x_n \cos n\theta$$

this time with $\theta=\pi$ we obtain

B_{3u} Species

$$S_1=x_{C_2}, \qquad S_1(\pi)=+x_{C_1}+x_{C_2}+x_{C_5}+x_{C_8}$$

$$S_2=\frac{1}{\sqrt{2}}(x_{H_3}+x_{H_4}), \qquad S_2(\pi)=\frac{1}{\sqrt{2}}(x_{H_3}+x_{H_4}+x_{H_6}+x_{H_7})$$

$$S_3=\frac{1}{\sqrt{2}}(y_{H_3}-y_{H_4}), \qquad S_3(\pi)=\frac{1}{\sqrt{2}}(y_{H_3}-y_{H_4}-y_{H_6}+y_{H_7})$$

B_{3g} Species

$$S_1=\frac{1}{\sqrt{2}}(z_{H_3}-z_{H_6}), \qquad S_1(\pi)=\frac{1}{\sqrt{2}}(z_{H_3}-z_{H_4}+z_{H_6}-z_{H_7})$$

B_{2u} Species

$$S_1 = y_{C_2}, \qquad S_1(\pi) = +y_{C_1} + y_{C_2} + y_{C_5} + y_{C_8}$$

$$S_2 = \frac{1}{\sqrt{2}}(x_{H_3} - x_{H_4}), \qquad S_2(\pi) = \frac{1}{\sqrt{2}}(x_{H_3} - x_{H_4} - x_{H_6} + x_{H_7})$$

$$S_3 = \frac{1}{\sqrt{2}}(y_{H_3} + y_{H_4}), \qquad S_3(\pi) = \frac{1}{\sqrt{2}}(y_{H_3} + y_{H_4} + y_{H_6} + y_{H_7})$$

B_{2g} Species

$$S_1 = z_{C_2}, \qquad S_1(\pi) = -z_{C_1} + z_{C_2} - z_{C_5} + z_{C_8}$$

$$S_2 = \frac{1}{\sqrt{2}}(z_{H_3} + z_{H_4}), \qquad S_2(\pi) = \frac{1}{\sqrt{2}}(z_{H_3} + z_{H4} - z_{H_6} - z_{H_7})$$

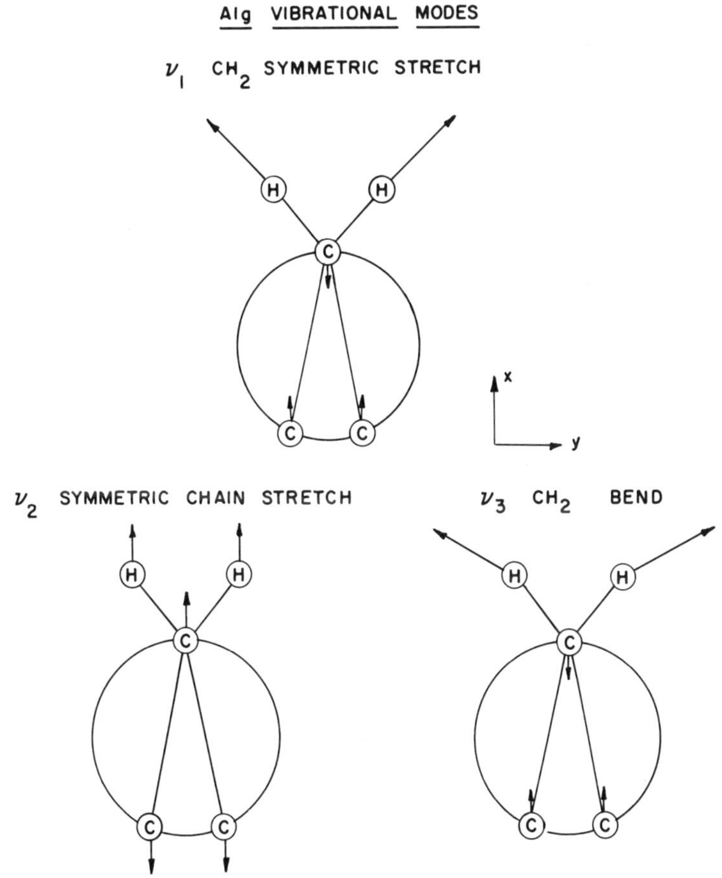

Figure 13.10. Eigenvectors for the A_{1g} vibrational modes of polyethylene.

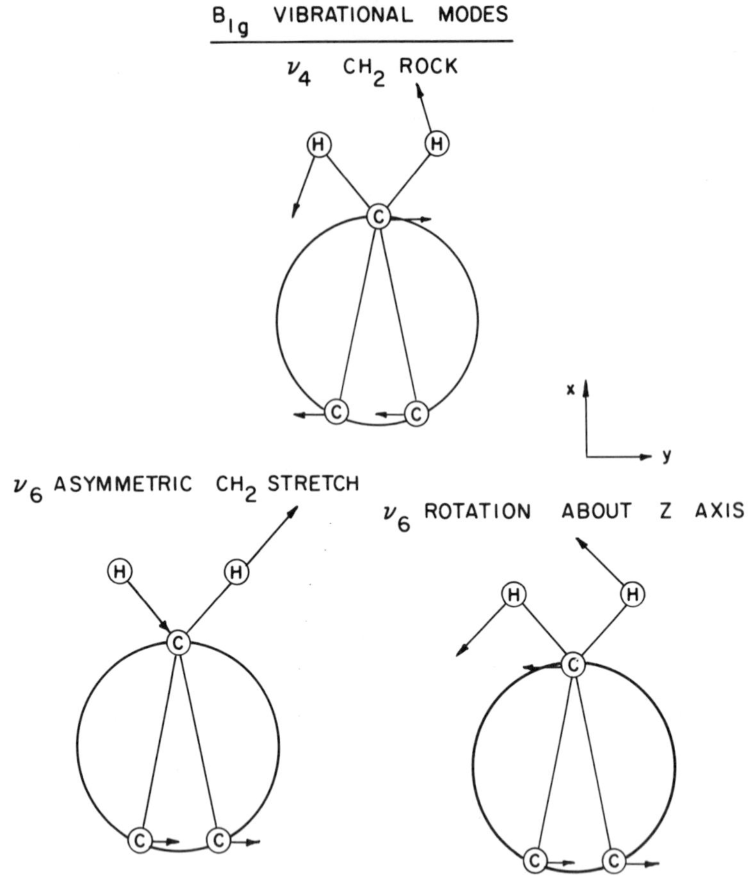

Figure 13.11. Eigenvectors for the B_{1g} vibrational modes of polyethylene.

If we wish to calculate the frequencies as a function of phase angle in order to plot the dispersion curve, we have to consider the complex component of equation (13.6), as discussed in Chapter 12. For such nonfactor group frequencies (values of θ other than 0 and π) one cannot factor the secular equation and it is therefore not necessary to calculate the local symmetry coordinates. We simply take the x, y, and z coordinates for each atom in the chemical repeat unit and apply the polymer symmetry (phasing). For polyethylene this is, of course, the 2_1 screw axis. Usually the phasing for the "A modes" ($\theta=0$) is specified and the computer program will calculate the appropriate symmetry coordinates for the "E modes" ($\theta \neq 0$) for each phase angle specified in the program input.

Using the data input specified, the frequencies shown in Table 13.14 were calculated and compared to the results of Snyder. The results are in excellent

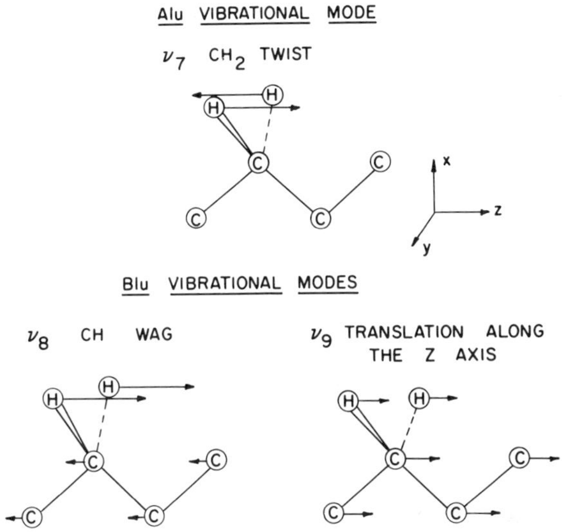

Figure 13.12. Eigenvectors for the A_{1u} and B_{1u} vibrational modes of polyethylene.

Table 13.15 Description of the Motion for the Active Polyethylene Vibrations

Vibration	Approx. Description of Vibration	Symmetry Species	Calc. Frequency (cm^{-1})	Potential Energy Distribution (%)
$\theta=0$				
ν_1	Symmetric CH_2 stretch	A_{1g}	2864	$d(98)$
ν_2	Chain stretch	A_{1g}	1133	$R(54)$, $\omega(42)$, $\gamma(8)$
ν_3	Asymmetric CH_2 stretch	B_{1g}	2928	$d(99)$
ν_4	CH_2 wagging motion	B_{1u}	1176	$\gamma(100)$
ν_5	CH_2 rocking motion	B_{1g}	1170	$\gamma(79)$
ν_6	CH_2 bend	A_{1g}	1439	$\delta(80)$, $\gamma(18)$
ν_7	CH_2 twist	A_{1u}	1063	$\gamma(99)$
ν_8	Translation in z direction, T_z	B_{1u}	0	
ν_9	R_z Rotation about z axis,	B_{1g}	0	
$\theta=\pi$				
ν_1	Symmetric CH_2 stretch	B_{3u}	2850	$d(99)$
ν_2	Chain stretch	B_{2g}	1060	$R(63)$, $\gamma(18)$
ν_3	Asymmetric CH_2 stretch	B_{2u}	2910	$d(100)$
ν_4	CH_2 wagging motion	B_{2g}	1385	$\gamma(89)$, $R(52)$
ν_5	CH_2 rocking motion	B_{2u}	716	$\gamma(96)$, $\tau(16)$
ν_6	CH_2 bend	B_{3u}	1472	$\delta(76)$, $\gamma(24)$
ν_7	CH_2 twist	B_{3g}	1301	$\gamma(100)$
ν_8	Translation in x direction, T_x	B_{3u}	0	
ν_9	Translation in y direction, T_y	B_{2u}	0	

agreement, as are the description of the motion in terms of eigenvectors, shown in Figures 13.10, 13.11, and 13.12, and the assignment in terms of the potential energy distribution, presented in Table 13.15.

13.4 Dispersion Curves and Two-Phonon Effects

As we have mentioned in the preceding sections, the dispersion curve can be plotted from experimental observations of the frequencies of the *n*-paraffins and compared to that calculated for polyethylene. We will defer a direct comparison of theoretical and calculated results until we consider the added complication of interchain interactions in the crystal. However, it is useful at this point to consider briefly the alternate ways of plotting the dispersion curve and the relationship between frequency dispersion and two-phonon effects.

The dispersion curve for a polymer can be plotted in two forms. The more usual way is a plot of frequency against θ, the phase angle between vibrations in adjacent chemical units. For simplicity we have shown such a dispersion curve for the skeletal vibrations of a polyethylene molecule (i.e., the vibrations of the carbon backbone alone) in Figure 13.13. For the two acoustic branches of this isolated chain the frequencies are reduced to $\nu=0$ for a phase difference of $\theta=0$. These vibrational modes are a translation along the chain axis and a rotation about the chain axis. The frequencies of these two acoustic branches are again reduced to zero at $\theta=\pi$ where we again have optical activity. The planar zigzag conformation of polyethylene is, in fact, a 2_1 helix allowing activity at $\theta=\pi$, as discussed in Chapter 10. To reiterate, the selection rules

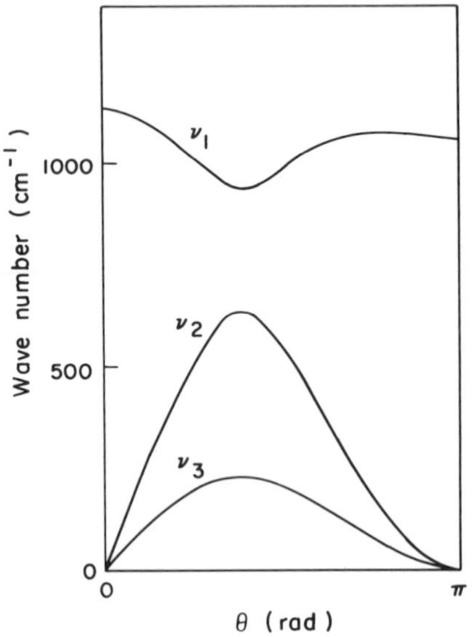

Figure 13.13. Dispersion curves for the skeletal vibrations of the polyethylene (PE) molecule. θ is the phase difference between adjacent CH_2 groups.

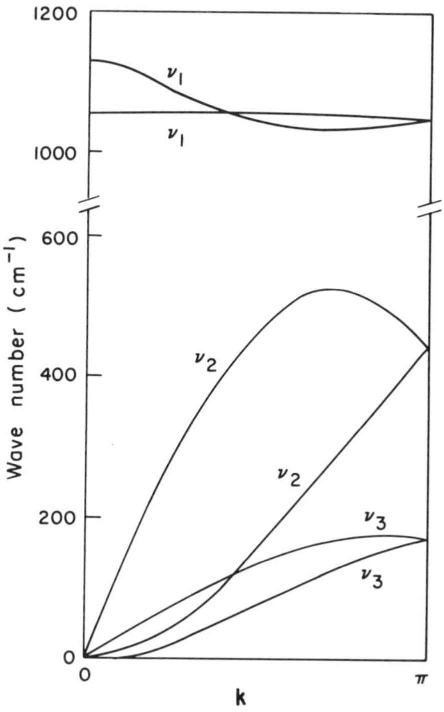

Figure 13.14. Dispersion curves for the skeletal vibrations of the PE molecule: k is the phase difference between adjacent unit cells.

require that vibrations in each *translational* repeat unit be totally in phase (i.e., the wave vector $\mathbf{k}=0$). This is obviously satisfied by the condition $\theta=0$. But if the phase difference between adjacent chemical units is π, then the phase difference between alternate chemical units is 2π, which is also equivalent to the $\mathbf{k}=0$ condition. In other words, because polyethylene has two chemical units per translational repeat unit, a phase difference of $\theta=0$ or $\theta=\pi$ will satisfy the condition that equivalent chemical units in adjacent translational repeat units vibrate in phase. The modes at $\theta=\pi$ differ in form from those on the same branch at $\theta=0$; they are reduced to translations perpendicular to the chain axis.

If we now consider dispersion curves plotted as a function of $\mathbf{k}$, the phase difference between vibrations in adjacent translational units, we effectively double the size of the unit cell relative to our plot of ν against θ. Consequently, the length of the Brillouin zone is half that discussed earlier and we can obtain the dispersion curve of ν as a function of $\mathbf{k}$ by "folding back" the right-hand half of the plot shown in Figure 13.13 to give the plot shown in Figure 13.14. Optical activity is allowed only for $\mathbf{k}=0$, but we now have the four acoustic branches corresponding to T_x, T_y, T_z, and R_z, reaching $\nu=0$ at $\mathbf{k}=0$. At the point $\mathbf{k}=\pi$ each branch degenerates into two vibrational modes, illustrated for ν_2 and ν_3 of Figure 13.14 in Figure 13.15.

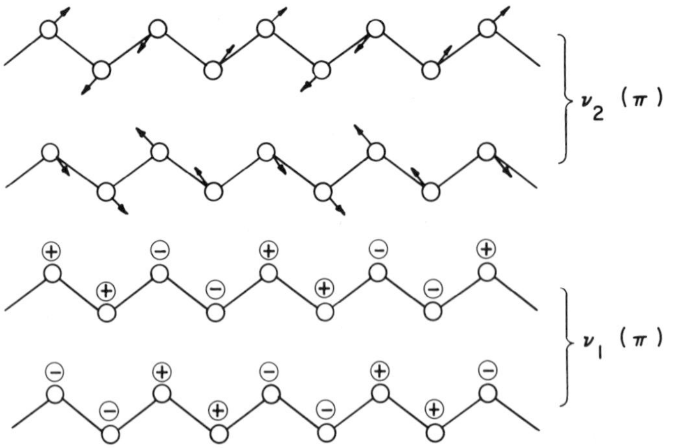

Figure 13.15. Examples of doubly degenerate vibrations of PE chain (the modes at $k=\pi$ in Figure 13.14.

We will discuss other aspects of dispersion curves when we discuss polymer crystals. Now we turn our attention to the relationship between frequency dispersion and the appearance of combinations or two-phonon effects (9).

In the spectra of small molecules, observed infrared bands or Raman lines can often be assigned to binary combinations of fundamental modes. When considering weak features in the spectra of polymers one is often tempted to do the same. It must be kept in mind, however, that the observed fundamentals of an ordered polymer represent only the $\theta=0$, ψ, or 2ψ (or wave vector $\mathbf{k}=0$) selection rules. Other points on the dispersion curve can contribute to a combination band. The necessary condition is that the wave vector must be the same for both components. For polyethylene the two branches, A and B, that

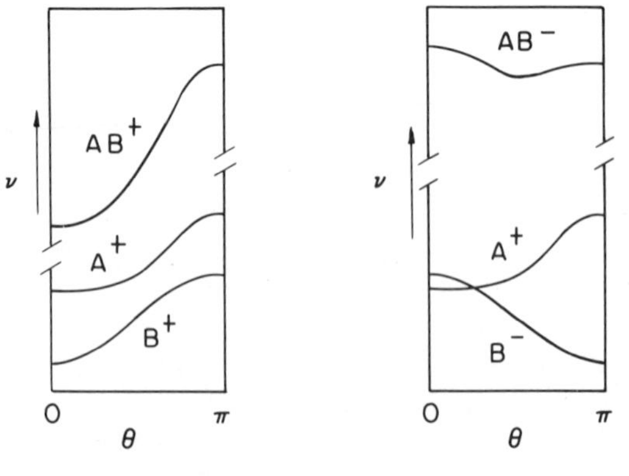

Figure 13.16. Schematic representation of dispersion curve combinations. (Reprinted by permission from ref. 9.)

combine can then do so in two ways:

$$\nu_{AB^+}(\theta)=\nu_A(\theta)+\nu_B(\theta) \tag{13.8}$$

$$\nu_{AB^-}(\theta)=\nu_A(\theta)+\nu_B(\pi-\theta) \tag{13.9}$$

where angle θ in these equations refers to the phase angle of the combining branches. Naturally, for the combination state the wave vector is zero at all points on the combined dispersion curve and all these points are allowed. We would then expect to see singularities or peaks at so-called critical points, where the combined dispersion curve is flat, that is, where there are a large number of points at a particular frequency. Figure 13.16 illustrates this designation of the binary states.

13.5 The Tetrahedral Approximation in Normal Coordinate Calculations of High Polymers

In many normal coordinate calculations of polymers, tetrahedral geometry is assumed, as in the polyethylene example discussed in this chapter. Opaskar and Krimm (10) pointed out that for many hydrocarbon polymers this is not true, and in order to illustrate the effect on nontetrahedral geometry these authors calculated the frequencies of polyethylene using the force field of Schachtschneider and Snyder (3), but with the planar C—C—C bond angle put equal to 112° and the H—C—H angle equal to 107°. No significant differences were found in the potential energy distributions of the modes, but it was seen that a change in the **G** matrix alone had resulted in one or two large frequency shifts and two instances in which the order of the frequencies is inverted. It is to be expected that this problem would be more acute where, rather than simply transferring a force field from a set of oligomers, the polymer frequencies are included in a refinement. We have made the point previously that this is not the best way to obtain a reasonable valence force field, but in some cases where there are few model compounds available there is little choice. Clearly, in such cases great caution must be used when considering the results of normal mode calculations, and wherever possible other evidence (group frequencies, etc.) should be used to support band assignments.

Finally, although the apparent sensitivity of the vibrational spectrum to the geometry of the polymer chain has certain negative aspects from the point of view of calculating normal coordinates, Zerbi (11) has pointed out some balancing positive elements. The sensitivity of the vibrational spectrum, particularly in the low frequency region, to changes in geometry (especially torsional angles) and the corresponding sensitivity of normal coordinate calculations to such effects should (theoretically) increase the importance of vibrational analysis in structural work. The authors of this book are a little less optimisitic in view of the essentially circular nature of the arguments often needed to obtain a good force field in such studies. However, we have no doubt that as long as suitable oligomers with an appropriate geometry are available, a force field should be directly transferable.

13.6 The Calculation of the Infrared and Raman Intensities of Polyethylene

The problem in the calculation of intensities revolves around the selection of a proper set of electro-optical parameters, eop's. It was demonstrated in Chapter 9 that within the framework of the electro-optical valence theory, the variation of the molecular dipole moment with respect to the ith normal mode can be written in terms of two kinds of parameters, equilibrium bond dipole moments and variations of bond dipole moments with the internal coordinates. These eop's should be transferable for molecules of similar classes and a set of eop's has been obtained for a series of n-paraffins by an overlay calculation (11–13).

Table 13.16 Values of the IR Electro-Optical Parameters Used for the Calculation of IR Intensities of Polyethylene from Least-Squares Overlay Refinement on Ethanes and Propanes[a]

Parameter	Calc. Value	Unit
$\frac{\partial m}{\partial r}$	0.86	Debye/Å
$\frac{\partial m}{\partial r'}$	0.06	Debye/Å
$\frac{\partial M}{\partial r}$	0.011	Debye/Å
$\frac{\partial M}{\partial \omega}$	0.023	Debye/rad
$\frac{\partial M}{\partial \gamma}$	−0.098	Debye/rad
$\frac{\partial M}{\partial \gamma'}$	0.060	Debye/rad
$\frac{\partial m}{\partial \omega}$	−0.023	Debye/rad
$\frac{\partial m}{\partial \gamma}$	−0.022	Debye/rad
$\frac{\partial m}{\partial \gamma'}$	0.039	Debye/rad
μ^0_{CH}	−0.245	Debye
μ^0_{CC}	0.000	Debye

[a]From Abbate et al. (13). The bond dipole moments of C—C are denoted by M, those of C—H bonds by m.

Unfortunately, the earlier work (13) on which the published intensity calculations of polyethylene are based has been superseded by revised calculations of the eop's of the alkanes (11). These were reported in Chapter 9. The earlier values of the infrared eop's are reproduced in Table 13.16. They differ from the more recently reported values principally in the sign of the parameters (see Table 9.4). The parameter set listed in Table 13.16 has the following features. The equilibrium C—C bond moment $\mu^0_{\mathrm{C-C}}$ is zero by symmetry. The C—H bond moment $\mu^0_{\mathrm{C-H}}$ is positive in the sense $\mathrm{C}^{(-)}$—$\mathrm{H}^{(+)}$. The dipole moment derivative $\delta\mu_{\mathrm{C-H}}/\delta r$ decreases while stretching the bond. The rate of change of $\mu_{\mathrm{C-H}}$ with angular displacement is smaller by an order of magnitude than that with the stretching of the bond.

In order to apply the eop's obtained from the paraffins to a calculation of the intensities of polyethylene, the equations have to be modified. As was shown in Chapter 9, the variation of the total dipole moment $\boldsymbol{\mu}$ with the normal coordinate Q_i is given by

$$\frac{\partial \boldsymbol{\mu}}{\partial Q_i}=\sum_m^M \sum_k^K \left(\mathbf{e}^{k0}\frac{\partial \mu^k}{\partial R_m}+\mu^{k0}\frac{\partial \mathbf{e}^k}{\partial R_m}\right)L_{mi} \tag{13.10}$$

where

$$R_m=\sum_i L_{mi}Q_i$$

For a polymer we have to "phase" our parameters between repeat units n. If θ is the phase angle between vibrations in adjacent units,

$$\mathbf{R}(\boldsymbol{\theta})=\frac{1}{\sqrt{N}}\sum_n^N R_n e^{is\theta} \tag{13.11}$$

where N is the number of units in the chain and s is the number of units between a given repeat unit and the chosen reference unit. We can now write

$$\mathbf{R}(\boldsymbol{\theta})=\mathbf{L}(\boldsymbol{\theta})\mathbf{Q}(\boldsymbol{\theta}) \tag{13.12}$$

If we assume a perfectly ordered chain, then the only optically active modes are those for which $\boldsymbol{\theta}=0$, so that we need to calculate only the eigenvectors $\mathbf{L}(0)$.

$$R_n=\frac{1}{\sqrt{N}}L(0)Q(0) \tag{13.13}$$

The variation of the total dipole moment for the mode Q_i is then

$$\frac{\partial \boldsymbol{\mu}}{\partial Q_i(0)}=\frac{1}{\sqrt{N}}\sum_{nn'}^N \sum_m^M \sum_k^K \left(\mathbf{e}^{k0}\frac{\partial \mu^{kn'}}{\partial R_{mn}}+\mu^{k0}\frac{\partial \mathbf{e}^{kn'}}{\partial R_{mn}}\right)L_{mi} \tag{13.14}$$

Table 13.17 Matrices of the IR Valence Electro-Optical Parameters for a Single Chain of Polyethylene[a]

$\frac{\partial\mu}{\partial R(0)}$	R_1^0	R_2^0	r_3^0	r_4^0	r_5^0	r_6^0	ω_1^0	ω_2^0	δ_1^0	δ_2^0	γ_3^0	γ_4^0	$\gamma_3'^0$	$\gamma_4'^0$	γ_5^0	γ_6^0	$\gamma_5'^0$	$\gamma_6'^0$	τ_1^0	τ_2^0
1	0*	0*	$\frac{\partial M}{\partial r}$	$\frac{\partial M}{\partial r}$	$\frac{-\partial M}{\partial r}$	$\frac{-\partial M}{\partial r}$	$\frac{\partial M}{\partial \omega}$	$\frac{-\partial M}{\partial \omega}$	$\frac{\partial M}{\partial \delta}$	$\frac{-\partial M}{\partial \delta}$	$\frac{\partial M}{\partial \gamma}$	$\frac{\partial M}{\partial \gamma}$	$\frac{\partial M}{\partial \gamma'}$	$\frac{\partial M}{\partial \gamma'}$	$\frac{-\partial M}{\partial \gamma'}$	$\frac{-\partial M}{\partial \gamma'}$	$\frac{-\partial M}{\partial \gamma}$	$\frac{-\partial M}{\partial \gamma}$	0*	0*
2	0*	0*	0	0	$\frac{\partial M}{\partial r}$	$\frac{\partial M}{\partial r}$	0	$\frac{\partial M}{\partial \omega}$	0	$\frac{\partial M}{\partial \delta}$	0	0	0	0	$\frac{\partial M}{\partial \gamma}$	$\frac{\partial M}{\partial \gamma}$	$\frac{\partial M}{\partial \gamma'}$	$\frac{\partial M}{\partial \gamma'}$	0*	0*
3	$\frac{\partial m}{\partial R}$	0	$\frac{\partial m}{\partial r}$	$\frac{\partial m}{\partial r'}$	0	0	$\frac{\partial m}{\partial \omega}$	0	$\frac{\partial m}{\partial \delta}$	0	$\frac{\partial m}{\partial \gamma}$	$\frac{\partial m}{\partial \gamma'}$	$\frac{\partial m}{\partial \gamma}$	$\frac{\partial m}{\partial \gamma'}$	0	0	0	0	0	0
4	$\frac{\partial m}{\partial R}$	0	$\frac{\partial m}{\partial r'}$	$\frac{\partial m}{\partial r}$	0	0	$\frac{\partial m}{\partial \omega}$	0	$\frac{\partial m}{\partial \delta}$	0	$\frac{\partial m}{\partial \gamma'}$	$\frac{\partial m}{\partial \gamma}$	$\frac{\partial m}{\partial \gamma'}$	$\frac{\partial m}{\partial \gamma}$	0	0	0	0	0	0
5	$\frac{\partial m}{\partial R}$	$\frac{\partial m}{\partial R}$	0	0	$\frac{\partial m}{\partial r}$	$\frac{\partial m}{\partial r'}$	0	$\frac{\partial m}{\partial \omega}$	0	$\frac{\partial m}{\partial \delta}$	0	0	0	0	$\frac{\partial m}{\partial \gamma}$	$\frac{\partial m}{\partial \gamma'}$	$\frac{\partial m}{\partial \gamma}$	$\frac{\partial m}{\partial \gamma'}$	0	0
6	$\frac{\partial m}{\partial R}$	$\frac{\partial m}{\partial R}$	0	0	$\frac{\partial m}{\partial r'}$	$\frac{\partial m}{\partial r}$	0	$\frac{\partial m}{\partial \omega}$	0	$\frac{\partial m}{\partial \delta}$	0	0	0	0	$\frac{\partial m}{\partial \gamma'}$	$\frac{\partial m}{\partial \gamma}$	$\frac{\partial m}{\partial \gamma'}$	$\frac{\partial m}{\delta \gamma}$	0	0
$\frac{\partial\mu}{\partial R(-1)}$	R_1^{-1}	R_2^{-1}	r_3^{-1}	r_4^{-1}	r_5^{-1}	r_6^{-1}	ω_1^{-1}	ω_2^{-1}	δ_1^{-1}	δ_2^{-1}	γ_3^{-1}	γ_4^{-1}	$(\gamma_3')^{-1}$	$(\gamma_4')^{-1}$	γ_5^{-1}	γ_6^{-1}	$(\gamma_5')^{-1}$	$(\gamma_6')^{-1}$	τ_1^{-1}	τ_2^{-1}
1	0	0*	0	0	0	0	0	0	0	0	0	0	0	0	0	0	0	0	0	0

2	0	0	0	0	0	0	0	0	0	0	0	0	0	0	0	0	0	0	0	0
3	0	$\frac{\partial m}{\partial R}$	0	0	0	0	0	0	0	0	0	0	0	0	0	0	0	0	0	0
4	0	$\frac{\partial m}{\partial R}$	0	0	0	0	0	0	0	0	0	0	0	0	0	0	0	0	0	0
5	0	0	0	0	0	0	0	0	0	0	0	0	0	0	0	0	0	0	0	0
6	0	0	0	0	0	0	0	0	0	0	0	0	0	0	0	0	0	0	0	0
$\frac{\partial \mu}{\partial R(+1)}$	R_1^1	R_2^1	r_3^1	r_4^1	r_5^1	r_6^1	ω_1^1	ω_2^1	δ_1^1	δ_2^1	γ_3^1	γ_4^1	$(\gamma_3')^1$	$(\gamma_4')^1$	γ_5^1	γ_6^1	$(\gamma_5')^1$	$(\gamma_6')^1$	τ_1^1	τ_2^1
1	0	0	0	0	0	0	0	0	0	0	0	0	0	0	0	0	0	0	0	0
2	0*	0	$-\frac{\partial M}{\partial r}$	$-\frac{\partial M}{\partial r}$	0	0	$-\frac{\partial M}{\partial \omega}$	0	$-\frac{\partial M}{\partial \delta}$	0	$-\frac{\partial M}{\partial \gamma'}$	$-\frac{\partial M}{\partial \gamma'}$	$-\frac{\partial M}{\partial \gamma}$	$-\frac{\partial M}{\partial \gamma}$	0	0	0	0	0	0
3	0	0	0	0	0	0	0	0	0	0	0	0	0	0	0	0	0	0	0	0
4	0	0	0	0	0	0	0	0	0	0	0	0	0	0	0	0	0	0	0	0
5	0	0	0	0	0	0	0	0	0	0	0	0	0	0	0	0	0	0	0	0
6	0	0	0	0	0	0	0	0	0	0	0	0	0	0	0	0	0	0	0	0

[a] Rows refer to the bonds of the central unit and columns to the internal coordinates (Figure 13.19). Superscripts 0, −1, 1 refer to the translational repeat unit. The asterisk indicates elements that vanish because of symmetry.

The summation over the repeat units n is limited to those units whose internal coordinates produce a variation in the dipole moment of the reference unit and/or a distortion of its bond directions. If the electro-optical interaction is from $-s$ to $+s$ around the central unit and if we substitute the factor N (equal to the number of repeat units in the chain) for the summation over n', then in matrix notation equation (13.14) becomes

$$\frac{\partial \boldsymbol{\mu}}{\partial \mathbf{Q}_i(0)} = \sqrt{\mathbf{N}}\left[\mathbf{e}\cdot\mathbf{0}^T \frac{\partial \boldsymbol{\mu}}{\partial \mathbf{R}}(0) + \boldsymbol{\mu}^{0T}\frac{\partial \mathbf{e}}{\partial \mathbf{R}}(0)\right] L_i \tag{13.15}$$

where

$$\frac{\partial \boldsymbol{\mu}}{\partial \mathbf{R}}(0) = \sum_{n=-s}^{+s} \frac{\partial \mu}{\partial R_n} \tag{13.16}$$

and

$$\frac{\partial \mathbf{e}}{\partial \mathbf{R}}(0) = \sum_{n=-s}^{s} \frac{\partial \mathbf{e}}{\partial R_n} \tag{13.17}$$

These matrices are the variation of the bond dipole moments and bond directions of the central unit ($n=0$) generated by the internal coordinate of the nth unit ($n=-1,\ldots,-s$, $n=+1,\ldots,+s$). The calculation for polyethylene has been made with the assumption that interactions extend only to adjacent units.

The calculations of infrared intensities for polyethylene by Abbate et al. (13) can be summarized in the following tables. The $\partial\boldsymbol{\mu}/\partial\mathbf{R}(\mathbf{0})$ matrices are shown in Table 13.17 based on the definition of internal coordinates designated in Figure 13.17 (cf. Figure 13.19). The numbers 0, -1, 1 refer to the translational repeat unit. The force field used is described earlier and the

Table 13.18 Eigenvectors Calculated for IR Frequencies of Polyethylene

Frequency (cm^{-1})			Polyethylene
ν_{obs}	ν_{calc}	Sym. Species	Eigenvectors
2917	2910	B_{2u}	$1.048d^{-} - 0.025P - 0.189\tau$
2848	2850	B_{3u}	$1.030d^{+} + 0.68T - 0.134\delta$
1473	1473	B_{3u}	$0.028d^{+} - 0.676T + 1.352\delta$
1463			
. . .[a]	1176	B_{1u}	$1.205\ W$
731	716	B_{2u}	$0.004d^{-} + 0.659P + 2.425\tau$
719			

[a]Presumably not observed.

Table 13.19 Matrix of the IR Valence Electro-Optical Parameters in Phonon Coordinates ($q=0$ wave vector) for Polyethylene.[a]

$\frac{\partial\mu}{\partial R}(0)$	R_1	R_2	γ_3	γ_4	γ_5	γ_6	ω_1	ω_2	δ_1	δ_2	γ_3	γ_4	γ_3'	γ_4'	γ_5	γ_6	γ_5'	γ_6'	τ_1	τ_2
1	0	0	$\frac{\partial M}{\partial r}$	$\frac{\partial M}{\partial r}$	$-\frac{\partial M}{\partial r}$	$-\frac{\partial M}{\partial r}$	$\frac{\partial M}{\partial \omega}$	$-\frac{\partial M}{\partial \omega}$	$\frac{\partial M}{\partial \delta}$	$-\frac{\partial M}{\partial \delta}$	$\frac{\partial M}{\partial \gamma}$	$\frac{\partial M}{\partial \gamma}$	$\frac{\partial M}{\partial \gamma'}$	$\frac{\partial M}{\partial \gamma'}$	$-\frac{\partial M}{\partial \gamma'}$	$-\frac{\partial M}{\partial \gamma'}$	$-\frac{\partial M}{\partial \gamma}$	$-\frac{\partial M}{\partial \gamma}$	0	0
2	0	0	$-\frac{\partial M}{\partial r}$	$-\frac{\partial M}{\partial r}$	$\frac{\partial M}{\partial r}$	$\frac{\partial M}{\partial r}$	$-\frac{\partial M}{\partial \omega}$	$\frac{\partial M}{\partial \omega}$	$-\frac{\partial M}{\partial \delta}$	$\frac{\partial M}{\partial \delta}$	$-\frac{\partial M}{\partial \gamma'}$	$-\frac{\partial M}{\partial \gamma'}$	$-\frac{\partial M}{\partial \gamma}$	$-\frac{\partial M}{\partial \gamma}$	$\frac{\partial M}{\partial \gamma}$	$\frac{\partial M}{\partial \gamma}$	$\frac{\partial M}{\partial \gamma'}$	$\frac{\partial M}{\partial \gamma'}$	0	0
3	$\frac{\partial m}{\partial R}$	$\frac{\partial m}{\partial R}$	$\frac{\partial m}{\partial r}$	$\frac{\partial m}{\partial r'}$	0	0	$\frac{\partial m}{\partial \omega}$	0	$\frac{\partial m}{\partial \delta}$	0	$\frac{\partial m}{\partial \gamma}$	$\frac{\partial m}{\partial \gamma'}$	$\frac{\partial m}{\partial \gamma}$	$\frac{\partial m}{\partial \gamma'}$	0	0	0	0	0	0
4	$\frac{\partial m}{\partial R}$	$\frac{\partial m}{\partial R}$	$\frac{\partial m}{\partial r'}$	$\frac{\partial m}{\partial r}$	0	0	$\frac{\partial m}{\partial \omega}$	0	$\frac{\partial m}{\partial \delta}$	0	$\frac{\partial m}{\partial \gamma'}$	$\frac{\partial m}{\partial \gamma}$	$\frac{\partial m}{\partial \gamma'}$	$\frac{\partial m}{\partial \gamma}$	0	0	0	0	0	0
5	$\frac{\partial m}{\partial R}$	$\frac{\partial m}{\partial R}$	0	0	$\frac{\partial m}{\partial r}$	$\frac{\partial m}{\partial r'}$	0	$\frac{\partial m}{\partial \omega}$	0	$\frac{\partial m}{\partial \delta}$	0	0	0	0	$\frac{\partial m}{\partial \gamma}$	$\frac{\partial m}{\partial \gamma'}$	$\frac{\partial m}{\partial \gamma}$	$\frac{\partial m}{\partial \gamma'}$	0	0
6	$\frac{\partial m}{\partial R}$	$\frac{\partial m}{\partial R}$	0	0	$\frac{\partial m}{\partial r'}$	$\frac{\partial m}{\partial r}$	0	$\frac{\partial m}{\partial \omega}$	0	$\frac{\partial m}{\partial \delta}$	0	0	0	0	$\frac{\partial m}{\partial \gamma'}$	$\frac{\partial m}{\partial \gamma}$	$\frac{\partial m}{\partial \gamma'}$	$\frac{\partial m}{\partial \gamma}$	0	0

[a]Rows refer to the bonds and columns to the internal coordinates.

Table 13.20 Observed and Calculated Values of the Relative Intensities of Polyethylene[a]

Observed Frequency (cm^{-1})	Description	I_{obs}	I_{calc}
2917	d^-	21.50	25.99
2848	d^+	8.67	10.38
1473–1463	δ	2.25	2.37
731–719	P	1	1

[a]From Abbate et al. (13).

calculated eigenvector corresponding to each symmetry coordinate is given in Table 13.18. It is important to note that the intensities calculated depend on the eigenvectors **L** and can therefore vary with the force field selected. Table 13.19 gives the matrix elements of the infrared valence electro-optical parameters after phasing. The calculated infrared intensities are reported in Table 13.20 for the single-chain model. The calculated intensities are also compared with the experimental values in Table 13.20. It is apparent that the experimental and theoretical infrared intensities are in good agreement.

The problem of calculating Raman intensities is complicated by the lack of experimental data required to derive a set of Raman eop's. There are four equilibrium eop's for polyethylene: $\bar{\alpha}^0_{C-C}$, $\bar{\alpha}^0_{C-H}$, γ^0_{C-C}, γ^0_{C-H}. However, $\bar{\alpha}^0_{C-C}$ and $\bar{\alpha}^0_{C-H}$ do not affect the intensities. The calculated results (13) available in the literature utilize data from CH_4 and cyclohexane. Table 13.21 gives the Raman valence eop's required for polyethylene using the definition of internal

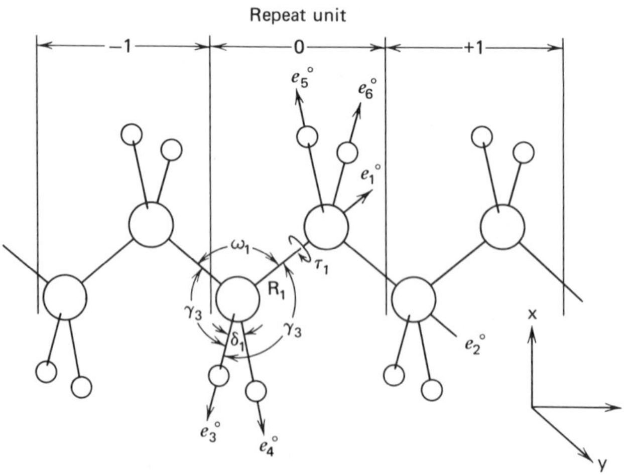

Figure 13.17. Model of polyethylene single-chain molecule. Definition of the cartesian reference system, of internal coordinates, and of unit bond vectors. Geometry adopted: $R^{eq} = 1.54$ Å; $r^{eq} = 1.09$ Å; all angles tetrahedral. (Reprinted by permission from ref. 13.)

Table 13.21 Matrices of the Raman Valence Electro-Optical Parameters in Phonon Coordinates ($q=0$ Wave Vector) for Polyethylene[a]

	R_1	R_2	r_3	r_4	r_5	r_6	ω_1	ω_2	δ_1	δ_2	γ_3	γ_4	γ_3'	γ_4'	γ_5	γ_6	γ_5'	γ_6'	τ_1	τ_2
$\frac{\partial\bar{\alpha}_{CC}}{\partial R}(0)$ *Bond Mean Polarizability Derivatives*																				
1 $\frac{\partial\bar{\alpha}_{CC}}{\partial R}$	0	0	0	0	0	$\frac{\partial\bar{\alpha}_{CC}}{\partial\omega}$	$\frac{\partial\bar{\alpha}_{CC}}{\partial\omega}$	$\frac{\partial\bar{\alpha}_{CC}}{\partial\delta}$	$\frac{\partial\bar{\alpha}_{CC}}{\partial\delta}$	0	0	0	0	0	0	0	0	0	0
2 0	$\frac{\partial\bar{\alpha}_{CC}}{\partial R}$	0	0	0	0	$\frac{\partial\bar{\alpha}_{CC}}{\partial\omega}$	$\frac{\partial\bar{\alpha}_{CC}}{\partial\omega}$	$\frac{\partial\bar{\alpha}_{CC}}{\partial\delta}$	$\frac{\partial\bar{\alpha}_{CC}}{\partial\delta}$	0	0	0	0	0	0	0	0	0	0
3 $\frac{\partial\bar{\alpha}_{CH}}{\partial R}$	$\frac{\partial\bar{\alpha}_{CH}}{\partial R}$	$\frac{\partial\bar{\alpha}_{CH}}{\partial r}$	0	0	0	$\frac{\partial\bar{\alpha}_{CH}}{\partial\omega}$	0	$\frac{\partial\bar{\alpha}_{CH}}{\partial\delta}$	0	0	0	0	0	0	0	0	0	0	0
4 $\frac{\partial\bar{\alpha}_{CH}}{\partial R}$	$\frac{\partial\bar{\alpha}_{CH}}{\partial R}$	0	$\frac{\partial\bar{\alpha}_{CH}}{\partial r}$	0	0	$\frac{\partial\bar{\alpha}_{CH}}{\partial\omega}$	0	$\frac{\partial\bar{\alpha}_{CH}}{\partial\delta}$	0	0	0	0	0	0	0	0	0	0	0
5 $\frac{\partial\bar{\alpha}_{CH}}{\partial R}$	$\frac{\partial\bar{\alpha}_{CH}}{\partial R}$	0	0	$\frac{\partial\bar{\alpha}_{CH}}{\partial r}$	0	0	$\frac{\partial\bar{\alpha}_{CH}}{\partial\omega}$	0	$\frac{\partial\bar{\alpha}_{CH}}{\partial\delta}$	0	0	0	0	0	0	0	0	0	0
6 $\frac{\partial\bar{\alpha}_{CH}}{\partial R}$	$\frac{\partial\bar{\alpha}_{CH}}{\partial R}$	0	0	0	$\frac{\partial\bar{\alpha}_{CH}}{\partial r}$	0	$\frac{\partial\bar{\alpha}_{CH}}{\partial\omega}$	0	$\frac{\partial\bar{\alpha}_{CH}}{\partial\delta}$	0	0	0	0	0	0	0	0	0	0
$\frac{\partial\gamma_{CC}}{\partial R}(0)$ *Bond Anisotropy Derivatives*																				
1 $\frac{\partial\gamma_{CC}}{\partial R}$	0	0	0	0	0	0	0	$\frac{\partial\gamma_{CC}}{\partial\delta}$	$\frac{\partial\gamma_{CC}}{\partial\delta}$	$\frac{\partial\gamma_{CC}}{\partial\gamma}$	$\frac{\partial\gamma_{CC}}{\partial\gamma}$	$\frac{\partial\gamma_{CC}}{\partial\gamma'}$	$\frac{\partial\gamma_{CC}}{\partial\gamma'}$	$\frac{\partial\gamma_{CC}}{\partial\gamma'}$	$\frac{\partial\gamma_{CC}}{\partial\gamma'}$	$\frac{\partial\gamma_{CC}}{\partial\gamma}$	$\frac{\partial\gamma_{CC}}{\partial\gamma}$	0	0
2 0	$\frac{\partial\gamma_{CC}}{\partial R}$	0	0	0	0	0	0	$\frac{\partial\gamma_{CC}}{\partial\delta}$	$\frac{\partial\gamma_{CC}}{\partial\delta}$	$\frac{\partial\gamma_{CC}}{\delta\gamma'}$	$\frac{\partial\gamma_{CC}}{\delta\gamma'}$	$\frac{\partial\gamma_{CC}}{\delta\gamma}$	$\frac{\partial\gamma_{CC}}{\partial\gamma}$	$\frac{\partial\gamma_{CC}}{\partial\gamma}$	$\frac{\partial\gamma_{CC}}{\partial\gamma}$	$\frac{\partial\gamma_{CC}}{\partial\gamma'}$	$\frac{\partial\gamma_{CC}}{\partial\gamma'}$	0	0
3 $\frac{\partial\gamma_{CH}}{\partial R}$	$\frac{\partial\gamma_{CH}}{\partial R}$	$\frac{\partial\gamma_{CH}}{\partial r}$	0	0	0	0	0	$\frac{\partial\gamma_{CH}}{\partial\delta}$	0	$\frac{\partial\gamma_{CH}}{\partial\gamma}$	$\frac{\partial\gamma_{CH}}{\partial\gamma'}$	$\frac{\partial\gamma_{CH}}{\partial\gamma}$	$\frac{\partial\gamma_{CH}}{\partial\gamma'}$	0	0	0	0	0	0
4 $\frac{\partial\gamma_{CH}}{\partial R}$	$\frac{\partial\gamma_{CH}}{\partial R}$	0	$\frac{\partial\gamma_{CH}}{\partial r}$	0	0	0	0	$\frac{\partial\gamma_{CH}}{\partial\delta}$	0	$\frac{\partial\gamma_{CH}}{\partial\gamma'}$	$\frac{\partial\gamma_{CH}}{\partial\gamma}$	$\frac{\partial\gamma_{CH}}{\partial\gamma'}$	$\frac{\partial\gamma_{CH}}{\partial\gamma}$	0	0	0	0	0	0
5 $\frac{\partial\gamma_{CH}}{\partial R}$	$\frac{\partial\gamma_{CH}}{\partial R}$	0	0	$\frac{\partial\gamma_{CH}}{\partial r}$	0	0	0	0	$\frac{\partial\gamma_{CH}}{\partial\delta}$	0	0	0	0	$\frac{\partial\gamma_{CH}}{\partial\gamma}$	$\frac{\partial\gamma_{CH}}{\partial\gamma'}$	$\frac{\partial\gamma_{CH}}{\partial\gamma}$	$\frac{\partial\gamma_{CH}}{\partial\gamma'}$	0	0
6 $\frac{\partial\gamma_{CH}}{\partial R}$	$\frac{\partial\gamma_{CH}}{\partial R}$	0	0	0	$\frac{\partial\gamma_{CH}}{\partial r}$	0	0	0	$\frac{\partial\gamma_{CH}}{\partial\delta}$	0	0	0	0	$\frac{\partial\gamma_{CH}}{\partial\gamma'}$	$\frac{\partial\gamma_{CH}}{\partial\gamma}$	$\frac{\partial\gamma_{CH}}{\partial\gamma'}$	$\frac{\partial\gamma_{CH}}{\partial\gamma}$	0	0

[a] Rows refer to the bonds and columns to the internal coordinates.

Table 13.22 Values of the Raman Electro-Optical Parameters Used for the Calculation of Raman Intensities of Polyethylene from Least-Squares Refinements on Cyclohexane

Parameter	Value	Unit
$\frac{\partial \bar{\alpha}_{CC}}{\partial R}$	0.939	Å^2
$\frac{\partial \bar{\alpha}_{CC}}{\partial \omega}$	0.044	Å^3/rad
$\frac{\partial \bar{\alpha}_{CH}}{\partial r}$	1.300	Å^2
$\frac{\partial \bar{\alpha}_{CH}}{\partial \omega}$	0.000	Å^3/rad
$\frac{\partial \gamma_{CC}}{\partial R}$	1.458	Å^2
$\frac{\partial \gamma_{CC}}{\partial \gamma}$	−0.043	Å^3/rad
$\frac{\partial \gamma_{CC}}{\partial \gamma'}$	0.019	Å^3/rad
$\frac{\partial \gamma_{CH}}{\partial r}$	2.200	Å^2
$\frac{\partial \gamma_{CH}}{\partial \gamma}$	−0.089	Å^3/rad
$\frac{\partial \gamma_{CH}}{\partial \gamma'}$	0.113	Å^3/rad
γ_{CC}	0.050	Å^3
γ_{CH}	0.318	Å^3

coordinates given in Figure 13.17. The bond mean polarizability and bond anisotropy derivatives are given in Table 13.22. Because of the redundancy in the internal coordinates, general relationships exist between the eop derivatives:

$$\frac{\partial \bar{\alpha}_{C-C}}{\partial \delta} = -\frac{\partial \bar{\alpha}_{C-C}}{\partial \omega}$$

$$\frac{\partial \bar{\alpha}_{C-H}}{\partial \delta} = -\frac{\partial \bar{\alpha}_{C-H}}{\partial \omega}$$

$$\frac{\partial \gamma_{C-C}}{\partial \delta} = -2\frac{\partial \gamma_{C-C}}{\partial \gamma} - 2\frac{\partial \gamma_{C-C}}{\partial \gamma}$$

$$\frac{\partial \gamma_{C-H}}{\partial \delta} = -2\frac{\partial \gamma_{C-H}}{\partial \gamma} - 2\frac{\partial \gamma_{C-H}}{\partial \gamma}$$

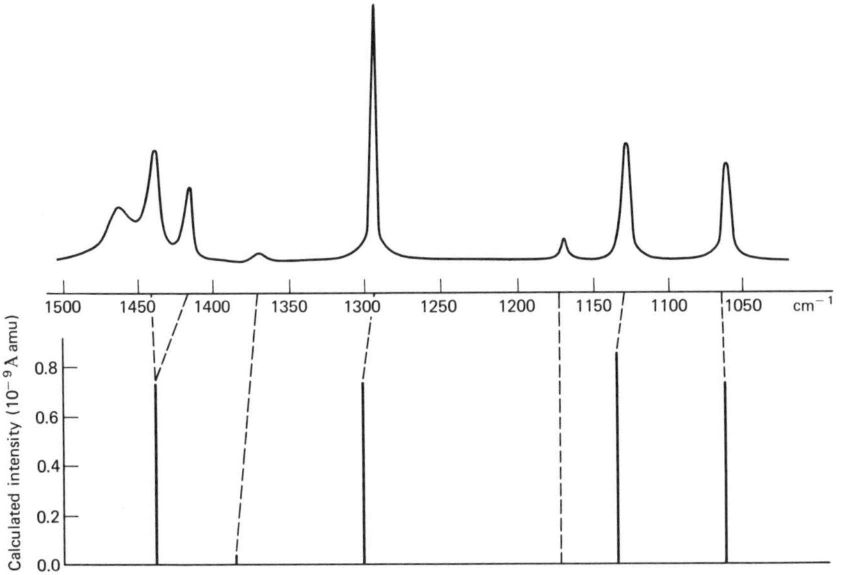

Figure 13.18. Top: hydrogenated polyethylene Raman spectrum in the CH_2 deformation and skeletal stretching region. Exciting line 5145 Å; power at the sample 140 mW; spectral slit width 1 cm^{-1}; gain×10; time constant 2 sec; scan speed 5 cm^{-1}/min. Bottom: graphical representation of the calculated intensities reported in Table 13.23 column C. (Reprinted by permission from ref. 13.)

Table 13.23 Observed and Calculated Values of the Raman Intensities of Polyethylene.

$\nu_{obs}(cm^{-1})$	Exptl. Band Areas (cm^2)	I_{calc}
1440 1416	~13.0	0.75
1370	0.8	0.05
1295	12.5	0.74
1170	1.4	0.00
1130	7.7	0.99
1062	8.0	0.76

The values of the eop's used for the calculation, shown in Table 13.22, are based on a least-squares refinement of cyclohexane. The calculated intensities are compared with the experimental intensities in Figure 13.18 and Table 13.23. The agreement obtained is very encouraging and suggests the potential of future efforts on this direction, providing that more experimental data from low molecular weight analogues are obtained.

References

1 M. Tasumi, in "Vibrational Spectroscopy—Modern Trends," A. J. Barnes and W. J. Orville-Thomas, Eds., Elsevier, New York, 1977, Chapter 23, p. 365.

2 R. G. Snyder and J. H. Schachtschneider, *Spectrochim. Acta* **19**, 85 (1963).

3 J. H. Schachtschneider and R. G. Snyder, *Spectrochim. Acta* **19**, 117 (1963).

4 R. G. Snyder and J. H. Schachtschneider, *Spectrochim. Acta* **21**, 169 (1965).

5 R. G. Snyder, *J. Mol. Spect.* **23**, 224 (1967).

6 R. G. Snyder, *J. Chem. Phys.* **47**, 1316 (1967).

7 J. Barnes and B. Fanconi, *J. Phys. Chem. Ref. Data* **7**(4), 1309 (1978).

8 K. Holland-Moritz and E. Sausen, *J. Polymer Sci. Polymer Phys. Ed.* **17**, 1 (1979).

9 R. G. Snyder, *J. Chem. Phys.* **68**, 4156 (1978).

10 C. G. Opaskar and S. Krimm, *Spectrochim. Acta* **21**, 1165 (1965).

11 M. Gussoni, S. Abbate, and G. Zerbi, *J. Chem. Phys.* **71**(8), 3428 (1979).

12 M. Gussoni, S. Abbate, and G. Zerbi, in "Vibrational Spectroscopy—Modern Trends," A. W. J. Barnes and J. Orville-Thomas, Eds., Elsevier, New York, 1977.

13 S. Abbate, M. Gussoni, G. Masetti, and G. Zerbi, *J. Chem. Phys.* **67**, 1519 (1977).

14

VIBRATIONAL ANALYSIS OF POLYMER CRYSTALS

Order leads to all the virtues, but what leads to order?
G. C. Lichtenberg

The investigation of the structure of polymer crystals has been a source of employment for polymer scientists for many years and promises to remain so for some time to come, particularly since the arguments concerning various proposed models have recently taken on an interesting degree of rancor. The controversy predominantly concerns the concept and nature of chain folding, the conformation and spatial arrangement of the chains within the bulk of the crystal being well established for many polymers by X-ray diffraction methods. Vibrational spectroscopy is very sensitive to the degree of order prevailing in a particular sample, but the information present in a spectrum is not always easy to classify or interpret. In a simplistic sense a particular sample could be considered a "mixture" of various phases: ordered conformations arranged in a regular three-dimensional array, regular or irregular chain folds, and a bulk amorphous phase. Naturally, all of these components will contribute to the spectrum, but there will also be effects due to vibrational coupling and complications due to factors such as the presence of defects in the crystal.

Experimentally, one of the major problems is separating the spectral contributions of the various phases. Although the complex structure of semi-crystalline polymers and the possibility or even probability of the perturbation of the vibrations of one phase by another would seem to make this problem intractable, the recent application of computerized instruments has led to a breakthrough in this area (1, 2). It is not the main purpose of this chapter nor indeed of this book to discuss such recent advances in instrumentation. However, a brief consideration of the separation of the spectral contribution of the amorphous and crystalline phases of *trans*-1, 4-polychloroprene is of some use in that it provides a framework in which we can define terms that are often confused when the spectra of polymer crystals are discussed.

Figure 14.1 compares the infrared spectrum of *trans*-1, 4-polychloroprene obtained as an amorphous film to the spectrum of the same sample subsequent

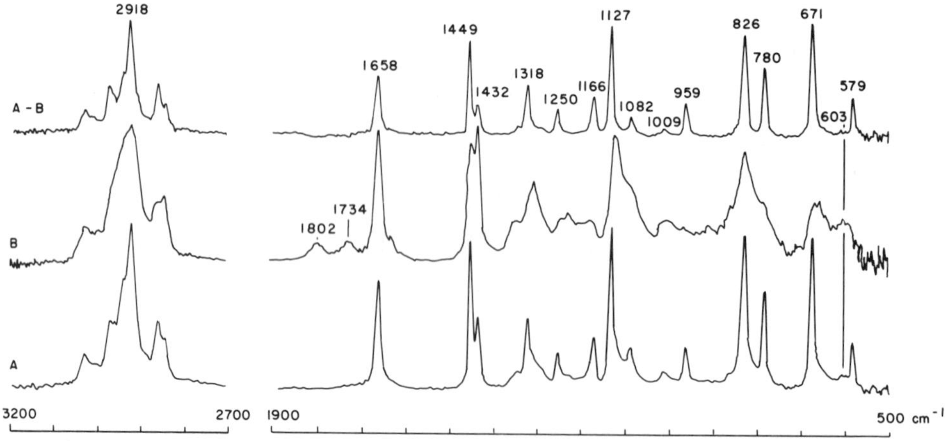

Figure 14.1. The infrared spectra of *trans*-1,4-polychloroprene. (*A*) Absorbance spectrum at room temperature (*B*) Absorbance spectrum at 80° (*A*−*B*) Difference spectrum. (Reprinted by permission from ref. 3.)

to a degree of crystallization. Spectra were obtained on Digilab FTS system (3). It can be seen from this figure that upon crystallization the bands become sharper and in certain cases more intense (similar results can be observed using Raman spectroscopy). Nevertheless, there is still an apparent contribution from the amorphous phase in the spectrum of the semicrystalline material. The computer manipulation capabilities of FT-IR systems are now routinely applied to such systems, so that by subtracting the spectrum of the amorphous component, scaled appropriately, from that of the semicrystalline material, a difference spectrum is obtained. This difference spectrum is also shown in Figure 14.1. Such spectra are often referred to as characteristic of the crystalline phase, but this designation is not entirely accurate. The sharp isolated bands in this difference spectrum are, in fact, characteristic of a single polymer chain in the preferred conformation, because effects due to interchain interactions in the crystal are apparently too small to be observed. In addition, this polymer has only translational symmetry elements relating chemical units along the chain, so that only $\theta=0$ modes are observed. In helical polymers, splitting into $\theta=0$ and $\theta=\psi$ modes are observed in the infrared, where ψ is the rotation angle between adjacent units, as discussed in Chapter 11. Zerbi et al. (4) have defined modes that are generated by the rototranslational symmetry of the polymer chain as “regularity bands.” Although ordered polymer chains are usually found in a three-dimensional lattice, this is not always the case (e.g., isotactic polypropylene can be obtained in a “smectic” form that has two-dimensional order). Consequently, the phrase “crystalline bands” (or lines) is used to designate those modes for which factor group splitting and correlation field splitting of line group modes are identified with certainty. On this basis,

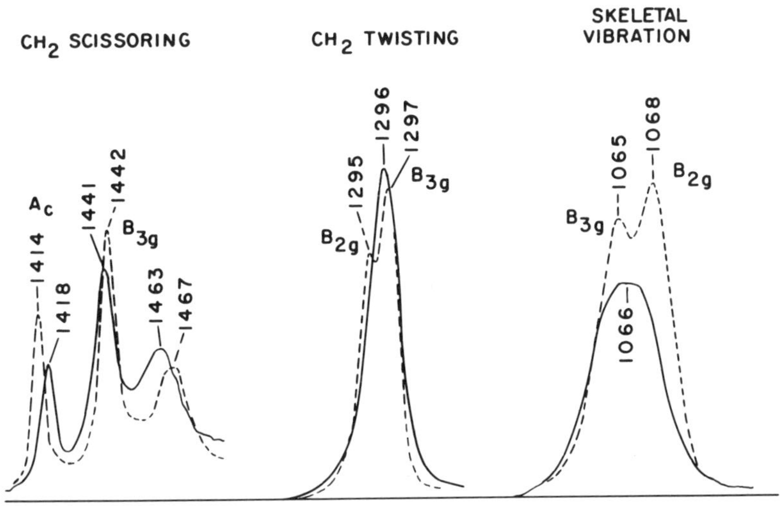

Figure 14.2. Raman spectra of polyethylene at 25°C (_____) and at −160°C (---).

the modes appearing in the difference spectrum shown in Figure 14.1 cannot be classified as regularity bands (since there is not splitting due to intrachain interactions, only $\theta=0$ modes appear) or crystalline bands. Nevertheless, this spectrum is clearly characteristic of the ordered polymer chain in the crystal lattice.

An example of the "splitting" of infrared bands due to interchain interactions in the crystal was presented earlier when we discussed a space group analysis of polyethylene. A similar splitting of lines in the Raman spectrum was observed by Boerio and Koenig (5), as shown in Figure 14.2, although not without going to liquid nitrogen temperatures. It was argued that cooling the sample brings the chains closer together and therefore increases the magnitude of intermolecular forces. In this chapter we will consider in some detail the interchain interactions in polymer crystals that can give rise to such splittings. First, we will discuss the theory of the molecular vibrations of polymer crystals, which uses modifications to the secular equation that allow the calculation of the normal modes and dispersion of a three-dimensional system. We will then proceed to examine in detail calculations for two polymer systems, polyethylene and polyglycine. Finally, we will turn our attention to lattice modes. These modes, although experimentally difficult to observe, are extremely useful in characterizing the structure of crystalline polymers. Unfortunately, such lattice modes, and indeed the splitting of line group modes due to interchain interactions, have been observed for very few polymers. For most highly ordered macromolecules the single-chain treatment discussed in previous chapters gives a practically complete description of the observed vibrational spectrum.

14.1 The Lattice Dynamics of Crystalline Polymers

The normal coordinate treatment of a single chain discussed in Chapter 12 can be readily extended to a three-dimensional crystal. As might be intuitively expected, the form of the equation is the same but we now have a dependence on the wave vector **k** that, is no longer a one-dimensional dispersion relation, but has components k_a, k_b, and k_c along the crystallographic axes of the unit cell. Consequently, although the concepts involved in performing a normal vibrational analysis are relatively straightforward, the actual calculations are complicated by the sheer physical size of the problem.

The derivation of the dynamical matrix that we will present is that of Piseri and Zerbi (6). These authors pointed out that the application of the **GF** method to a crystalline system is difficult, especially when phase coupling between equivalent units is included in order to calculate the dispersion curves. As pointed out previously in this text, the **GF** matrix is not symmetric and must be reduced to Hermitian form by proper transformation in order to allow solution by computer methods (see Chapter 8). It was demonstrated in Chapter 3 (although a formal derivation was not presented there) that the symmetric cartesian coordinate description of the dynamical matrix can be written in terms of the **F** matrix expressed in internal coordinates. We will consider a derivation of this form of the secular equation here.

As before, we first define the kinetic and potential energy terms.

$$2T=\sum_{n,n'}\dot{\mathbf{x}}_n^T\mathbf{M}\dot{\mathbf{x}}_{n'}(n=n') \tag{14.1}$$

$$2V=\sum_{n,n'}\mathbf{r}_n^T\mathbf{F}_{n,n'}\mathbf{r}_{n'} \tag{14.2}$$

where $\mathbf{x}_n$ and $\mathbf{r}_n$ are the vectors of the cartesian displacement coordinates and internal coordinates of the nth unit cell respectively. Each unit cell n is labeled by a triplet of integers n_1, n_2, n_3. Each unit cell contains N atoms. The location of each unit cell is given by the vector $t(n)$ referred to a suitably chosen origin.

In equation (14.1) **M** is a diagonal $3N\times3N$ matrix of the masses of the atoms in the unit cell appropriately arranged, while $\mathbf{F}_{nn'}$ in equation (14.2) is the $3N\times3N$ matrix of harmonic force constants describing the interaction between internal coordinates of the unit cell n with those of the unit cell n'. The periodicity of the crystal requires that $\mathbf{F}_{nn'}$ depends only on the difference $|n-n'|=s$, so that we can define $\mathbf{F}_{nn'}=\mathbf{F}^s$ and $\mathbf{F}_{n'n}=\mathbf{F}_s^T$. As in Chapter 12, we can now effect a transformation to symmetry coordinates based on the translational symmetry of the lattice. These symmetry coordinates, or cartesian and internal phonon coordinates, are

$$\mathbf{S}_x(\mathbf{k})=(2\pi)^{-1/2}\sum_{n=-\infty}^{+\infty}\mathbf{x}_n e^{-i\mathbf{k}\cdot t(n)} \tag{14.3}$$

$$\mathbf{S}_R(\mathbf{k})=(2\pi)^{-1/2}\sum_{n=-\infty}^{+\infty}\mathbf{r}_n e^{-i\mathbf{k}\cdot t(n)} \tag{14.4}$$

where **k** is the wave vector. Equations (14.3) and (14.4) transform (14.1) and (14.2) into

$$2T = \int \dot{\mathbf{S}}_x(\mathbf{k})^T \mathbf{M} \dot{\mathbf{S}}_x(\mathbf{k})\, d\mathbf{k} \tag{14.5}$$

$$2V = \int \mathbf{S}_R(\mathbf{k})^T \mathbf{F}_R(\mathbf{k}) \mathbf{S}_R(\mathbf{k})\, d\mathbf{k} \tag{14.6}$$

where the integration is extended over the first Brillouin zone. It will be recalled from Chapter 12 that this is equivalent to the derivation of Higgs for the single chain and that $\mathbf{F}_R(\mathbf{k})$ takes the form

$$\mathbf{F}_R(\mathbf{k}) = \mathbf{F}^0 + \sum_s \left[(\mathbf{F}^s)^T e^{-i\mathbf{k}\cdot\mathbf{t}(s)} + \mathbf{F}^s e^{i\mathbf{k}\cdot\mathbf{t}(s)} \right] \tag{14.7}$$

Similarly, the $\mathbf{B}(\mathbf{k})$ matrix can be obtained from

$$\mathbf{S}_R(\mathbf{k}) = \sum_{l=-m}^{m} \mathbf{B}_l e^{-i\mathbf{k}\cdot t(l)} \mathbf{S}_x(\mathbf{k}) \tag{14.8}$$

so that

$$\mathbf{B}(\mathbf{k}) = \sum_{l=-m}^{m} \mathbf{B}_l e^{-i\mathbf{k}\cdot t(l)} \tag{14.9}$$

The dynamical matrix can then be written

$$|\mathbf{D}(\mathbf{k})^T \mathbf{F}_R(\mathbf{k}) \mathbf{D}(\mathbf{k}) - \boldsymbol{\lambda}(\mathbf{k})\mathbf{E}| = \mathbf{0} \tag{14.10}$$

where $\mathbf{D}(\mathbf{k}) = \mathbf{B}(\mathbf{k})\mathbf{M}^{-1/2}$, as defined in a similar form for simple molecules in Chapter 3, and for the one-dimensional polymer chain in Chapter 12. It is important to reiterate that the quantity **k** is the wave vector with three components, k_a, k_b, and k_c, which have to be specified for a three-dimensional crystal.

Even though the form of equation (14.10) is not more complex than previously discussed secular equations describing single polymer chains or even small molecules, the size of the problem can have a daunting magnitude if dispersion curves are to be calculated. Some simplification is made possible by confining the calculation to the optically active modes given by the condition $k_a = k_b = k_c = 0$. But if it is desired to interpret the results of thermodynamic or neutron scattering measurements, a calculation of the dispersion curves for the enormous number of points in three-dimensional k-space necessary for their accurate definition can only be contemplated with dismay. Not surprisingly, the most detailed work has concerned polyethylene and the dispersion curves for this polymer have most often been determined for $k_a = k_b = 0$. Even so, certain approximations have been used in calculating the normal modes of crystalline materials. Tasumi and Shimanouchi (7) in their calculations of intermolecular forces in polyethylene essentially separated the internal and external vibrations, so that the potential energy could be written as a sum of the force constants associated with a single chain and a set of force constants accounting for intermolecular interactions. Cross terms between internal and external coordinates were neglected. Methods used to approximately separate high and low frequencies then allow the calculation of a set of force constants that account for the observed splittings. Similar approximations have been used by Kitagawa and Miyazawa (8) to calculate the dispersion curves in all **k**-space for the lattice modes of polyethylene. This treatment is presented in a general form for crystals by Turrell (9). We will now examine these calculations in more detail, since they provide an example of one of the two types of intermolecular forces that have been applied to an analysis of the vibrational spectra of polymers, namely, atom-atom interactions (i.e., central forces acting between pairs of atoms). The second type of interaction we will consider is based on dipole-dipole interactions, which we will discuss in detail later in the context of an analysis of the amide I mode (C=O stretching) of polyglycine I.

14.2 Interchain Interactions in Polyethylene

Although the vibrational problem as set up by Tasumi and Shimanouchi (7) is basically straightforward, there is a degree of "bookkeeping" complexity introduced by the need to define two different cartesian coordinate systems, one fixed to the chain and the other fixed to the crystal. This complication is introduced essentially because the potential energy is assumed to be the sum of the intramolecular forces associated with the individual chains plus a perturbation due to intermolecular interactions. The arrangement of the chains in the crystal is shown in Figure 14.3, while the Bravais unit cell of the crystal, consisting of two chains, is shown in Figure 14.4. Each Bravais unit cell is labeled by the letters l, m (in Figure 14.3) where l and m refer to the numbering along the a and b axes, respectively. A cartesian coordinate system is fixed to each chain in the conventional manner with the z axis coincident with that of the chain, as shown in Figure 14.4. The cartesian displacement coordinates $\mathbf{X}_p(l, m)$ consist of the displacements of the atoms in the chain $p(p=1$ or 2) of the Bravais cell (l, m).

$$\mathbf{X}_p^T(l,m)=$$

$$\left(\ldots, \Delta x_n^{Cp}, \Delta y_n^{Cp}, \Delta z_n^{Cp}, \Delta x_n^{Hp1}, \Delta y_n^{Hp1}, \Delta z_n^{Hp1}, \Delta x_n^{Hp2}, \Delta y_n^{Hp2}, \Delta_n^{Hp2}, \ldots\right) \tag{14.11}$$

A set of cartesian displacement coordinates $\mathbf{Y}_p(l, m)$ fixed to the crystal and

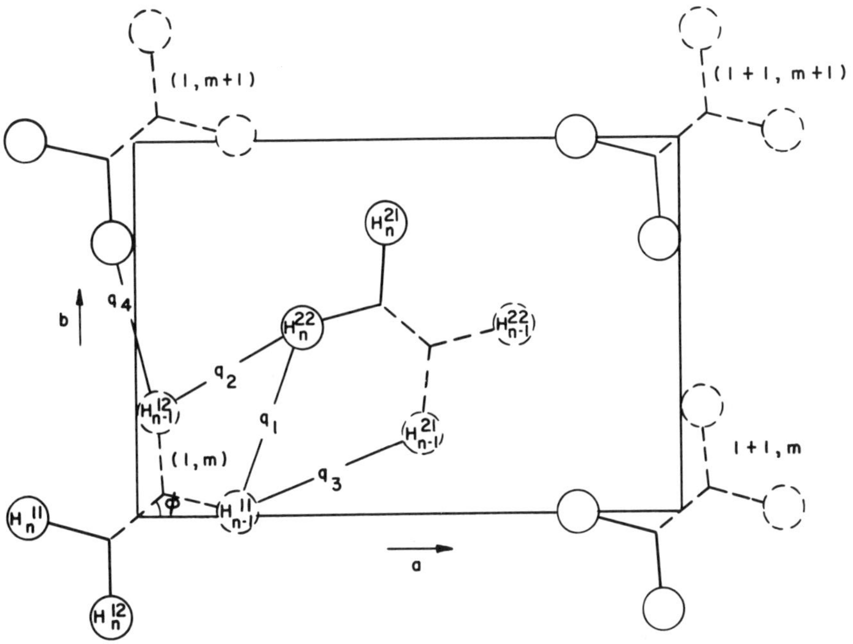

Figure 14.3. Structure of the orthorhombic crystalline form of polyethylene. (Reprinted by permission from ref. 7.)

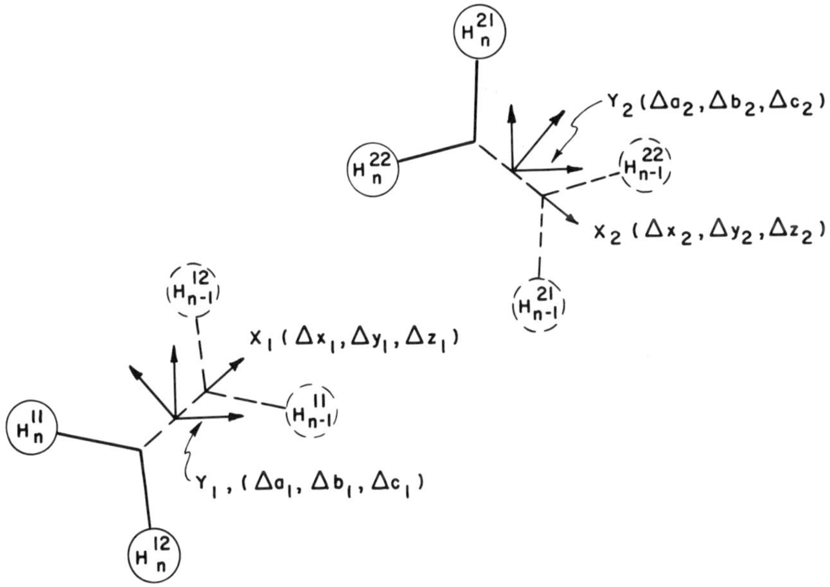

Figure 14.4. Bravais unit cell and coordinate systems of polyethylene. (Reprinted by permission from ref. 7.)

using the axes a, b, and c (see Figure 14.3) are defined in a similar fashion. The relationship between the internal coordinates and these cartesian coordinate systems can then be defined as

$$\mathbf{R}=\mathbf{B}_x\mathbf{X}_p \tag{14.12}$$

$$\mathbf{R}=\mathbf{B}_{yp}\mathbf{Y}_p \tag{14.13}$$

The coordinates $\mathbf{X}_p$ and $\mathbf{Y}_p$ are related by

$$\mathbf{Y}_p=\mathbf{T}_p\mathbf{X}_p\left(\mathbf{X}_p=\mathbf{T}_p^T\mathbf{Y}_p\right) \tag{14.14}$$

where for any given atom $\mathbf{T}_p$ depends on the setting angle ϕ_p between the skeletal plane of the chain p and the a axis:

$$\mathbf{T}_p=\begin{bmatrix} \cos\phi_p & -\sin\phi_p & 0 \\ \sin\phi_p & \cos\phi_p & 0 \\ 0 & 0 & 1 \end{bmatrix} \tag{14.15}$$

Consequently, $\mathbf{B}_x$ and $\mathbf{B}_{yp}$ are related by

$$\mathbf{B}_{yp}=\mathbf{B}_x\mathbf{T}_p^T \tag{14.16}$$

Tasumi and Shimanouchi (7) then defined a set of cartesian symmetry coordinates $\mathbf{S}_x(\theta)$ and internal symmetry coordinates $\mathbf{S}_R(\theta)$ for each chain, as discussed in Chapter 12, where θ is the phase difference between adjacent CH_2 units along each polymer chain. The internal symmetry coordinates are related to the cartesian (or external) symmetry coordinates by an appropriately phased **B** matrix:

$$\mathbf{S}_R(\theta)=\mathbf{B}_x(\theta)\mathbf{S}_x(\theta) \tag{14.17}$$

Using cartesian coordinates fixed to the crystal, we can also construct symmetry coordinates $\mathbf{S}_{yp}(\boldsymbol{\theta})$ with the appropriately phased **B** matrix:

$$\mathbf{S}_R(\theta)=\mathbf{B}_{yp}(\theta)\mathbf{S}_{yp}(\theta) \tag{14.18}$$

The relationship between the **B** matrices is simply,

$$\mathbf{B}_{yp}(\theta)=\mathbf{B}_x(\theta)\mathbf{T}_p^T \tag{14.19}$$

Tasumi and Shimanouchi combined the intramolecular force constants of the polyethylene chain into local symmetry force constants, $\mathbf{F}_s(\theta)$. These force constants are transformed into the $\mathbf{F}_{yp}(\theta)$ matrix expressed in terms of the

$\mathbf{Y}(\boldsymbol{\theta})$ cartesian system in the usual manner:

$$\mathbf{F}_{yp}(\theta)=\mathbf{B}_{yp}(\theta)\mathbf{F}_s(\theta)\mathbf{B}_{yp}(\theta) \tag{14.20}$$

The secular equation

$$|\mathbf{M}^{-1/2}\mathbf{F}_{yp}(\theta)\mathbf{M}^{-1/2}-\boldsymbol{\lambda}(\theta)\mathbf{E}| \tag{14.21}$$

is equivalent to equation (12.21), and may be solved to give the frequencies or eigenvalues of an isolated chain.

This procedure, which is essentially the same as that discussed in Chapter 12 with the additional complication of converting to a cartesian system **Y** based on the unit cell, may seem a long and pointless exercise at this stage. However, this form of the equation makes possible the simple inclusion of intermolecular forces. Tasumi and Shimanouchi neglected dipole-dipole interactions and assumed only short-range intermolecular H---H interactions. A quadratic potential for the H---H distances q_1, q_2, q_3, and q_4 shown in Figure 14.3 (and those equivalent to them by symmetry) was assumed. The intermolecular potential V' can then be written

$$2V'=\sum_{\substack{\text{whole}\\ \text{crystal}}}\sum_i f_i(\Delta_{qi})^2, \qquad i=1,2,3,\text{ and }4 \tag{14.22}$$

where f_i is the force constant. In terms of the **Y** coordinates the change of distance Δq_{AB} between atoms A and B can be expressed as

$$\Delta q_{\mathrm{AB}}=\frac{1}{q_{\mathrm{AB}}}\big[(a_{\mathrm{A}}-a_{\mathrm{B}})(\Delta a_{\mathrm{A}}-\Delta a_{\mathrm{B}})+(b_{\mathrm{A}}-b_{\mathrm{B}})(\Delta b_{\mathrm{A}}-\Delta b_{\mathrm{B}})$$
$$+(c_{\mathrm{A}}-c_{\mathrm{B}})(\Delta c_{\mathrm{A}}-\Delta c_{\mathrm{B}})\big] \tag{14.23}$$

The distances q_1, q_2, q_3, and q_4 defined in terms of the atoms shown in Figure 14.3 are listed in Table 14.1.

The Raman- and infrared-active vibrations are those corresponding to $k_a=k_b=k_c=0$, with $k_c=0$ condition being equivalent to $\theta=0$ or π. It is assumed that the potential energy of the crystal can be written as a sum of the intra- and intermolecular potential matrices for such active coordinates. However, because modes for $0\neq\theta\neq\pi$ become active in crystalline paraffins, the force constant matrix is constructed with just $k_a=k_b=0$ and the dependence on θ (k_c) still included. In terms of the cartesian coordinates **Y** this force constant matrix for the crystal $\mathbf{F}_{cx}(\boldsymbol{\theta})$ is then a simple sum of the corresponding elements of the intra- and interchain **F** matrices.

$$\mathbf{F}_{cx}(\theta)=\begin{bmatrix}\mathbf{F}_{y1}(\theta)+\mathbf{F}'_{y1}(\theta) & \mathbf{F}'^{T}_{y12}(\theta)\\ \mathbf{F}'_{y12}(\theta) & \mathbf{F}_{y2}(\theta)+\mathbf{F}'_{y2}(\theta)\end{bmatrix} \tag{14.24}$$

Table 14.1 Intermolecular H---H Distances and Force Constants

H---H Distance (Å)	No. per Unit Cell	Corresponding Force Constants (mdyn/Å)	
		Set 1	Set 2
$q_1=2.943$	8	$f_1=0.0045$	0.0038
$q_2=2.743$	8	$f_2=0.0133$	0.0153
$q_3=2.755$	4	$f_3=0.0080$	0.0120
$q_4=2.575$	4	$f_4=0.02$ (assumed)	
q_1: $H_n^{22}(l,m)-H_{n\pm1}^{11}(l,m)$		q_3: $H_{n\pm1}^{11}(l,m)-H_{n\pm1}^{21}(l,m)$	
$H_n^{22}(l,m)-H_{n\pm1}^{11}(l,m+1)$		$H_n^{22}(l,m)-H_n^{12}(l,m+1)$	
$H_{n\pm1}^{22}(l,m)-H_n^{11}(l+1,m)$		$H_{n\pm1}^{22}(l,m)-H_{n\pm1}^{12}(l+1,m)$	
$H_{n\pm1}^{22}(l,m)-H^{11}(l+1,m+1)$		$H_n^{21}(l,m)-H_n^{11}(l+1,m+1)$	
q_2: $H_n^{22}(l,m)-H_{n\pm1}^{12}(l,m)$		q_4: $H_{n\pm1}^{12}(l,m)-H_n^{12}(l,m+1)$	
$H_n^{21}(l,m)-H_{n\pm1}^{11}(l,m+1)$		$H_n^{21}(l,m)-H_{n\pm1}^{21}(l,m+1)$	
$H_{n\pm1}^{21}(l,m)-H_n^{11}(l+1,m)$			
$H_{n\pm1}^{22}(l,m)-H_n^{12}(l+1,m+1)$			

where $\mathbf{F}_{yp}(\theta)$ of equations (14.20) and (14.21) is now written separately for the chains 1 and 2 as $\mathbf{F}_{y1}(\theta)$ and $\mathbf{F}_{y2}(\theta)$. Element $\mathbf{F}'_{y12}(\theta)$ comes from the intermolecular potential between chains 1 and 2 and therefore determines the splitting. The term $\mathbf{F}'_{yp}(\theta)$ accounts for frequency shifts on taking the isolated chain and placing it in the crystal. Naturally, because the force constants for the isolated chain $\mathbf{F}_{yp}(\theta)$ are essentially determined from the observed frequencies of the crystalline oligomers and polymer, there is an implicit assumption in the single-chain model that the $\mathbf{F}'_{yp}(\theta)$ term can be accounted for by adjustments of the intramolecular force constants in $\mathbf{F}_{yp}(\theta)$.

From equation (14.24), the secular equation for the two chains in the Bravais cell can be written

$$|\mathbf{M}^{-1/2}\mathbf{F}_{cy}(\theta)\mathbf{M}^{-1/2}-\boldsymbol{\lambda}(\theta)\mathbf{E}|=0 \tag{14.25}$$

Accordingly, this equation can be solved for $\theta=0$ and π and the intermolecular force constants f_1, f_2, f_3, and f_4 adjusted to obtain a best fit between the observed and calculated values of splitting. However, prior to this refinement Tasumi and Shimanouchi (7) obtained approximate initial values of three of the force constants from splittings of three modes using a perturbation treatment. The fourth constant was given an assumed value. When intermolecular force constants are small compared to intramolecular forces, the splitting $\Delta\nu_i$ can be estimated from (7, 9)

$$\Delta\nu_i=2\left[\mathbf{L}_{y2}(\theta)\mathbf{F}'_{y12}(\theta)\mathbf{L}_{y1}(\theta)\right]_{ii} \tag{14.26}$$

where the eigenvectors $\mathbf{L}_{y1}(\theta)=\mathbf{L}_{y2}(\theta)$ are determined from the normal coordinate treatment of the isolated chain. The values of f_1, f_2, and f_3 were determined from the splittings of the modes $\nu_2(\pi)$ (1473 and 1463 cm^{-1}), $\nu_8(0)$ (6.4 and 1050 cm^{-1}), and $\nu_8(\pi)$ (731 and 720 cm^{-1}) and then these force constants were adjusted by a least-squares procedure so that the splitting of the frequencies calculated from the secular equation [equation (14.25)] gave a best fit to those observed experimentally. These force constants, together with the assumed value of f_4, are presented as Set 1 in Table 14.1. To obtain Set 2 force constants, intermolecular force constants were adjusted at $\theta=0$, $2\pi/9$, $5\pi/9$, and π, presumably using frequencies observed in the spectra of *n*-paraffins.

The intermolecular force constants determined by Tasumi and Shimanouchi (7) were compared with those obtained from the second derivative of the potential function:

$$V_{\mathrm{HH}}=-Ar^{-6}+B\exp(-Cr) \qquad (14.27)$$

where r is the distance between two hydrogen atoms. Various authors have given values of the coefficients A, B, and C, as tabulated by Coulson and Haigh (10). It was found that only those proposed by Bartell ($A=3.573$, $B=10.51$, $C=2.160$) and Muller ($A=4.496$, $B=10.09$, $C=2.645$) could be compared with the results of the normal coordinate calculations, as shown in Figure 14.5.

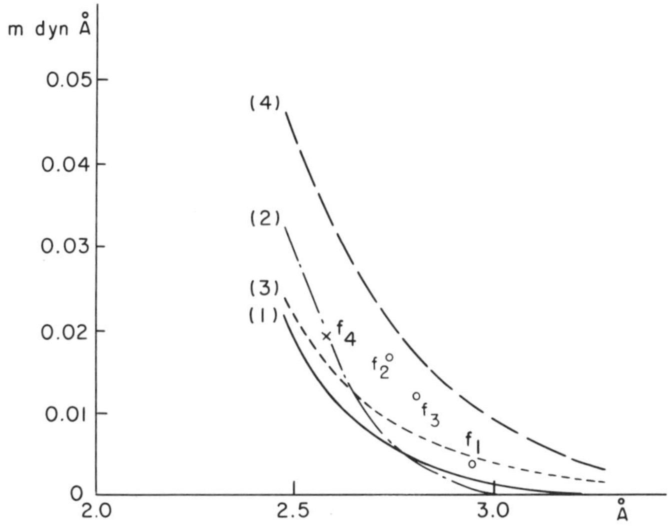

Figure 14.5. Intermolecular H · · · H force constants; ●, Set 1; ○ Set 2; ×, assumed. (1) Bartelle function; (2) Muller potential; (3) de Boer repulsive potential; (4) Buckingham potential ($E_{3/2}$). (Reprinted by permission from ref. 7.)

Table 14.2 Molecular and Crystallographic Structural Parameters of Polyethylene

Molecular Structure[a]	
C—C	0.1532 nm
C—H	0.1058 nm
C—C—C	112.01°
HCH	109.29°
Orthorhombic Lattice Parameters[b]	
$a=0.7161$ nm, $b=0.4866$ nm, $c=0.25412$ nm, $\Phi=41°$	
Triclinic[a]	
$a=0.422$ nm, $b=0.479$, $c=0.25412$ nm, $\alpha=89.545°$, $\beta=71.497°$ nm, $c=0.25412$ nm, $\alpha=89.545°$	

[a] H. Mathisen, N. Norman, and B. F. Pedersen, *Acta Chem. Scand.* **21**, 127–135 (1967).
[b] G. Avitabile, R. Napolitano, B. Pirozzi, K. D. Rouse, M. W. Thomas, and B. T. M. Willis, *J. Polymer Sci.* **B13**, 351 (1975).

On the basis of these results, Tasumi and Shimanouchi (7) concluded that the H---H repulsive potential plays the major role in determining intermolecular forces in paraffin crystals. Tasumi and Krimm (11) investigated this further by calculating dipole-dipole interactions, which they found make an insignificant contribution to the translational lattice frequencies and only small contributions to the splitting of internal modes. These authors also attempted to derive the interaction force constants from a common potential function of the type given in equation (14.27) and also considered the effects of changes in lattice parameters and the setting angle (ϕ in Figure 14.3). Because of the difficulty of using assumed potential functions to determine the best setting angle, a value of $\phi=48°$ was assumed and a "best" set of force constants determined. Both Tasumi and Shinanouchi (7) and Tasumi and Krimm (11) calculated dispersion curves (as a function of θ), which we will consider in the following section. We conclude this section by considering the more recent calculations of Barnes and Fanconi (12), which included additional structural and vibrational data.

Table 14.3 Interatomic Potential Energy Parameters

Type	a(mdyn/Å^7)	b(mdyn/Å)	c(Å^{-1})
H---H	0.250	17.77	3.74
H---C	2.014	99.49	3.67
C---C	3.716	1036.4	3.60

In order to utilize the data from n-alkanes, Barnes and Fanconi (12) used two different subcell geometries for the force field refinement, the usual orthorhombic lattice of polyethylene and the triclinic subcell geometry of the even n-alkanes. The lattice parameters are given in Table 14.2 and it can be seen that the setting angle, 41°, is significantly different from that used by Tasumi and Krimm (11). The vibrational frequencies were collected according to dispersion curves, 705 frequencies in all being used in the calculation. The intramolecular force field, defined in terms of local symmetry coordinates, was discussed in Chapter 13. Of more relevance to this discussion are the intermolecular forces used in the lattice dynamical calculation. Again, the potential energy term given in equation (14.27) was used, but instead of calculating only the second derivative of this nonbonded potential energy expression (the quadratic term), the first derivative or linear term was also determined. Barnes and Fanconi (12) pointed out that the linear term can be appreciable for

Table 14.4 Experimental and Calculated Frequencies of Polyethylene (Orthorhombic Structure)

Symmetry	Calc. (cm^{-1})	Obs. (cm^{-1})	Assignment
A_g		Raman Active	
	141.2	136	Libration
	1137.5	1133	Optical skeletal
	1175.2	1170	Methylene rocking
	1445.0	1442	Methylene scissors
	2854.6	2848	Symmetric C—H stretch
	2884.8	2883	Asymmetric C—H stretch
A_u			
	54.9	inactive	Translatory
	1040.6	inactive	Methylene twisting
	1180.8	inactive	Methylene wagging
B_{1g}		Raman Active	
	1065.8	1065	Optical skeletal
	1296.8	1297	Methylene twisting
	1370.7	1370	Methylene wagging
B_{1u}		Infrared Active	
	0.0		
	85.9	81	Translatory
	736.5	734	Methylene rocking
	1477.8	1473	Methylene scissors
	2851.5	2851	Symmetric C—H stretch
	2921.5	2919	Asymmetric C—H stretch

Table 14.4 Continued

B_{2g}	Raman Active		
	1069.6 1293.4 1373.2	1068 1295 1370	Optical skeletal Methylene twisting Methylene wagging
B_{2u}	Infrared Active		
	0.0 109.1 720.7 1472.3 2852.9 2921.7	 106 721 1463 2851 2919	 Translatory Methylene rocking Methylene scissors Symmetric C—H stretch Asymmetric C—H stretch
B_{3g}	Raman Active		
	101.5 1137.0 1176.1 1458.3 2849.9 2888.7	108 1133 1170 1442 2848 2883	Libration Optical skeletal Methylene rocking Methylene scissors Symmetric C—H stretch Asymmetric C—H stretch
B_{3u}	Infrared Active		
	0.0 1042.9 1177.3	 1050 1175	 Methylene twisting Methylene wagging

distances in the repulsive part of the potential. Furthermore, H---C and C---C terms in addition to H---H interactions were included in the interatomic potential energy, and the final values of the parameters are reproduced in Table 14.3. It was found that the linear term had a major effect on the librational lattice frequencies and explained why in previous calculations these frequencies were calculated to be too large. The experimental and calculated Brillouin zone center frequencies of orthorhombic polyethylene are compared in Table 14.4 and it can be seen that Barnes and Fanconi obtained an outstanding agreement.

14.3 Dispersion Curves of Polyethylene

The advent of large computers made possible the extension of normal mode calculations from the spectroscopically active modes at the Brillouin zone center ($\mathbf{k}=0$) to the nonactive modes corresponding to the intermediate phase

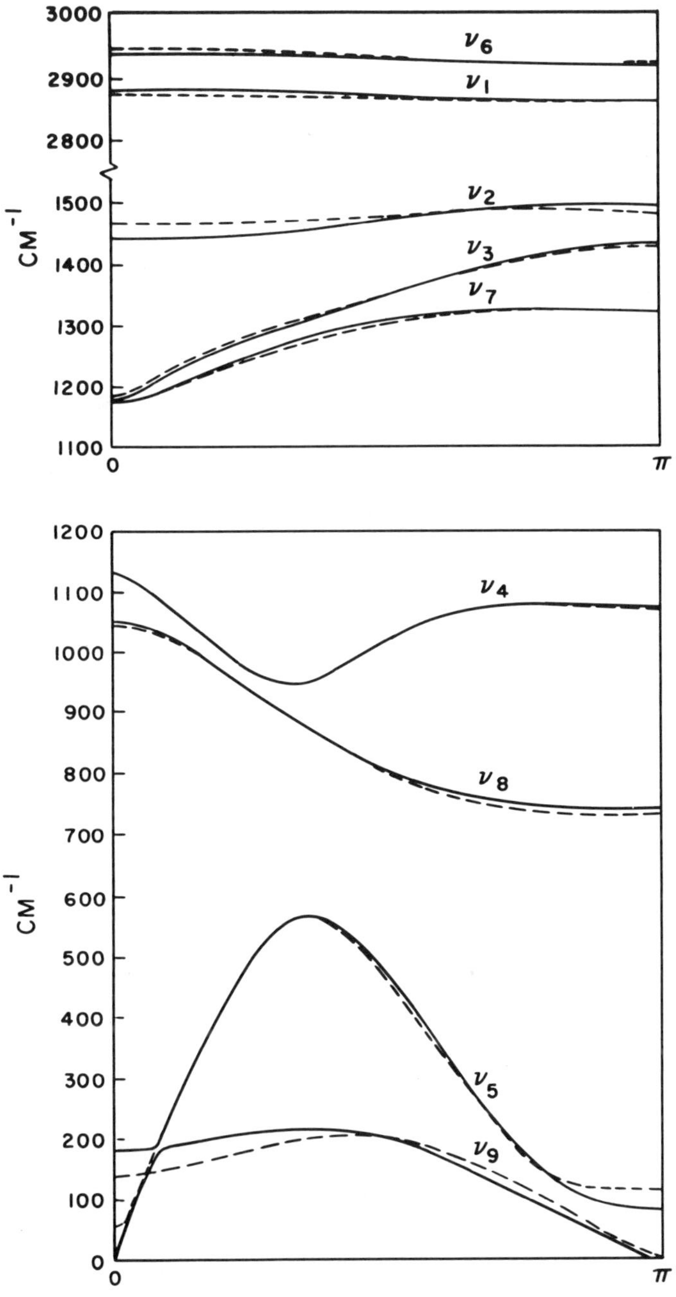

Figure 14.6. Dispersion curves for crystalline polyethylene; ——, ν^a branch; ---, ν^b branch. In-plane modes: ν_1, CH symmetric stretching; ν_2, CH_2 scissors; ν_3, CH_2 wagging; ν_4, CC stretching–CCC bending; ν_5, CCC bending–CC stretching; Out-of-plane modes; ν_6, CH_2 antisymmetric stretching; ν_7, CH_2 rocking–CH_2 twisting; ν_8, CH_2 twisting–CH_2 rocking; ν_9, torsional. (Reprinted by permission from ref. 11.)

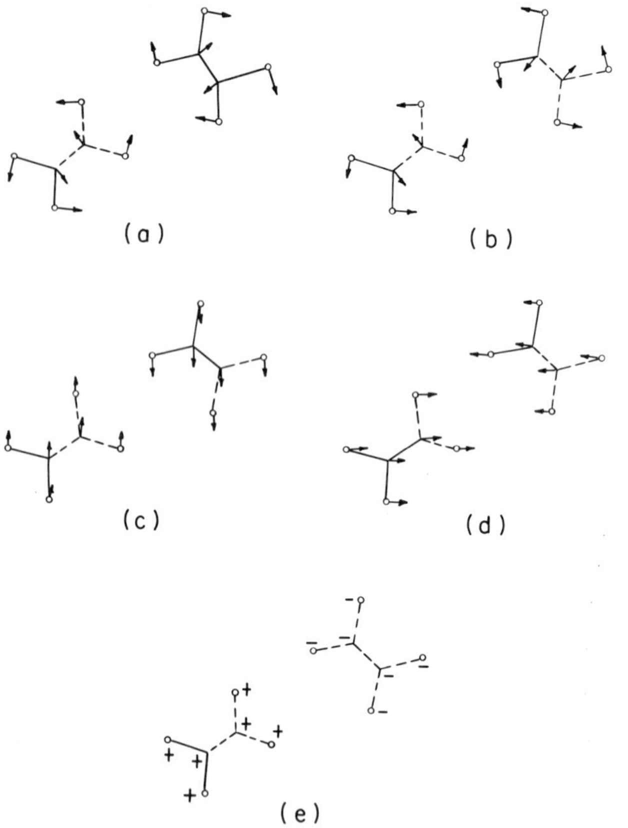

Figure 14.7. Lattice vibration of orthorhombic polyethylene crystal: (*a*) A_g rotational mode; (*b*) B_{3g} rotational mode; (*c*) B_{1u} translational mode; (*d*) B_{2u} translational mode; (*e*) A_u translational mode. (Reprinted by permission from ref. 7.)

coupling between adjacent chemical units. The calculation of the dispersion curves is useful for a number of different purposes. Several physical properties (e.g., specific heat) can be calculated from a knowledge of polymer dynamics. Analysis of inelastic neutron scattering data depends on a knowledge of the dispersion relation. Furthermore, nonactive modes in the theoretically infinite perfect chain become active in a chain containing defects, a topic we will consider in Chapter 15.

For some purposes the dispersion relation in all *k*-space is required, which as we noted earlier would necessitate an astronomical computer budget. Nevertheless, dispersion curves for the lattice modes of polyethylene have been determined by using some simplifying assumptions. We will consider these at the end of this section. First we will discuss the dispersion relationship determined simply as a function of k_c or θ, since this follows directly from the work discussed in the preceding section.

Tasumi and Shimanouchi (7), Tasumi and Krimm (11), and Barnes and Fanconi (12) all determined dispersion curves for $k_a = k_b = 0$ and $0 \leqslant \theta \leqslant \pi$. To our eyes these dispersion curves appear very similar and we have arbitrarily chosen to reproduce those calculated by Tasumi and Krimm (11) in Figure 14.6. Each of the nine branches, labeled ν_1 and ν_9 and keyed in the figure caption, is split into two subbranches ν_i^a and ν_i^b. For example, the splitting of the ν_8 CH_2 rocking mode at $\theta = \pi$ accounts for the characteristic doublet observed at 731 and 720 cm^{-1} in the infrared spectrum of crystalline polyethylene. The modes ν_5 and ν_9 are drastically affected by the intermolecular forces (7). In the isolated chain ν_5 is the skeletal deformation vibration and ν_9 is the torsional vibration around the C—C bond. In the crystal these modes mix. In the single chain ν_5 and ν_9 have zero frequencies at $\theta = 0$ and $\theta = \pi$ (corresponding to three translations and a rotation about the chain axis), but in the **crystal** ν_5 and ν_9 correspond to five lattice vibrations and three translational modes [$\nu_9^a(\pi)$, $\nu_9^b(\pi)$, and $\nu_9^c(0)$] of the whole crystal (null frequency). The modes $\nu_5^a(0)$ and $\nu_5^b(0)$ are the A_g and B_{3g} rotatory or librational lattice vibrations; $\nu_5^a(\pi)$ and $\nu_5^b(\pi)$ are the B_{1u} and B_{2u} translational lattice modes, and $\nu_9^b(0)$ is the A_u translational lattice vibration. The form of these modes, as calculated by Tasumi and Shimanouchi, is illustrated in Figure 14.7. In the original crystal lattice dynamic calculations (7, 11) good agreement between the observed and calculated lattice translatory frequencies was obtained, but the librational lattice frequencies were calculated to be too large. The linear term in the interaction potential introduced by Barnes and Fanconi (12) had a large effect on the librational modes and resulted in a much better fit of calculated and observed frequencies. The results of some of these calculations are summarized in Table 14.5.

Table 14.5 Lattice Modes of *n*-Alkanes and Polyethylene

Type	Symmetry	Frequency (cm^{-1}) Obs.	Calc.[a]	Calc.[b]
Orthorhombic Structure				
Librational	A_g	136	141	181.7
Translatory	A_u	Inactive	55	57.5
Translatory	B_{1u}	81	86	80.6
Translatory	B_{2u}	106	109	113.3
Librational	B_{3g}	108	102	138.8
Triclinic Structure				
Librational	A_g	60	63	

[a]From Barnes and Fanconi (12).
[b]From Tasumi and Krimm (11).

The vibrations at $\theta=0$ and π, discussed above, are special in the sense that they can be classified according to the factor group of the space group of the crystal (D_{2h}). This is an obvious advantage when solving the dynamical matrix in that it can be factored into a simpler block diagonal form by the appropriate transformation. In general, however, $k\neq 0$ modes do not have this symmetry. Nevertheless, for crystal lattices of high space group symmetry, crystal vibrations of special wave vectors **k** can be grouped into symmetry classes that form a subgroup of the factor group. This is not only useful in factoring the dynamical matrix, but greatly aids the construction of dispersion curves, since branches belonging to the same symmetry species do not cross (8). For a general phase difference δ, defined by Kitagawa and Miyazawa (8) so that $\delta_a=a_0k_a$, $\delta_b=b_0k_b$, and $\delta_c=c_0k_c$, a k-group can be defined that satisfies the condition

$$\alpha_u\delta=\delta+2\pi m \tag{14.28}$$

where α_u is a symmetry operation and m is a vector. For $\delta=0$ all the operations of the factor group of the space group satisfy equation (14.28), and so the factor group itself is the k-group. At the other extreme, a general phase difference $\delta(\delta_a\neq 0, \delta_b\neq 0, \delta_c\neq 0)$, only the identity operation constitutes the k-group and the dynamical matrix cannot be factored into a simpler form. There are regions of k-space that lie between these two conditions. For example, for δ values $(0,0,\delta_c)$ four operations, the identity, the twofold screw axis along the c axis, and glide planes normal to the a and b axes satisfy equation (14.28) ($m=0$) and the k-group is C_{2v}. The k-groups for the space group of polyethylene, as listed by Kitagawa and Miyazawa (8), are reproduced in Table 14.6.

Kitagawa and Miyazawa (8) applied k-groups and used certain simplifying assumptions (principally the separation of high and low frequencies) to determine the frequency dispersion curves of eight vibrational branches below

Table 14.6 ***k*-Groups of P_{nam}**

Phase Difference Vector (δ)	k-Group
$(\delta_a, \delta_b, \delta_c)$	C_1
$(0, \delta_b, \delta_c)$, $(\delta_a, 0, \delta_c)$, $(\delta_a, \delta_b, 0)$	C_s
$\Sigma(\delta_a,0,0)$, $\Lambda(0,\delta_b,0)$, $\Delta(0,0,\delta_c)$	C_{2v}
$\Gamma(0,0,0)$	D_{2h}
$H(0,\delta_b,\pi)$, $B(0,\pi,\delta_c)$, $D(\pi,0,\delta_c)$	C_{2v}
$X(\pi,0,0)$, $Z(0,\pi,0)$, $Y(0,0,\pi)$, $T(0,\pi,\pi)$	D_{2h}
(π,δ_b,δ_c), (δ_a,π,δ_c), (δ_a,δ_b,π)	C_s
$A(\delta_a,\pi,0)$, $C(\delta_a,0,\pi)$, $E(\delta_a,\pi,\pi)$, $G(\pi,\delta_b,0)$, $F(\pi,\pi,\delta_c)$	C_{2v}
$U(\pi,\pi,0)$	D_{2h}
$Q(\pi,\delta_b,\pi)$	C_{2v}
$S(\pi,0,\pi)$, $R(\pi,\pi,\pi)$	D_{2h}

600 cm^{-1}; some examples of the dispersion curves are shown in Figures 14.8 and 14.9. In Figures 14.8*a* and 14.8*b* the normal modes corresponding to phase differences $(\delta_a, 0, 0)$ and $(0, \delta_b, 0)$ belong to a k-group C_{2v}, with four symmetry species $(\Sigma_1, \ldots, \Sigma_4$ or $\Lambda_1, \ldots, \Lambda_4)$. The highest branches Σ_1 and Σ_2 (or Λ_1 and Λ_4) are due to rotational modes about the chain axis, while the next highest Σ_1 and Σ_2 (or Λ_1 and Λ_4) branches are due to translational modes in the c plane.

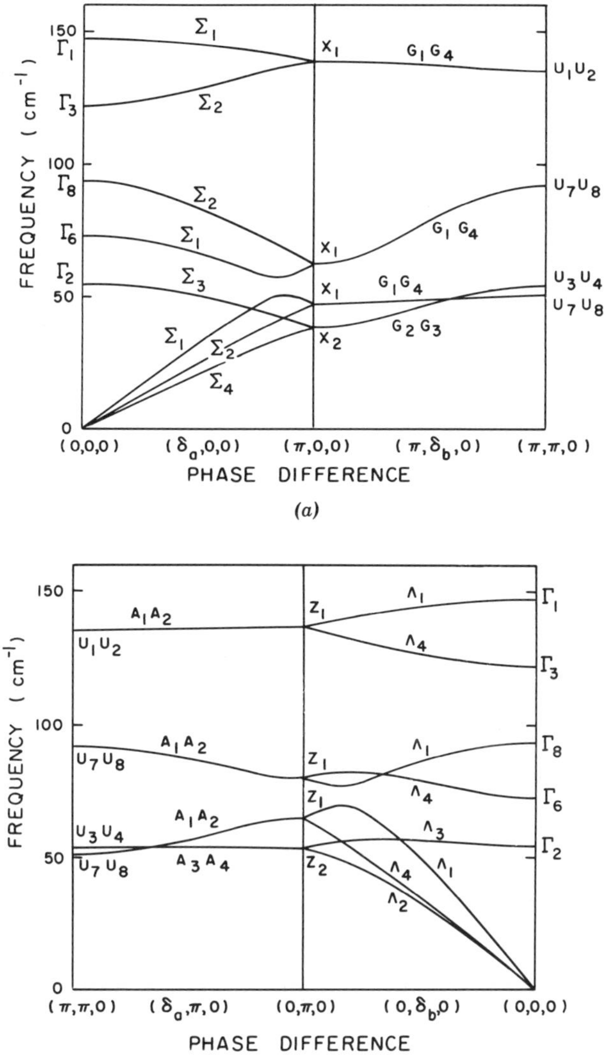

(a)

(b)

Figure 14.8. Dispersion curves for orthorhomic polyethylene. (*a*) Phase difference $(\delta_a,0,0)$, $(\pi, \delta_b, 0)$. (*b*) Phase difference $(\delta_a, \pi, 0)$, $(0, \delta_b, 0)$. (Reprinted by permission from ref. 8.)

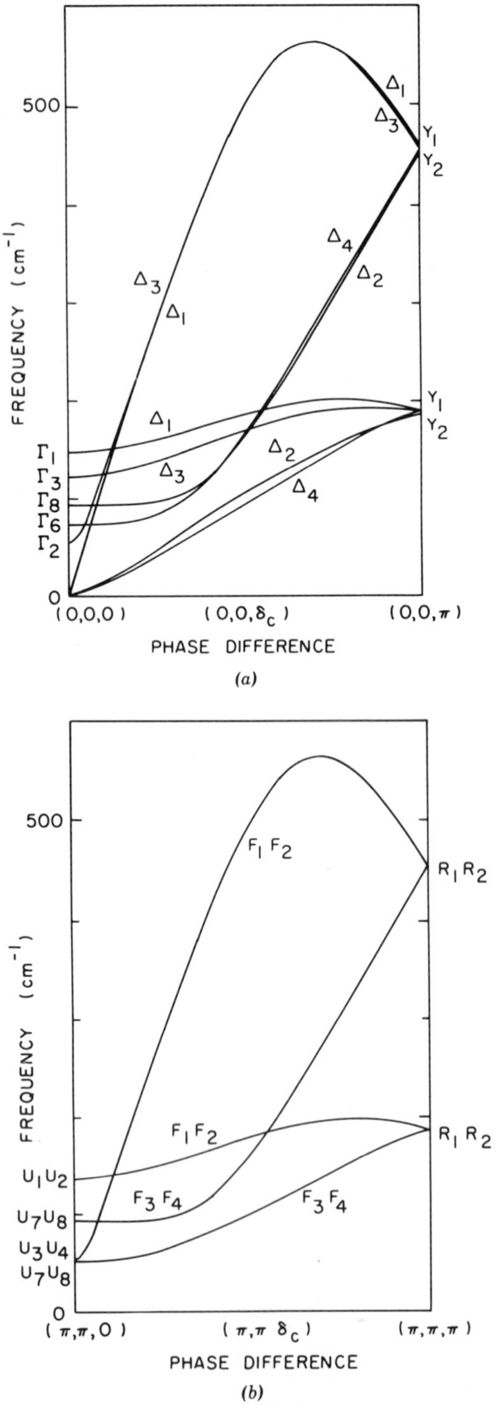

Figure 14.9. Dispersion curves for orthorhombic polyethylene. (a) Phase difference $(0, 0, \delta_c)$, (b) Phase difference (π, π, δ_c). (Reprinted by permission from ref. 8.)

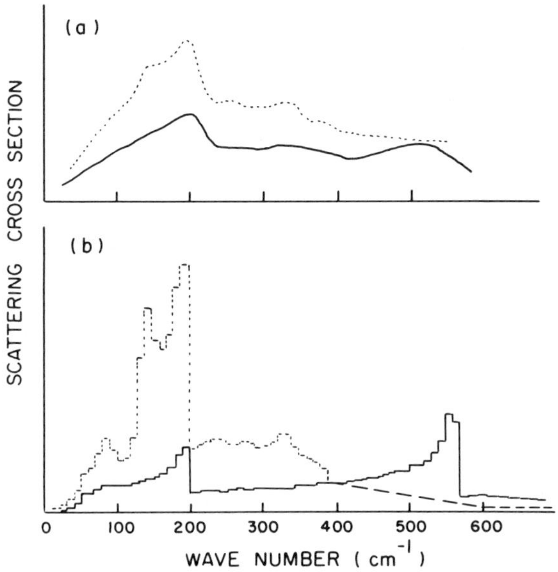

Figure 14.10. (*a*) Distribution of the cross section of neutron incoherent inelastic scattering by a uniaxially oriented orthorhombic PE crystal (100°K); ——, momentum transfer vector parallel to the *c* axis; ---, momentum transfer vector perpendicular to the *c* axis. (*b*) Theoretical curve obtained from the frequency distribution by weighting the squares of the amplitudes of the hydrogen atoms. (Reprinted by permission from ref. 8.)

The Σ_3 (or Λ_3) branch is due to translational modes along the *c* axis. As the phase difference δ_a approaches π in Figure 14.8*a* the Σ_1 and Σ_2 (or Σ_3 and Σ_4) branches approach each other and finally meet at the zone boundary $(\pi,0,0)$ to form doubly degenerate X_1 or X_2 vibrations. Similarly, in Figure 14.8*b* the Λ_1 and Λ_4 (and Λ_2 and Λ_3) branches meet to form doubly degenerate Z_1 (or Z_2 vibrations). A complete discussion of the $\delta \neq 0$ dispersion curves has been given by Kitagawa and Miyazawa, who from these curves determined the frequency distribution. This is in good agreement with the experimental results of inelastic neutron scattering, as shown in Figure 14.10.

14.4 Transition Dipole Coupling in Polypeptides

In the preceding sections we have examined the use of atom-atom potentials in determining intermolecular interactions in polyethylene. In this section we will turn our attention to the other type of intermolecular force that has been used in calculations of interchain interactions, dipole-dipole interactions or transition dipole coupling. This approach has been central to attaining an understanding of the vibrational spectra of polypeptides, so we will examine it in some detail.

The effective application of infrared spectroscopy to the characterization of protein structure dates from 1950, when Elliott and Ambrose (15) made the seminal observation that the frequency of the amide I band (C=0 stretching) of α-helical polypeptides is about 20 cm^{-1} higher than that of their β-sheet counterparts. It has subsequently been established that the amide I mode of highly ordered polypeptides is split as a consequence of inter- and intramolecular interactions. A significant advance in our understanding of the observed splittings was made by Miyazawa (16) and Miyazawa and Blout (17), who developed a perturbation treatment for the interaction of amide group vibrations. This analysis assumes that the observed frequencies can be equated to an unperturbed frequency plus the contribution of inter- and intramolecular interaction terms. The approach is similar to the coupled oscillator model discussed in Chapter 10 and can best be illustrated by reference to polyglycine I. This polypeptide has an antiparallel β-sheet structure, illustrated in Figure 14.11. The four possible amide I modes are also illustrated in this figure. These four active vibrations are labeled by a phase relationship (δ, δ'), where δ is the phase angle between vibrations in adjacent peptide units in the same chain (i.e., an intramolecular interaction term) and δ' is the phase angle between vibrations of adjacent units in different chains (i.e., an intermolecular term). The four optically active vibrations are $\nu(0,0)$, $\nu(\pi,0)$, $\nu(0,\pi)$, and $\nu(\pi,\pi)$. The original treatment of Miyazawa essentially assumed that the amide I was localized as a C=0 stretching vibration in the amide group, so that the amide I mode could be modeled as a chain of antiparallel coupled oscillators. If the force constant describing interactions between oscillators is then small relative to the force constant for the oscillator itself, the interactions can be introduced via a first-order classical perturbation treatment (16). The perturbed frequency of an isolated chain of such oscillators can then be expressed as

$$\nu(\delta) = \nu_0 + \sum_s D_s \cos(s\delta) \tag{14.29}$$

where $\nu(\delta)$ is the observed frequency, ν_0 the unperturbed frequency of the localized amide group, δ the phase angle between the motions of adjacent oscillators, s the separation of interacting groups, and D_s the interaction constant between oscillators separated by s units. For an isolated planar zigzag chain δ can be either 0 or π for optically active modes. Interactions were further limited to only nearest neighbor terms, so that

$$\nu(0) = \nu_0 + D_1 \tag{14.30a}$$

$$\nu(\pi) = \nu_0 - D_1 \tag{14.30b}$$

Miyazawa (16) generalized equation (14.29) to include both intermolecular and intramolecular interactions in the two-dimensional crystal of the β-sheet:

$$\nu(\delta, \delta') = \nu_0 + \sum_{s,t} D_{st} \cos(s\delta)\cos(t\delta') \tag{14.31}$$

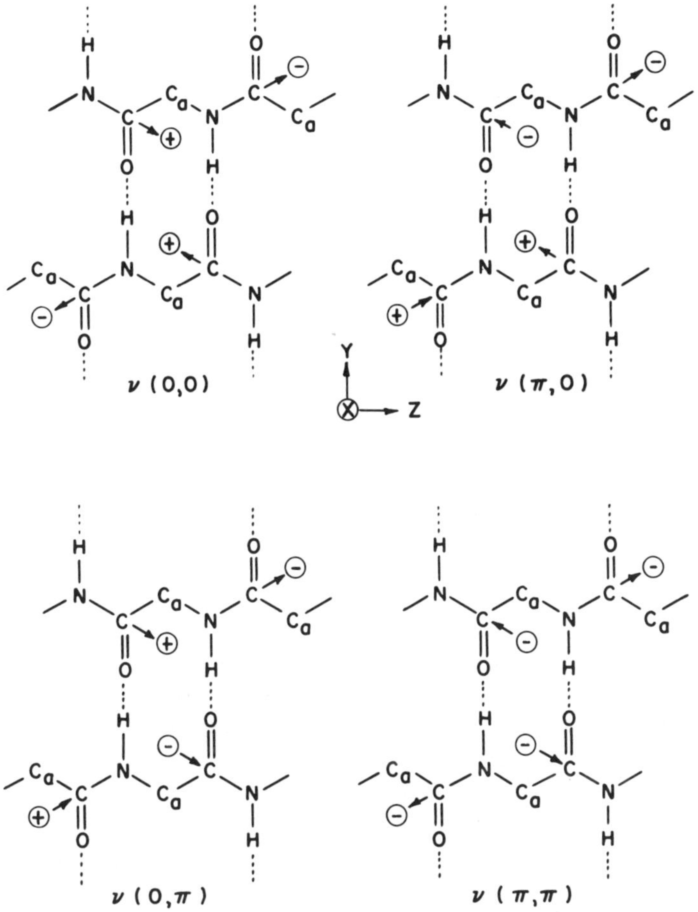

Figure 14.11. The four amide I vibrational modes of polyglycine. (Reprinted by permission from T. Miyazawa, in "Polyamino Acids, Polypeptides and Proteins," M. A. Stahmann, Ed., University of Wisconsin Press, Madison, 1962.)

where δ and δ' are the respective intrachain and interchain phase angles and D_{st} is the interaction constant between peptide groups separated by t chains and s groups along the tth neighboring chain. Only nearest neighbor interactions D_{01} and D_{10}, illustrated in Figure 14.12, were included in Miyazawa's treatment, which basically used the observed frequencies in the infrared spectra of β-sheet polypeptides and nylons to calculate values of the interaction terms.

In the early applications of this perturbation treatment the origin of the intra- and intermolecular forces was not really considered. Furthermore, the treatment itself as well as the assumptions made have been criticizied on a number of grounds (18–21), perhaps the most important being that normal coordinate calculations of polyglycine I (20) could not reproduce the large

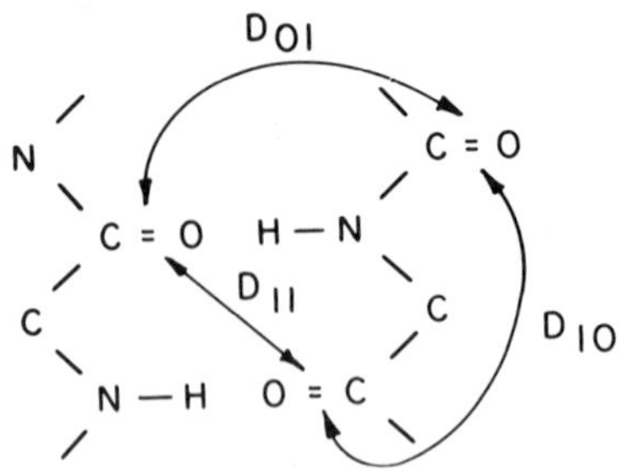

Figure 14.12. Adjacent interaction constants for β-sheet polypeptides.

values of the D_{10} term predicted on the basis of Miyazawa's original equations. Krimm and Abe (21) proposed a modified form of the theory that contained a D_{11} term (also illustrated in Figure 14.12). This interaction constant was considered to have its origin in transition dipole couplings, and Miyazawa's original equation was modified to

$$V(\delta,\delta') = V_0 + D_{10}\cos\delta + D_{01}\cos\delta' + D_{11}\cos\delta\cos\delta' \tag{14.32}$$

A common feature of the original treatment and this initial modification is the assumption that only neighboring interactions are important. This assumption has now been abandoned and the effective "sphere of influence" for transition dipole coupling has been extended to groups within a radius of 30 Å (22–24). In this more generalized form, equation (14.32) is further modified to account for inter- as well as intrasheet interactions and a "rippled sheet" rather than a pleated sheet structure was considered for polyglycine. We will not discuss this extensive and important body of work in detail, but will simply consider the calculation of the interaction terms D_{st} from the assumption of transition dipole coupling.

The change in potential energy due to any two interacting dipoles is given by

$$\Delta V = \frac{1}{\varepsilon}\left|\frac{\partial\mu_1}{\partial S_1}\right|\left|\frac{\partial\mu_2}{\partial S_2}\right|\Delta S_1\,\Delta S_2\cdot X \quad \text{(ergs)} \tag{14.33}$$

where ΔS is in angstroms, $\partial\mu/\partial S$ is in Debye per angstrom, ε is the dielectric constant (assumed to be 1), and X is a geometry term that depends on the location and relative orientation of the transition dipole moments (23). If the directions of the two dipole moments i, j, which are separated by a distance R angstrom, are represented by unit vectors $\overline{e_i}$, $\overline{e_j}$, as illustrated in Figure 14.13, then

$$X = \left[\overline{e_i}\,\overline{e_j} - \frac{3\left(\overline{e_i}\,\overline{R_{ij}}\right)\left(\overline{e_j}\,\overline{R_{ij}}\right)}{R_{ij}^2}\right]\frac{1}{R_{ij}^3} \tag{14.34}$$

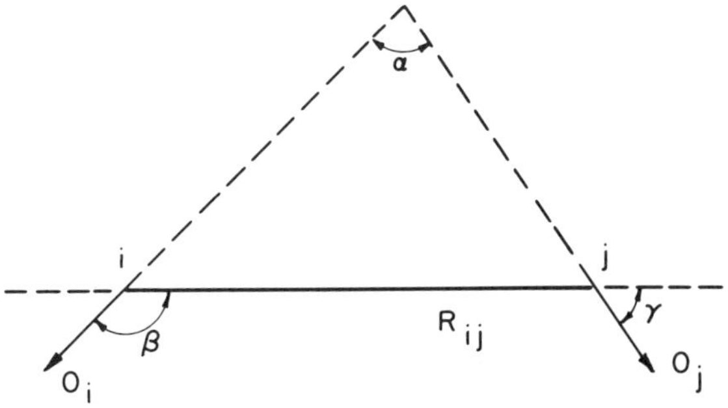

Figure 14.13. Vector representation of geometry terms for transition dipole coupling.

where $\overline{R_{jj}}$ is the vector connecting groups i and j. For the simpler planar case illustrated in Figure 14.13

$$X = \frac{\cos X - 3\cos\beta\cos\gamma}{R^3} \tag{14.35}$$

Since both transition dipoles in the interaction term are associated with C=0 stretching, equation (14.33) can be written in the following form

$$\Delta V_{s,t} = 5.035 \times 10^3 \left(\frac{\partial\mu}{\partial S}\Delta S\right)^2 X \tag{14.36}$$

where the units of ΔV have been converted to cm^{-1} and $(\partial\mu/\partial S)\Delta S$ is interpreted as an "effective dipole moment" (23). The energy terms $\Delta V_{s,t}$ represent the coefficients D_{st} in the perturbation equation. Originally, an interaction force constant F_{st} was defined (21) such that

$$F_{st} = 0.1\left(\frac{\partial\mu}{\partial S}\right)^2 X \tag{14.37}$$

and D_{st} was determined from

$$D_{st} = \left(\frac{\partial\nu}{\partial F_{C=0}}\right) F_{st} \tag{14.38}$$

By assuming values of $\partial\mu/\partial S$, F_{st} could be calculated. The term $\partial\nu/\partial F$ was found approximately from the Jacobian matrix derived in the normal coordinate calculations, thus allowing D_{st} to be calculated. In subsequent work (23, 24), however, the experimental frequencies (or more precisely, ratios of the splittings between modes) were used to calculate values of the transition dipole, its location, and its orientation. The calculated values agreed well with experimental values taken from measurements of the polarization and intensity of infrared amide I bands. The introduction of transition dipole coupling to account for the splittings of amide modes is an important advance, since it potentially allows details of polypeptide structure to be determined from the observed vibrational spectrum.

14.5 The Longitudinal Acoustic Mode

In 1949 Mizushima and Shimanouchi (25) made the important observation of a low frequency line in the Raman spectra of *n*-paraffins whose frequency was inversely proportional to the number of carbon atoms in the chain. This mode is now known as the longitudinal acoustic mode or LAM, and has been used in a number of studies of the morphology of polyethylene crystals. In many such studies the observed LAM frequency has been interpreted in terms of mechanical models. These we will briefly discuss later. First we will consider the vibrational character of this mode in both polyethylene and the *n*-paraffins.

Zerbi and co-workers (26) have made the point that, strictly speaking, the term "longitudinal acoustic mode" refers to the mode at the beginning of the longitudinal acoustic branch of the dispersion relation with $\mathbf{k}=0$ and zero frequency. If we consider the dispersion curves calculated for an infinite isolated polyethylene chain, the acoustic branches of which are shown in Figure 14.14, this is equivalent to ν_5 at $\theta=0$ and corresponds to the translation T. Of course, polymers may be many things, but they are not so remarkable as to be infinite. For a finite chain of N coupled oscillators the frequencies can be expressed as a function of the phase difference $s\pi/(N+1)$, as discussed in Chapter 10. The form of the normal modes for a finite chain that "map" the dispersion curves shown in Figure 14.14 are illustrated schematically in Figure 14.15, together with the modes for the infinite chain ($s=0$). The longitudinal acoustic mode corresponds to a translation for $s=0$, as mentioned earlier. For $s=1$ the mode is an accordion-like motion with a node in the middle, as illustrated in more detail in Figure 14.16. Although in polyethylene it is this mode that is referred to as the longitudinal acoustic mode, Zerbi et al. (26) have pointed out that the term "longitudinal accordion motion" is a more accurate description (which fortunately allows us to maintain the acronym LAM). However, in keeping with general usage in the literature we will refer to other modes on this branch of the dispersion curve as LAM-3, LAM-5, etc. (meaning $s=3$ and $s=5$, respectively). In a finite three-dimensional crystal there should also be analogous transverse acoustic modes, as illustrated in Figure 14.17, but these have not been observed in polyethylene.

As in other studies of polyethylene, the availability of the *n*-paraffins has proved indispensable to a detailed analysis of the observed spectrum. For example, the low frequency Raman spectrum of $C_{36}H_{74}$ is shown in Figure 14.18, as reported by Schaufele and Shimanouchi (27). The band progression for the acoustic vibrations is reproduced in Figure 14.19 so as to show the corresponding points on the dispersion curve. We will discuss the exact form of

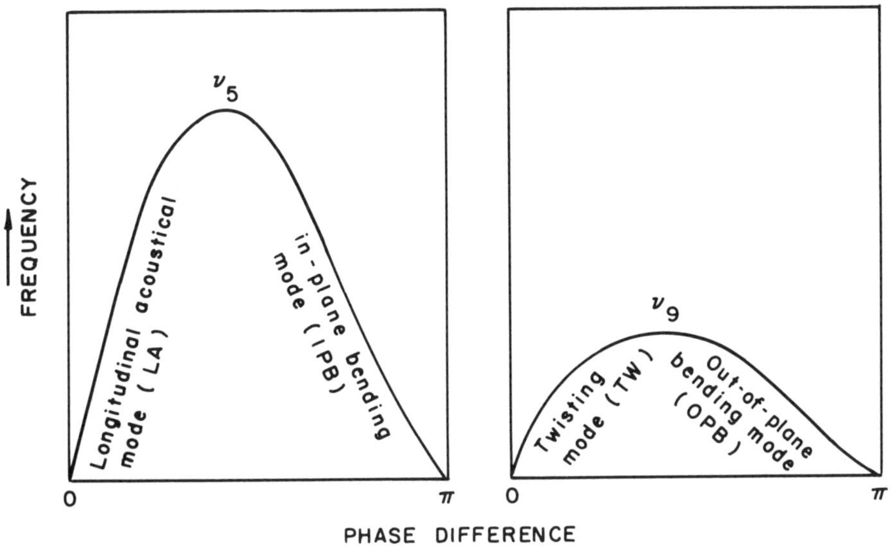

Figure 14.14. Frequency versus phase difference curves for acoustic vibrations of the polyethylene chain. (Reprinted by permission from T. Shimanouchi, in "Structural Studies of Macromolecules by Spectroscopic Methods," K. J. Ivin, Ed.)

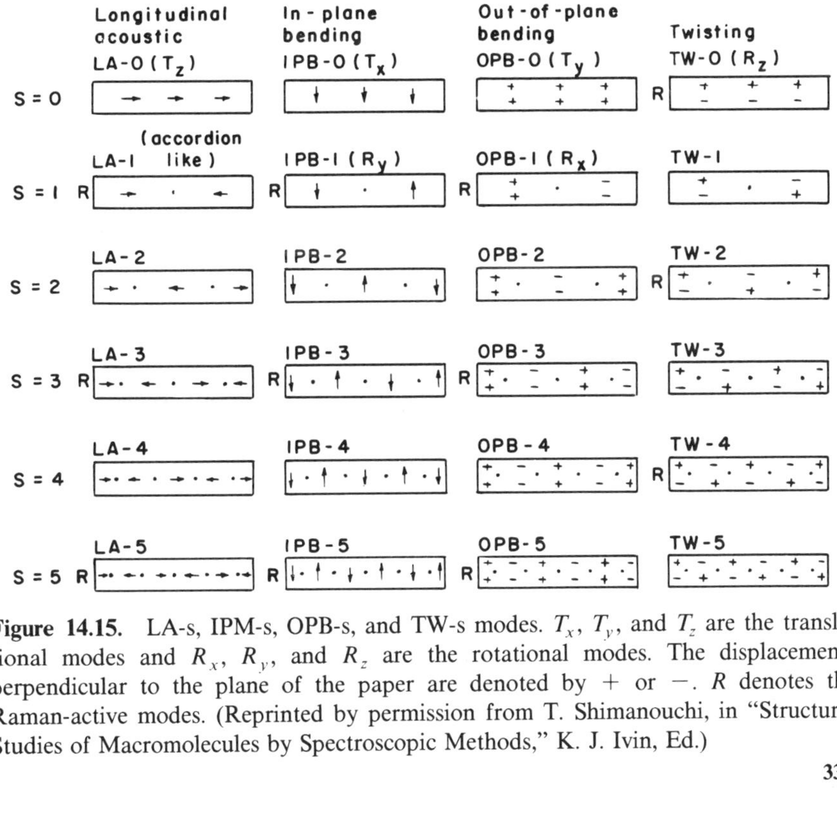

Figure 14.15. LA-s, IPM-s, OPB-s, and TW-s modes. T_x, T_y, and T_z are the translational modes and R_x, R_y, and R_z are the rotational modes. The displacements perpendicular to the plane of the paper are denoted by + or −. R denotes the Raman-active modes. (Reprinted by permission from T. Shimanouchi, in "Structural Studies of Macromolecules by Spectroscopic Methods," K. J. Ivin, Ed.)

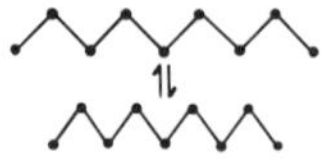

Figure 14.16. Schematic representation of longitudinal acoustic mode for planar zigzag chain.

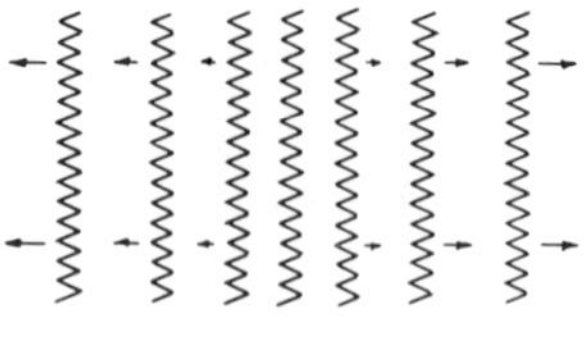

Figure 14.17. Schematic representation of transverse acoustic mode.

the phase relationship later. First it should be emphasized that only the lowest frequency mode (corresponding to a phase difference of $1/36\pi$ in Figure 14.19) is the LAM in the sense of being an accordion-like motion; the other modes on the dispersion curve for ν_5 are related to other kinds of overall vibrations. Second, there is a significant difference between the dispersion curves calculated for a single chain and those calculated for a crystal (with $k_b = k_c = 0$) for the branches ν_5 and ν_9. If Figure 14.6, discussed earlier, is examined carefully, it can be seen that at low values of $\mathbf{k}$ ν_5 and ν_9 approach each other and "swap" character, so that $\nu_5^a(0)$ and $\nu_5^b(0)$ are the A_g and B_{3g} rotational lattice vibrations and $\nu_9^b(0)$ becomes the longitudinal acoustic mode.

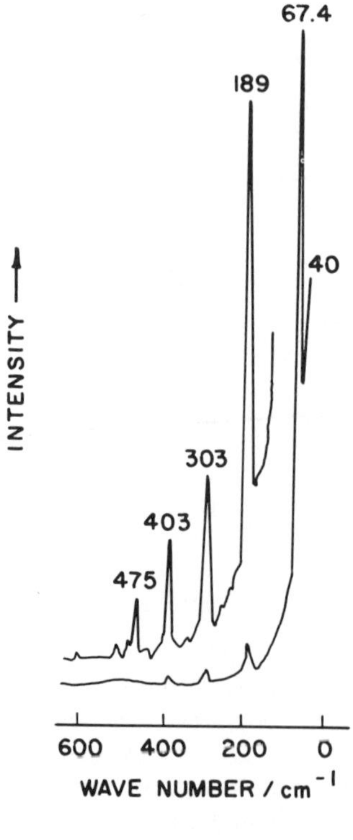

Figure 14.18. Low frequency Raman spectrum of $C_{36}H_{74}$. (Reprinted by permission form ref. 27.)

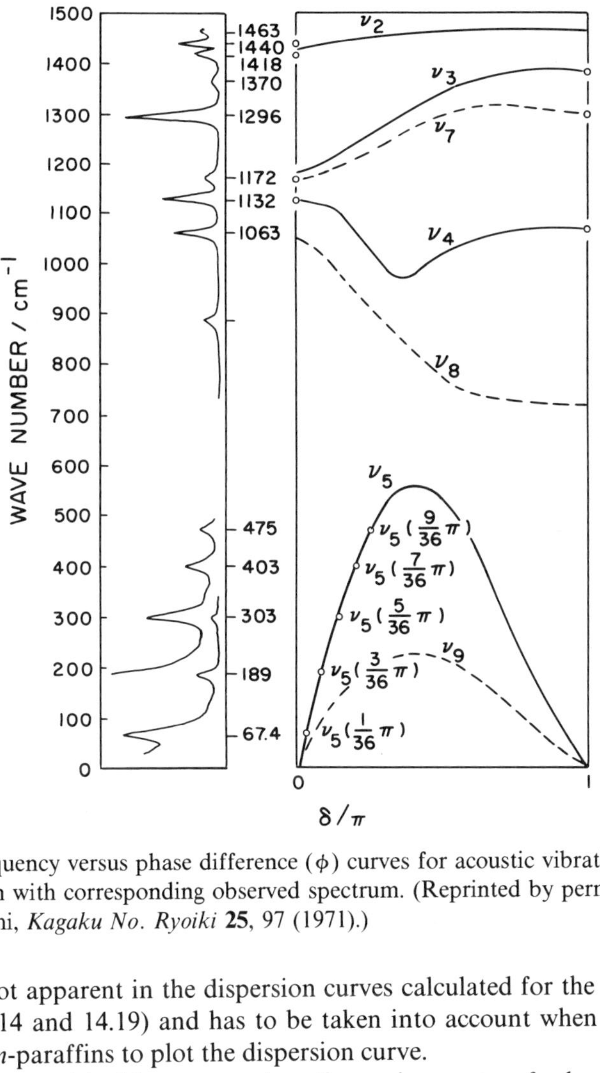

Figure 14.19. Frequency versus phase difference (ϕ) curves for acoustic vibrations of polymethylene chain with corresponding observed spectrum. (Reprinted by permission from T. Shimanouchi, *Kagaku No. Ryoiki* **25**, 97 (1971).)

This behavior is not apparent in the dispersion curves calculated for the single chain (Figures 14.14 and 14.19) and has to be taken into account when using the data from the *n*-paraffins to plot the dispersion curve.

There are some ambiguities concerning the assignments of phase differences to the low frequency progression bands observed in the infrared and Raman spectra of *n*-paraffins. As we noted earlier, for a chain of N coupled oscillators we can use $\phi = s\pi/(N+1)$ $(s = 1, 2, 3, \ldots, N)$ and this relationship has been successfully applied to analysis of CH_2 wagging, twisting, and rocking vibrations. (We will maintain the use of ϕ to describe the phase relationship in a model system of coupled oscillators and will use θ to describe the phase relationship between adjacent chemical units in a polymer chain.) For these high frequency vibrations, however, the vibrations of the end methyl groups are considered separately, so that if n_C is the number of carbon atoms in the chain, $N = n_C - 2$. For the low frequency modes the CH_3 group is treated

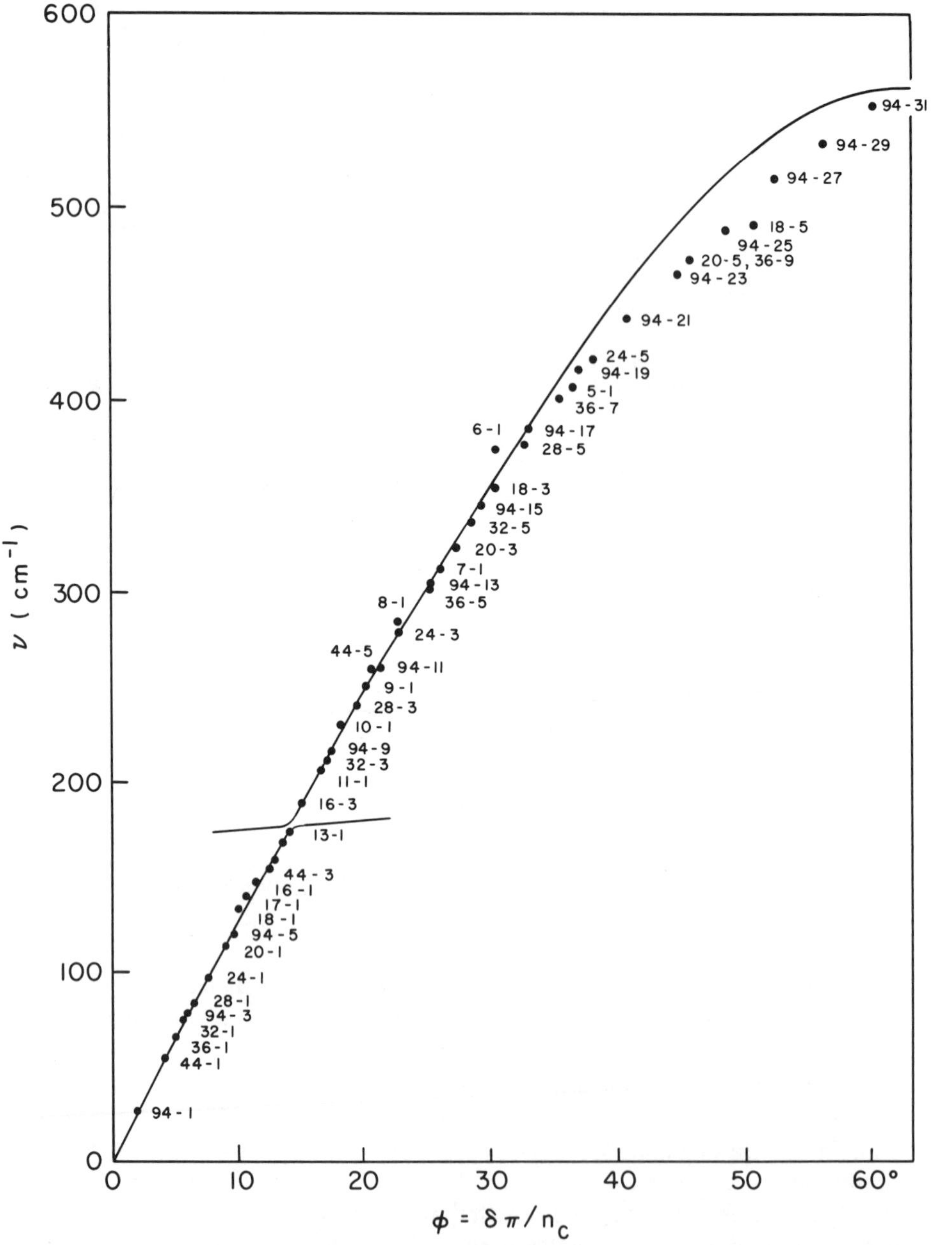

Figure 14.20. Observed Raman frequencies of *n*-paraffins versus $\phi = s\pi/n_c$. The assignment is indicated by N_{c-s}, where N_C is the carbon number and s is the vibration order. The continuous curve indicates the dispersion relation calculated for an infinite polyethylene crystal. (Reprinted by permission from ref. 28.)

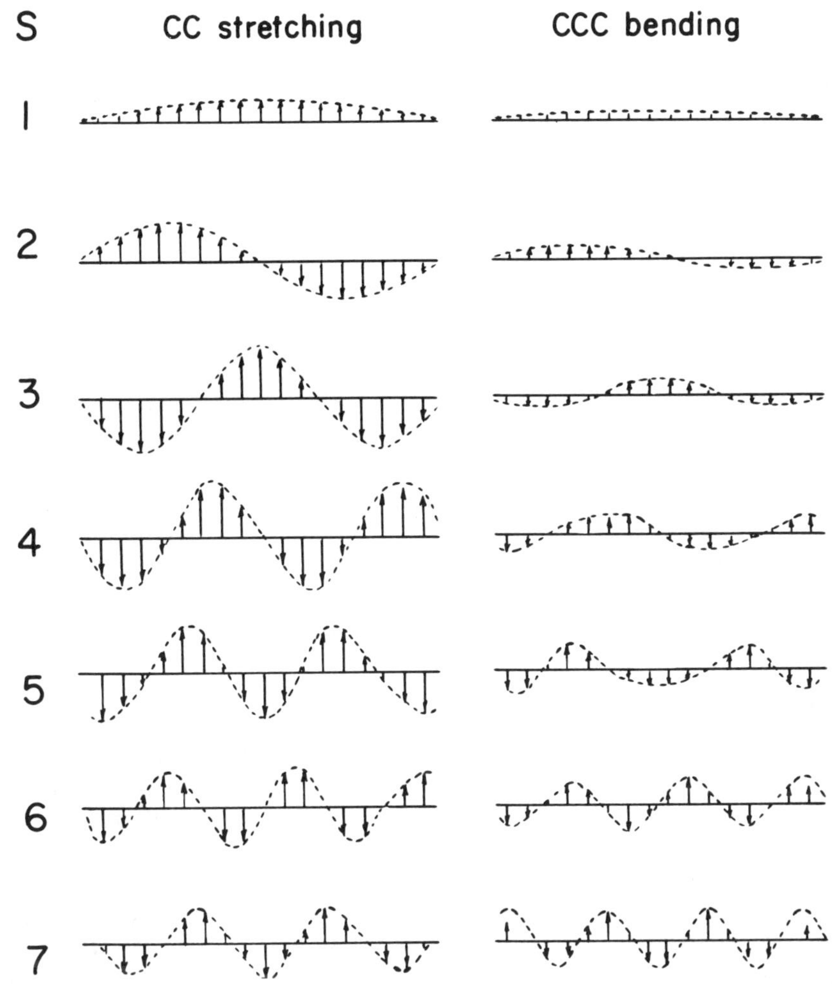

Figure 14.21. Schematic representation of the contributions of CC stretching and CC bending coordinates (*not* the atomic displacements) to the normal coordinates of the LAM ($s=1$) and its overtones ($s=2,\ldots,7$) (Reprinted by permission from ref. 28.)

as being identical to the CH_2 units and the ambiguity arises from the definition of the dynamical unit involved in the vibration. Normal coordinate calculations (28) have demonstrated that both C—C stretching and C—C—C bending coordinates are involved in the LAM-s. In a chain of n_C carbon atoms we have (n_C-1) C—C "oscillators" and (n_C-2) CCC "oscillators," implying that the phase relationship could be $\phi=s\pi/n_C$ or $\phi=s\pi/n_C-1$. It was mentioned at the beginning of this section that the former phase relationship was found empirically to give a better fit to the observed frequencies, and this dispersion plot is shown in Figure 14.20. This observation was subsequently explained by a normal coordinate analysis of $n-C_{18}H_{38}$, which demonstrated that the contribution of C—C stretching increased relative to C—C—C bending with decreasing vibrational order, so that the normal modes for small s number ($s<n_C/3$) could be described essentially as C—C stretching modes with a phase difference $\phi=s\pi/n_C$ (28). For larger s values ($s>n_C/3$) the normal modes may be described as C—C—C bending vibrations. This change in character of the skeletal modes with increasing values of s is illustrated by the displacement coordinates shown in Figure 14.21 for $s=1,\ldots,7$. The change in character is also evident from the potential energy distribution shown in Table 14.7.

This change in character of the normal modes helps to explain the deviation of some of the points from the dispersion curve shown in Figure 14.20. However, there are other factors involved, such as coupling between modes belonging to the same symmetry species and coupling of these vibrations with the translational and rotational lattice vibrations in the crystal. Apparently this coupling is not a serious problem for the longitudinal acoustic modes, but is more serious for other modes.

The inverse dependence of the frequency of the LAM on the length of the all-trans chain has obvious potential for studying the morphology of polymer crystals. Peticolas et al. (29) compared the fold period determined from the LAM frequency of polyethylene single crystals to that determined by small-angle x-ray diffraction. A range of values of the crystal thickness was obtained by annealing, and the results are reproduced in Figure 14.22. The difference in the value of the fold period as determined by the two techniques can be attributed to a number of factors. Perhaps the most important is that the fold period measured by Raman spectroscopy essentially measures the length of the all-trans sequence in the crystal. This will differ from the x-ray measurement according to the degree of "tilt" of the chains in the crystal and the dimensions and effect of the fold or amorphous layer at the crystal surface. Recently, attempts have been made to account for such "end effects" by the use of appropriately modified mechanical models. Originally (25, 27) the observed inverse dependence of the lower harmonics of the LAM-s on the number of carbon atoms (or length of the chain), suggested an analogy to a continuous elastic rod. For such a rod the frequencies of the LAM-s would be given by

$$\nu_s=\frac{s}{2l}\left(\frac{E}{\rho}\right)^{1/2} \qquad (14.39)$$

where l, E, and ρ are the length of the rod, Young's modulus, and the density of the rod, respectively. (The integer s is odd, since only centrosymmetric modes are Raman active.) This model gave reasonably good results but has subsequently been modified in an attempt to obtain a more precise and hence more useful relationship (30–32). Various measurements on polyethylene (33) together with attempts to fit the newly observed LAM frequency of polypropylene (34) to the simple mechanical model resulted in the proposal of various

Table 14.7 Skeletal Vibrations and Potential Energy Distributions (PED) of *n*-Octadecane

Mode	Sym. Species	Frequency		PED[a]	
		Obs. (cm^{-1})	Calc. (cm^{-1})	CC stretching (%)	CC bending (%)
1	A_g	133	129	58	42
2	B_u		248	58	42
3	A_g	355	358	49	51
4	B_u		452	35	65
5	A_g	493	524	20	80
6	B_u		568	11	89
7	A_g		557	8	92
8	B_u		492	7	93
9	A_g		410	4	96
10	B_u		326	3	97
11	A_g		246	3	97
12	B_u		175	2	98
13	A_g		115	2	98
14	B_u		68	1	99
15	A_g		34	1	99
16	B_u		12	0	100

[a]PED are defined for ν_m as follows:

$$\text{PED(CC stretching)} = \frac{100a}{a+b}$$

$$\text{PED(CCC bending)} = \frac{100b}{a+b}$$

where

$$a = \sum_i L_{im}^2 F_{ii}$$

(i refers to the CC stretching coordinates) and

$$b = \sum_j L_{jm}^2 F_{jj}$$

(j refers to the CCC bending coordinates).

composite rod models. As an example, the model proposed by Hsu et al. (32) is reproduced in Figure 14.23. Essentially, it consists of a crystalline core of length D, elastic modulus E_C, density ρ_C, and two "amorphous" ends of length $(L-D)/2$, elastic modulus E_a, and density ρ_a. At each end of this composite rod is assumed a point mass M and an elastic force f. We will not discuss further aspects of the application of this model, since at present it is still being

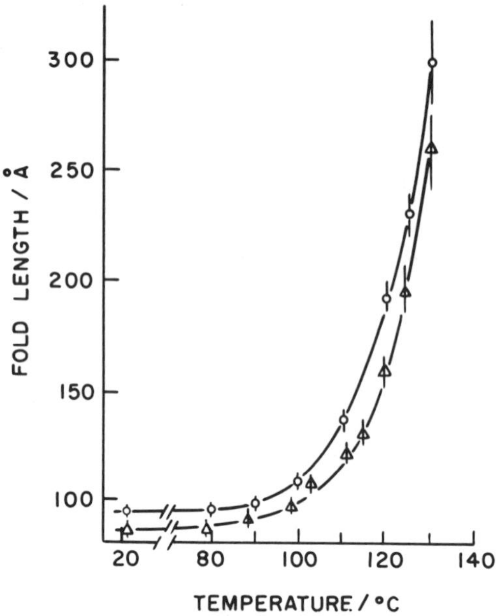

Figure 14.22. Dependence of fold length of polyethylene single crystals on the annealing temperature. ●, x-ray long spacings; ▲, predicted from low frequency Raman spectra. (Reprinted by permission from ref. 29.)

developed and applied. In concluding, however, we would like to draw attention to the proposal of Zerbi et al. (26) that a mechanical model may not necessarily be the only approach to obtaining a detailed understanding of the relationship of the LAM to polymer morphology. These authors have calculated the dynamics of a chain of trans sequences containing a defect or fold consisting of other conformations. Calculations of the dynamics of this system by methods discussed in the following chapter indicate that the LAM frequencies of the trans sequences are perturbed by coupling through the defect. Fortunately, this work is also just appearing at the time of writing this book, so that we have not been put in the position of having to compare the merits of these two approaches.

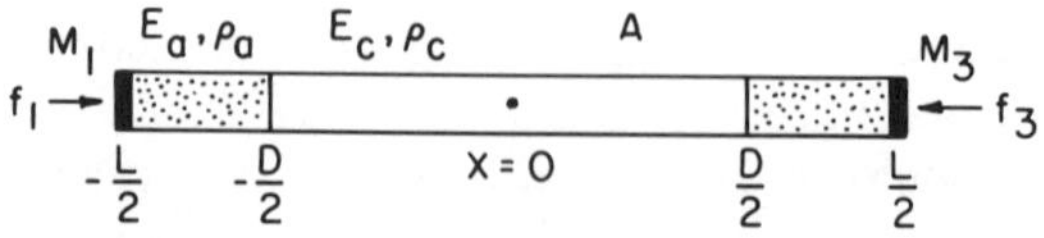

Figure 14.23. Model of composite elastic rod with general perturbing forces and masses at ends. (Reprinted by permission from ref. 34.)

References

1 J. L. Koenig, *Appl. Spect.* **29**, 293 (1975).

2 M. M. Coleman and P. C. Painter, *J. Macromol. Sci. Macromol. Chem.* **C16**(2), 197 (1978).

3 M. M. Coleman, P. C. Painter, D. L. Tabb, and J. L. Koenig, *J. Polymer Sci. Polymer Lett. Ed.* **12**, 577 (1974).

4 G. Zerbi, F. Ciampelli, and V. Zamboni, *J. Polymer Sci.* **C7**, 141 (1965).

5 J. Boerio and J. Koenig, *J. Chem. Phys.* **52**, 3425 (1970).

6 L. Piseri and G. Zerbi, *J. Mol. Spect.* **26**(2), 254 (1968).

7 M. Tasumi and T. Shimanouchi, *J. Chem. Phys.* **43**(4), 1245 (1965).

8 T. Kitagawa and T. Miyazawa, *Adv. Polymer Sci.* **9**, 335 (1972).

9 G. Turrell, "Infrared and Raman Spectra of Crystals," Academic Press, New York, 1972.

10 C. A. Coulson and C. W. Haigh, *Tetrahedron* **19**, 527 (1963).

11 T. Tasumi and S. Krimm, *J. Chem. Phys.* **46**(2), 755 (1967).

12 J. Barnes and B. Fanconi, *J. Phys. Chem. Ref. Data* **7**(4), 1309 (1978).

13 W. Myers, G. C. Summerfield, and J. S. King, *J. Chem. Phys.* **44**, 184 (1966).

14 T. Miyazawa and T. Kitagawa, *J. Polymer Sci.* **B2**, 395 (1964).

15 A. Elliott and E. J. Ambrose, *Nature* **165**, 921 (1950).

16 T. Miyazawa, *J. Chem. Phys.* **32**, 1647 (1960).

17 T. Miyazawa and E. R. Blout, *J. Am. Chem. Soc.* **83**, 712 (1961).

18 A. Elliott and E. M. Bradbury, *J. Mol. Biol.* **5**, (1962).

19 E. M. Bradbury and A. Elliott, *Polymer* **4**, 47 (1963).

20 Y. Abe and S. Krimm, *Biopolymers* **11**, 1817 (1972).

21 S. Krimm and S. Abe, *Proc. Natl. Acad. Sci.* **69**, 2788 (1972).

22 W. H. Moore and S. Krimm, *Proc. Natl. Acad. Sci.* **72**, 4933 (1975).

23 W. H. Moore and S. Krimm, *Biopolymers* **15**, 2439 (1976).

24 W. H. Moore and S. Krimm, *Biopolymers* **15**, 2465 (1976).

25 S. I. Mizushima and T. Shimanouchi, *J. Am. Chem. Soc.* **71**, 1320 (1949).

26 G. Zerbi, P. Graziari, and M. Gussoni, Private communication, to be published.

27 R. J. Schaufele and T. Schimanouchi, *J. Chem. Phys.* **47**, 3605 (1967).

28 T. Shimanouchi and M. Tasumi, *Indian J. Pure Appl. Phys.* **9**, 958 (1971).

29 W. L. Peticolas, G. W. Hibler, J. L. Lippert, A. Peterlin, and H. Olf, *Appl. Phys. Lett.* **18**, 87 (1971).

30 H. G. Olf, A. Peterlin, and W. L. Peticolas, *J. Polymer Sci. Polymer Phys. Ed.* **12**, 359 (1974).

31 S. L. Hsu and S. Krimm, *J. Appl. Phys.* **47**, 4265 (1976).

32 S. L. Hsu, G. W. Ford, and S. Krimm, *J. Polymer Sci. Polymer Phys. Ed.* **15**, 1769 (1977).

33 A. Peterlin, H. Olf, W. L. Peticolas, G. W. Hibler, and J. L. Lippert, *J. Polymer Sci.* **B9**, 583 (1971).

34 S. Hsu, S. Krimm, S. Krause, and G. S. Y. Yeh, *J. Polymer Sci. Polymer Lett. Ed.* **14**, 195 (1976).

15

THE INFLUENCE OF DEFECTS AND DISORDER ON THE VIBRATIONAL SPECTRA OF POLYMERS

A sweet disorder in the dress
kindles in clothes a wantonness:
A lawn about the shoulders thrown
Into a fine distraction. . . .
A careless shoe-string in whose tie
I see a wild civility:
Do more bewitch me, than when Art
Is too precise in every part.

Robert Herrich

15.1 The Nature of Defects in Polymers

For an infinite, ordered polymer chain the vibrational analysis can be reduced through symmetry to an analysis of the translational repeat unit. Similar reductions in the size of the vibrational problem can be made for polymer crystals. In the case of high molecular weight polymeric materials, the assumption of infinite chain length does not introduce significant errors and allows a satisfactory fit of observed and calculated frequencies. However, there are usually a number of secondary infrared bands or Raman lines in the spectra of even the most highly crystalline macromolecule that cannot be assigned to fundamentals, combinations, and overtones, or even to end groups. In addition, the vibrational modes of highly crystalline polymers demonstrate a temperature dependence that cannot be explained in terms of an ideal translationally invariant structure (1). For *trans*-1,4-polychloroprene it has also been demonstrated that certain "crystalline" bands are sensitive to the temperature of polymerization of the polymer (2,3). These experimentally observed effects are due to the semicrystalline nature of even the most highly regular polymer. Even polymer single crystals grown from dilute solution have both a crystalline and an amorphous component, and defects of one kind or another can be included in (or excluded from) the crystalline lattice. Defects can be categorized into four major classes:

1 **Chemical defects** Consider the vinyl monomer:

$$\underset{\text{tail}}{CH_2} = \underset{\text{head}}{CX_2}$$

Usually, polymer chains consist of predominantly head-to-tail placements:

$$-CH_2-CX_2-CH_2-CX_2-CH_2-CX_2-$$

However, in certain polymers [e.g., free radically synthesized poly(vinyl fluoride), poly(vinylidene fluoride), polychloroprene] a monomer unit is occasionally incorporated "backward," giving head-to-head and tail-to-tail placements:

$$-CH_2-\underbrace{CX_2-CX_2}_{\text{h-h}}-\underbrace{CH_2-CH_2}_{\text{t-t}}-CX_2-$$

Similarly, defects due to isotopic substitution, chain branching (short or long), and cross-linking can occur. A particularly intriguing combination of geometric and configurational isomers is produced by the synthesis of diene polymers. For example, in the polymerization of chloroprene the trans-1,4 structure dominates but there is significant incorporation of cis-1,4, −1,2, and −3,4 structures into the polymer backbone, as illustrated in Figure 15.1.

2 **Conformational Defects** It is now accepted that conformational defects are always present in polymer chains. For example, the minimum energy

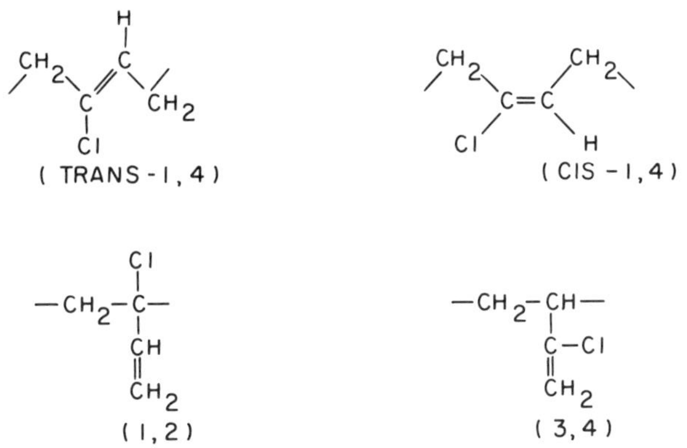

Figure 15.1. *trans*-1,4-Polychloroprene and its configurational and sequence isomers.

conformation of a polyethylene chain is the all-trans conformation. However, the incorporation of gauche units, obtained by a rotation of 120° around a C—C bond, can result in a folding or twisting of this chain that yields the characteristic chain-folded crystalline morphology. In certain cases [e.g., *trans*-1,4-polyisoprene, poly(vinylidene fluoride) and the poly(α-olefins)] there is more than one conformationally stable state, each characterized by different bond angles and rotations.

3 **Stereochemical Defects** Vinyl polymers that contain asymmetric carbon atoms, such as polypropylene,

$$-\!\!(CH_2-\underset{\displaystyle CH_3}{\underset{|}{CH}}\!\!)_n$$

can be synthesized to give distinct structures known as stereoisomers. Stereoregular isotactic and syndiotactic structures are compared to the stereoirregular atactic structure in Figure 15.2. The synthesis of stereoregular polymers is seldom perfect, so that the polymer chain consists of stereoregular sequences interrupted by defect units.

4 **Packing Defects** Most structural defects are usually excluded from the crystal lattice. Accordingly, as the concentration of defects increases, the degree of crystallinity of the polymer decreases and there is a larger contribution to the vibrational spectrum from the amorphous regions. However, certain types of crystalline packing defects, such as those described as kinks and jogs, may be incorporated into the lattice, perturbing the vibrational spectrum of the theoretically perfect crystal. Additionally, defects that are chemically and structurally similar to the host lattice may also be incorporated (3).

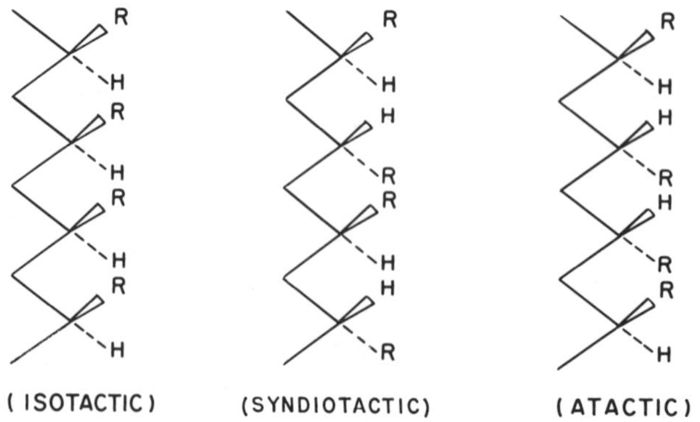

Figure 15.2. Stereoisomers of a $(-CH_2-CHR-)$ polymer.

In order fully to interpret the vibrational spectrum of polymers it is necessary to achieve an understanding of the effect of defects and conformational disorder. This is not just an interesting (and exacting) theoretical problem but one of practical significance as well. Many of the properties of polymeric materials depend on the amorphous component, and vibrational spectroscopy is one of the few tools that is potentially capable of analyzing the conformational structure of chains in the noncrystalline state. Unfortunately, the theory of the vibrations of unordered chains, or even of chains with a low concentration of defects, can become extraordinarily complex. As a result, many approaches rely on simplifying assumptions. We will now consider a preliminary qualitative description of the problem and then critically discuss various approaches and models that have been applied so far.

15.2 Preliminary Analysis of the Problem

The dynamics of a lattice containing defects is a familiar problem in solid state science. However, the mathematical treatments used to obtain precise solutions (Green's function methods) depend on the system being simple and the defect concentration small (4, 5). The problem that is capable of solution is essentially a near-perfect system containing an impurity. In order to describe more complex systems, brute-force numerical methods have been employed (6–12).

One of the characteristics of polymeric systems is that they exhibit a range of complexity, so that a semicrystalline polymer such as isotactic polystyrene can be obtained in a completely amorphous form by rapid quenching from the melt; in the form of a semicrystalline plastic material by annealing the quenched material at temperatures above the glass transition but below the melting point; or as highly ordered "single-crystal" lamellae obtained by crystallization from appropriate dilute solutions. In order to analyze the vibrational modes of such systems it is necessary first to examine various structural models and to discuss how the nature of the defects would be expected to affect the spectrum. We can then proceed to examine methods that have been applied to specific problems.

It was demonstrated in Chapter 12 that the translational symmetry of a theoretically perfect infinite chain greatly simplifies the vibrational problem by restricting allowed frequencies to those corresponding to values of the wave vector **k** equal to zero. As the simplest case of a partially disordered polymer chain we can consider the effect of randomly including a low concentration of defects in such a chain. If the geometry and force constants of the defect unit are such that its modes do not couple with those of the host lattice, then we can assume that the vibrational spectrum of such a system is a composite of two contributions. First, there is the effect of the localization of chain vibrations. In the infinite perfect chain such vibrations are delocalized along the entire length of the chain and involve all monomer units. Randomly introducing a small concentration of defects truncates this chain into units of finite length, so that we have to consider the changes in the spectrum due to the localization of vibrations in these sequences. In effect, the $\mathbf{k}=0$ selection rule no longer applies and the entire density of states of the perfect lattice becomes activated, so that all points on the dispersion curves are allowed. The second effect is the appearance of localized vibrational modes due to the impurity. If the structure or mass of the defect unit is sufficiently different from the repeat unit that makes up the perfect chain, these localized modes may appear outside the ideal phonon bands of the host lattice, that is, in the frequency gaps of the dispersion curve. In such cases the modes of the defect cannot exchange energy with the modes of the perfect lattice. These gap modes are termed localized since their neighbors cannot follow the vibrations of the defects and the amplitudes of the oscillations decays exponentially with distance from the defects. Highly localized defect modes can also occur within the frequency bands, so that relatively sharp infrared bands or Raman lines appear superimposed on the density of states of the host lattice. Furthermore, if there is more than one defect unit per chain, the possibility of splitting within the vibrational energy levels of the localized mode occurs; this possibility increases with decreasing distance between defects.

The next level of complexity is the case of a low concentration of defects with in-band defect modes coupled to some extent with the normal modes of the host lattice. In such cases we would expect a perturbation of both the displacements and the frequencies of the phonons of the theoretically infinite perfect chain. Such perturbations will increase with increasing concentration of defects, and additional complexities due to coupling between the defect modes will also become apparent. Kozyrenko et al. (13) have illustrated these changes in the density of states of a polymer chain due to the introduction of an increasing number of defects, as shown in Figure 15.3. The dispersion curves of the ideal polymer chain are shown in Figure 15.3*a*. The allowed frequency bands are shaded. Figure 15.3*b* shows the density of vibrational states of the perfect chain together with impurity bands appearing in the frequency gaps. It can be seen that near the edges of the frequency branches the density of states tends to infinity and so these critical points give rise to singularities in the density of states and probably appear as strong absorption bands in the vibrational spectrum, depending on intensity factors. As the concentration of defects increases, the perturbation of the phonons of the original ideal lattice is such that the boundaries of the allowed regions become blurred and the density of states, $g(\omega^2)$ or $g(\nu)$, has nonzero values in the normally forbidden frequency gap regions. The singularities in the density of states broaden around the impurity frequencies to form impurity bands. These progressive changes are illustrated in Figure 15.3*c* and 15.3*d*. Note that the concept of frequency branches has at this stage lost its meaning and the features of the spectra of such disordered systems are best described in terms of the density of vibrational states.

The various approaches to the problem of interpreting the vibrational spectra of polymer chains with defects can be categorized in terms of the

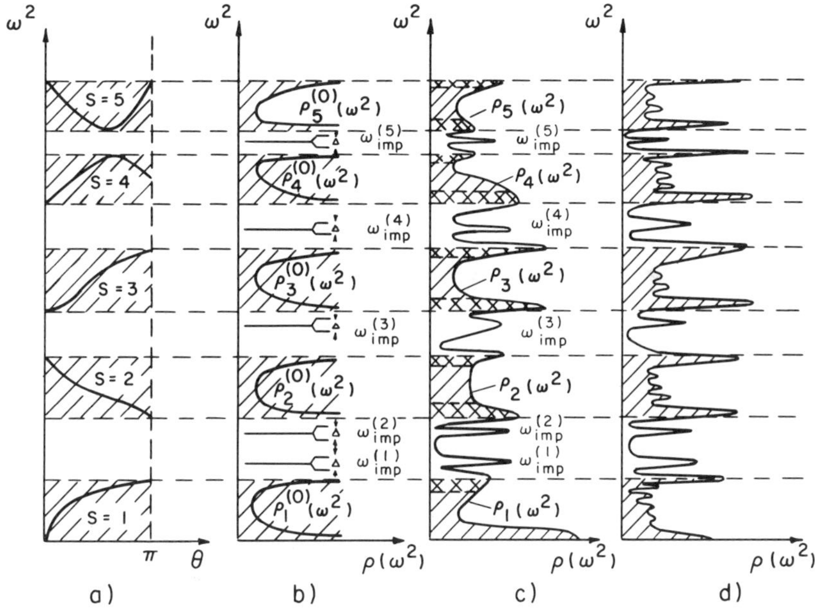

Figure 15.3. Dispersion curves (a) and changes in the density of states of a regular infinite polymer chain due to defects: (b) one or two defects; (c, d) polymer chain with large number of defects. (Reprinted by permission from ref. 13.)

degrees of increasing complexity discussed earlier. At the simplest level of a low concentration of defects whose modes are not coupled to the phonons of the ideal lattice, the spectrum can be simply interpreted as a superposition of defect modes on the effects of finite chain length. At the next level of complexity, coupling between the vibrational modes of the defect and the host lattice has to be considered. One way of estimating the degree of mixing between local and chain vibrations is the application of transfer matrices (14). However, numerical methods using the negative eigenvalue theorem have been more widely applied, and we will discuss the application of this theory in detail. Finally, at the level of completely amorphous polymers, adequate mathematical tools are only now being developed (13). Theoretically, Green's functions are capable of solving this problem, but the mathematical complexities are such that this method will probably be effectively applied only to the situation of an ordered chain with a small concentration of defects.

15.3 The Effect of Finite Chain Length

For many semicrystalline polymers we can assume a model consisting of conformationally ordered sequences of different length separated by "defect" groups or regions. The defect groups could simply be a conformational

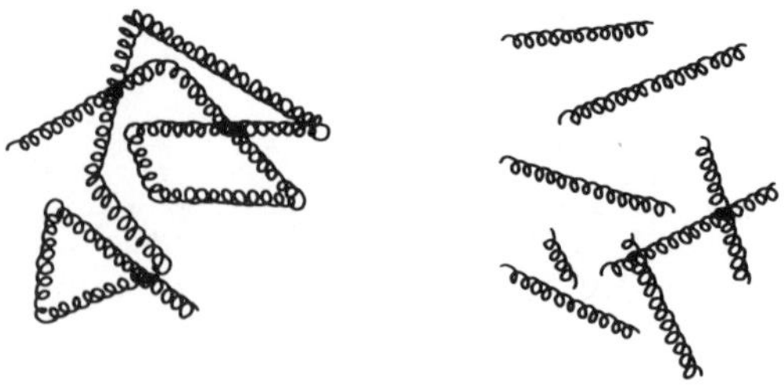

Figure 15.4. Model of polymer chain consisting of uncoupled finite ordered sequences.

disturbance, or a chemical or configurational defect of the type discussed in Section 15.1. The ordered sections of the chain need not be aligned, as in a crystal, but could be randomly arranged, as illustrated in Figure 15.4*a*. A number of tactic polymers in solution can have partial order of this type. However, since for most polymers the intermolecular forces are extremely small compared to the intramolecular interactions, this model can also be applied to situations of higher order, for example, a polymer crystal with included defects. The simplest way to treat such a system is to assume that the defect units do not couple or interact with the ordered chain segments. In effect, we are "breaking" the chain at the defect points to give various segments of finite length, as illustrated in Figure 15.4*b*. Obviously, the validity of this approximation will depend on the specific polymer under consideration and the type of defect unit present in the chain.

We solved the secular equation for a simple finite lattice of N units in Chapter 10. Since the noninteracting defect units constitute the boundaries of each finite chain, then provided that each finite section is long, we can neglect end effects. The phase relationship between vibrations in adjacent units in the same finite segment is given by

$$\phi = \frac{s\pi}{N+1} \tag{15.1}$$

where N is the number of monomer units in the specific finite chain segment under consideration. Consider the case where each monomer unit has p degrees of freedom. We will observe a dispersion relationship consisting of p frequency branches. There will be, in the general case of n units in the chain, n points on each of these dispersion curves. Theoretically, we should then observe pn optical modes in the vibrational spectrum. However, if the interaction between repeating units is negligibly small, which is often the case in polymers with large monomer units, then the frequencies of all n units will be the same and the dispersion curve will be horizontal. Only p vibrational modes will be observed, and this is represented schematically by horizontal lines in Figure 15.5 (I). However, if we consider interactions between the monomer units, each of these p bands will be split into n components, as illustrated in Figure 15.5

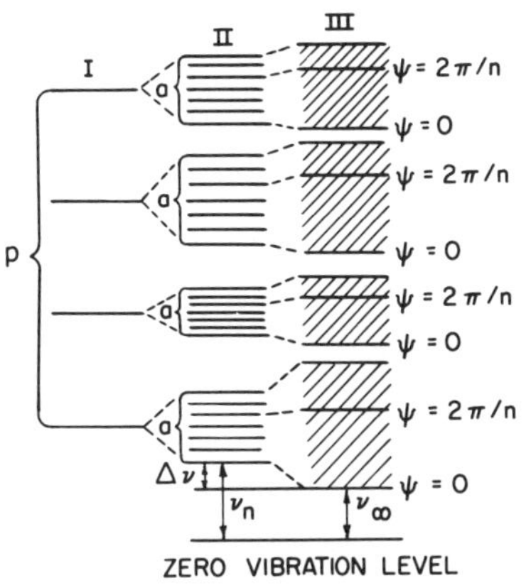

Figure 15.5. Spectrum of the frequencies of the normal vibrations of chains of finite (II) and infinite (III) lengths with mutually noninteracting (I) and interacting (II and III) monomers.

(II). In the limiting case of an infinite chain the dispersion curves become continuous as the number of frequencies in each branch tends to infinity. The branches can then be represented by continuous bands, as shown in Figure 15.5 (III). Of course, the selection rules imposed by the translational symmetry of the infinite chain limit the observable infrared bands to values of the phase angle corresponding to $\phi=0$ and $\phi=\psi$, where $\psi=2\pi m/n$, while the allowed Raman lines correspond to $\phi=0$, ψ, and 2ψ. Consequently, on transition from the theoretically perfect infinite chain to one of finite length n we are activating n vibrational modes in each branch. However, we demonstrated in Chapter 10 that the intensity distribution in such band series is not equal. In the infrared spectrum the lowest frequency band of a chain of parallel coupled oscillators is intense relative to the other bands in the series. The weaker bands on the higher frequency side in effect make the strong lowest frequency band appear asymmetrically broadened, since for large n the weaker bands will be increasingly difficult to resolve. A system having a distribution of chain lengths will naturally give rise to a composite of such band series.

Although this model allows a qualitative understanding of the vibrational spectrum of a polymer chain containing a small concentration of defects, in more precise treatments the effect of coupling between the vibrations of the end groups and the normal modes of the chain has to be considered. It is no surprise that comparisons of the data obtained from n-paraffins with those of n-fatty acid salts demonstrates that the nature of the end group (which in our model corresponds to the defect) produces appreciable differences in the spectra. The method of transfer matrices has been applied to finite polymethylene chains with different end groups (14). The degree of mixing of local and chain vibrations can be estimated from the following expression for the phase angle ϕ:

$$\phi=\left(\frac{\pi}{N}\right)k+\left(\frac{\delta}{N}\right) \tag{15.2}$$

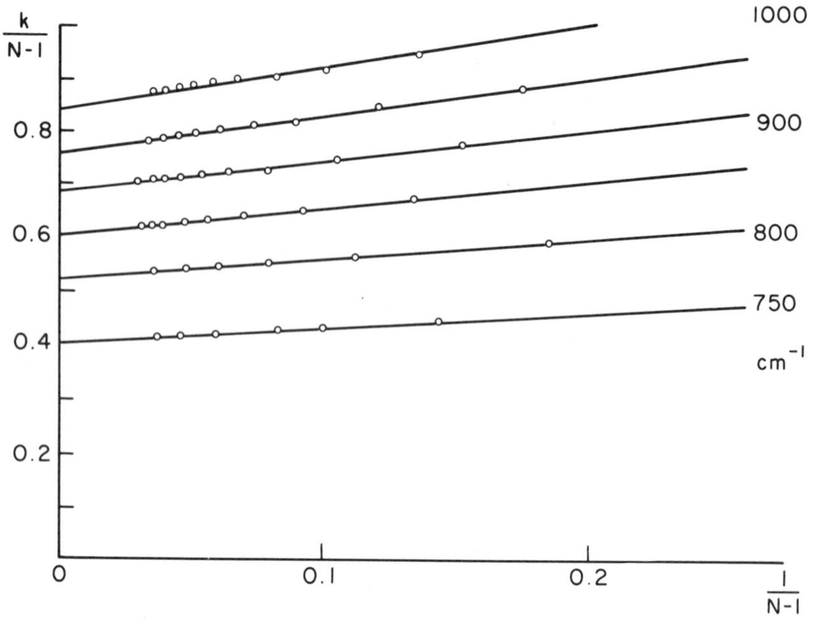

Figure 15.6. Phase shift analysis for methylene rocking and twisting modes. (Reprinted by permission from ref. 14.)

where δ depends on the structure of the end groups. This expression successfully accounts for deviations in the frequency phase curve observed in the paraffins of short chain length. For example, the array of methylene rocking modes for C_3H_8 through n-$C_{30}H_{62}$ as determined by Snyder and Schachtschneider was shown previously in Figure 10.21. The frequency phase curve was shown in Figure 10.22. Reorganizing equation (15.2), we can obtain

$$\frac{k}{N-1}=\frac{\phi}{\pi}-\frac{\delta}{(N-1)\pi} \tag{15.3}$$

and the plot of $k/(N-1)$ against $1/(N-1)$ at several frequencies, shown in Figure 15.6, is practically a straight line, demonstrating that this theoretical approach is in good agreement with experiment. The gradient of the straight line is equal to $(\phi-\delta)/\pi$, which is a measure of the deviation in the phase relationship due to end effects. Unfortunately, it would be extremely difficult to apply this method to the analysis of polymer chains containing defects, since the progression band series required to make such plots are not experimentally obtainable. Instead, an asymmetric broadening of the fundamentals of the perfect chain, together with some frequency shifts, is observed. Consequently, numerical methods that allow a determination of the density of states appear to be a superior approach. We will discuss these later.

15.4 Localized Conformationally Sensitive Modes

Empirical correlations of infrared bands and Raman lines to vibrations of the partially ordered polymers has been reported for a number of polymers. An example is the use of the observed frequencies of CH_2 rocking bands as a measure of the sequence length of these units in ethylene–propylene copolymers. Such bands are useful as group frequencies to the extent that their corresponding normal modes are localized. Snyder (15) has argued that certain modes in a polyethylene chain containing conformational defects are localized to the extent that an analysis of the structures present can be based on a detailed normal coordinate analysis of several rotamers of normal hydrocarbons. This theory is in marked contrast to the approach used by Zerbi and co-workers (6–12), who applied numerical methods to calculate the density of states of chains containing a random distribution of defects. The density of states $g(\nu)$ (as illustrated, e.g., in Figure 15.3) is plotted as a histogram and compared to the observed spectrum. We will consider this method in detail in the following section, but it is important to note that this analysis assumes that the spectrum is determined by the dynamics of the whole chain. Obviously, contributions from both the density of states and localized vibrations will be present, but which of these will predominate will depend on intensity factors. In this section we will outline the arguments of Snyder (15) and also discuss an analysis of partially deuterated polyethylene by Snyder and Poore (16). This work probably represents the most detailed analysis of localized conformationally sensitive modes yet achieved.

The infrared spectrum of polyethylene in the molten state is compared to the spectrum taken at room temperature in Figure 15.7. Snyder (15) has compared the spectra of polymers with vinyl and methyl end groups in order to identify the end group modes in the latter specimens. At first glance the spectrum of the molten state seems to be simpler than that of the semicrystalline material. This apparent simplicity is due to the absence of intermolecular effects characteristic of the orthorhombic crystal. It can also be seen that the characteristic fundamental modes broaden considerably while at the same time certain bands increase considerably in intensity. Note in particular the modes between 1250 and 1400 cm^{-1}, since these have been the focus of much attention in attempts to correlate observed bands to fold structures. Koenig (17) has reported that the Raman spectrum of molten polyethylene displays corresponding features.

As mentioned earlier, Snyder (15) has determined a valence force field for both nonplanar and planar *n*-paraffins. The calculated force constants are listed in Table 15.1. The internal coordinates correspond to those defined in Chapter 13. We will consider in detail only a few of the band assignments that are based on this analysis. The frequencies observed between 1500 cm^{-1} and 600 cm^{-1} in the infrared spectrum, together with the appropriate band assignments, are more completely summarized in Table 15.2.

The asymmetric broadening of the CH_2 rocking mode near 720 cm^{-1} can be interpreted in terms of the superposition of contributions of all-trans

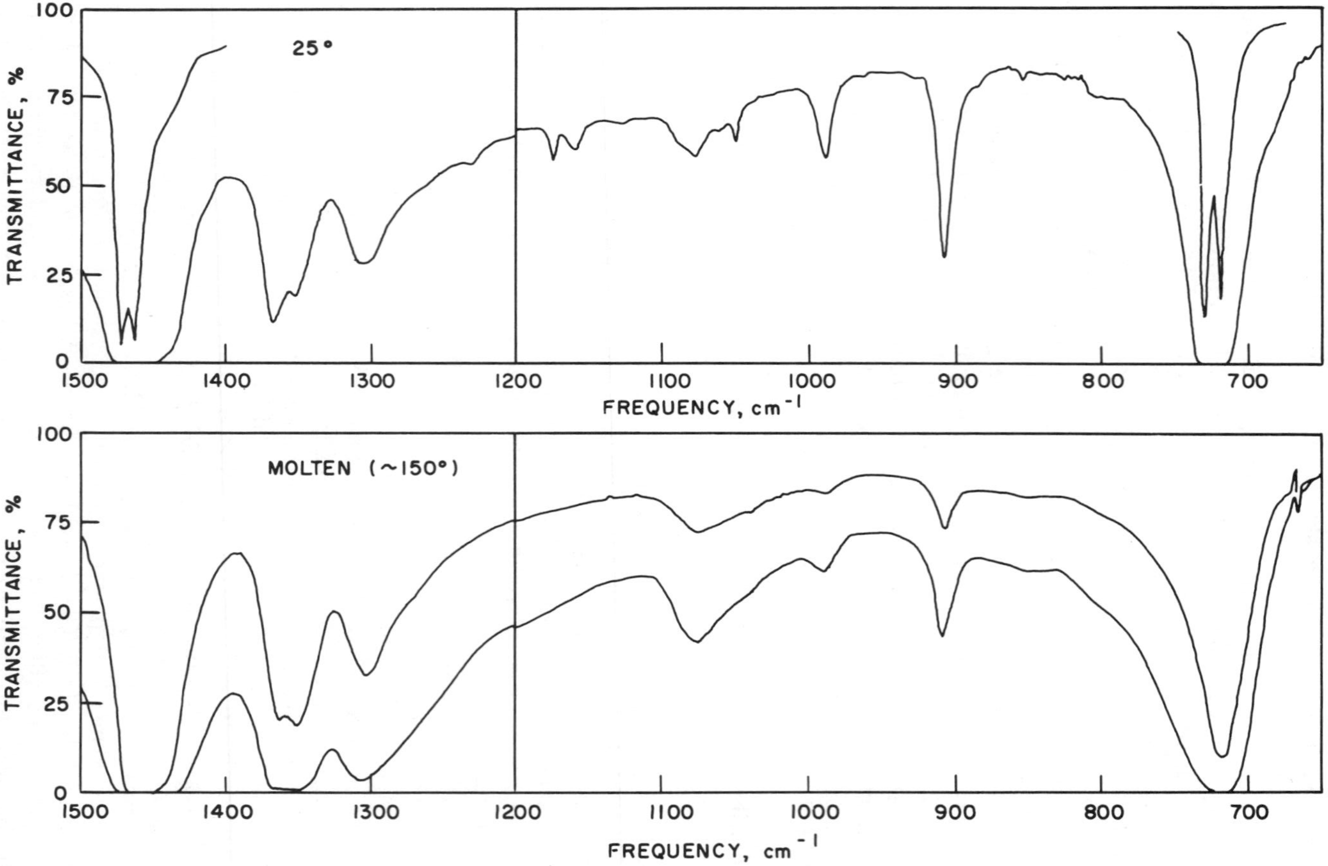

Figure 15.7. Infrared spectra of polyethylene (vinyl end groups). (Reprinted by permission from ref. 15.)

Table 15.1 Force Constants for trans and gauche *n*-Paraffins

Constant[a]	Group	Coordinate(s) Involved	Atoms Common to Interacting Coordinates	Calc. Value[b] Φ_i	Error[c] $\sigma(\Phi_i)$
K_r	CH_3	C—H		4.699	0.003
K_d	CH_2	C—H		4.538	0.003
K_R	CC	C—C		4.532	0.041
H_α	CH_3	∠HCH		0.539	0.001
H_β	CH_3	∠HCC		0.618	0.005
H_δ	CH_2	∠HCH		0.533	0.009
H_γ	CH_2C	∠HCC		0.663	0.003
H_ω	CCC	∠CCC		1.032	0.085
H_r	\ / — CC — / \	C—C		0.024[d]	
H_τ	CC	C—C		0.024[d]	
F_r	CH_3	C—H, C—H	C	0.032	0.002
F_d	CH_2	C—H, C—H	C	0.019	0.003
F_R	CCC	C—C, C—C	C	0.083	0.027
$F_{R\gamma}$	HCC	C—C, ∠HCC	C—C	0.174	0.014
$F'_{R\gamma}$	H CCC	C—C, ∠HCC	C	−0.097	0.016
$F_{R\omega}$	CCC	C—C, ∠CCC	C—C	0.303	0.018
F_β	CH_3	∠HCC, ∠HCC	H—C	−0.031	0.004
F_γ	CCH_2	∠HCC, ∠HCC	C—C	−0.019	0.003
F'_γ	CCH_2C	∠HCC, ∠HCC	H—C	0.021	0.002
$F_{\gamma\omega}$	H CCC	∠HCC, ∠CCC	C—C	−0.022	0.050
f_γ^t	HCCH	∠HCC, ∠HCC	C—C	0.073	0.006
f_γ^g	HCCH	∠HCC, ∠HCC	C—C	−0.058	0.006
$f_{\gamma\omega}^{t'}$	H CCCH	∠HCC, ∠HCC	C	−0.009	0.005
$f_\gamma^{g'}$	H CCCH	∠HCC, ∠HCC	C	−0.004	0.005
$f_\gamma^{t''}$	HH CCCC	∠HCC, ∠HCC		0.010	0.006
$f_\gamma^{g''}$	HH CCCC	∠HCC, ∠HCC		0.012	0.006
$f_{\gamma\omega}^t$	HCCC	∠HCC, ∠CCC	C—C	0.073	0.011
$f_{\gamma\omega}^g$	HCCC	∠HCC, ∠CCC	C—C	−0.064	0.008
f_ω^t	CCCC	∠CCC, ∠CCC	C—C	0.097	0.018
f_ω^g	CCCC	∠CCC, ∠CCC	C—C	−0.005	0.021

[a]For definition of the internal coordinates, see Chapter 13; for a more complete definition of these force constants, see ref. 3.
[b]Stretch constants are in units of millidynes per angstrom; stretch-bend interaction constants are in units of millidynes per radian; bending and torsion constants are in units of millidyne·angstrom per square radian.
[c]$\sigma(\Phi_i)$ is the standard error in Φ_i estimated from the standard error in the frequency parameters and the variance-covariance matrix.
[d]Value taken to fit the torsion frequency (280 cm^{-1}) of ethane.

Table 15.2 Observed Infrared Bands of Polyethylene and Their Assignment[a]

Polyethylene				
Solid (25°C)		Melt (>130°C)		
I	II	I	II	Assignment
1472, vs	1472, vs			δ, cryst
		1463, vs	1463, vs	δ, $-GT_mG^*-$, m large
1462, vs	1462, vs			δ, cryst
		1455, s, sh	1455, s, sh	δ, $-GT_mG^*-$, m small
~1438, m, b, sh	~1440, m, b, sh	~1438, s, b, sh	~1438, s, b, sh	δ, —GG—
1413, vvw				
	1378, m		1378, m	U
1366, m	1368, m	1368, s	1367, s	W, —GTG*—
1352, m	1353, m	1349, s	1352, s	W, —GG—
	1344, w, sh		1344, m, sh	W, —TG
1338, w, sh				W, —GTTG*—
1309, m	1308, m	1303, s	1305, s	W, —GTG*—
~1270, w, b, sh	~1270, w, b, sh	≈1270, m, b, sh	≈1270, m, b, sh	W, $-GT_mG^*-$, $m \geq 3$
1229, vw	1220, vvw			
1175, w	1176, vw			W, cryst
1160, w				
1129, vw		1125, vvw?		
~1085, w, b, sh	~1087, vw, b, sh	~1087, w, sh, b	~1087, w, sh	$R+W$, $-TG_mT-$, $m \geq 2$
1078, w, b	1079, w, b	1076, m, b	1073, m, b	$R+W$, —TGT—
1062, vw	1062, vw			T, cryst
1050, w	1050, vw			T, cryst
989, w		989, w		$RCH{=}CH_2$
	966, vvw, b		968, vvw, b	
953, vvw				
908, m	909, vvw	907, m	910, w	$RCH{=}CH_2$
888, vw	888, w		888, w	β
~ 858, vw, b		~ 855, vw, b	~ 850, vw, b	$-TG_mT-$, $m>2$, (?)
~ 808, vw, b		~ 805, vw, b, sh	~ 800, vw, b, sh	$-T_mGT_n$, m, n large, (?)
	~ 775, w, b, sh		~ 775, w, b, sh	P, —TG
~ 745, w, b, sh		~ 745, m, b, sh	~ 745, m, b, sh	P, —GTG*—
730, vs	731, vs			P, cryst
		718, s	720, s	P, $-GT_mG^*-$, $m>2$
719, vs	719, vs			P, cryst
620, vw, b		~ 625, vw, b		

[a] Polyethylene I has vinyl end groups; polyethylene II has methyl end groups. δ, W, T, and P are methylene bending, wagging, twisting, and rocking; U and β are methyl symmetric bending and rocking; R is C—C stretching. G* means that the bond can be either right or left gauche or that it leads to a methyl group terminating the chain. Abbreviations: v, very; s, strong; m, medium; b, broad; sh, shoulder; w, weak.

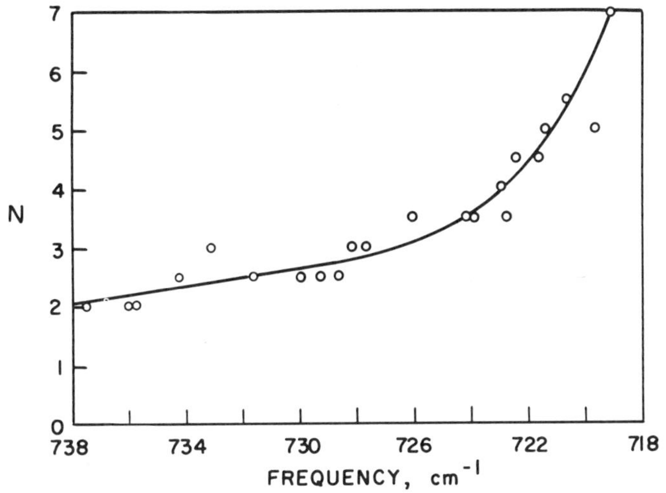

Figure 15.8. Number of methylenes in a trans sequence versus the calculated frequency of the most in-phase rocking mode. (Reprinted by permission from ref. 15.)

sequences of various sequence length. This analysis depends on a localization of vibrations in these sequences, and the calculations of Snyder (15) have indicated that this is the case. The observed frequency of the longest trans sequence of various paraffin rotamers is plotted against sequence length in Figure 15.8. Only the lowest frequency mode of these chains of finite length, which are the most intense, is plotted. The observed band at 719 cm^{-1} in molten polyethylene indicates that trans sequences of four or more methylene units predominate in the melt (at 150°C). However, the asymmetric broadening of this band is not solely due to the weaker bands of the finite trans sequences. Alternating trans-gauche sequences absorb near 745 cm^{-1} and in the limiting case of $(TG)_\infty$ and $(TGTG')_\infty$ chains the rocking mode has calculated frequencies of 738 and 736 cm^{-1}, respectively. The high frequency tail is therefore due to overlapping contributions from both localized defect vibrations, the lowest frequency bands of very short trans sequences, and the numerous weak-intensity superimposed higher frequency bands of longer trans sequences.

In contrast to the methylene rocking modes, where the evidence indicates that there is a localization of vibrations in ordered trans sequences, the strongly coupled C—C stretching vibrations might be expected to be more amenable to an analysis based on the determination of the density of states of a long chain containing defects (see Section 15.5). In fact, a broad band is observed spanning the frequency range of the dispersion curve of the fully extended chain. However, Snyder (15) has pointed out that, for a number of reasons, the actual shape of the absorption band may not correspond to the density of states derived from the dispersion curve. The most important factor

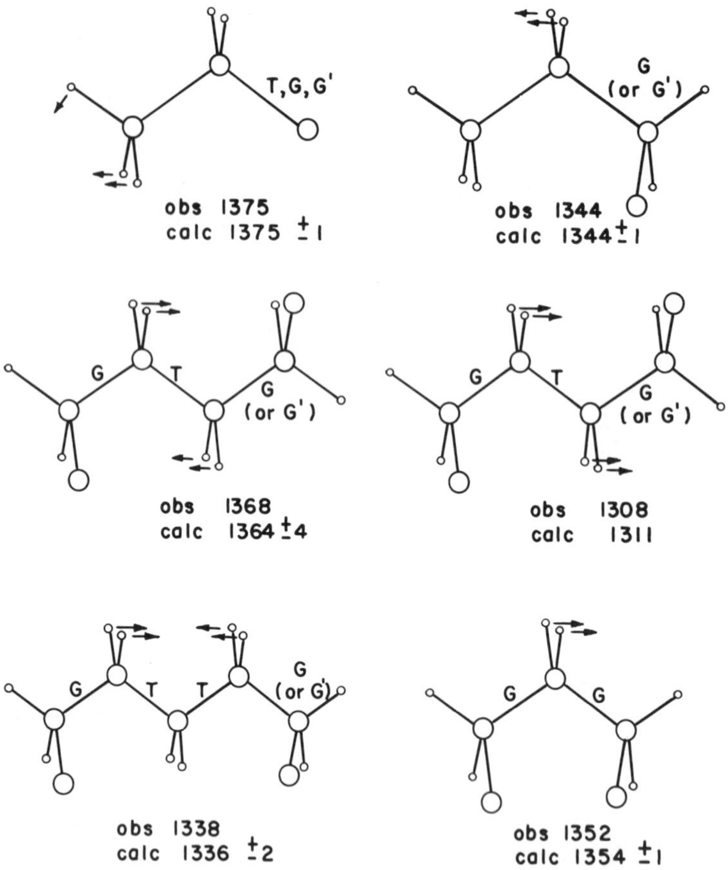

Figure 15.9. Approximate vibrational motion or bands observed in the spectrum of polyethylene in the region form 1380 to 1300 cm^{-1} (Reprinted by permission from ref. 15.)

is probably the variable mixing of the C—C stretching modes with other coordinates. For example, in the *n*-paraffins much of the intensity of the 1134 cm^{-1} band is due to strong mixing with the in-plane methyl rocking mode. In polyethylene, mixing of this mode with other kinds of normal coordinates (e.g., methylene wagging and twisting modes) depends on the conformations present, so that these other modes will contribute to the intensity of the observed bands in a frequency-dependent fashion. Consequently, the modes identified as being predominantly C—C stretching are not very useful for determining conformation, because vibrations are not localized in specific structures and analysis based on the calculation of the density of states is to some degree suspect.

Unlike the C—C stretching modes the methylene wagging modes of certain conformations are localized sufficiently to give characteristic bands. This is because the contribution to the dipole moment derivative from the wagging of CH_2 units adjoining trans C—C bonds is small, so that the intensity of methylene wagging modes is derived predominantly from units adjacent to gauche bonds. However, methylene units connected by short sequences of trans units are more strongly coupled than those connected by gauche units, so that localized modes due to gauche-isolated trans sequences of one, two, or three methylene units would be expected to be the most prominent in the spectrum. Based on an analysis of rotamers of *n*-paraffins, Snyder (15) has assigned the prominent wagging bands observed at 1368 and 1350 cm^{-1} in the spectrum of polyethylene to GTG (or GTG′) and GG conformations, respectively. Two modes are possible for the GTG structure, one with motion symmetric and the other antisymmetric to the center of the trans bond. The former is assigned to the 1368 cm^{-1} band while the latter is calculated to have a frequency near 1310 cm^{-1}. There is a broad conformationally sensitive band observed in this region of the spectrum. However, this frequency also corresponds to a peak in the density of states of the ideal chain, which becomes activated when there is a low concentration of defects. Consequently, assignments of bands near 1300 cm^{-1} to localized vibrations in specific structures are not certain. In the spectra of the paraffins a well-defined band observed near 1338 cm^{-1} can be assigned to GTTG′ conformations where normal mode calculations indicate a significant contribution from displacements of the methylenes adjoining and isolated by the gauche bonds. The vibrational motions in the gauche-isolated methylene units discussed earlier are illustrated in Figure 15.9. Also shown in this figure are modes associated with the methyl end group. The symmetric methyl bending mode is observed near 1375 cm^{-1} in the spectrum of the *n*-paraffins and in polyethylene samples having this end group, although in high molecular weight material the intensity of this band is weak. There is in addition a band at 1344 cm^{-1} in the paraffins which demonstrates intensity variations corresponding to the 1375 cm^{-1} band and is therefore most likely associated with a vibration at the end of the molecule. When a TG conformation is located at the end of the molecule, the wagging of the terminal methylene group is localized and the TG frequency is calculated to be 1346 cm^{-1}. The assignments of these localized methylene wagging vibrations are summarized in Table 15.2. As mentioned earlier, the broad band centered near 1308 cm^{-1} cannot be unambiguously assigned to a localized mode. Snyder has argued that gauche-isolated trans sequences of more than three units will tend to absorb in less well-defined bands and contribute to a broad region of absorption on the low frequency side of the 1308 cm^{-1} band. The origin of this band may be best understood in terms of density of states calculations.

Because of the problems associated with identifying and assigning bands to localized vibrations, particularly when there is an overlap of such bands in a specific spectral region and the vibrations are not entirely localized (so that there is an additional unknown intensity contribution from the density of states of the disordered chain), Snyder and Poore (16) investigated the use

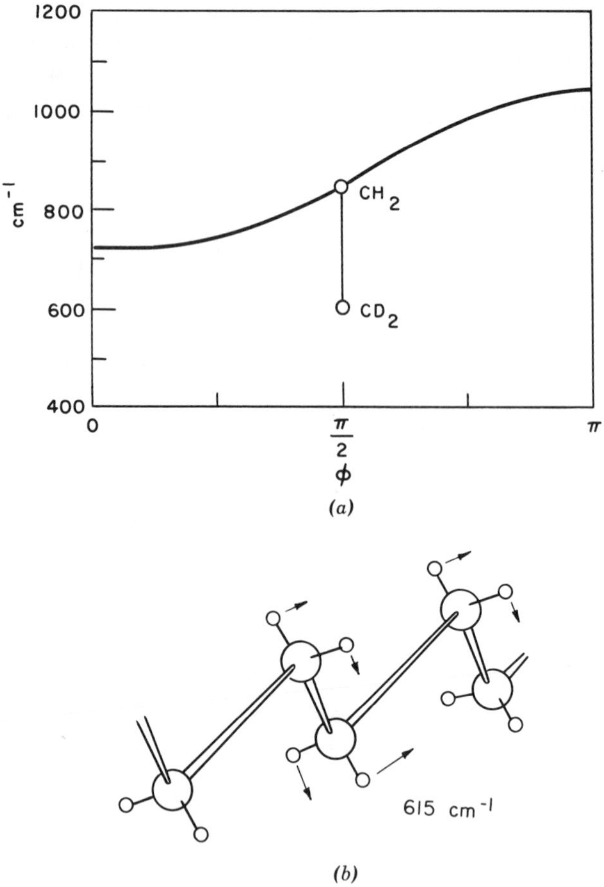

Figure 15.10. (*a*) Isolated CD_2 rocking mode frequency shown relative to the (CH_2) rocking-twisting mode dispersion curve. (*b*) Form of the CD_2 rocking mode in a *trans*-polyethylene chain. (Reprinted by permission from ref. 16.)

of partially deuterated chains in the analysis of the conformational structure of polyethylene. The rationale of this approach is based on the following arguments. Figure 15.10*a* shows the dispersion curve for the methylene rocking-twisting mode. The rocking mode for an isolated CH_2 unit occurs near $\phi=\pi/2$ (i.e., near 850 cm^{-1}). If this group were replaced by a deuterated methylene unit, corresponding CD_2 rocking vibration would be shifted to about 600 cm^{-1}. Since this frequency is outside the lattice band of the ideal chain, there should be little coupling with CH_2 rocking vibrations. Nevertheless, there is still a significant contribution from vibrations in CH_2 units adjacent to the CD_2 groups, as shown by the approximate form of the rocking mode calculated

Table 15.3 Calculated Frequencies (cm^{-1}) of the CD_2 Rocking Modes of 4,4-*n*-Heptane-d_2 and 5,5-*n*-Nonane-d_2

$CH_3CH_2CH_2CD_2$-$CH_2CH_2CH_3$		$CH_3CH_2CH_2CH_2$-$CD_2CH_2CH_2CH_2CH_3$	
(T)TT(T)	616	(TT)TT(TT)	615
(G)TT(T)	615		
(G)TT(G)	615		
(G)TT(G′)	615		
(T)TG(T)	649		
(G)TG(G)	649		
(G)TG(T)	654		
(G′)TG(T)	649		
(T)GG(T)	678	(TT)GG(TT)	689
(T)GG(G)	677		
(G)GG(G)	676		

by Snyder and Poore illustrated in Figure 15.10*b*. This is by virtue of large kinetic energy coupling between neighboring methylene groups. As a result, the CD_2 rocking frequency is affected by the conformation of the adjoining CH_2 groups and can be used as a conformational probe. Band assignments were based on calculations of the normal modes of the *n*-paraffin 4,4-*n*-heptane-d_2. The calculated frequencies of the CD_2 rocking mode for most of the possible conformations of the bonds adjacent to the CD_2 group are listed in Table 15.3. The normal coordinates of the TTTT, TGTT, and TGGT conformations are listed in Table 15.4. Although the vibration is significantly localized, the contribution of displacements from adjacent CH_2 units is apparent. This contribution varies with conformation, depending on the proximity of the frequency of the CD_2 rocking mode to the dispersion curve of the polymer. The calculated frequencies of the CD_2 modes are shown superimposed upon the dispersion curve for the rocking mode of polyethylene in Figure 15.11. It can be seen that the coupling interaction is greatest for the vibration of the GG bond pair, whose frequency most closely approaches the 720 cm^{-1} lower limit

Table 15.4 Normal Coordinates of the CD_2 Rocking Modes of 4,4-*n*-Heptane-d_2

Conformation	$\nu_{calc.}$ (cm^{-1})	L_{ik}				
		CH_2	CH_2	CD_2	CH_2	CH_2
TTTT	615	0.015	0.202	0.489	0.202	0.015
TGTT	650	−0.033	−0.065	0.435	0.298	0.060
TGGT	678	−0.094	−0.201	0.420	−0.201	−0.094

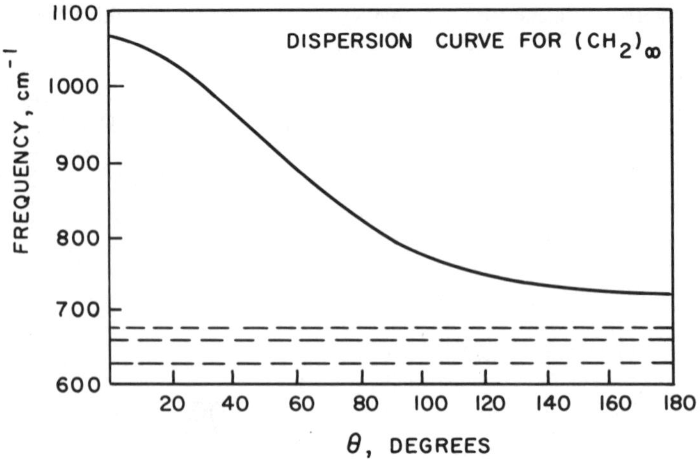

Figure 15.11. Relationship between the methylene rocking-twisting dispersion curve and the "characteristic" frequencies for an isolated CD_2 group. (Reprinted by permission from ref. 16.)

of the dispersion curve. This effect was confirmed by comparing the frequencies of 5,5-*n*-nonane-d_2 with those of 4,4-*n*-heptane-d_2, as shown in Table 15.3. The additional CH_2 groups of the former molecule bring it closer to the polymer and it can be seen that the GG conformation is shifted appreciably in frequency but the TT conformation is practically the same as in 4,4-*n*-heptane-d_2. The infrared spectrum of a pressed film of polyethylene containing 5% CD_2 groups is shown in Figure 15.12. Band assignments are given in Table 15.5. It

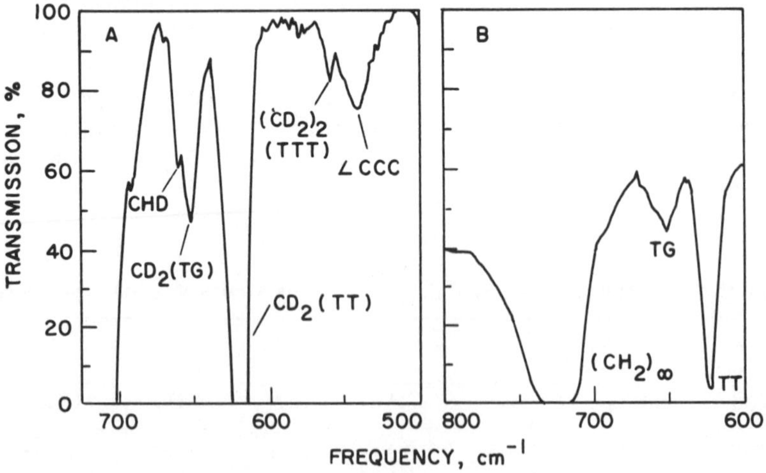

Figure 15.12. Infrared spectra of a pressed film of polyethylene (at 77°K) containing 5% CD_2 groups. (Reprinted by permission from ref. 16.)

Table 15.5 Frequencies and Assignments of Bands Observed in the Infrared Spectrum of 5% CD_2 Polyethylene in the Region 730–500 cm^{-1}

Frequency (cm^{-1})	Assignment
725 (vvs)	$(CH_2)_\infty$, $(t)_\infty$, CH_2 rocking (ν_8)
690 (vvw, sh)	$(CH_2)_\infty$, cis C=C
661 (vvw, sh)	CHD, TT (calc. 658)
652 (vw)	CD_2, TG (calc. 650)
622 (w)	CD_2, TT (calc. 615)
562 (vvw)	CD_2CD_2, TTT (calc. 556)
540 (vw)	$(CH_2)_\infty$, ∠CCC bend cutoff (ν_5)

can be seen that the strong band at 622 cm^{-1} can be confidently assigned to TT conformations while the band at 651 cm^{-1} can be identified with the TG bond pair. Bands due to GG structures were not detected. However, even in the melt (at 140°C) it is estimated (16) that this bond pair is only about 10% of the total, so that the rocking band associated with this conformation could be too weak to detect.

It can be seen in Figure 15.12 that there is a weak band at 562 cm^{-1}. This band has been assigned to a CD_2CD_2 pair, since about 2.5% of CD_2 groups will have CD_2 neighbors and the $(CD_2)_2$ group in the trans conformation has a calculated frequency of 556 cm^{-1}. In fact, polyethylene synthesized with a

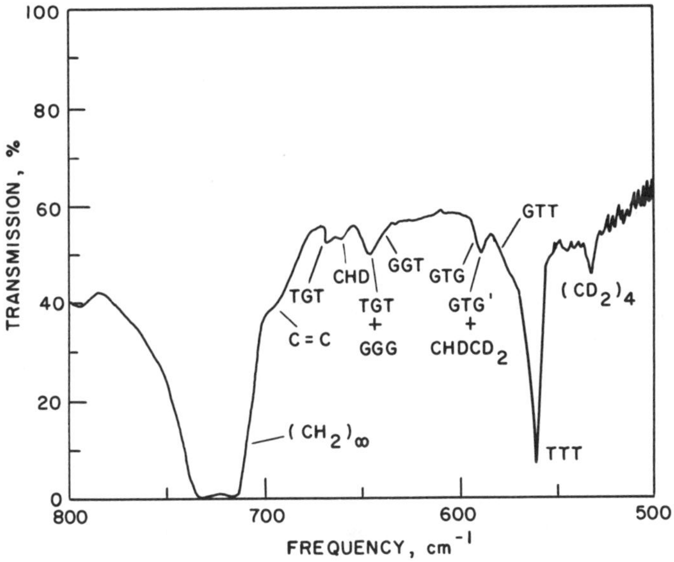

Figure 15.13. Infrared spectrum of a pressed film of polyethylene (at 77°K) containing ~5% CD_2CD_2 groups. (Reprinted by permission from ref. 16.)

Table 15.6 Frequencies and Assignments of Bands Observed in the Infrared Spectrum of 5% $(CD_2)_2$ Polyethylene in the Region 730–500 cm^{-1}

Freq. Rel. Intensity	Assignment	Rel. Integrated Intensity	
		Film	Cryst.
725 (vvs)	$(CH_2)_\infty$, CH_2 rocking (ν_8)		
690 (vvw, sh)	$(CH_2)_\infty$, cis C=C	0.05_5	0.06_9
666 (vvw, sh)	$(CD_2)_2$, TGT (calc. 662)	0.02_0	0.00_7
662 (vvw)	CHD, TTT (calc. 658)	0.02_9	0.03_4
645 (vw)	$(CD_2)_2$ {TGT (calc. 640); GGG (calc. 640)}	0.05_5	0.03_4
640 (vvw, sh)	$(CD_2)_2$, GGT (calc. 636)	0.02_9	0.00
625 (vvw)			
600 (vvvw)	$(CD_2)_2$, GTG (calc. 595)		
	$(CD_2)_2$, GTG′ (calc. 581)		
589 (vw)	$CHDCD_2$, TTT (calc. 583)	0.07_9	0.07_2
575 (vvw, sh)	$(CD_2)_2$, GTT (calc. 569)	0.06_7	0.00
562 (vw)	$(CD_2)_2$, TTT (calc. 556)	1.00	1.00
~542 (vvw, b)	$(CH_2)_\infty$, ∠CCC bend cutoff (ν_5)		
531 (vvw)	$(CD_2)_4$, T_5 (calc. 527)		

small amount of CD_2CD_2 units shows a wealth of conformationally sensitive bands in the rocking region of the spectrum, as shown in Figure 15.13. Band assignments have been made based on a vibrational analysis of 4,4,5,5-*n*-octane-d_4, and are summarized in Table 15.6. The normal coordinates of this molecule indicate that the two rocking modes expected for the $(CD_2)_2$ groups is only a fair description. There is some mixing of this vibration with carbon skeletal modes, and serious perturbations occur as the frequencies of certain conformations approach 720 cm^{-1} and the vibrations tend to become delocalized. Fortunately, there is a local center of inversion in the $(CD_2)_2$ bond for TTT and GTG′ triads, such that the higher frequency symmetric mode will tend to be infrared inactive. Conversely, the lower frequency antisymmetric rocking mode is infrared active and less subject to complicating interactions.

The introduction of isotopic impurities seems to be a powerful tool in the investigation of conformation, since it is much easier to identify bands that are effectively due to localized vibrations. However, the rocking mode of polyethylene is particularly suitable for such an analysis, since the frequencies associated with the dyad and triad conformations fall in a region of the spectrum that is relatively free of other absorption bands. The major experimental difficulties noted were the presence of isotopic impurities and the low intensities of the bands (16).

15.5 Numerical Analysis of Imperfect Chains

Although specific bands can occasionally be identified with localized vibrations, for example, when mass defects result in such modes appearing outside the lattice band of the ideal chain, it is necessary to consider the lattice dynamics of the disordered system in order to obtain a clearer understanding of the complete vibrational spectrum. It has been shown that the absorption intensity of an impure crystal can be written as

$$I(\omega)=M^2(\omega)g(\omega) \tag{15.4}$$

where $M^2(\omega)$ is the transition moment for the mode of frequency ω and $g(\omega)$ is the density of states. Even for simple molecules *a priori* calculations of $M^2(\omega)$ are at present impossible. Consequently, we encounter the first major approximation to the vibrations of disordered systems. It is assumed that the term $M^2(\omega)$ is constant, so that the observed vibrational spectrum is equated directly to the density of states, $g(\omega)$, without dipole weighting. What remains is the all but trivial problem of calculating $g(\omega)$! Zerbi and co-workers (6–12) have applied numerical methods based on the negative eigenvalue theorem (NET). Naturally, for a largely disordered system or even a polymer with a low concentration of defects, the reduction in the vibrational problem allowed by the translational symmetry of the repeat unit of a theoretically perfect infinite chain vanishes. Chains of finite length have to be considered. It has been shown, however, that provided that the chain segments considered are not too short, the features of $g(\omega)$ are apparently reliable. Calculations on models of 100, 200, 400, and 600 chemical units demonstrate the same features in $g(\omega)$, provided that the statistical distribution of conformational sequences in the chains are the same.

The density of states is determined by dividing the spectrum into frequency intervals, say ω_1, ω_2 where $\omega_2>\omega_1$, and counting the number of eigenvalues of the dynamical matrix $\mathbf{D}_m$ that lie in each interval. Note that $\mathbf{D}_m$ is not the same as the $\mathbf{D}$ matrix defined in Chapter 3, but is in fact equal to $\mathbf{D}^T\mathbf{F}_s\mathbf{D}$. The number of eigenvalues in the frequency interval of interest, $n(\omega_2-\omega_1)$, is given by

$$n(\omega_2-\omega_1)=\eta(\mathbf{D}_m-\omega_2\mathbf{E})-\eta(\mathbf{D}_m-\omega_1\mathbf{E}) \tag{15.5}$$

where $\eta(\mathbf{D}_m-\omega_i\mathbf{E})$ is the number of negative eigenvalues of $(\mathbf{D}_m-\omega_i\mathbf{E})$. This number is of course a "count" of the number of eigenvalues whose values are less than ω_i. It follows directly that the right-hand side of equation (15.5) is a measure of the number of eigenvalues in the interval (ω_1, ω_2). If there are N atoms in each unit cell and we are considering as our model a chain of p such units, $\mathbf{D}_m$ is a symmetric $3Np\times3Np$ matrix and $\mathbf{E}$ is the $3Np\times3Np$ unit matrix.

The negative eigenvalue theorem applies to a partioned symmetric matrix $\mathbf{M}$ of dimensions $r \times r$ that takes the form

$$\mathbf{M}=\begin{bmatrix} \mathbf{A}_1 & \mathbf{B}_2 & & & & \\ \mathbf{B}_2^T & \mathbf{A}_2 & \mathbf{B}_3 & & \mathbf{0} & \\ & \mathbf{B}_3^T & \mathbf{A}_3 & \mathbf{B}_4 & & \\ & & \ddots & \ddots & & \\ & \mathbf{0} & & & & \\ & & & & \mathbf{B}_m^T & \mathbf{A}_m \end{bmatrix} \qquad (15.6)$$

where $\mathbf{A}_i$ is a symmetric square matrix whose dimensions are $r_i \times r_i$, $\mathbf{B}_i$ has dimensions $r_{i-1} \times r_i$, and $\sum_{i=1}^{m} r_i = r$. All other elements of $\mathbf{M}$ apart from $\mathbf{A}_i$, $\mathbf{B}_i$, and $\mathbf{B}_i^T$ are zero. The negative eigenvalue theorem states that

$$\eta(\mathbf{M}-\mathbf{xE}) = \sum_{i=1}^{m} \eta(\mathbf{U}_i)$$

where

$$\mathbf{U}_1 = \mathbf{A}_1 - \mathbf{xE}$$

$$\mathbf{U}_i = \mathbf{A}_i - \mathbf{xE}_1 - \mathbf{B}_i^T \mathbf{U}_{i-1}^{-1} \mathbf{B}_i$$

At this stage it is not easy to see how the application of NET allows us to determine the number of negative eigenvalues of $\mathbf{M}-\mathbf{xE}$. The problem has apparently only changed form, so that we now have to determine the number of negative eigenvalues of the submatrices $\mathbf{U}_i$ and then sum them. However, if it could be arranged so that $\mathbf{U}_1 = \mathbf{A}_1 - \mathbf{xE}_1$ were of order unity, $\eta(\mathbf{U}_1)$ would simply be determined from the sign of this scalar quantity. The successive partitioning of the matrix $\mathbf{D}_m - \mathbf{w}_i\mathbf{E}$ into such scalar submatrices allows us to determine the number of negative eigenvalues through the application of equation (15.5).

Consider the simple partition of the dynamical matrix $\mathbf{D}_m$ into just four submatrices:

$$\mathbf{D}_m = \begin{bmatrix} \mathbf{A}_1 & \mathbf{B}_2 \\ \mathbf{B}_2^T & \mathbf{A}_2 \end{bmatrix} \qquad (15.7)$$

where $\mathbf{A}_2$ is the submatrix of $\mathbf{D}_m$ obtained by striking out the first row and column, and can be written

$$\mathbf{A}_2 = \begin{bmatrix} d_{22} & d_{23} & \cdots & d_{zn} \\ d_{32} & d_{33} & \cdots & \\ \vdots & \vdots & & \\ & & \cdots & d_{nn} \end{bmatrix} \tag{15.8}$$

where $n = 3Np$. The submatrix $\mathbf{A}_1$ is equal to d_{11} and is of order unity (i.e., scalar). $\mathbf{B}_2$ is given by

$$\mathbf{B}_2 = (d_{12}, d_{13}, \ldots, d_{1n})$$

For convenience we will now write

$$\mathbf{X}_1 = \mathbf{A}_1 - \omega_i \mathbf{E}_1, \qquad \mathbf{Z}_1 = \mathbf{A}_2 - \omega_i \mathbf{E}_2, \qquad \text{and} \qquad \mathbf{Y}_1 = \mathbf{B}_2$$

so that the matrix $\mathbf{D}_m - \omega_i \mathbf{E}$ can be partitioned into

$$\mathbf{D}_m - \omega_i \mathbf{E} = \begin{bmatrix} \mathbf{X}_1 & \mathbf{Y}_1 \\ \mathbf{Y}_1^T & \mathbf{Z}_1 \end{bmatrix} \tag{15.9}$$

Applying the NET, we obtain the result

$$\eta(\mathbf{D}_m - \omega_i \mathbf{E}) = \eta(\mathbf{X}_1) + \eta(\mathbf{Z}_1 - \mathbf{Y}_1^T \mathbf{X}_1^{-1} \mathbf{Y}_1) \tag{15.10}$$

The matrix $\mathbf{Z}_1 - \mathbf{Y}_1^T \mathbf{X}_1^{-1} \mathbf{Y}_1$, of order $n-1$, can be partitioned in a corresponding fashion, so that

$$\mathbf{Z}_1 - \mathbf{Y}_1^T \mathbf{X}_1^{-1} \mathbf{Y}_1 = \begin{bmatrix} \mathbf{X}_2 & \mathbf{Y}_2 \\ \mathbf{Y}_2 & \mathbf{Z}_2 \end{bmatrix}$$

and

$$\eta(\mathbf{Z}_1 - \mathbf{Y}_1^T \mathbf{X}_1^{-1} \mathbf{Y}_1) = \eta(\mathbf{X}_2) + \eta(\mathbf{Z}_2 - \mathbf{Y}_2^T \mathbf{X}_1^{-1} \mathbf{Y}_2) \tag{15.11}$$

where $\mathbf{X}_2$ is scalar and $\mathbf{Z}_2$ is a matrix of order $n-2$. The rationale behind the notation changes introduced here should now be clear. The procedure outlined in equations (15.7)–(15.11) can be repeated $n-1$ times and the number of negative eigenvalues determined from

$$\eta(\mathbf{D}_m - \omega_i \mathbf{E}) = \sum_{i=1}^{n} \mathbf{X}_i \tag{15.12}$$

where the $\mathbf{X}_i$ are simple scalar quantities and $\eta(\mathbf{X}_i)$ is determined simply from the sign of $\mathbf{X}_i$. Zerbi has noted that the time required for computation is reduced by the fact that the row matrix $\mathbf{Y}$ has only $C-1$ nonzero elements, where C is the number of codiagonals of $\mathbf{D}_m$. In computing the density of states of various polymer chains, intervals of $\omega_2-\omega_1=5\ \text{cm}^{-1}$ have been used (8). Eigenvectors can also be determined by using the "inverse iteration method," but to perform such calculations $\omega_2-\omega_1$ has to be restricted to the point at which it contains only one eigenvalue.

We will consider an example of the application of the NET shortly, but first we will briefly discuss the problem of generating a suitable large dynamical matrix. This has been effectively treated by Zerbi and co-workers (6–12). Although a symmetric matrix is required to apply the NET, it is convenient and useful to maintain force constants defined in terms of internal coordinates. Consequently, it is necessary to define the **B** matrix for large finite chains that include defects. It will be recalled (Chapter 3) that the linear transformation between the internal displacement coordinates **R** and the cartesian displacement coordinates **X** is given by

$$\mathbf{R}=\mathbf{B}\mathbf{X} \tag{15.13}$$

The potential energy with force constants $\mathbf{F}_R$ defined in terms of internal coordinates is

$$\begin{aligned} 2V&=\mathbf{R}^T\mathbf{F}_R\mathbf{R} \\ &=\mathbf{X}^T\mathbf{B}^T\mathbf{F}_R\mathbf{B}\mathbf{X} \end{aligned} \tag{15.14}$$

The symmetric dynamical matrix can then be derived in the form

$$\mathbf{D}_m=\mathbf{M}^{-1/2}\mathbf{B}^T\mathbf{F}_R\mathbf{B}\mathbf{M}^{1/2} \tag{15.15}$$

where **M** is a diagonal matrix whose elements are the atomic masses. The major part of the problem is setting up the **B** matrix. When internal coordinates are defined for a particular chemical unit it is often necessary to include atoms from adjacent chemical units (see Chapter 3). However, even when defining torsions, each internal coordinate involves at the most atoms from four adjacent monomer units. Consequently the transformation matrix from cartesian to internal coordinates can be written

$$\mathbf{R}_i=\left[\mathbf{B}_i^{-1}\ \vdots\ \mathbf{B}_i^{0}\ \vdots\ \mathbf{B}_i^{1}\ \vdots\ \mathbf{B}_i^{2}\ \vdots\right]\mathbf{X}_i \tag{15.16}$$

where $\mathbf{B}_i^{-1}$ gives the contribution to $\mathbf{R}_i$ of the atoms belonging to the $(i-1)$th chemical unit, $\mathbf{B}_i^0$ that of unit i, and so on; $\mathbf{X}_i$ is the vector of the cartesian components of the displacement of the atoms belonging to the $i-1, i, i+1, i+2$ chemical units in a right-handed reference system fixed at a given skeletal atom n.

In this reference system, the x axis is oriented from the nth to the $(n+1)$th skeletal atoms and the y axis is taken to form an angle of less than $\pi/2$ with the bond joining the $(n-1)$th and nth skeletal atoms. It is assumed that N is the number of atoms contained in a chemical repeat unit and the reference system is rigidly connected with the first skeletal atom of the chain. The formulas that generate all the cartesian coordinates of the atoms in the polymer are then the following:

$$\mathbf{X}_i^1 = \mathbf{X}_{i-1}^1 + d\mathbf{t}_{i-1}$$

$$\mathbf{X}_i^\alpha = \prod_{n=1,i-1} \mathbf{\Gamma}_n \mathbf{X}_1^\alpha + \mathbf{X}_i^1, \qquad (\alpha = 2, N) \qquad (15.17)$$

$$\mathbf{t}_i = \prod_{n=1,i-1} \mathbf{\Gamma}_n \mathbf{t}_1$$

with

$$\mathbf{X}_1^1 = \begin{bmatrix} 0 \\ 0 \\ 0 \end{bmatrix}, \qquad \mathbf{X}_1^\alpha = \begin{bmatrix} x \\ y \\ z \end{bmatrix}, \qquad \text{and} \qquad \mathbf{t}_1 = \begin{bmatrix} 1 \\ 0 \\ 0 \end{bmatrix}$$

where $\mathbf{X}_i^1$ are the cartesian components of the skeletal atom of the ith chemical unit, $\mathbf{X}_i$ those of the remaining $N-1$ atoms of the ith chemical unit, $\mathbf{t}_i$ a unit vector fixed at the ith skeletal atom and oriented toward the $(i+1)$th atom, d the bond length, and $\mathbf{\Gamma}_n$ the rotation matrix that transforms the reference system fixed at the $(i+1)$th skeletal atom into that fixed at the ith atom. The expression for the rotation matrix $\mathbf{\Gamma}_i$ is

$$\mathbf{\Gamma}_i = \begin{bmatrix} -\cos\vartheta & -\sin\vartheta & 0 \\ \sin\vartheta\cos\tau_i & -\cos\vartheta\cos\tau_i & -\sin\tau_i \\ \sin\vartheta\sin\tau_i & -\cos\vartheta\sin\tau_i & \cos\tau_i \end{bmatrix} \qquad (15.18)$$

where ϑ is the valence angle between adjacent skeletal bonds and τ_i is the dihedral angle between the two planes defined by skeletal atoms $i-1, i+1$, and by $i, i+1, i+2$. With the use of equations (15.17) and (15.18) it is easy to generate by digital computer the elements of the transformation matrix for each chemical unit and obtain the complete **B** matrix for the chain. This matrix takes into account all the internal rotations required to describe the complete geometry for the chain considered. The complete **B** matrix has the form

(15.19)

B	$i-2$	$i-1$	i	$i+1$	$i+2$	$i+3$	
...	...	...	...				
R_{i-1}	$\mathbf{B}_{i-1}^{-1}$	$\mathbf{B}_{i-1}^{0}$	$\mathbf{B}_{i-1}^{1}$	$\mathbf{B}_{i-1}^{2}$			
R_i		$\mathbf{B}_{i}^{-1}$	$\mathbf{B}_{i}^{0}$	$\mathbf{B}_{i}^{1}$	$\mathbf{B}_{i}^{2}$		
R_{i+1}			$\mathbf{B}_{i+1}^{-1}$	$\mathbf{B}_{i+1}^{0}$	$\mathbf{B}_{i+1}^{1}$	$\mathbf{B}_{i+1}^{2}$	
...				...	...	...	...

and the dynamical matrix (when end effects are neglected) takes the form

$$\begin{bmatrix} \cdots & \cdots & \cdots & \cdots & & & \\ \mathbf{D}^0_{i-1} & \mathbf{D}^1_{i-1} & \mathbf{D}^2_{i-1} & \mathbf{D}^3_{i-1} & \mathbf{D}^4_{i-1} & & \\ & \mathbf{D}^0_i & \mathbf{D}^1_i & \mathbf{D}^2_i & \mathbf{D}^3_i & \mathbf{D}^4_i & \\ & & \mathbf{D}^0_{i+1} & \mathbf{D}^1_{i+1} & \mathbf{D}^2_{i+1} & \mathbf{D}^3_{i+1} & \mathbf{D}^4_{i+1} \\ & & & \cdots & \cdots & \cdots & \cdots \end{bmatrix} \tag{15.20}$$

where

$$\mathbf{D}^0_i = \mathbf{M}_i^{-1/2}\big(\tilde{\mathbf{B}}^2_{i-2}\mathbf{F}_0\mathbf{B}^2_{i-2} + \tilde{\mathbf{B}}^1_{i-1}\mathbf{F}_0\mathbf{B}^1_{i-1} + \tilde{\mathbf{B}}^0_i\mathbf{F}_0\mathbf{B}^0_i + \tilde{\mathbf{B}}^{-1}_{i+1}\mathbf{F}_0\mathbf{B}^{-1}_{i+1}$$

$$+ \tilde{\mathbf{B}}^2_{i-2}\mathbf{F}_{i-2}\mathbf{B}^1_{i-1} + \tilde{\mathbf{B}}^1_{i-1}\tilde{\mathbf{F}}_{i-2}\mathbf{B}^2_{i-2} + \tilde{\mathbf{B}}^1_{i-1}\mathbf{F}_{i-1}\mathbf{B}^0_i + \tilde{\mathbf{B}}^0_i\tilde{\mathbf{F}}_{i-1}\mathbf{B}^1_{i-1}$$

$$+ \tilde{\mathbf{B}}^{-1}_{i+1}\tilde{\mathbf{F}}_i\mathbf{B}^0_i + \tilde{\mathbf{B}}^0_i\mathbf{F}_i\mathbf{B}^{-1}_{i+1}\big)\mathbf{M}_i^{-1/2}$$

$$\mathbf{D}^1_i = \mathbf{M}_i^{-1/2}\big(\tilde{\mathbf{B}}^{-1}_{i+1}\mathbf{F}_0\mathbf{B}^0_{i+1} + \tilde{\mathbf{B}}^0_i\mathbf{F}_0\mathbf{B}^1_i + \tilde{\mathbf{B}}^1_{i-1}\mathbf{F}_0\mathbf{B}^2_{i-1} + \tilde{\mathbf{B}}^2_{i-2}\mathbf{F}_{i-2}\mathbf{B}^2_{i-1}$$

$$+ \tilde{\mathbf{B}}^1_{i-1}\mathbf{F}_{i-1}\mathbf{B}^1_i + \tilde{\mathbf{B}}^0_i\tilde{\mathbf{F}}_{i-1}\mathbf{B}^2_{i-1} + \tilde{\mathbf{B}}^0_i\mathbf{F}_i\mathbf{B}^0_{i+1} + \tilde{\mathbf{B}}^{-1}_{i+1}\tilde{\mathbf{F}}_i\mathbf{B}^1_i$$

$$+ \tilde{\mathbf{B}}^{-1}_{i+1}\mathbf{F}_{i+1}\mathbf{B}^{-1}_{i+2}\big)\mathbf{M}_i^{-1/2}$$

$$\mathbf{D}^2_i = \mathbf{M}_i^{-1/2}\big(\tilde{\mathbf{B}}^0_i\mathbf{F}_0\mathbf{B}^2_i + \tilde{\mathbf{B}}^{-1}_{i+1}\mathbf{F}_0\mathbf{B}^1_{i+1} + \tilde{\mathbf{B}}^1_{i-1}\mathbf{F}_{i-1}\mathbf{B}^2_i + \tilde{\mathbf{B}}^0_i\mathbf{F}_i\mathbf{B}^1_{i+1}$$

$$+ \tilde{\mathbf{B}}^{-1}_{i+1}\tilde{\mathbf{F}}_i\mathbf{B}^2_i + \tilde{\mathbf{B}}^{-1}_{i+1}\mathbf{F}_{i+1}\mathbf{B}^0_{i+2}\big)\mathbf{M}_i^{-1/2}$$

$$\mathbf{D}^3_i = \mathbf{M}_i^{-1/2}\big(\tilde{\mathbf{B}}^{-1}_{i+1}\mathbf{F}_0\mathbf{B}^2_{i+1} + \tilde{\mathbf{B}}^0_i\mathbf{F}_i\mathbf{B}^2_{i+1} + \tilde{\mathbf{B}}^{-1}_{i+1}\mathbf{F}_{i+1}\mathbf{B}^1_{i+2}\big)\mathbf{M}_i^{-1/2}$$

$$\mathbf{D}^4_i = \mathbf{M}_i^{-1/2}\tilde{\mathbf{B}}^{-1}_{i+1}\mathbf{F}_{i+1}\mathbf{B}^2_{i+2}\mathbf{M}_i^{-1/2}$$

In these equations $\mathbf{F}_0$ is the matrix of the quadratic force constants describing the interactions of the atoms within the same chemical unit and $\mathbf{F}_i$ accounts for the interactions between the ith and $(i+1)$th chemical units. The desired randomness of the model adopted is translated into the dynamical matrix through equation (15.18) where τ_i is the internal rotational angle, which may vary along the chain. The density of states of a chain generated by this procedure is then obtained by applying the negative eigenvalue theorem, as discussed earlier.

As an example of the application of this numerical method we will consider polyethylene (8, 9), since doing so will permit a comparison with the results of the preceding section, which were based on the assumption of localized modes. Zerbi et al. (8, 9) considered the application of the NET to three situations: (a) The polyethylene chain is largely in the all-trans preferred conformation and the concentration of conformational defects is small. In addition to the $\mathbf{k}=0$ modes characteristic of the infinite chain, weak bands due to $g(\omega)$ of the host

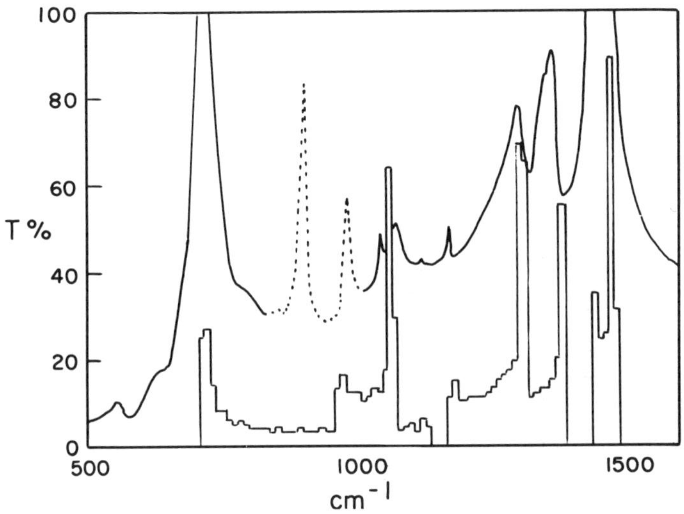

Figure 15.14. Comparison of the infrared spectrum (at room temperature) of crystalline polyethylene with the density of states calculated for a segment of 200 Ch_2 units in the trans configuration. (Reprinted by permission from ref. 8.)

lattice are expected. The density of states of 200 CH_2 units in the trans conformation has been determined as a model for this situation. (b) The concentration of defects is small but not negligible. In addition to the activation of $g(\omega)$ for the host trans lattice, local defect modes will appear. Models of the type —$(T)_m$—X—$(T)_n$— have been generated for this situation, where X is equal to defect conformations such as G, GG, GTG, and GTTG with a regular $(m=n)$ or irregular $(m \neq n)$ distribution of trans units. (c) The concentration of defects is large. The normal modes of the host lattice are significantly perturbed and $g(\omega)$ of an amorphous material has to be determined. This will include singularities due to specific defects; in addition, interactions between certain modes of such defects can occur. As a model of such a highly disordered system Zerbi et al. (8, 9) have generated a chain of 200 CH_2 units with a random sequence of the most common internal rotation angles, T, G, and G′. The random sequence of internal rotation angles was extracted as a "well-balanced" section of a sequence of 5000 units generated with an Ising model. Sequences corresponding to several different temperatures have been generated.

The force field used by Zerbi et al. in their application of the NET to these model systems is the one determined by Snyder (15). The density of states of the host lattice, the first situation listed in the preceding paragraph, is compared to the infrared spectrum of polyethylene at room temperature in Figure 15.14. In addition to the relatively strong spectroscopically active $\mathbf{k}=0$

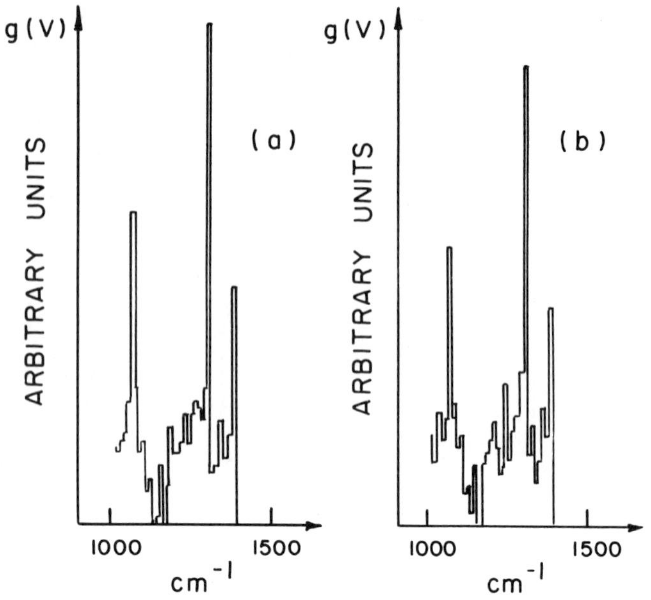

Figure 15.15. Density of states in the 1000–1400 cm^{-1} region (a) for a model -(T)$_n$—GTG—(T)$_m$—with random values of n and m; (b) for a model —(T)$_n$—GTTG —(T)$_m$—with random values of n and m. (Reprinted by permission from ref. 8.)

frequencies of the crystal, there are weaker infrared bands at 1075, 1128, 1300, 1375, and 1450 cm^{-1}, which correspond to peaks or singularities in the density of states. However, the observed 1375 cm^{-1} band is characteristic of localized vibrations of the methyl end group, so at least one or two of the assignments made by Zerbi et al. (8, 9) have to be considered with caution. Nonetheless, it was demonstrated that the asymmetry toward lower frequencies of the bands near 550 and 1300 cm^{-1}, together with the asymmetric broadening toward higher frequencies of the rocking mode at 720 cm^{-1}, coincides nicely with the calculated $g(\nu)$. The calculation of the $g(\nu)$ for model systems of the type ⟨T) —$_n$X ⟨T ⟩$_m$ results in assignments that are in more dispute. The density of states of two such systems is reproduced in Figure 15.15. Zerbi et al. (8, 9) concentrated their attention on the region 1300–1400 cm^{-1}, focusing particularly on the observed bands near 1368 and 1351 cm^{-1}. The application of the NET to the models listed earlier gave the following singularities in $g(\nu)$:

X=G	no peak
X=GG	1352 cm^{-1} (sharp)
X=GTG	1350 cm^{-1} (broad)
X=GTTG	1370 cm^{-1}

The assignment of the band near 1350 cm^{-1} to localized wagging modes in GG structures is in agreement with the analysis of Snyder (15) summarized in

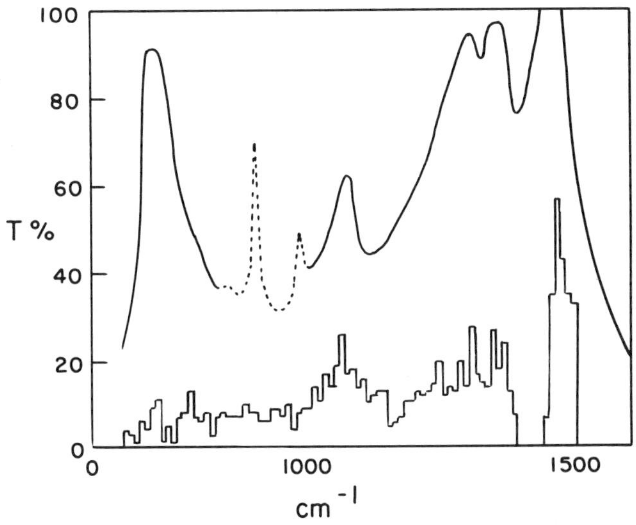

Figure 15.16. Comparison of the infrared spectrum of polyethylene at 160°C with the density of states calculated for a segment of 200 CH_2 units containing a random distribution of *T*, *G*, and *G′* conformations. (Reprinted by permission from ref. 8.)

the preceding section. However, the assignment of the band at 1368 cm^{-1} differs. Zerbi (8, 9) argues that the CH_3 end groups of short-chain paraffins give rise to intramolecular coupling throughout the chain different from that generated by conformational defects in a long chain.

Furthermore, bands in the spectrum of polyethylene at room temperature assigned by Snyder to localized modes (e.g., near 1300 cm^{-1}) are interpreted by Zerbi as being due to the activation of the density of states of the perfect lattice. Both factors could well contribute to the spectrum, but a clear distinction is not possible without a knowledge of intensity factors.

Finally, the $g(\nu)$ of a system with a large concentration of defects is compared to the infrared spectrum of polyethylene in the melt in Figure 15.16. Peaks due to specific localized modes can be identified. However, certain bands observed in the spectrum also correspond to singularities of the trans chain, leading Zerbi et al. to suggest the existence of trans segments in the melt. Zerbi et al. (18) have also treated the various chemical and conformational defects in poly(vinyl chloride). A discussion of these results is given in Chapter 16.

References

1 R. G. Brown, *J. Appl. Phys.* **34**, 2382 (1963).

2 M. M. Coleman, P. C. Painter, D. L. Tabb, and J. L. Koenig, *J. Polymer Sci. Polymer Lett. Ed.* **12**, 577 (1974).

3 D. L. Tabb, J. L. Koenig, and M. M. Coleman, *J. Polymer Sci. Polymer Phys. Ed.* **13**, 1145 (1975).

4 A. A. Maradudin, E. W. Montroll, and G. H. Weiss, "Theory of Lattice Dynamics in the Harmonic Approximation," Academic, New York, 1963.

5 C. Schmid and K. Holzl, *J. Phys.* **C7**, 2970 (1974).

6 G. Zerbi, 10th Microsymposium on Conformational Structure of Polymers, Prague, 1972, *Pure Appl. Chem.* **36**, 35 (1973).

7 G. Zerbi, Enrico Fermi Summer School: Lattice Dynamics and Intermolecular Forces, 1972, Varenna, *Nuovo Cimento*, in press.

8 G. Zerbi, L. Piseri, and F. Cabassi, *Mol. Phys.* **22**, 241 (1971).

9 G. Zerbi, *Pure Appl. Chem.* **236**, 499 (1971).

10 G. Zerbi, in "Phonons," M. A. Nusimovici, Ed., Flammarion, Paris, 1972.

11 G. Zerbi and M. Sacchi, *Macromolecules* **6**, 692 (1973).

12 G. Masetti, F. Cabassi, G. Morelli, and G. Zerbi, *Macromolecules* **6**, 700 (1973).

13 V. N. Kozyrenko, I. V. Kumparenko, and I. D. Mikhailov, *J. Polymer Sci. Polymer Phys. Ed.* **15**, 1721 (1977).

14 H. Matsuda, K. Okada, T. Takese, and T. Yamamoto, *J. Chem. Phys.* **41**, 1527 (1964).

15 R. G. Snyder, *J. Chem. Phys.* **47**, 1316 (1967).

16 R. G. Snyder and M. W. Poore, *Macromolecules* **6**, 708 (1973).

17 J. L. Koenig, *Appl. Spect. Rev.* **4**, 233 (1971).

18 A. Rubcic and G. Zerbi, *Macromolecules* **6**, 751 (1974); **6**, 759 (1974).

16

SELECTED EXAMPLES OF THE APPLICATION OF NORMAL COORDINATE ANALYSIS TO POLYMERS

Give all thou canst; high Heaven rejects the lore
of nicely-calculated less or more.
William Wordsworth

In this chapter we will review published results from selected normal coordinate calculations for a variety of polymeric materials. We have made no attempt to cover all of the large number of studies that have been reported. Rather, we have decided to illustrate, with selected examples (chosen primarily on the basis of transferability of force fields and inherent bias on the part of the authors!), the type of information that can be obtained from normal coordinate analysis. Accordingly, we will restrict our discussions to the polyolefins, haloethylene polymers, polydienes and polymers containing aromatic rings, and the amide group.

16.1 Polyolefins

Polyolefins are a general class of polymers containing only saturated carbon and hydrogen atoms in the chain. The simplest polyolefin is polyethylene, or structurally polymethylene, which contains the chemical repeat unit $-(CH_2)_n-$. The vibrational analysis of this polymer was discussed in detail in Chapter 13 and will not be considered further. Substitution of alkyl groups on the α-carbon of ethylene ($CH_2{=}CRR'$ where R and R′ can be a hydrogen atom or methyl, ethyl, propyl, etc. groups) leads, upon polymerization, to a wide variety of polyolefins. Of particular significance are the monosubstituted polymers

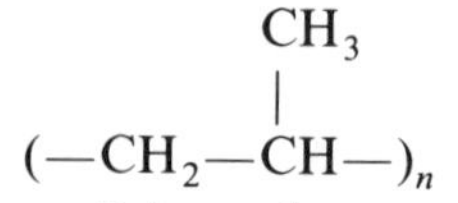

$$CH_3$$
$$|$$
$$(-CH_2-CH-)_n$$

Polypropylene

$$C_2H_5$$
$$|$$
$$(-CH_2-CH-)_n$$

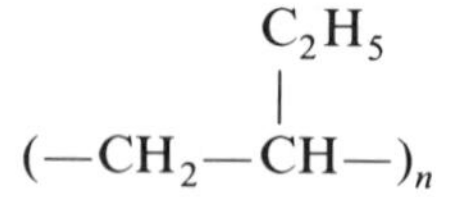

Poly(butene-1)

alternatively called poly(ethyl ethylene)

$$C_3H_7$$
$$|$$
$$(-CH_2-CH-)_n$$

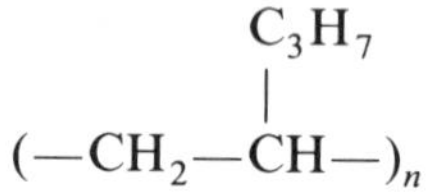

Poly(propyl ethylene)

$$R$$
$$|$$
$$(-CH_2-\ CH-)_n$$

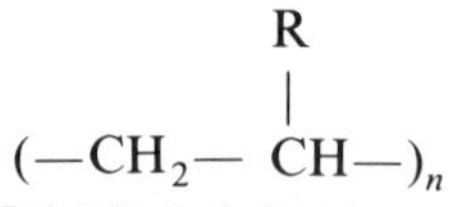

Poly(alkyl ethylenes)

which contain longer side chains

In all these cases, owing to the presence of an asymmetric α-carbon atom, the polymers synthesized may contain a complex sequence distribution of stereo-isomeric placements. In general, polymerization via Zeigler-Natta catalysis yields predominantly isotactic polymers. However, syndiotactic polypropylene has been synthesized from specific catalysts (1). Disubstituted ethylene poly-mers, for example

$$CH_3$$
$$|$$
$$+CH_2-C+_n$$
$$|$$
$$CH_3$$

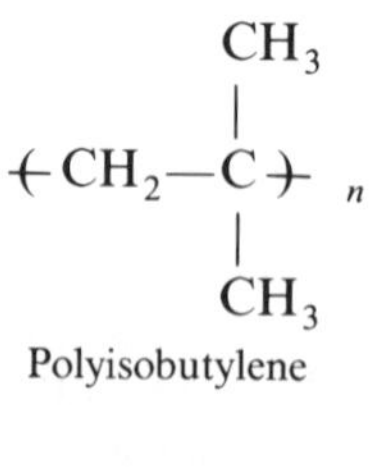

Polyisobutylene

have been synthesized by cationic polymerization.

In the following section, we will consider the normal coordinate calcula-tions that have been reported for polypropylene, polyisobutylene, and poly(al-kyl ethylenes) containing longer alkyl side groups.

Polypropylene

Normal coordinate calculations of polypropylene have been performed by several groups using a variety of force fields (2–7). In the following discussion, we will restrict ourselves primarily to the studies of Snyder and Schachtschneider (2, 3), since these authors employed the valence force field described in Chapter 13. Calculations by these authors have been reported for both isotactic (IPP) and syndiotactic (SPP) polypropylenes.

The preferred chain conformation of crystalline IPP is in the form of a 3_1 helical structure (8), illustrated in Figure 16.1. Three chemical repeat units are involved in the translational repeat unit and the 77 optically active normal modes are classified under the line group isomorphous to the point group C_3 and are both infrared and Raman active. The 25 A modes exhibit parallel dichroism in the infrared, while the 26 doubly degenerate E modes exhibit perpendicular dichroism. The A modes are polarized in the Raman whereas the E modes are depolarized. SPP exists in two polymorphic crystalline forms, denoted forms I and II, as shown in Figure 16.2. The preferred chain conformation of the crystalline form I is a twofold helix with four chemical repeat units per translational repeat unit (9). This polymer chain may be described as TGGTTG′G′T (where T, G, and G′ represent dihedral angles of 180°, 60°, and $-60°$, respectively) and the line group is isomorphous to the

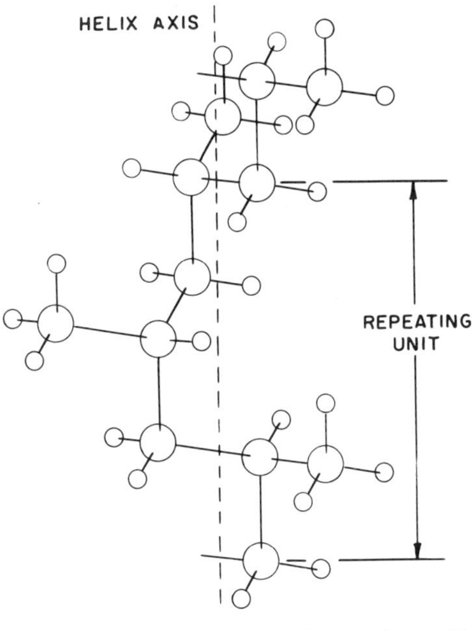

Figure 16.1. Structure of crystalline isotactic polypropylene. (Reprinted by permission from ref. 2.)

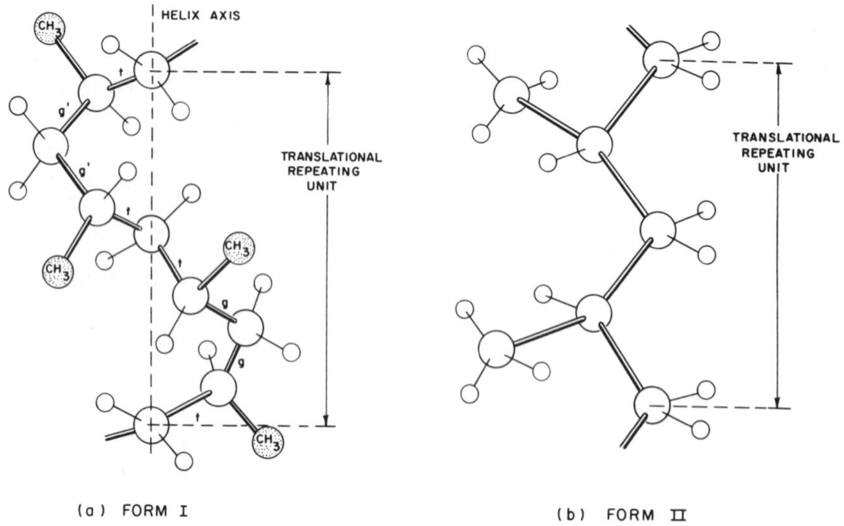

Figure 16.2. Structure of crystalline syndiotactic polypropylene. (Reprinted by permission from ref. 3.)

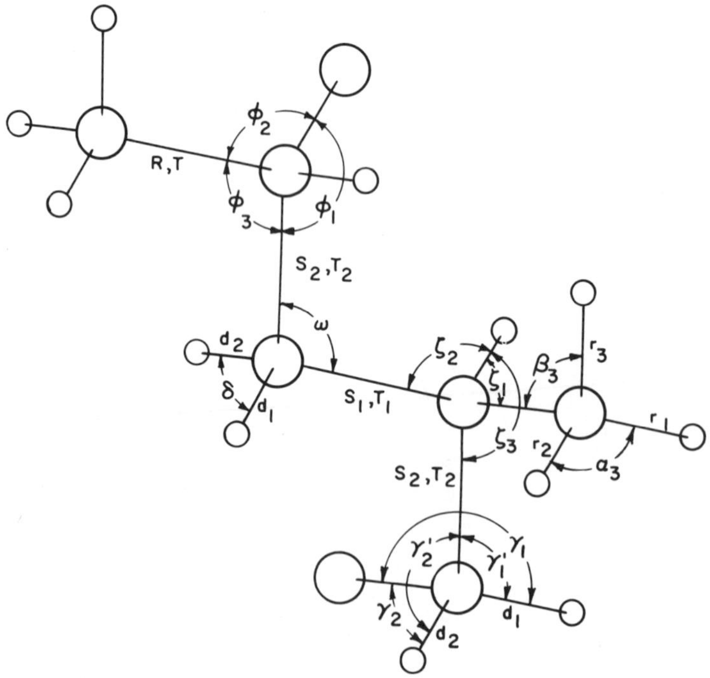

Figure 16.3. Internal coordinates for polypropylene. (Reprinted by permission from ref. 3.)

point group D_2. Twenty-six optically active modes are predicted for each of the A, B_1, B_2, and B_3 species. The A modes are only Raman active (polarized), while the B modes are both infrared and Raman (depolarized) active. B_1 modes exhibit perpendicular infrared dichroism and the B_2 and B_3 display parallel dichroic behavior. The preferred conformation of the crystalline form II of SPP is essentially planar zigzag (10). Two chemical repeat units are contained in the translational repeat unit and the line group of the isolated chain is isomorphous to the point group C_{2v}. There are 50 optically active fundamental vibrations, which are distributed among the symmetry species in the following manner: 14 A_1 (Raman active, infrared active, perpendicular), 11 A_2 (Raman active), 11 B_1, and 14 B_2 (Raman active, infrared active, parallel).

Table 16.1 Group Coordinates (—CH(CH_3)—CH_2—)

	Coordinate	Symbol	Description
1	$2r_1-r_2-r_3$	r_a^-	CH_3 asymm. C—H stretching
2	r_2-r_3	r_b^-	CH_3 asymm. C—H stretching
3	$r_1+r_2+r_3$	r^+	CH_3 sym. C—H stretching
4	d_1-d_2	d^-	CH_2 asymm. C—H stretching
5	d_1+d_2	d^+	CH_2 sym. C—H stretching
6	s	s	CH C—H stretching
7	S_1+S_2	S^+	C—C skeletal stretching
8	S_1-S_2	S^-	C—C skeletal stretching
9	R	R	C—CH_3 stretching
10	$\alpha_1+\alpha_2+\alpha_3-\beta_1-\beta_2-\beta_3$	U^-	CH_3 sym. bending
11	$2\alpha_1-\alpha_2-\alpha_3$	α_a	CH_3 asymm. bending
12	$\alpha_2-\alpha_3$	α_b	CH_3 asymm. bending
13	Υ	Υ	CH_3 torsional
14	$2\beta_1-\beta_2-\beta_3$	β_a	CH_3 rocking
15	$\beta_2-\beta_3$	β_b	CH_3 rocking
16	δ	δ	CH_2 bending
17	$\gamma_1+\gamma_2+\gamma_1'+\gamma_2'$	Γ	CH_2 bending
18	$\gamma_1+\gamma_2-\gamma_1'-\gamma_2'$	W	CH_2 wagging
19	$\gamma_1-\gamma_2-\gamma_1'+\gamma_2'$	T	CH_2 twisting
20	$\gamma_1-\gamma_2+\gamma_1'-\gamma_2'$	P	CH_2 rocking
21	$2\zeta_1-\zeta_2-\zeta_3$	ζ_a	CH bending
22	$\zeta_2-\zeta_3$	ζ_b	CH bending
23	$\zeta_1+\zeta_2+\zeta_3$	ζ^+	∠C—C—C bending
24	ω	ω	∠C—C—C bending
25	$2\varphi_1-\varphi_2-\varphi_3$	φ_a	∠C—C—C bending
26	$\varphi_2-\varphi_3$	φ_b	∠C—C—C bending
27	$\varphi_1+\varphi_2+\varphi_3$	φ^+	∠C—C—C bending
28	$\tau_1+\tau_2$	τ^+	CH_2—CH torsional
29	$\tau_1-\tau_2$	τ^-	CH_2—CH torsional
30	$\alpha_1+\alpha_2+\alpha_3+\beta_1+\beta_2+\beta_3$	U^+	Redundant

Table 16.2 Frequencies of Isotactic Polypropylene

A Block			*E* Block		
ν_{calc} (cm^{-1})	ν_{obs}[a, b] (cm^{-1})	Approx. PED[c] (%)	ν_{calc} (cm^{-1})	ν_{obs}[a, d] (cm^{-1})	Approx. PED (%)
Polypropylene					
2962	2956 (s)	$r^-(100)$	2962	2956	$r^-(100)$
2962		$r^-(100)$	2962	2951 (s)	$r^-(100)$
2928		$d^-(99)$	2930	2925 (sh)	$d^-(94)$
2906		$s(98)$	2904	2907 (sh)	$s(94)$
2882		$r^+(98)$	2882	2880 (s)	$r^+(98)$
2855	2843 (s)	$d^+(99)$	2856	2868 (s)	$d^+(99)$
1465		$\alpha(78)$	1463	1459 (sh)	$\alpha(88)$
1462		$\alpha(87)$	1462		$\alpha(78)$
1453	1454 (s)	$\delta(69)$, $\alpha(11)$	1454		$\delta(69)$
1380	1378 (m)[e]	$W(27)$, $S(26)$, $\zeta(25)$, $U(20)$	1377	1377 (s)	$U(38)$, $W(21)$, $\zeta(21)$, $S(16)$
1369	1365 (vw)[e]	$U(89)$	1367	1359 (m)	$U(69)$, $\zeta(19)$
1348		$\zeta(65)$, $S(12)$	1352	1329 (w)	$\zeta(52)$, $T(15)$
1284	1305 (w)	$W(37)$, $T(36)$	1292	1297 (w)	$W(35)$, $\zeta(25)$, $T(13)$
1242	1257 (w)	$\zeta(37)$, $T(19)$, $\beta(15)$	1201	1219 (w)	$T(26)$, $\zeta(20)$, $R(19)$
1170	1167 (s)	$S(30)$, $\beta(22)$	1155	1153 (sh)	$S(23)$, $R(15)$, $\zeta(13)$, $\beta(13)$

1066	1044 (w)	$R(51)$, $S(31)$	1121	1103 (w)	$R(32)$, $\beta(24)$, $\zeta(16)$
1004	997 (s)	$\beta(46)$, $\zeta(19)$	1024		$S(30)$, $R(18)$, $\zeta(15)$, $T(14)$
969	973 (s)	$\beta(51)$, $S(28)$	940	941 (w)	$\beta(45)$, $R(40)$
853 }	841 (s)	$P(43)$, $R(23)$	901	900 (w)	$\beta(49)$, $P(17)$, $\zeta(17)$
829 }		$P(29)$, $S(32)$, $\beta(20)$	817	809 (m)	$R(32)$, $P(30)$, $S(23)$
456	456[e]	$\omega(81)$	513	528[e]	$\omega(63)$, $R(10)$, $P(9)$
392	398[e]	$\omega(62)$, $\zeta(15)$	429	408[f]	$\omega(58)$, $\zeta(29)$, $P(17)$
267		$\omega(58)$, $\tau(15)$	311	321[e]	$\omega(74)$
195		$\tau(88)$	198		$\tau(94)$
140		$\omega(50)$, $\tau(30)$	147		$\omega(77)$, $\tau(16)$
			63		$\tau(86)$

[a]Abbreviations: s, m, w, v, and sh mean strong, medium, weak, very, and shoulder, respectively. Unless otherwise indicated, frequencies are from ref. 12.
[b]All bands that have been measured have a polarization direction parallel to the helix axis.
[c]Potential energy distributions are given in percentage of contribution to λ by the diagonal force constants for the following group coordinates: r, CH_3 stretching; d, CH_2 stretching; s, CH stretching; α, antisymmetric CH_3 bending; δ, CH_2 bending; U, symmetric CH_3 bending; W, CH_2 wagging; T, CH_2 twisting; P, CH_2 rocking; ζ, CH bending; β, CH_3 rocking; S, C—C chain stretching; R, C—CH_3 stretching; ω, C—C—C bending; τ, C—C torsion. Only contributions $>8\%$ are included in the table.
[d]All bands that have been measured have a polarization direction perpendicular to the helix axis.
[e]From ref. 13.
[f]Frequencies from ref. 14; polarization not measured.

Table 16.3 Observed and Calculated Frequencies for the D_2 Form of Syndiotactic Polypropylene

	ν_{calc} (cm^{-1})	$\nu_{obs}{}^{a}$ (cm^{-1})	Polarization[a]	Potential Energy Distribution
A block	2962			$r_a^-(91), r_b^-(8)$
(IR inactive)	2962			$r_b^-(91), r_a^-(8)$
	2906			$s(98)$
	2882			$r^+(99)$
	2856			$d_1^+(50), d_2^+(50)$
	2855			$d_1^+(50), d_2^+(50)$
	1463			$\alpha_b(77), \alpha_a(7), \beta_b(8)$
	1462			$\alpha_a(71), \alpha_b(9), \beta_a(7)$
	1454			$\delta(74), \Gamma(17)$
	1452			$\delta(70), \Gamma(15), \alpha_a(8)$
	1372			$U(99)$
	1353			$\zeta_b(45), T(30), S(10)$
	1339			$\zeta_a(48), T(25), R(14)$
	1259			$T(50), \zeta_a(11), \beta_a(11), S(10)$
	1191			$T(30), \zeta_a(14), S(13), \zeta_b(7), \beta_b(7)$
	1168			$S(36), \beta_b(18), \zeta_b(14), \varphi(12)$
	1041			$T(36), R(31), \zeta_b(22), S(15)$
	996			$\beta_a(38), T(19), \zeta_a(18), \beta_b(11), R(11)$
	918			$S(48), \beta_b(43)$
	839			$S(38), \beta_a(24), R(17), \omega(14)$
	542			$\omega(32), \varphi_a(26), R(8), S(8)$
	347			$\varphi_b(33), \varphi^+(27), \zeta^+(18)$
	315			$\varphi_b(22), \varphi^+(10), S(29), \omega(10)$
	199			$\Upsilon(84), \varphi_b(8)$
	176			$\omega(44), \varphi_b(27), \Upsilon(14)$
	66			$\varphi_a(44), \tau(29)$
B_1 block	2962	2960–2950	(g)	$r_a^-(54), r_b^-(45)$
	2962			$r_b^-(54), r_a^-(45)$
	2929	2927	(g)	$d^-(96)$
	2904	2915	(g)	$s(95)$
	2882	2882, 2866	(g)	$r^+(99)$
	2855	2843	(b)	$d^+(99)$
	1465	1465	(d)	$\alpha_b(50), \alpha_a(26), \delta(8)$
	1462	1460	(d)	$\alpha_a(50), \alpha_b(36)$
	1453	1432	(d)	$\delta(65), \alpha_a(14)$
	1380	1379	(b)	$U(65), W(18), S(8)$
	1357			$U(36), W(24), S(13), R(13)$
	1352	1332	(b)	$\zeta_b(56), S(15), T(14)$
	1275	1287	(b)	$W(40), \zeta_a(37), \beta_b(8)$
	1223	1242	(b)	$T(45), S(11), \beta_a(10)$
	1167	1152	(b)	$S(28), R(12), P(11), \varphi_a(11), \zeta_a(10)$
	1113	1088, 1083	(b)	$S(37), \zeta_b(24), \beta_b(17), R(7)$
	980	1006	(b)	$R(25), \beta_a(20), \beta_b(16), T(11), S(9), \zeta_a(8)$
	971	976	(f), (d?)	$\beta_b(34), S(25), R(19), \beta_a(10)$
	903	906	(b)	$\beta_a(36), R(23), P(22), \zeta_a(10)$
	786	776	(b)	$P(41), S(38), R(13), \omega(8), \zeta_b(7)$

Table 16.3 Continued

	ν_{calc} (cm^{-1})	$\nu_{obs}{}^{a}$ (cm^{-1})	Polarization[a]	Potential Energy Distribution
B_1	537	535	(b)	$\varphi_b(45)$, $\omega(20)$, $\beta_b(10)$
	403			$\varphi^+(40)$, $\zeta^+(27)$, $\varphi_a(13)$
	264			$\varphi_a(55)$, $\tau(12)$, $\Upsilon(10)$, $S(8)$
	196			$\Upsilon(74)$, $\varphi_b(13)$
	170			$\omega(40)$, $\varphi_b(29)$, $\Upsilon(15)$
	62			$\tau(85)$
B_2 block	2962 }	2960–2950	(g)	$r_b^-(59)$, $r_a^-(39)$
	2962 }			$r_a^-(59)$, $r_b^-(40)$
	2929	2923	(g), (c)	$d^-(95)$
	2904	2915	(g)	$s(95)$
	2882	2882, 2867	(g)	$r^+(99)$
	2855	2837	(c)	$d^+(99)$
	1464	1463	(d)	$\alpha_b(78)$, $\beta_b(8)$
	1461	1455	(g)	$\alpha_a(86)$, $\beta_a(8)$
	1454			$\delta(73)$, $\Gamma(20)$
	1374	1378?	(g)	$\zeta_b(34)$, $S(29)$, $W(26)$, $T(12)$
	1372	1373	(c)	$U(100)$
	1331	1360	(c)	$\zeta_a(57)$, $W(11)$, $R(12)$
	1301	1311	(c)	$W(40)$, $T(29)$
	1194	1202	(c)	$\zeta_b(39)$, $S(14)$, $T(11)$, $P(10)$
	1159	1153	(c)	$S(22)$, $\beta_b(17)$, $\beta_a(10)$, $W(10)$, $R(10)$
	1115			$R(40)$, $T(13)$, $S(10)$, $\beta_b(10)$, $\zeta_b(10)$
	1044	1035	(c)	$S(30)$, $T(22)$, $\beta_a(16)$, $\zeta_a(12)$, $R(8)$
	931	935	(c)	$\beta_b(45)$, $S(38)$
	878	~870	(d)	$\beta_a(43)$, $S(30)$, $P(21)$, $\zeta_a(15)$
	825	812	(c)	$P(36)$, $S(23)$, $R(18)$, $\zeta_b(7)$
	476	483	(c)	$\varphi_a(45)$, $\omega(20)$, $R(11)$, $S(10)$
	433			$\varphi^+(37)$, $\zeta^+(25)$, $\varphi_b(14)$, $P(12)$
	280			$\varphi_b(67)$, $S(8)$
	202			$\Upsilon(84)$
	171			$\omega(33)$, $\varphi_a(28)$, $\tau(19)$, $\Upsilon(14)$
	58			$\tau(78)$, $\varphi_a(10)$, $\omega(8)$
B_3 block	2962 }	{ 2959	(h), (a)	$r_b^-(60)$, $r_a^-(38)$
	2962 }	{ 2948	(h)	$r_a^-(60)$, $r_b^-(39)$
	2930 }	~2925	(h)	$d_2^+(52)$, $d_1^+(41)$
	2928 }			$d_1^+(55)$, $d_2^+(44)$
	2903	2912	(h), (a)	$s(93)$
	2882	2882, 2867	(h)	$r^+(99)$
	1464 }	{ 1460	(h)	$\alpha_b(87)$, $\beta_b(9)$
	1462 }	{ 1455	(h)	$\alpha_a(89)$, $\beta_a(8)$
	1378 }	{ 1377	(h)	$U(70)$, $W(15)$, $S(10)$
	1373 }	{ 1374	(h), (a)	$W(35)$, $S(34)$, $\zeta_b(30)$
	1352	1346	(a)	$W(34)$, $\zeta_a(29)$, $U(22)$, $R(13)$

Table 16.3 Continued

	ν_{calc} (cm^{-1})	$\nu_{obs}{}^{a}$ (cm^{-1})	Polarization[a]	Potential Energy Distribution
B_3	1300	1293	(a)	$W(44)$, $\zeta_b(27)$, $\zeta_a(7)$
	1265	1264	(a)	$\zeta_a(39)$, $W(33)$, $\zeta_b(10)$
	1174	1167	(a)	$S(25)$, $\beta_b(15)$, $P(14)$, $\zeta_b(12)$
	1096			$S(38)$, $W(14)$, $\beta_a(14)$, $P(14)$, $\beta_b(9)$
	1068	1060	(a)	$R(57)$, $S(17)$, $W(11)$, $P(8)$
	974	977	(a)	$\beta_b(53)$, $S(29)$
	902	901	(a)	$\beta_a(53)$, $P(21)$, $\zeta_a(16)$, $S(9)$
	867	867	(a)	$P(52)$, $S(20)$, $\zeta_b(10)$
	829			$P(43)$, $R(35)$, $S(11)$
	436	468	(a)	$\varphi^+(46)$, $\zeta^+(32)$, $P(17)$
	410			$\varphi_b(86)$, $\varphi_a(9)$, $\beta_b(9)$
	358			$\varphi_a(73)$, $\varphi_b(7)$, $P(8)$
	200			$\Upsilon(98)$
	74			$\tau(88)$, $P(7)$
	60			$\tau(86)$, $P(7)$

[a] From ref. 15.
(a) Pol. ∥ to chain axis.
(b) Pol. ↕◇◇.
(c) Pol. ↔◇◇.
(d) Pol. ✥; i.e., ⊥.
(e) Probably pol. ⊥.
(f) May be an amorphous band.
(g) Frequency measured when radiation polarized ⊥ to the chain axis.
(h) Frequency measured when radiation polarized ∥ to the chain axis.

Table 16.4 Observed and Calculated Frequencies for the C_{2v} Form of Syndiotactic Polypropylene

	ν_{calc} (cm^{-1})	$\nu_{obs}{}^{a}$ (cm^{-1})	Polarization[a]	Potential Energy Distribution
A_1 block	2962	2959	(d)	$r_a(99)$
	2906	2916, 2905	(b)	$s(98)$
	2882	2880, 2868	(b)	$r^+(99)$
	2856			$d^+(99)$
	1464	1466	(b), (d)	$\alpha_a(66)$, $\delta(18)$, $\beta_a(7)$
	1453	1450	(d)	$\delta(59)$, $\alpha_a(24)$, $\Gamma(17)$
	1373	1381	(b)	$U^-(87)$, $T(9)$
	1364	1375	(c)	$\zeta_a(37)$, $T(32)$, $R(17)$, $U^-(15)$
	1242	1200	(b)	$T(34)$, $\zeta_a(27)$, $\beta_a(17)$
	1152	1154	(c)	$S^+(39)$, $R(22)$, $\omega(16)$, $\varphi_a(16)$, $\zeta_a(13)$
	971	972	(b)	$\beta_a(47)$, $R(25)$, $T(20)$, $\zeta_a(21)$
	863	867	(b)	$R(45)$, $\beta_a(20)$, $S^+(17)$, $\omega(7)$
	404			$\varphi^+(38)$, $\zeta^+(26)$, $\varphi_a(25)$
	66			$\tau(97)$

Table 16.4 Continued

	ν_{calc} (cm^{-1})	$\nu_{obs}{}^{a}$ (cm^{-1})	Polarization[a]	Potential Energy Distribution
A_2 block	2962			$r_b^-(99)$
(IR inactive)	2855			$d^+(99)$
	1463			$\alpha_b(88)$, $\beta_b(9)$
	1452			$\delta(78)$, $\Gamma(21)$
	1345			$\zeta_b(59)$, $T(16)$, $S^-(8)$
	1185			$T(50)$, $S^-(22)$, $\beta_b(19)$
	1082			$T(35)$, $\zeta_b(33)$, $\beta_b(18)$, $S^-(16)$
	920			$S^-(60)$, $\beta_b(41)$
	574			$\omega(54)$, $\varphi_b(28)$, $\zeta_b(9)$, $\beta_b(8)$
	215			$\varphi_b(61)$, $\omega(25)$, $\Upsilon(14)$
	192			$\Upsilon(84)$, $\omega(8)$
B_1 block	2962	2959	(a)	$r_b^-(99)$
	2928	2926	(e)	$d^-(99)$
	1464	~1450	(a)	$\alpha_b(88)$, $\beta_b(9)$
	1370	1379	(e)	$S^-(41)$, $\zeta_b(38)$, $W(36)$
	1228	1233	(a)	$\zeta_b(49)$, $W(40)$, $S^-(7)$
	1154	1130	(a)	$\beta_b(27)$, $W(26)$, $S^-(20)$, $\varphi_b(10)$
	949	962	(a)	$\beta_b(50)$, $S^-(40)$
	847	831	(a)	$P(81)$, $\zeta_b(15)$
	386			$\varphi_b(83)$, $\beta_b(8)$, $P(8)$
	198			$\Upsilon(97)$
	114			$\tau(90)$
B_2 block	2962	2959	(d)	$r_a^-(98)$
	2930	2925	(b)	$d^-(93)$
	2903	2916, 2905	(b)	$s(93)$
	2882	2880, 2868	(b)	$r^+(99)$
	1462	1466	(b)	$\alpha_a(89)$, $\beta_a(9)$
	1372	1381	(b)	$U(100)$
	1355	~1350	(d)	$W(49)$, $\zeta_a(28)$, $R(7)$
	1330	1322	(b)	$\zeta_a(39)$, $W(37)$, $\zeta^+(18)$, $R(7)$
	1162	1153	(c)	$P(26)$, $\beta_a(25)$, $R(18)$, $\varphi^+(7)$
	1088	1095	(b)	$S^+(53)$, $R(26)$, $\zeta_a(14)$
	903	899	(b)	$\beta_a(56)$, $\zeta_a(17)$, $P(15)$, $S^+(13)$
	827	828	(b)	$R(47)$, $P(27)$, $S^+(13)$
	447	492	(b)	$\varphi^+(44)$, $\zeta^+(30)$, $W(14)$, $P(10)$
	372			$\varphi_a(79)$, $P(7)$

[a]From ref. 15.
(a) Pol. ∥ to chain axis.
(b) Pol. ⊥ to chain axis.
(c) May be an amorphous band.
(d) Frequency measured when radiation polarized ⊥ to the chain axis.
(e) Frequency measured when radiation polarized ∥ to the chain axis.

Snyder and Schachtschneider employed essentially the same internal and group coordinates for both IPP and SPP. The internal coordinates are reproduced in Figure 16.3 and the group coordinates (internal symmetry coordinates) are reproduced in Table 16.1. Some minor changes in the symbols employed is apparent, but these should not cause confusion. Symmetry coordinates were constructed from the group coordinates listed in Table 16.1 and the reader is referred to refs. 2 and 3 for a complete description. In all cases for the construction of the **G** matrix, the bond angles were assumed to be tetrahedral and the lengths of the C—H and C—C bonds were taken to be 1.093 and 1.54 Å, respectively. The force constants employed were those determined from the calculations performed on model paraffin molecules (11) that were given in Chapter 13 (Table 13.6) and illustrated in Figures 13.7 and 13.8. The results of the normal coordinate analysis of IPP and the two polymorphic forms of SPP are shown in Tables 16.2, 16.3, and 16.4, respectively. The agreement between the calculated frequencies and observed vibrational data is very good in all three cases. It must be emphasized that Snyder and Schachtschneider transferred the force field derived from model paraffins to the polymers **without adjustment** of any of the force constant values. The authors pointed out that they did not wish to bias their final calculated frequencies by forcing them to some observed values from a zero-order assignment. If the C—H stretching frequencies are excluded, the average error in the calculated frequencies in all three cases (i.e., IPP and the two forms of SPP) in the range of 0.8–1.1%, which is roughly equivalent to that determined for paraffins using the same force field. There are marked differences in the calculated infrared frequencies of the two forms of SPP, as illustrated in the schematic diagram reproduced in Figure 16.4, and it is evident that strong coupling between the groups in the polymer

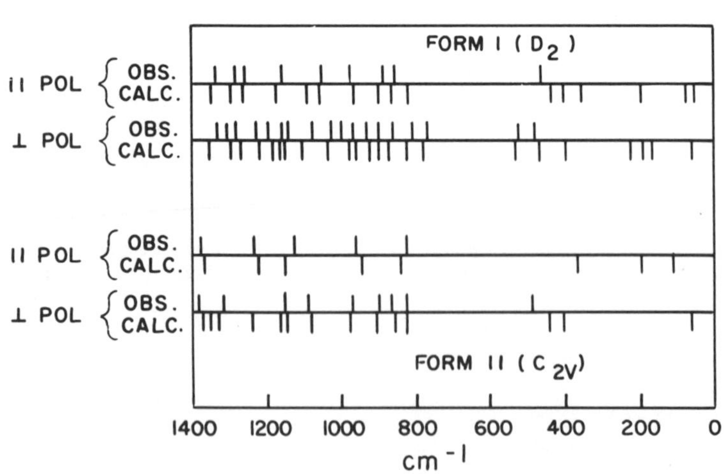

Figure 16.4. Observed and calculated frequencies of syndiotactic polypropylene. (Reprinted by permission from ref. 3.)

leads to vibrations exhibiting conformational sensitivity. In fact, the excellent correlation between the observed and calculated frequencies is verification of the proposed structures of these polymers. Snyder and Schachtschneider also calculated the frequencies of several deuterated polypropylenes and compared them to the experimental data reported in the literature. In general there is good agreement between the observed and calculated frequencies. These results demonstrate the difficulties inherent in the group frequency approach to selective deuteration. Upon deuteration, there is a drastic alteration in the form of many of the normal modes resulting from groups of atoms being strongly coupled. Accordingly, correlations of the vibrational frequencies between protonated and deuterated analogues can be specious. The interested reader is referred to refs. 2 and 3 for further details of the calculations of the deuterated polypropylenes.

Finally, Zerbi et al. (16) studied the dynamics of a "growing" chain of IPP by calculating the normal modes of vibration of finite chains. A simplified model with point masses for the CH_3, CH_2, and CH groups was employed, together with Urey-Bradley force constants taken from the work of Miyazawa (17). From this work the authors were able to conclude that ordered segments of at least five propylene units exist in the molten state of IPP. The results have also been verified by Raman studies of the molten state.

Poly(ethyl ethylene)

Normal coordinate studies of isotactic poly(ethyl ethylene) (PEE), or as it is alternatively described, poly(butene-1), were initially reported by Ukita (18). This analysis was based on a 3_1 helical model and a modified Urey-Bradley force field, derived from the model compounds C_2H_6, C_3H_8, and *i*-C_4H_{10}, was employed. Subsequent normal coordinate calculations have been performed by Cornell and Koenig (19) and more recently by Holland-Moritz and Sausen (20). In the following discussion we will concentrate primarily on the results of the latter two groups, since they both employed modified valence force fields transferred from the studies of Snyder and Schachtschneider (2, 11, 21) on model paraffins, polyethylene, and polypropylene.

Isotactic PEE exists in three distinct polymorphic forms, designated forms I, II, and III. Wide-angle x-ray diffraction studies have shown that form I of PEE contains six 3_1 helices in a hexagonal unit cell (22–26); form II contains four 11_3 helices in a tetragonal unit cell (24, 26–28); and form III contains two 4_1 helices in an orthorhombic unit cell (22–26, 28–30).

The infrared and Raman spectra of the three polymorphic forms of PEE are markedly different (19, 31, 32) and are an excellent illustration of the effect of polymer chain conformation on the vibrational spectra. An example of the Raman spectra of the three forms in the "helix-sensitive region" of the spectrum (700–1000 cm^{-1}) is reproduced in Figure 16.5 (19). Cornell and Koenig (19), who first reported the Raman spectra of the three forms of PEE, performed normal coordinate calculations based on modifications of the

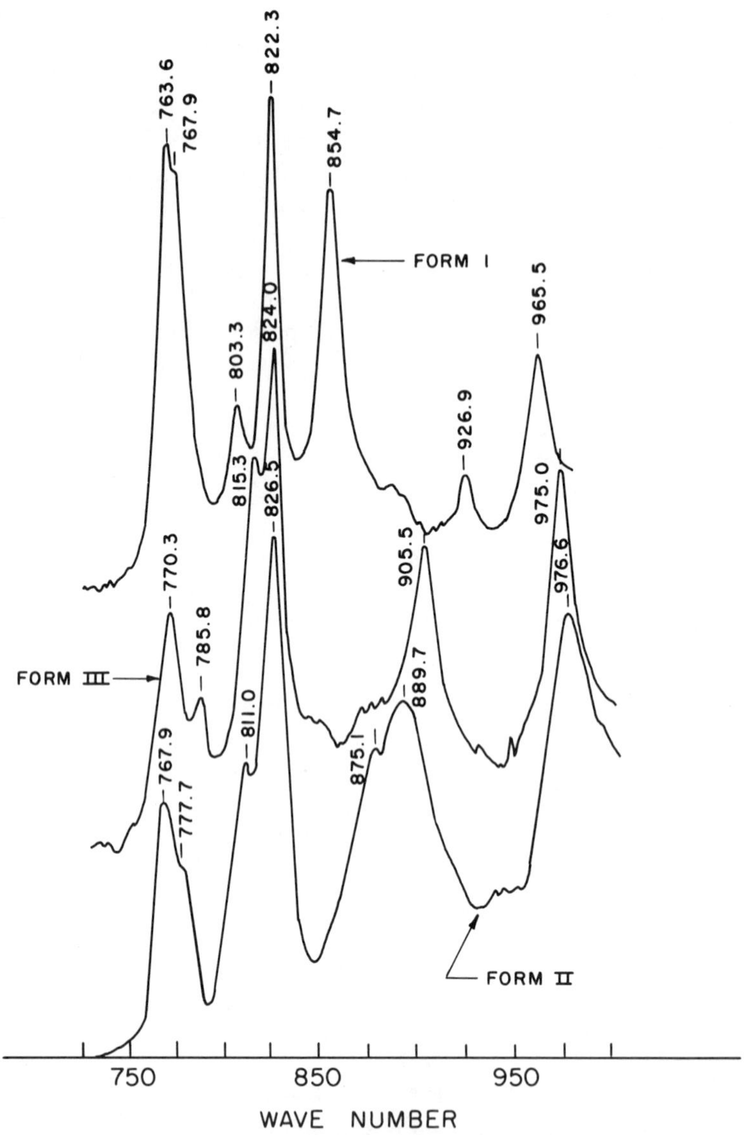

Figure 16.5. High resolution scan of helix-sensitive region of poly(ethyl ethylene) Raman spectrum (Reprinted by permission from ref. 19.)

valence force field of Snyder and Schachtschneider (2, 21) for saturated hydrocarbons. All bond angles were assumed to be tetrahedral and the C—C and C—H bond lengths were taken to be 1.54 and 1.093 Å, respectively. The cartesian coordinates were calculated for five regular helices, namely, the 3_1, 10_3, 7_2, 11_3, and 4_1 helical conformations. The line group of the isolated helical chain is isomorphous to the point group $C(2\pi m/n)$ where m is the number of

Table 16.5 Poly(Ethyl Ethylene) *A* Mode Vibrations Sensitive to Helix Angle

3_1 120°	10_3 108°	7_2 103°	11_3 98°	4_1 90°	Helix ψ Assignment[a]	
		Wavenumber				
64	63	63	63	62	τ (84)	
114	114	112	110	105	γ_{CC} (CH_2) (59)	τ(33)
187	182	179	177	172	τ (49)	γ_{CC} (CH_2) (30)
212	211	211	211	210	γ_{CC} (CH_2) (44)	τ (36)
259	248	243	238	230	γ_{CC} (CH) (50)	τ (22)
347	354	358	361	367	γ_{CC} (CH) (43)	
532	516	509	503	492	γ_{CC} (CH_2) (51)	γ_{CC} (CH) (22)
774	781	784	786	790	ν_C (CH_2) (40)	ν_C (CH—CH_2) (41)
824	822	820	819	817	ν_C (CH_2) (62)	ν_C (CH—CH_2) (19)
875	885	889	893	901	γ_C (CH_2) (41)	γ_C (CH_3) (23)
982	986	988	990	993	γ_C (CH_3) (29)	ν_C (CH—CH_2) (21)
1031	1032	1032	1033	1033	γ_C (CH_3) (33)	ν_C (CH_2—CH_3) (34)
1039	1039	1038	1038	1038	γ_C (CH_3) (35)	γ_C (CH_2) (20)
1095	1095	1095	1094	1092	γ_C (CH_2) (59)	
1136	1134	1133	1133	1133	γ_C (CH_2) (33)	ν_C (CH—CH_2) (32)
1181	1179	1178	1178	1176	γ_C (CH_2) (46)	γ_C (CH) (37)
1296	1295	1294	1292	1289	γ_C (CH_2) (62)	γ_C (CH) (30)
1307	1301	1299	1298	1296	γ_C (CH_2) (75)	
1362	1363	1363	1363	1364	γ_C (CH_3) (39)	γ_H (CH_3) (33)

[a]Key: ν=stretching, γ=bending, τ=torsional; $\nu_H=\nu$ (CH), $\nu_C=\nu$ (CC), $\gamma_H=\gamma$ (HCH), $\gamma_C=\gamma$ (HCC), $\gamma_{CCC}=\gamma$ (CCC).

turns of the helix for *n* units in the translational repeat unit. In Chapter 11 we discussed the optical activity of the normal modes of helical polymer molecules. In any event, the *A* modes are infrared and Raman active and Cornell and Koenig (19) calculated the frequencies of the *A* modes for the five molecular models. At that time, because of the limited computer memory size available, the calculation of the *E* modes was not possible. Nevertheless, the authors reasoned that the *A* modes should be strongest in the Raman spectra and are therefore of primary importance. Those calculated *A* modes that were found to be sensitive to the helix angle are shown in Table 16.5. On the basis of these calculated frequencies, the authors concluded that form III of PEE had a 10_3 helical conformation, since the results were more consistent with the experimental Raman data. In retrospect, this conclusion is subject to considerable doubt. It will be recalled that the Snyder-Schachtschneider force field was derived for paraffins with only gauche and trans conformations. The force field may reasonably be transferred to the 3_1 and 4_1 helical models but significant errors can be introduced in applying it to the other three models where torsional angles exist that are significantly different from pure trans and gauche conformations. In effect, Cornell and Koenig had assumed that the calculated frequency changes are derived solely from corresponding changes in

Table 16.6 Calculated and Observed Frequencies of IR and Raman Spectra of Poly(propyl Ethylene) (Modification I) and Their Assignments ($\sigma = 1.06\%$)

Calc. Freq. *A*	Calc. Freq. *E*	Poly(propyl ethylene) *P*-5-1, Obs. Freq. IR	Obs. Freq. Raman	Error (%)	H—C(—CH_2—)—CH_2—CH_2—CH_3 Assignment and Potential Energy Distribution	Type of Vibration
	114				42% H_τ, 31% H_ω(m′)	
131					37% H_τ, 28% H_ω(m)	
167			164 w	1.82	21% H_ω(m), 50% H_ω(s)	
	207		200 w	3.50	64% H_τ	
210					93% H_τ	
	216		220	1.82	50% H_τ, 19% H_ω(s)	
261			264 w	1.13	53% H_ω(m′)	
	281		280 vw	0.35	37% H_ω(s), 23% H_ω(m)	δ(CCC), τ
288			292 w	1.36	24% H_ω(m′), 21% H_ω(s)	
			(320 w)			
	390		375	4.00	28% H_ω(s), 25% H_ω(m′)	
418			425 vw	1.75	41% H_ω(m′), 30% H_ω(s)	
	461	460 w(?)	460 w	0.21	38% H_ω(m′), 19% H_ζ(m)	
	532	510 vw(⊥)	525 vw	4.31	27% H_ω(m′), 26% H_τ	
550		550 w(‖)	549 vw	0.18	38% H_ω(m), 20% H_ω(s)	
746		735 ms(‖)		1.50	76% H_γ(s)	
			(760 w)			$r_s(CH_2)$
	764		771 w	0.90	88% H_γ(s)	
813		825 m(‖)	828 m	1.81	54% H_γ(m)	

	823	806 m(⊥)		2.10	25% H_γ(m), 23% K_S(m)	$r_s(CH_2)$, ν_m(CC)
840		856 m(‖)	855 m	1.86	22% H_γ(s), 27% $F_{R\omega}$(m), 22% K_S(m)	$t_s(CH_2)$, ν_m(CC)
906	904	905 m(⊥)	906 mw	0.11	26% K_S(s), 29% H_β(s)	ν_s(CC), $w(CH_3)$
					27% H_β(s), 27% K_S(s)	$w(CH_3)$, ν_s(CC)
968	925	925 w(⊥)	923 vw	0.21	30% H_γ(s), 23% H_ζ(m)	} $r_s(CH_2)$, δ(CH)
					44% H_β(s), 21% H_γ(s)	
	979	970 w(⊥)		0.92	79% H_β(s)	$t(CH_3)$
993		990 vw(‖)	988 w	0.50	50% H_β(s), 43% H_γ(s)	$t(CH_3)$, $t_s(CH_2)$
	1008		1022 m	1.36	39% H_γ(s)	$t_s(CH_2)$
1035		1041 m(‖)		0.57	62% K_S(s)	ν_s(CC)
	1037		1050 m	1.23	47% $F_{R\beta}$(s), 14% H_γ(s)	$t_s(CH_2)$
	1077	1075 m(⊥)		0.18	25% K_S(s), 18% K_S(m)	ν_s(CC), ν_m(CC)
1079		1065 m(‖)	1065 m	1.31	42% K_S(s), 18% H_γ(s)	ν_s(CC), $w(CH_2)$
	1088				38% H_γ(m), 28% H_ζ(m)	$t_m(CH_2)$, δ(CH)
1091			1090 m	0.09	19% K_S(m), 18% K_S(s),	δ(CH), $w_s(CH_2)$
1114					66% H_γ(s)	$t_s(CH_2)$
				0.54		
	1114	1108 m(⊥)	1110 w		47% H_γ(s), 27% H_ζ(m)	$t_s(CH_2)$, δ(CH)
	1172	1160 sh(⊥)	1162 sh	1.03	22% K_S(m), 22% H_γ(m), 18% H_γ(s), 18% H_ζ(m)	ν_m(CC), $t_m(CH_2)$, $t_s(CH_2)$, δ(CH)
1185		1171 m(‖)	1171 ms	1.53	25% H_ζ(m)	δ(CH), $w(CH_2)$
	1193	1199 m(?)		0.50	27% H_γ(m), 23% H_γ(s), 20% H_ζ(m)	$t_m(CH_2)$, $t_s(CH_2)$, δ(CH)
1251		1247 m(‖)	1246 vw	0.40	51% H_γ(s), 21% H_γ(m)	$t_s(CH_2)$, δ(CH)
	1262	1260 wm(⊥)	1259 w	0.23	55% H_γ(s), 24% H_γ(m)	$w_s(CH_2)$, $t_m(CH_2)$
1272		1264 m(‖)		0.63	46% H_γ(m), 32% H_ζ(m), 17% H_γ(s)	$t_m(CH_2)$, δ(CC), $w_s(CH_2)$
1291		1280 w(‖)	1279 vw	0.93	49% H_γ(s), 32% H_γ(m)	$w_s(CH_2)$, $t_m(CH_2)$

Table 16.6 Continued

Poly(propyl ethylene) *P*-5-1

$H{-}C(-CH_2-){-}CH_2{-}CH_2{-}CH_3$

Calc. Freq. *A*	Calc. Freq. *E*	Obs. Freq. IR	Obs. Freq. Raman	Error (%)	Assignment and Potential Energy Distribution	Type of Vibration
1308	1297		1292 w	0.38	67% H_γ(s)	$t_s(CH_2)$
					67% H_γ(s)	
		1305 m(⊥)		0.84		
	1316				53% H_γ(s), 21% H_γ(m)	$t_m(CH_2)$, $w_m(CH_2)$
1332		1330 sh(‖)		0.15	51% H_γ(s), 19% H_ζ(m)	$w_s(CH_2)$, $\delta(CH)$
	1340	1340 sh(⊥)	1336 m	0.29	46% H_γ(s), 27% H_γ(m), 18% $F_{R\beta}$	$w_s(CH_2)$, $w_m(CH_2)$
	1367				33% H_ζ(m), 27% K_S(m), 20% $H_\gamma(m)$, 20% $F_{R\beta}$	$\delta(CH)$, $\nu(CC)$, $w_m(CH_2)$
			1356 w	0.88		
1368					24% H_γ(s), 20% H_β(s), 18% H_α(s)	$w_s(CH_2)$, $\delta(CH_3)$
	1375				27% H_β(s), 23% H_α(s)	$\delta(CH_3)$
1385	1388	1375 ms(?)	1373 m	1.00	31% H_β(s), 30% H_γ(s), 28% H_α(s)	$\delta(CH_3)$, $w_s(CH_3)$
					32% H_γ(s), 20% H_β(s), 19% H_α(s)	$w_s(CH_2)$, $\delta(CH_3)$
1409		1387 sh(?)	1395 vw	1.58	46% K_s(m), 25% H_γ(m), 21% H_ζ(m)	$\nu(CC)$, $w_m(CH_2)$, $\delta(CH)$
1448	1448	1440 sh	1440 sh	0.55	60% H_δ(m), 19% H_δ(s)	$\delta(CH_2)$, $\delta_s(CH_2)$
					62% H_δ(s), 18% H_γ(s)	$\delta_s(CH_2)$
1454	1461	1450 s		0.75	44% H_δ(s)	$\delta_m(CH_2)$
					47% H_δ(m)	

1467	1467	1455 s	1455 sh	0.82	75% H_α(s)	$\delta(CH_3)$
					69% H_α(s)	
1473	1473	1465 s	1465 sh	0.54	88% H_α(s)	
					88% H_α(s)	
1479	1479		1480 sh	0.06	69% H_δ(s), 21% H_γ(s)	$\delta_s(CH_2)$
					63% H_δ(s), 20% H_γ(s)	
2850	2850	2845 vs	2848 ms		97% K_d(m)	$\nu_m^s(CH_2)$
2855	2856				75% K_d(m), 24% K_d(s)	$\nu_m^s(CH_2)$, $\nu_s^s(CH_2)$
2856	2856				22% K_d(m), 77% K_d(s)	$\nu_m^s(CH_2)$, $\nu_s^s(CH_2)$
2881	2881	2875 vs	2875 vs		98% K_r	$\nu(CH_3)$
2904	2902		2900 sh		96% K_s	ν(CH)
2920	2920	2912 vs	2915 vs		98% K_d(s)	$\nu_s^{as}(CH_2)$
2928	2928		2935 ms		99% K_d(m)	$\nu_s^{as}(CH_2)$
2930	2930				98% K_d(s)	$\nu_a^{as}(CH_2)$
2961	2961	2960 vs	2960 ms		100% K_r	$\nu^{as}(CH_3)$
2963	2963				100% K_r	

the **G** matrix. However, Zerbi et al. (33) have emphasized that for hydrocarbon polymers the dynamics are strongly affected by both the force field (**F** matrix) and the geometry (**G** matrix). In fact, recent x-ray studies indicate that the chain conformation of form III is a 4_1 helix (30).

Normal coordinate calculations of form I of PEE have recently been reported by Holland-Moritz and Sausen (20). These authors obtained additional experimental data in the form of polarized infrared and Raman spectra of isotactic PEE and vibrational spectra of three deuterated analogues of PEE. From the polarized Raman data, the strongest polarizability tensor components (i.e., α_{yz}, α_{zz}, and α_{zx}) could be determined, which permitted a distinction between *A* and *E* modes in the Raman spectrum. In contrast to previous studies, Holland-Moritz and Sausen calculated the cartesian coordinates from the measured unit cell dimensions (22–26). Internal rotation angles of 180° and 60° for a CCC angle of 114° within the main chain were determined. The 40 internal coordinates shown schematically in Figure 13.9 were defined. In addition, the definitions and values of the force constants in terms of the internal coordinates were given in Tables 13.7A and 13.7B. It should be noted that the authors defined the force constants in such a way as to clearly differentiate between contributions of the characteristic groups belonging to the main chain (m) and side chain (s). This differentiation is of particular significance for the poly(alkyl ethylenes) containing longer alkyl side groups. The force constant values employed by Holland-Moritz and Sausen are also included in Tables 13.7A and 13.7B. These values deserve further comment. Since the cartesian coordinates were calculated from the measured unit cell dimensions, the basic Snyder-Schachtschneider force field (2, 21), which is based on tetrahedral geometry, is not entirely transferable. Nevertheless, from initial calculations, Holland-Moritz and Sausen transferred this force field to the 3_1 helical model of PEE and determined that only six calculated frequencies differed significantly from the observed frequencies. By inspection of the Jacobian matrix the authors were able to determine that four force constants, F'_γ(m), H_γ(m), f^t_γ(s), and f^g_γ(s), make major contributions to these six frequencies. Accordingly, a damped least-squares refinement of these four force constants was undertaken and reasonable agreement between the observed and calculated frequencies was obtained. Incidentally, the authors also attempted to use the revised hydrocarbon force field of Snyder (34), but because of the poor fit of the calculated to the observed frequencies for the deuterated polymers it was rejected. It was noted that the Snyder-Schachtschneider force field (2, 21) was derived from 17 protonated and some deuterated alkanes, which in the opinion of Holland-Moritz and Sausen gives this force field greater physical relevance to their calculations. The results of the normal coordinate analysis of PEE (form I) are reproduced in Table 16.6. In general, there is a very good agreement between the observed and calculated results. The potential energy distribution and the displacements of the atoms (**L** matrix) reveal that the majority of the normal modes are strongly coupled between the main and side chain. Calculations were also performed for the three deuterated PEEs with good results. (For a complete discussion of these calculations, see ref. 20.)

Poly(alkyl ethylenes)

The vibrational spectra and normal coordinate calculations of poly(propyl ethylene) (PPE) and two deuterated analogues have been reported by Holland-Moritz et al. (35). The force field employed was identical to that previously described for poly(ethyl ethylene). PPE also exists in three distinct polymorphic crystalline forms, and calculations of the normal coordinates were restricted to the 3_1 helical conformation (form I). The results are similar to those described earlier for poly(ethyl ethylene); see ref. 35 for details.

16.2 Haloethylene Polymers

In this section we will consider the results of normal coordinate calculations performed on the general class of polymers synthesized from vinyl ($CH_2{=}CHX$), vinylidene ($CH_2{=}CX_2$), and tetrasubstituted vinyl ($CX_2{=}CX_2$) halides where X can be fluorine, chlorine, bromine, or iodine atoms. Poly(vinyl halides) contain the chemical repeating unit $\leftarrow CH_2{-}CHX \rightarrow$ and because of the asymmetric center on the α-carbon atom, the polymer chain may contain a complicated sequence distribution of stereoisomeric placements. At the extremes, one can consider purely isotactic or syndiotactic poly(vinyl halides). Another microstructural complication may arise when a monomer unit is incorporated "backward" leading to sequence isomerism (i.e., head-to-head and tail-to-tail placements). This sequence isomerization is particularly significant in the case of poly(vinyl fluoride) (X=F). Free radical polymerization of vinyl halides generally results in the synthesis of predominantly atactic polymers. However, a predominantly syndiotactic poly(vinyl chloride) (X=Cl) may be synthesized by the urea canal technique (36).

Poly(vinylidene halides) contain $\leftarrow CH_2{-}CX_2 \rightarrow$ chemical repeat units in the chain. Stereoisomerism is not possible but significant sequence isomerism does occur in free radically polymerized poly(vinylidene fluoride) (X =F). Because of the symmetrical repeat unit, both poly(vinylidene chloride) (X=Cl) and poly(vinylidene fluoride) are semicrystalline polymers at room temperature.

Also included in this section is poly(tetrafluoroethylene), which has the chemical repeat unit $\leftarrow CF_2{-}CF_2 \rightarrow$. This is also a highly regular crystalline material.

In the following subsections we will consider the normal coordinate calculations that have been performed on models for poly(vinyl chloride), poly(vinylidene chloride), poly(vinylidene fluoride), and poly(tetrafluoroethylene).

Poly(vinyl chloride)

There have been several normal coordinate studies of poly(vinyl chloride) (PVC) employing a variety of force fields, notably by Krimm and co-workers (37–39), Rubĉić and Zerbi (40, 41), Shimanouchi and Tasumi (42, 43), and

Table 16.7 Force Constants for Secondary Chlorides

Name	Values[a]	Environment	Coordinates Coupled
$^*K_r$	4.6990	C(H,H,H)	(C,II)
$^*K_d$	4.5380	C(H,H)	(C,H)
$^\dagger K_{dCl}$	4.8460	C(H,Cl)	(C,H)
$^*K_R$	4.532	C(H,H)—C(H,H)	(C,C)
K_{RCl}	4.7047	C(H,H)—C(H,Cl)	(C,C)
K_{Cl}	2.7844	C(H,Cl)	(C,Cl)
$^*H_\alpha$	0.5390	$C(H_1,H_2,H_3)$	(H_1,C,H_2)
H_β	0.6213	$C(H_1,H_2,H_3)—C$	(C,C,H_1)
$^*H_\gamma$	0.663	$C_1—C_2(H_1,H_2)—C_3$	(C_1,C_2,H_1)
H_φ	0.6567	$C_1—C_2(H,Cl)$	(C_1,C_2,H)
H_θ	0.8518	C(H,Cl)	(H,C,Cl)
H_Ξ	1.1062	$C_1—C_2(H,Cl)$	(C_1,C_2,Cl)
$H_{1\omega}$	0.9792	$C_1—C_2(H,Cl)—C_3$	(C_1,C_2,C_3)
$^*H_{2\omega}$	1.032	$C_1—C_2(H,H)—C_3$	(C_1,C_2,C_3)
$^*H_{1\delta}$	0.533	$C_1(H,H)—C_2(H_1,H_2)—C_3(H,H)$	(H_1,C_2,H_2)
$H_{2\delta}$	0.5028	$C_1(H,Cl)—C_2(H_1,H_2)—C_3(H,Cl)$	(H_1,C_2,H_2)
$H_{3\delta}$	0.5198	$C_1(H,H)—C_2(H_1,H_2)—C_3(H,Cl)$	(H_1,C_2,H_2)
$H_{1\tau}$	0.1105	$C_1(H,H,H)—C_2(H,Cl)—C$	\ / $—C_1 — C_2—$ / \
$H_{2\tau}$	0.0775	$C_1(H,H,H)—C_2(H,H)—C$	\ / $—C_1 — C_2—$ / \
$H_{3\tau}$	0.1309	$C_1(H,H,H)—C_2(H_1,Cl)—C(H,H,H)$	\ / $—C_1 — C_2—$ / \
$H_{4\tau}$	0.0744	$C—C_1(H,H)—C_2(H,Cl)—C$	\ / $—C_1 — C_2—$ / \
$^*F_r$	0.032	$C(H_1,H_2,H_3)$	$(C,H_1)(C,H_2)$
$^*F_{2d}$	0.019	$C(H_1,H_2)$	$(C,H_1)(C,H_2)$
$^*F_{1R}$	0.083	$C_1(H,H)—C_2(H,H)—C_3(H,H)$	$(C_1,C_2)(C_2,C_3)$
F_{2R}	0.177	$C_1(H,Cl)—C_2(H,H)—C_3(H)$	$(C_1,C_2)(C_2,C_3)$
F_{3R}	0.4058	$C_1(H,H)—C_2(H,Cl)—C_3(H,H)$	$(C_1,C_2)(C_2,C_3)$
F_{RCl}	0.6401	$C_1—C_2(H,Cl)$	$(C_1,C_2)(C_2,Cl)$
$^*F_{1R\gamma}$	0.174	$C_1—C_2(H,H)$	$(C_1,C_2)(C_1,C_2,H)$
$F_{2R\gamma}$	0.1845	$C_1(H_1,Cl)—C_2(H,H)$	$(C_1,C_2)(C_1,C_2,H)$
$F_{3R\gamma}$	0.1701	$C_1(H_1,Cl)—C_2(H,H)$	$(C_1,C_2)(C_2,C_1,H_1)$
$F_{1R\omega}$	0.349	$C_1(H,Cl)—C_2—C_3$	$(C_1,C_2)(C_1,C_2,C_3)$[b]
$^*F_{2R\omega}$	0.303	$C_1—C_2—C_3$	$(C_1,C_2)(C_1,C_2,C_3)$
$F'_{1R\gamma}$	−0.0218	$C_1—C_2(H_1,H_2)—C_3(H,Cl)$	$(C_2,C_3)(C_1,C_2,H_1)$ or $(C_1,C_2)(C_3,C_2,H_1)$
$F'_{2R\gamma}$	−0.0868	$C_1—C_2(H,Cl)—C_3$	$(C_1,C_2$
$^*F'_{3R\gamma}$	−0.097	$C_1—C_2(H,H)—C_3$	$(C_1,C_2$
$F_{\omega Cl}$	−0.1117	$C_1—C_2(H,Cl)—C_3$	$(C_2,C$
$F_{R\Xi}$	0.1003	$C_1—C_2(H,Cl)$	$(C_1,C_2$
$F_{Cl\Xi}$	0.4005	$C_1—C_2(H,Cl)$	$(C_2,Cl$

Table 16.7 Continued

Name	Values[a]	Environment	Coordinates Coupled
$F_{\theta Cl}$	0.2399	C(H,Cl)	(C,Cl)(H,C,Cl)
$*F_{\beta\beta}$	−0.0272	$C(H_1,H_2,H_3)—C$	$(C,C,H_1)(C,C,H_2)$
$*F_{\gamma\gamma}$	−0.0190	$C(H_1,H_2)—C$	$(C,C,H_1)(C,C,H_2)$
$*F'_{\gamma\gamma}$	0.0210	$C_1—C_2(H_1,H_2)—C_3$	$(C_1,C_2,H_1)(C_3,C_2,H_1)$
$*F_{1\gamma\omega}$	−0.022	$C_1—C_2(H_1,H_2)—C_3$	$(C_1,C_2,C_3)(C_1,C_2,H_1)$
$F_{2\gamma\omega}$	−0.0903	$C_1—C_2(H,Cl)—C_3$	$(C_1,C_2,C_3)(C_1,C_2,H)$
$F_{\varphi\theta}$	0.1055	$C_1—C_2(H,Cl)$	$(C_1,C_2,H)(H,C_2,Cl)$
$*f^t_{1\gamma\gamma}$	0.073	$C_1(H_1,H_2)—C_2(H_3,H_4)$	$(H_1,C_1,C_2)(C_1,C_2,H_3)$[c]
$*f^g_{1\gamma\gamma}$	−0.058	$C_1(H_1,H_2)—C_2(H_3,H_4)$	$(H_1,C_1,C_2)(C_1,C_2,H_3)$[c]
$*f^g_{1\gamma\omega}$	−0.064	$C_1(H_1,H_2)—C_2—C_3$	$(H_1,C_1,C_2)(C_1,C_2,C_3)$
$*f^t_{1\gamma\omega}$	0.073	$C_1(H_1,H_2)—C_2—C_3$	$(H_1,C_1,C_2)(C_1,C_2,C_3)$
$*f^t_{1\omega\omega}$	0.097	$C_1—C_2—C_3—C_4$	$(C_1,C_2,C_3)(C_2,C_3,C_4)$
$*f^g_{1\omega\omega}$	−0.005	$C_1—C_2—C_3—C_4$	$(C_1,C_2,C_3)(C_2,C_3,C_4)$
$f^t_{2\omega\omega}$	0.095	$C_1—C_2(H,Cl)—C_3—C_4$	$(C_1,C_2,C_3)(C_2,C_3,C_4)$
$f^g_{2\omega\omega}$	0.0436	$C_1—C_2(H,Cl)—C_3—C_4$	$(C_1,C_2,C_3)(C_2,C_3,C_4)$
$f^t_{2\gamma\gamma}$	0.0803	$C_1(H_1,H_2)—C_2(H,Cl)$	$(H_1,C_1,C_2)(C_1,C_2,H)$
$f^g_{2\gamma\gamma}$	−0.0327	$C_1(H_1,H_2)—C_2(H,Cl)$	$(H_1,C_1,C_2)(C_1,C_2,H)$
$f^t_{2\gamma\omega}$	0.073	$C_1(H,Cl)—C_2—C_3$	$(H,C_1,C_2)(C_1,C_2,C_3)$
$f^g_{2\gamma\omega}$	−0.0077	$C_1(H,Cl)—C_2—C_3$	$(H,C_1,C_2)(C_1,C_2,C_3)$
$f^g_{\gamma\Xi}$	−0.1085	$C_1(H,Cl)—C_2(H_1,H_2)$	$(Cl,C_1,C_2)(C_1,C_2,H_1)$
$f^t_{\gamma\Xi}$	−0.1548	$C_1(H,Cl)—C_2(H_1,H_2)$	$(Cl,C_1,C_2)(C_1,C_2,H_1)$
$f^t_{\omega\Xi}$	−0.0774	$C_1(H,Cl)—C_2—C_3$	$(Cl,C_1,C_2)(C_1,C_2,C_3)$
$f^g_{\omega\Xi}$	−0.0193	$C_1(H,Cl)—C_2—C_3$	$(Cl,C_1,C_2)(C_1,C_2,C_3)$
$f'^t_{\gamma\gamma}$	0.0136	$C_1—C_2(H_1)—C_3(H_2)$	$(C_1,C_2,H_1)(C_2,C_3,H_2)$[c]
$f'^g_{\gamma\gamma}$	0.0519	$C_1—C_2(H_1)—C_3(H_2)$	$(C_1,C_2,H_1)(C_2,C_3,H_2)$
$f''^t_{\gamma\gamma}$	0.0034	$C_1—C_2(H_1)—C_3(H_2)—C_4$	$(C_1,C_2,H_1)(H_2,C_3,C_4)$[d]
$f''^g_{\gamma\gamma}$	0.0222	$C_1—C_2(H_1)—C_3(H_2)—C_4$	$(C_1,C_2,H_1)(H_2,C_3,C_4)$[d]
K_{Cl---H}	0.07	C—Cl---H—C	Cl---H
$H_{\alpha op}$[e]	0.01	C—Cl---H—C	C—Cl---H
$H_{\alpha ip}$[e]	0.03	C—Cl---H—C	C—Cl---H
$H_{\beta op}$[e]	0.01	C—Cl---H—C	Cl---H—C
$H_{\beta ip}$[e]	0.25	C—Cl---H—C	Cl---H—C
$f_{\tau\tau}$	0.008	$C_1—C_2—C_3—C_4$	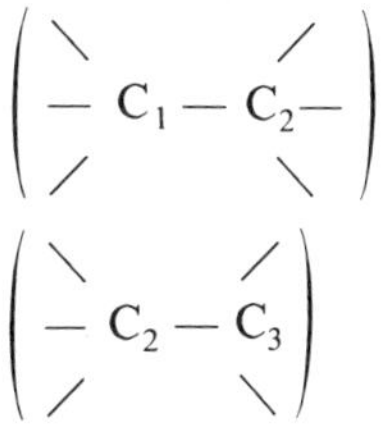

[a]Stretch constants are in units of mdyn/Å, stretch bend constants are in units of mdyn/rad, bend and torsion constants are in units of mdyn-Å/rad^2.

*Force constants transferred from ref. 11.

†Force constant transferred from ref. 46.

[b]A chlorine atom is attached to at least one of the carbon atoms involved.

[c]Superscripts *t*, *g* imply that extreme atoms are trans and gauche to one another, respectively.

[d]Trans implies that two angles are bisected by a common plane, gauche implies no such common bisector.

[e]op implies out-of-plane bending, ip implies in-plane bending.

Boitsov et al. (44). In the following discussion we will concentrate on the results of the former two groups, since they both employed similar modified valence force fields.

Crystalline PVC has a chain structure composed predominantly of syndiotactic chain placements. It is generally agreed that the preferred chain conformation is in the form of an extended planar zigzag structure. The crystal structure has been determined by Natta and Corradini (45) and contains two chains (i.e., four chemical repeat units) per unit cell. Bond lengths of 1.093, 1.54, and 1.798 Å were determined for the C—H, C—C, and C—Cl bonds, respectively. For the purposes of calculations, all angles were assumed to be tetrahedral. The space group of crystalline PVC is isomorphous to the point group D_{2h} (39) and the symmetry coordinates are classified into the following eight species:

Raman active	A_g, B_{1g}, B_{2g}, and B_{3g}
Infrared active	B_{1u}, B_{2u}, and B_{3u}
Forbidden	A_u

An initial force for secondary dichlorides was published by Opaskar and Krimm (46) and was subsequently modified by Moore and Krimm (47) to include torsional and intermolecular force constants. The latter authors employed their force field to calculate the normal vibrations of crystalline PVC. This valence force field, reproduced in Table 16.7, was determined by the refinement of 42 force constants to predict the greater than 90 experimentally observed frequencies of chloropropane, ${}_{\mathrm{H}}\mathrm{S}_{\mathrm{H}}$ and ${}_{\mathrm{H}}\mathrm{S}_{\mathrm{C}}$ 2-chlorobutane, the ${}_{\mathrm{H}}\mathrm{S}_{\mathrm{HH}}\mathrm{S}_{\mathrm{H}}$ form of DL-2,4-dichloropentane, and the ${}_{\mathrm{H}}\mathrm{S}_{\mathrm{HC}}\mathrm{S}_{\mathrm{H}}$ form of meso-2,4-dichloropentane. (In this notation S indicates the secondary chloride and the left- and right-hand subscripts describe the geometry to the immediate left and right, respectively, of the chlorine atom. Subscripts H and C denote, respectively, hydrogen and carbon atoms.) Additionally, experimental observations of low frequency intermolecular vibrations in secondary chlorides (48) and PVC (49) have led to the estimation of intermolecular force constants. The internal and symmetry coordinates used are consistent with those described in the publication of Opaskar and Krimm (39).

Initially, Moore and Krimm calculated the normal vibrations of an isolated chain of syndiotactic PVC, that has a line group isomorphous to the point group C_{2v}. The A_1, B_1, and B_2 species are both infrared and Raman active, while the A_2 species are forbidden in the infrared. The calculated frequencies and their potential energy distributions are compared to the observed frequencies and observed infrared dichroic behavior in Table 16.8. In general, there is excellent agreement with the calculated and observed frequencies. It should be reemphasized that Moore and Krimm transferred their valence force field derived from model secondary dichlorides without modification, thus increasing the confidence in the calculations. From correlation tables it may be determined that the C_{2v} point group associated with the isolated chain is

Table 16.8 Observed and Calculated Frequencies and Potential Energy Distributions (PED) of a Single Planar Zigzag Syndiotactic Poly(vinyl chloride) Chain

Obs. Frequency[b] (cm^{-1})	Obs. Dichroism[c]	Calc. Symmetry	Calc. Frequency[d] (cm^{-1})	PED[a] (%)
1437		A_2	1431	$\delta(73)^e$
1428s	σ	A_1	1422	$\delta(75)$
1387w	π	B_2	1384	$H_\pi(45)$, $R(39)$
1355w	σ	B_1	1356	$w(84)$
1338ms	σ	A_1	1320	$t(50)$, $\theta(31)$
1316*		A_2	1315	$H_\pi(62)$, $t(59)$
1258s	σ	B_1	1255	$\theta(79)$
1230mw	π	B_2	1226	$w(55)$, $H_\pi(41)$
1195w	σ	A_1	1192	$\theta(42)$, $R(31)$
		A_2	1151	$t(79)$, $H_\pi(19)$
1122w	π	B_2	1117(1118)	$R(55)$, $w(14)$
1105m	σ	A_1	1099	$R(36)$, $t(20)$
1090mw(sh)	σ	B_1	1079	$R(51)$, $r(24)$
1066*		A_2	1065	$R(68)$
960ms	σ	B_1	965	$r(50)$, $R(23)$
835mw	π	B_2	826	$r(88)$, $H_\pi(16)$
640ms	σ	A_1	638	$X(92)$
604s	σ	B_1	601	$X(89)$
544*		A_2	549	$W(68)$
492vw	σ	B_1	482	$W(38)$
358w	σ	A_1	357	$X_\pi(52)$
345w	π	B_2	330	$X_\pi(83)$
315m	σ	B_1	314	$X_\pi(47)$
		A_2	133	$X_\pi(68)$
89		B_2	84(105)	$\tau(85)$
		A_1	38(48)	$\tau(96)$

[a] Only values greater than 10% are given.
[b] Cf. ref. 50 for IR and ref. 51 for Raman bands (*).
[c] σ means perpendicular, π parallel.
[d] Frequencies in parentheses where obtained using force constant $f_{\tau\tau}=0.0744$.
[e] See ref. 47 for definition of coordinates.

related to the D_{2h} space group associated with the PVC crystal in the following manner:

Isolated Chain (line group)	Crystal (space group)
C_{2v}	D_{2h}

$$A_1 \quad < \quad \begin{array}{l} A_g \\ B_{2u} \end{array}$$

$$A_2 \quad < \quad \begin{array}{l} A_u \\ B_{2g} \end{array}$$

$$B_1 \quad < \quad \begin{array}{l} B_{1u} \\ B_{3g} \end{array}$$

$$B_2 \quad < \quad \begin{array}{l} B_{1g} \\ B_{3u} \end{array}$$

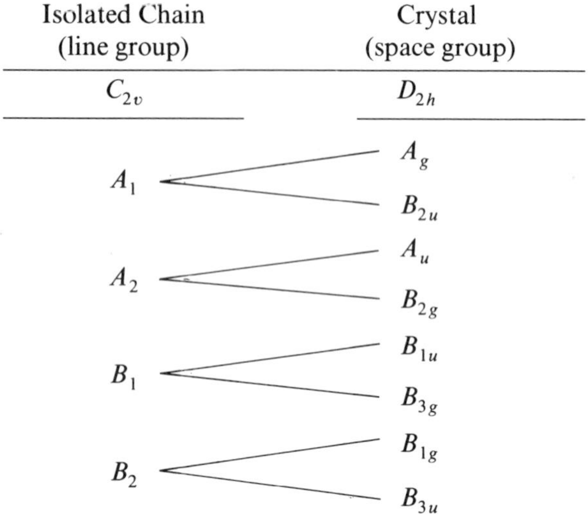

The results of the calculation based on the crystalline unit cell of syndiotactic PVC are given in Table 16.9 and are in good agreement with the observed experimental data. Assignments of the modes occurring between 300 and 1000 cm^{-1} are in general agreement with those previously proposed (37, 38, 43). However, reassignment of the bands in the 1030–1122 cm^{-1} region was necessary (see ref. 39 for further details). In fact, the splitting of the normal modes of PVC is difficult to observe, since each mode splits into an infrared- and a Raman-active band and little additional information is obtained from calculations of the crystal compared to the isolated chain above 80 cm^{-1}. However, introduction of the intermolecular force constants permits the calculation of five new frequencies. Thus one intermolecular infrared-active mode (associated with the rotation of chains in opposite directions) and four intermolecular Raman-active modes (associated with chain rotation in the same direction and with chains translating as units relative to one another in each of three directions) are calculated. Experimentally, only an infrared band at 67 cm^{-1} is observed, and Moore and Krimm assigned this band to a mode calculated at 64 cm^{-1}, whose potential energy is distributed between intramolecular torsional vibration and an intermolecular motion.

Moore and Krimm also applied their valence force field to calculate the frequencies of deuterated PVCs. The authors point out that because of differences in anharmonic components, transference of the force field will be less effective. Nevertheless, the agreement between observed and calculated frequencies for deuterated PVCs is reasonable. As expected, the assignment of the modes in the 700–1000 cm^{-1} region are the most difficult, owing to the contraction of the frequency range in which the CD_2 rocking, CD_2 twisting,

Table 16.9 Observed and Calculated Frequencies and Potential Energy Distributions (PED) of Crystalline Syndiotactic Poly(vinyl chloride)

Obs. Raman Frequency[b] (cm^{-1})	Obs. IR Frequency[c] (cm^{-1})	Calc. Frequency (cm^{-1})	Calc. Symmetry and Activity	PED[a] (%)
		12	B_{1g}, R[d]	$\beta(97)$[e]
		29	B_{1u}, IR	$\alpha(97)$
		39	A_g, R	$\tau(65)$, $\alpha(21)$
		45	B_{3g}, R	$\alpha(94)$
	67	64	B_{2u}, IR	$\tau_1(32)$, $\alpha(29)$
		80	B_{3g}, R	$X_P(93)$
		80	A_g, R	$X_P(85)$
		86	B_{1g}, R	$\tau_2(86)$
	89	90	B_{3u}, IR	$\tau_2(79)$
		134	B_{2g}, R	$X_\pi(66)$, $W_2(24)$
		135	A_u, f	$X_\pi(66)$, $W_2(24)$
310w		315	B_{3g}, R	$X_\pi(46)$, $W_1(33)$
	315m	321	B_{1u}, IR	$X_\pi(48)$, $W_1(28)$
	345w	330	B_{3u}, IR	$X_\pi(86)$
345sh		330	B_{1g}, R	$X_\pi(86)$
363		360	A_g, R	$X_\pi(50)$, $H_\pi(16)$
	358w	376	B_{2u}, IR	$X_\pi(47)$, $H_\pi(14)$
469vw		490	B_{3g}, R	$W_1(35)$, $H_\pi(19)$
	492vw	490	B_{1u}, IR	$W_1(40)$, $H_\pi(18)$
		550	A_u, f	$W_2(62)$, $X_\pi(26)$
544vw		550	B_{2g}, R	$W_2(62)$, $X_\pi(26)$
599m		602	B_{3g}, R	$X(105)$, $W_1(13)$
	604s	604	B_{1u}, IR	$X(105)$, $W_1(11)$
	640ms	637	B_{2u}, IR	$X(102)$, $X_\pi(21)$
638vs		642	A_g, R	$X(100)$, $X_\pi(21)$
	835mw	828	B_{3u}, IR	$r(89)$, $H_\pi(15)$
838w		828	B_{1g}, R	$r(89)$, $H_\pi(15)$
	960ms	969	B_{1u}, IR	$r(50)$, $R(24)$
964m		969	B_{3g}, R	$r(50)$, $R(24)$
1066w		1066	B_{2g}, R	$R(69)$, $H_\pi(13)$
		1066	A_u, f	$R(69)$, $H_\pi(13)$
	1090mw(sh)	1084	B_{1u}, IR	$R(52)$, $t(24)$
		1084	B_{3g}, R	$R(53)$, $t(23)$
1101m		1107	A_g, R	$R(41)$, $t(19)$
	1105m	1107	B_{2u}, IR	$R(41)$, $t(19)$
1119w		1118	B_{1g}, R	$R(55)$, $w(14)$
	1122w	1118	B_{3u}, IR	$R(55)$, $w(14)$
		1152	A_u, f	$t(81)$, $H_\pi(17)$
		1152	B_{2g}, R	$t(80)$, $H_\pi(17)$
	1195w	1200	B_{2u}, IR	$\theta(39)$, $t(31)$
1187w		1201	A_g, R	$\theta(39)$, $t(31)$
		1230	B_{1g}, R	$w(56)$, $H_\pi(39)$

Table 16.9 Continued

Obs. Raman Frequency[b] (cm^{-1})	Obs. IR Frequency[c] (cm^{-1})	Calc. Frequency (cm^{-1})	Calc. Symmetry and Activity	PED[a] (%)
	1230mw	1230	B_{3u}, IR	$w(56)$, $H_\pi(40)$
1257w		1277	B_{3g}, R	$\theta(78)$, $H_\pi(31)$
	1258s	1278	B_{1u}, IR	$\theta(78)$, $H_\pi(31)$
		1322	A_u, f	$H_\pi(63)$, $t(16)$
1316m		1321	B_{2g}, R	$H_\pi(63)$, $t(16)$
1335m		1332	A_g, R	$t(43)$, $\theta(36)$
	1338ms	1331	B_{2u}, IR	$t(44)$, $\theta(36)$
1357w		1356	B_{3g}, R	$w(84)$, $R(17)$
	1355w	1356	B_{1u}, IR	$w(84)$, $R(17)$
	1387w	1388	B_{3u}, IR	$H_\pi(46)$, $R(38)$
1379w		1388	B_{1g}, R	$H_\pi(46)$, $R(38)$
1430s		1422	A_g, R	$\delta(75)$
	1428s	1423	B_{2u}, IR	$\delta(75)$
1437m		1431	B_{2g}, R	$\delta(73)$
		1432	A_u, f	$\delta(72)$

[a] Only values greater than 10% are given.
[b] Cf. ref. 51.
[c] Cf. ref. 50.
[d] R≙Raman active, IR≙infrared active, f≙forbidden.
[e] Cf. ref. 47 for definition of coordinates, Approximately, β=Cl---H—C bending, α=C—Cl---H bending, τ=torsion, X_π=C—C—Cl bending, W=C—C—C bending, H_π=C—C—H bending, X_P=H---Cl stretching, X=C—Cl stretching, R=C—C stretching, r=CH_2 rocking, t=CH_2 twisting, δ=CH_2 bending, w=CH_2 wagging, θ=H—C—Cl bending.

CH bending, and CH wagging modes occur, resulting in a high degree of mixing. Further details of the assignment of the normal modes of deuterated PVCs are given in (39), to which the interested reader is referred.

Rubĉić and Zerbi (40) considered the problem of the effect of configurational and conformational defects on the vibrational spectra of PVC. Initially, these authors reanalyzed the perfect PVC chain, which was deemed essential as a reference point with which to compare disordered systems. Accordingly, the phonon dispersion curves $\omega(\mathbf{k})$, the vibrational density of states $g(\omega)$, the $\mathbf{k}=0$ phonon frequencies, and the corresponding form of the normal modes for PVC was calculated for three configurationally and conformationally most probable translationally symmetrical structures, these being (a) an extended syndiotactic planar zigzag (TTTT) that has a line group symmetry isomorphous to the point group C_{2v}; (b) an isotactic threefold helical structure (TGTGTG), C_3 point group; and (c) a folded syndiotactic structure (TTGGTTG′G′), D_2 point group. Geometric parameters were assumed as follows: C—Cl=1.795 Å, C—C=1.54 Å, C—H=1.09 Å; all bond angles were tetrahedral and the

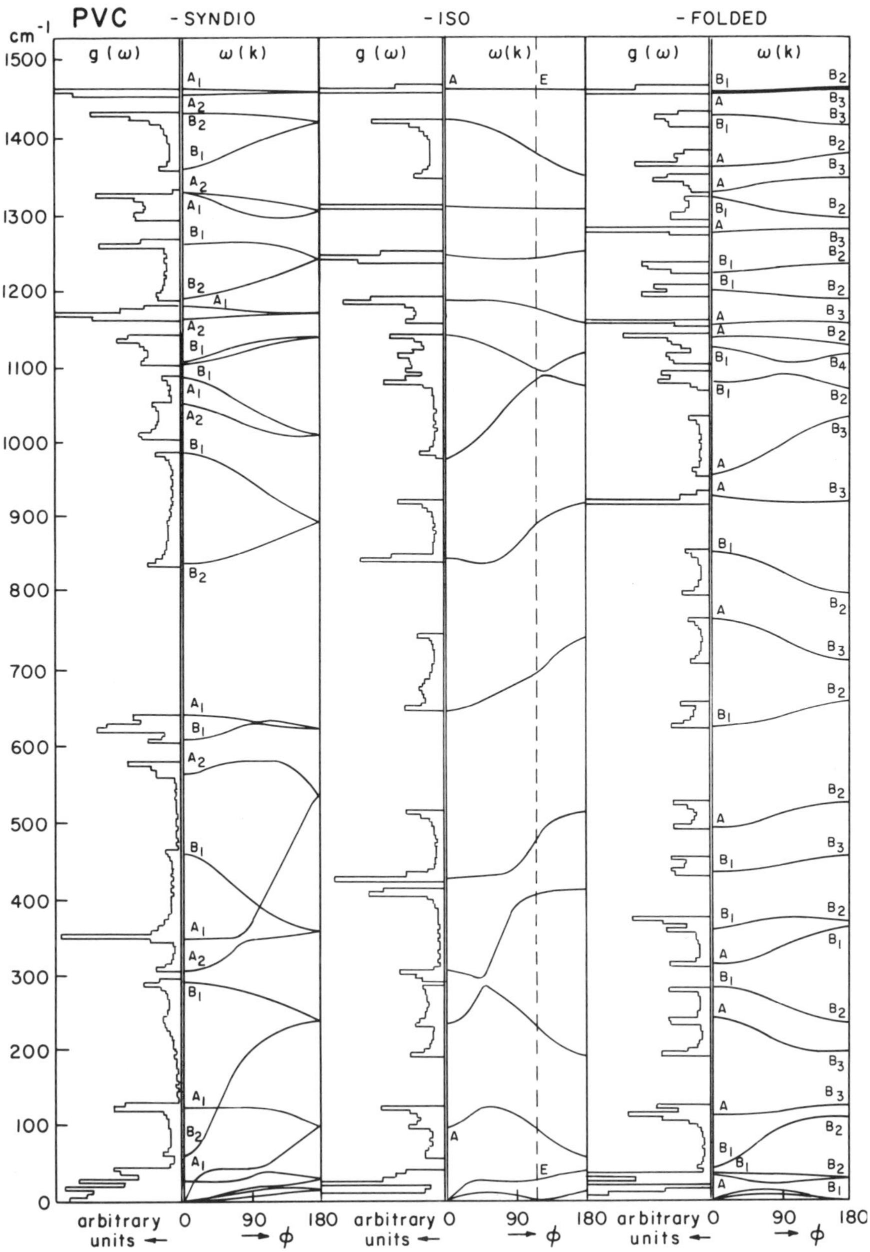

Figure 16.6. Dispersion curves $\omega(k)$ and density of states $g(\omega)$ from 0 to 1500 cm^{-1} of the three possible models of the PVC single chain. (Reprinted by permission from ref. 41.)

Table 16.10 $k=0$ Phonon Frequencies and Symmetry Species for Three Models of Poly(vinyl chloride)

Species	No.	Exptl. Freq. (cm^{-1})	Calc. Freq. (cm^{-1}) for $k=0$, OK Force Field	Description of Modes from Potential Energy Distribution
		I. Extended Syndiotactic PVC		
A_1	1	2970	2985	CH stretch
	2	2910	2856	CH_2 symm stretch
	3	1428	1460	CH_2 scissor
	4	1338	1324	CH_2 twist
	5		1177	CH bend
	6	1105	1084	Skeletal
	7	640	638	CCl stretch
	8	364	350	CCl bend
	9		26	Torsion
A_2	10		2855	CH_2 symm stretch
	11		1454	CH_2 scissor
	12		1326	CH bend
	13		1161	CH twist
	14		1047	CC stretch
	15		563	CCl bend
	16		125	CCl wag
B_1	17	2970	2985	CH stretch
	18	2930	2928	CH_2 asymm stretch
	19	1355	1356	CH_2 wag
	20	1258	1257	CH bend
	21	1030	1101	CC stretch, CH bend
	22	960	984	CH_2 rock, CC stretch
	23	604	606	CCl stretch
	24	490	462	CCC bend
	25	312	291	CCl bend
B_2	26		2928	CH_2 asymm stretch

27	1387	1427	CH bend
28	1230	1188	CH_2 wag
29	1090	1105	CC stretch
30	835	837	CH_2 rock
31	340	306	CCl wag
32		59	Torsion

Species	No.	Calc. Freq. (cm^{-1}), OK Force Field, $k=0$	Species	No.	Calc. Freq. (cm^{-1}), OK Force Field, $k=0$
			II. Isotactic PVC		
A	1	2985	*E*	17	2985
($\phi=0°$)	2	2929	($\phi=2\pi/3$)	18	2928
	3	2856		19	2857
	4	1460		20	1457
	5	1419		21	1376
	6	1308		22	1305
	7	1244		23	1241
	8	1185		24	1173
	9	1139		25	1093
	10	980		26	1083
	11	848		27	894
	12	646		28	700
	13	426		29	480
	14	306		30	400
	15	233		31	226
	16	95		32	92
				33	28

Table 16.10 Continued

No.	Calc. Freq. (cm^{-1}), $k=0$, for Species A and B ($\varphi=0°$) and B_2 and B_3 ($\varphi=\pi$) A	No.	B_1	No.	B_2	No.	B_3
			III. Folded PVC				
1	2986	18	2985	35	2985	52	2985
2	2929	19	2928	36	2928	53	2928
3	2856	20	2857	37	2856	54	2857
4	1457	21	1458	38	1461	55	1457
5	1361	22	1426	39	1376	56	1413
6	1327	23	1319	40	1293	57	1346
7	1274	24	1224	41	1233	58	1278
8	1154	25	1199	42	1187	59	1156
9	1137	26	1123	43	1126	60	1114
10	954	27	1077	44	1069	61	1029
11	926	28	852	45	796	62	917
12	763	29	622	46	655	63	707
13	491	30	434	47	524	64	454
14	311	31	358	48	368	65	659
15	239	32	280	49	232	66	195
16	111	33	40	50	107	67	123
17	30	34	35	51	26	68	25

torsional angles were taken as T = 180° and G = 120°. After critically considering the available force fields and with definite reservations, the authors decided to employ that proposed by Opaskar and Krimm (OK) (46) with minimal revision, although for the sake of completeness an attempt was made to improve the force field by a least-squares refinement of the protonated and deuterated PVCs. [Evidentally, the improved force field of Moore and Krimm (47), which was published in the same year, was not known at that time.] The dispersion curves and density of states for the three models are reproduced in Figure 16.6. It is apparent that many of the branches of the dispersion curves show sizable **k** dependence, indicating that significant intramolecular coupling occurs along the polymer chain. For syndiotactic PVC crossing and repulsions of the frequency branches occur. Although ordered forms of isotactic or folded

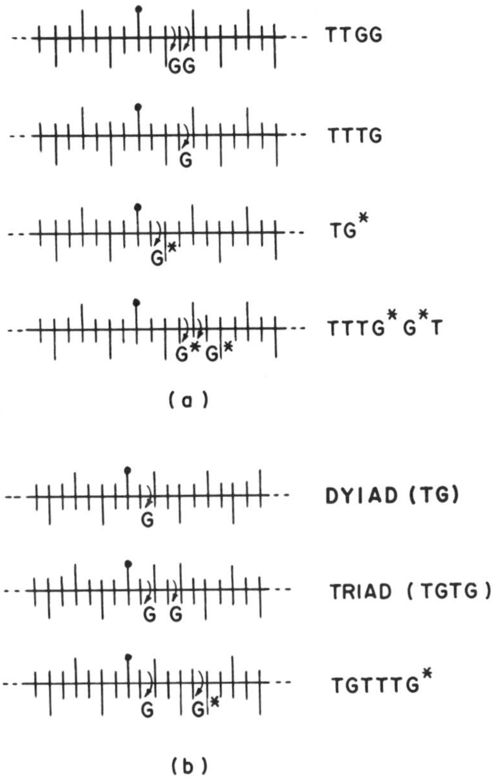

Figure 16.7. (*a*) Conformational defects introduced in an otherwise all-trans syndiotactic PVC chain. The dots refer to the starting point of the conformational sequence considered as defect. (*b*) Configurational defects introduced in an otherwise syndiotactic PVC chain. The conformational changes required by the introduction of an a configurational defect are also indicated. (Reprinted by permission from ref. 41.)

syndiotactic PVCs have not been synthesized, the calculated results are important because sequences of these structures could well be present in samples of PVC and the **k**=0 calculated phonon frequencies should appear in the optical spectrum. The calculated frequencies for all three models are reproduced in Table 16.10. In essence, from the foregoing studies, Rubĉić and Zerbi had obtained information concerning the precise position of the gaps and frequency bands over the entire vibrational spectrum and were in a position to search for characteristic gap or resonance modes in disordered PVC.

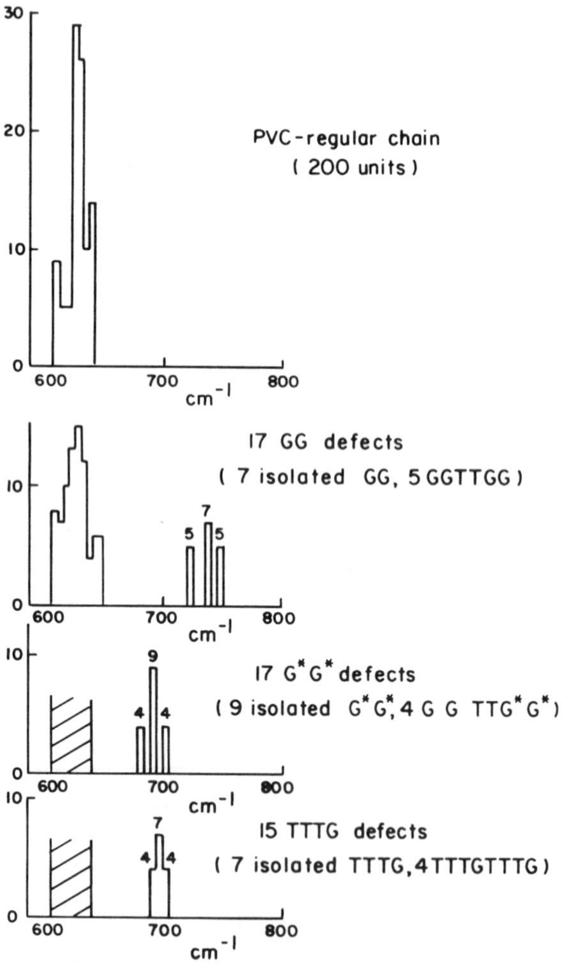

Figure 16.8. Comparison between the density of states in the C—Cl stretching region of a conformationally and configurationally regular syndiotactic PVC chain with that of the conformationally impure chains. Notice the correspondence between the population of defects and the number of gap frequencies indicated in the histograms. (Reprinted by permission from ref. 41.)

In a second paper, Rubĉić and Zerbi (41) considered the effects of defect structures in PVC and concentrated primarily on the 600–700 cm^{-1} region of the spectrum, where the characteristic C—Cl stretching modes occur. This region of the spectrum is sensitive to the local environment of the polymer chain (52). The approach taken by the authors was to use the negative eigenvalue theorem, which was described in Chapter 15 and which provides directly the density of states $g(\omega)$. Additionally, algorithms were employed to calculate approximate eigenvectors corresponding to the eigenvalues chosen in $g(\omega)$. The dynamics of PVC chains containing the following defects were considered:

1 Conformational isolated defects, as depicted in Figure 16.7*a*. Those considered were GG, G′G′, TTTG, and TTTG′ defect structures.
2 Configurational isolated defects; the introduction of a configurational defect (i.e., an isotactic placement) perturbs the chain conformation and thus there are additional conformational defects associated with a configurational defect. These are depicted in Figure 16.7*b* and are self-explanatory.

In calculations of the density of states, Rubĉić and Zerbi adopted chains of 200 "chemical units." Subsequently, the authors found that they could reduce computation time by calculating the eigenvectors for 50 chemical units without sacrificing accuracy. Resolution of the calculated $g(\omega)$ histograms and the eigenvectors was reported to be 5 cm^{-1} and 10^{-2} cm^{-1}, respectively.

Figures 16.8 and 16.9 show the comparison of the density of states in the C—Cl stretching region of a configurationally and conformationally pure syndiotactic PVC chain with that of PVC chains containing conformational and configurational defects. It is immediately apparent that clear gap frequencies are generated; furthermore, the shape of the density of states in the band is modified, which indicates that resonance modes are generated. Additionally, as the distance between defects along the chain is shortened, coupling between the gap modes gives rise to frequency splittings that are dependent on the distance. Finally, the gap frequencies for conformational and configurational defects occur very close to one another in the frequency range considered, leading to very complex spectral patterns, which are summarized in Table 16.11.

Rubĉić and Zerbi considered the nature of the gap and resonance modes from their calculations of the relevant eigenvectors. Accordingly, the authors expressed the total displacements of the atoms for GG conformational defects in an otherwise perfect syndiotactic chain. Additionally, they considered an isolated isotactic diad, two interacting isotactic diads, and an isotactic tetrad. An example of the GG conformational defect is shown in Figure 16.10, which illustrates the localization in space of the gap mode. The vibrational amplitudes decay very rapidly along the chain and the maximum amplitude is centered at the defect. It should be emphasized, however, that the number of chemical units involved in these gap modes is considerable (approximately 7–12).

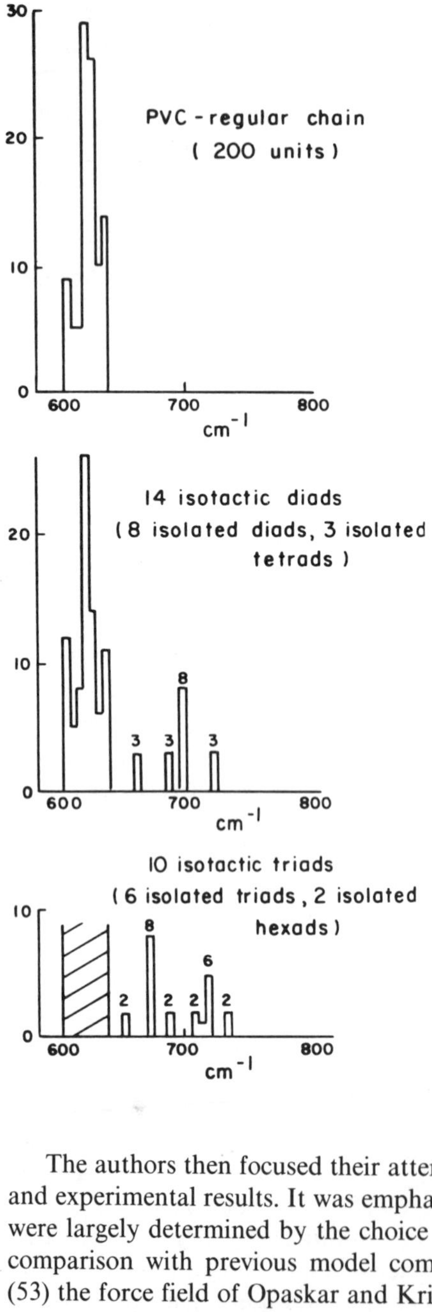

Figure 16.9. Comparison of the density of states in the C—Cl stretching region of a conformationally and configurationally regular syndiotactic PVC chain with that of configurationally impure chains. Notice the correspondence of the population of defects and the number of gap frequencies indicated in the histograms (Reprinted by permission from ref. 41.)

The authors then focused their attention on a comparison of the theoretical and experimental results. It was emphasized that the specific theoretical results were largely determined by the choice of force field, and in order to permit a comparison with previous model compound work reported in the literature (53) the force field of Opaskar and Krimm (46) was employed. Fair agreement was noted for gap frequencies. However, the calculations for the chain molecule containing defects suggest that frequencies occur within the C—Cl stretching band and these cannot be compared to model compound studies, since coupling with the phonons of the host lattice cannot be considered in

Table 16.11 Gap Frequencies in the C—Cl Stretching Region for Various Isolated or Interacting Conformational and Configurational Defects in a Trans Syndiotactic Host Lattice

Type of Defect	Position of Gap Freq. (cm^{-1}) in Histogram of Density of States[a]
	Conformation
GG	735–740
TTTG	695–700
G^+G^+	690–695
$TTTG^+$	650–655
GGTTGG	720–725; 745–750
$G^+G^+TTG^+G^+$	680–685; 700–705
TTTGTTTG	690–695; 700–705
$TTTG^+TTTG^+$	660–665
	Isoconfiguration
Diad	695–700
Triad	670–675; 715–720
Tetrad	660–665; 685–690; 720–725
Hexad	650–655; 670–675; 685–690 705–710; 730–735

[a] The histogram is calculated with a 5 cm^{-1} mesh.

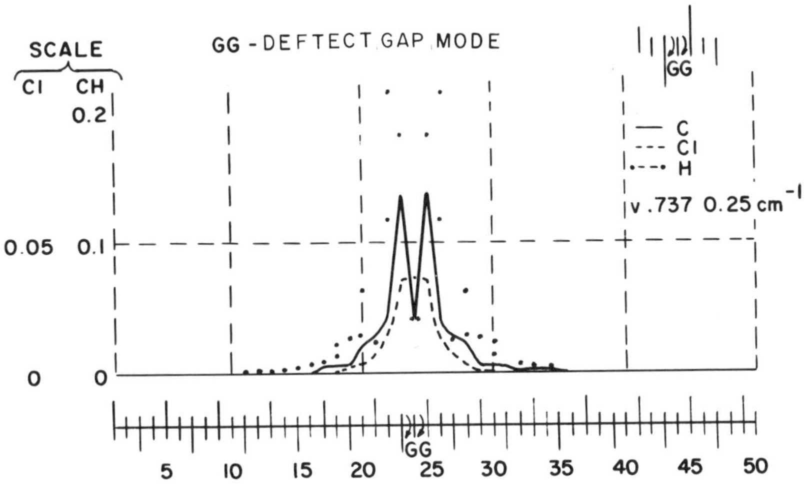

Figure 16.10. Gap mode (described in terms of total displacements of atoms) for a GG defect in a syndiotactic PVC chain. (Reprinted by permission from ref. 41.)

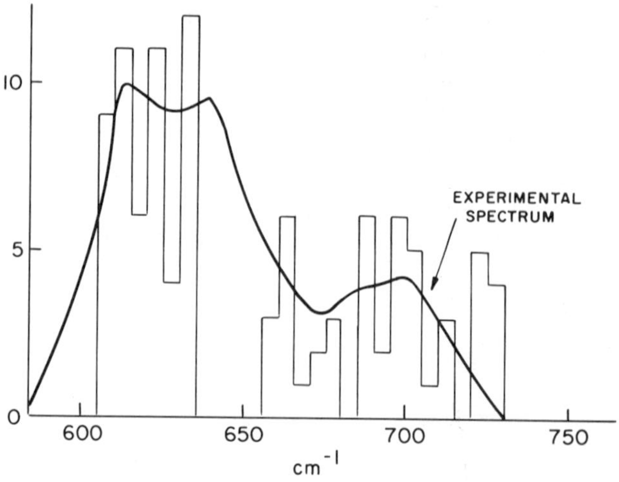

Figure 16.11. Density of vibrational states for a realistic model of configurationally disordered PVC. Comparison with the experimental spectrum. (Reprinted by permission from ref. 41.)

short-chain model compounds. Comparison of the theoretical and experimental results can be performed on the basis of frequency fitting alone or by the relative intensities of each normal mode. The latter requires a knowledge of the proper dipole or polarizability weighting of the calculated density of states. Unfortunately, our knowledge of the easier case, the calculation of the transition dipole moment, is strictly limited for even the simplest molecules. Nevertheless, Rubĉić and Zerbi decided to compare the dipole unweighted density of states for a highly disordered PVC to that of the experimental infrared spectrum in the C—Cl stretching region. The authors expected to find infrared absorptions due to (a) spectroscopically active **k**$=0$ modes arising from infinite or long sections of a translationally invariant chain; (b) gap modes from isolated or interacting defects (conformational and configurational); (c) activation of the density of states of band modes due to breakdown of selection rules; and (d) activation of characteristic in-band pseudolocalized motions. A comparison of the experimental spectrum and the unweighted density of states is shown in Figure 16.11.

In summary, Rubĉić and Zerbi have demonstrated that a study of the dynamics of conformationally and configurationally disordered PVC is feasible by numerical analysis. They have clearly shown that distinct gap frequencies and gap modes characteristic of structural defects occur in the C—Cl stretching region of the spectrum. Further, vibrational interactions between defects are predicted from their calculations. Finally, a reasonable agreement exists between the experimental spectrum and theoretical predictions for structurally disordered PVC. This excellent work has laid the foundation for a more complete understanding of the effect of conformational and configurational defects on the vibrational spectra of polymers in general.

Poly(vinylidene chloride)

Normal coordinate calculations of poly(vinylidene chloride) (PVDC) were first reported by Miyazawa and Ideguchi (54) in 1965 and more recently by Wu, Painter, and Coleman (55). The former authors proposed a TGTG′ conformation and performed a limited normal coordinate analysis utilizing a Urey-Bradley force field. Wu et al. obtained a valence force field from the model compounds 2,2-dichloropropane (DCP) and 2,2-dichlorobutane (DCB) and directly transferred this force field to a TXTX′ model of PVDC (where X and X′ are torsional angles of ±32°). In the following discussion, we will be emphasizing this work.

The polymer chain conformation of crystalline PVDC has been the subject of controversy for the past three decades. From x-ray studies it has been determined that the unit cell contains two polymer chains and four chemical repeat units (56, 57). The repeat distance is observed to be 4.68 Å, which effectively eliminates the planar zigzag conformation (2.34 Å). A number of conformational models containing two chemical repeat units per translational repeat have been proposed and are shown diagramatically in Figure 16.12. Wu et al. (58), employing a combination of geometric considerations, symmetry analysis, and predicted polarization and intensities, were able to eliminate specific conformational models and present arguments supporting a TXTX′ (X = ±32.5°) chain conformation for crystalline PVDC. This TXTX′ conformation is intermediate between the cis-planar and TGTG′ conformations and may be viewed as a compromise between repulsion of the eclipsed structure in the former and nonbonded adjacent CCl_2 units in the latter.

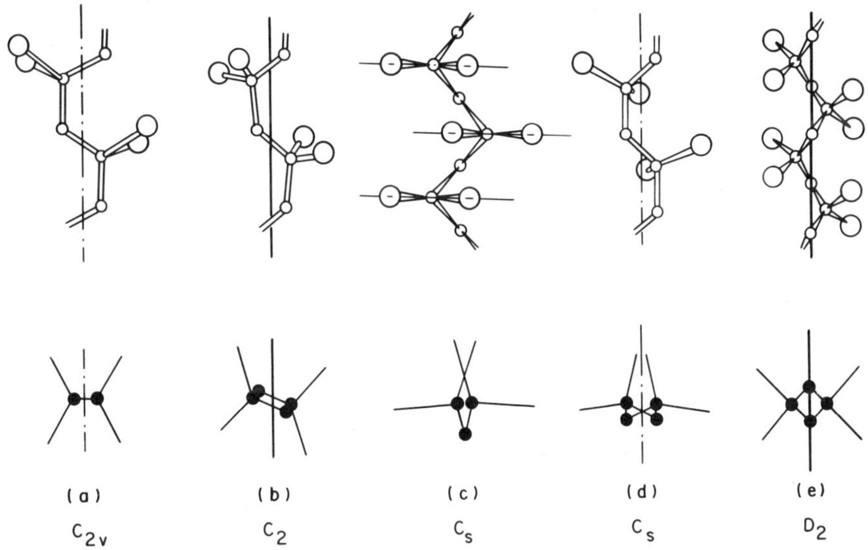

Figure 16.12. Conformational models of poly(vinylidene chloride). (Reprinted by permission from ref. 58.)

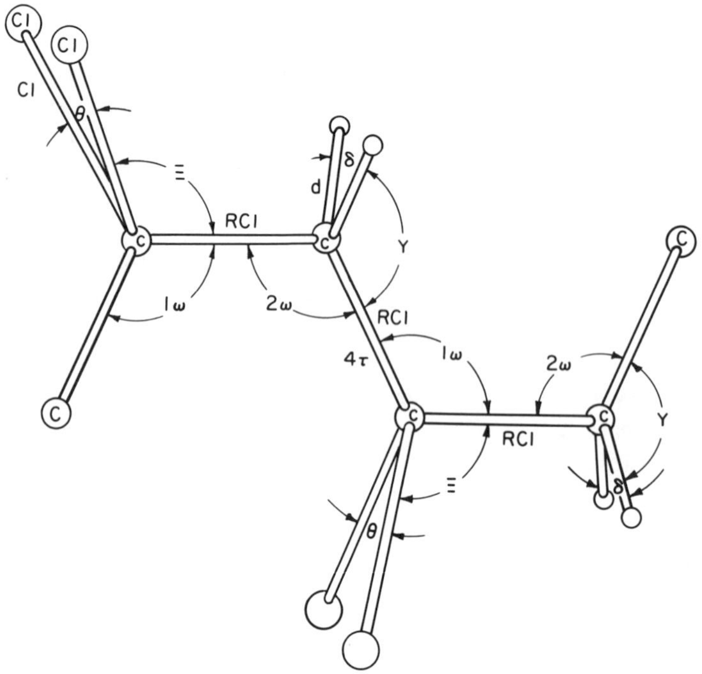

Figure 16.13. Internal coordinates of poly(vinylidene chloride). (Reprinted by permission from ref. 59.)

In a subsequent publication, Wu et al. (55) employed the TXTX′ conformational model in normal coordinate calculations. A schematic drawing of the model together with the nomenclature used for the internal coordinates is shown in Figure 16.13. In their calculations the authors used the following molecular parameters.

C—C = 1.54 Å	C—Ĉ—C = 120°
C—H = 1.09 Å	Cl—Ĉ—Cl = 110°
C—Cl = 1.795 Å	H—Ĉ—H = 110°

Internal rotation angle about the C—C—C bond = ±32.5°

The isolated chain TXTX′ model has a line group symmetry isomorphous to the point group C_s. This glide mirror plane is parallel to the polymer chain axis and the normal modes are distributed evenly between the two symmetry species A' and A''. The A'' modes exhibit σ polarization, while the A' modes have complex polarization with components of both σ and π, which is dependent on the internal torsional angle (X).

The valence force field employed by Wu et al. was directly transferred, without refinement, from normal coordinate studies of the model compounds

Table 16.12 Force Constant Data for PVDC

Description[a]	Φ_r^b	Description	Φ_r
K_d	4.5380	$F'_{R\Xi}$	−0.2395
K_{RCl}	4.7741	$F_{Cl\Xi}$	0.4716
K_{Cl}	2.7465	$F'_{CL\Xi}$	−0.0293
H_γ	0.6630	$F_{Cl\theta}$	0.2070
H_Ξ	1.3021	$F'_{1Cl\omega}$	−0.1248
H_θ	1.0693	$F_{\gamma\gamma}$	−0.0190
$H_{2\delta}{}^c$	0.5028	$F'_{\gamma\gamma}$	0.0210
$H_{2\omega}$	1.0320	$F_{2\gamma\omega}$	−0.0220
$H_{1\omega}$	0.9792	$F_{\Xi\Xi}$	0.1769
$H_{4\tau}{}^c$	0.0744	$F'_{\Xi\Xi}$	0.1963
F_d	0.0190	$F_{1\Xi\omega}$	−0.0114
F_{3R}	0.7123	$F'_{\theta\Xi}$	−0.1016
F_{2R}	0.2421	$f^t_{\gamma\Xi}$	−0.1211
F_{RCl}	0.2408	$f^g_{\gamma\Xi}$	−0.1544
F_{Cl}	0.3203	$f^t_{1\gamma\omega}{}^c$	0.0730
$F_{2R\gamma}$	0.1845	$f^g_{1\gamma\omega}$	−0.0640
$F'_{1R\gamma}$	−0.0218	$f^t_{2\Xi\omega}$	−0.0774
$F_{2R\omega}$	0.3490	$f^g_{2\Xi\omega}$	0.0094
$F_{1R\omega}$	0.3490	$f^t_{\omega\omega}$	0.0950
$F_{R\Xi}$	0.2578	$f^g_{\omega\omega}$	0.0436

[a]Force constants are defined as in ref. 59.
[b]Force constant units: stretching constants, mdyn/Å; bending and torsional constants, mdyn-Å/rad^2; stretch-bend interaction constants, mdyn/rad.
[c]Force constants transferred from ref. 47.

DCP and DCB (59) and is listed in Table 16.12. Those constants not applicable, such as $H_{2\delta}$, $f^t_{1\gamma\omega}$, $f^t_{2\Xi\omega}$, $f^g_{\omega\omega}$, and $H_{4\tau}$ were transferred from the valence force field of secondary chlorides (47). The authors emphasize that normal coordinate analysis alone cannot be used to differentiate between the various conformational models of PVDC. However, once a preferred conformation is selected from other considerations, it should be a relatively straight forward task to assign the normal modes of crystalline PVDC. One major difficulty with this approach was considered. The model force field was derived from secondary dichlorides containing trans and gauche conformations. For intermediate conformations there is the distinct possibility of variations in the calculated frequencies due to the inadequacies of the force field. Nevertheless, the authors point out that the interaction force constants between C—C—Cl and C—C—H bending coordinates are relatively small and the calculated

Table 16.13 Normal Coordinate Calculations for the TXTX′ Conformational Model

Symmetry Species	Calc. Freq. (cm^{-1})	Potential Energy Distribution (%)[a]
A'	2920	$K_d(100)$
	2853	$K_d(99)$
	1391	$H_\gamma(63)$, $H_{2\delta}(33)$
	1377	$H_\gamma(50)$, $H_{2\delta}(48)$
	1258	$H_\gamma(96)$
	1173	$K_{RCl}(132)$, $H_\Xi(15)$
	1053	$H_\gamma(45)$, $H_\Xi(31)$, $K_{Cl}(21)$, $K_{RCl}(12)$
	920	$K_{RCl}(53)$, $K_{Cl}(22)$, $H_\gamma(14)$, $H_\Xi(10)$
	670	$K_{Cl}(72)$, $H_\Xi(25)$, $H_\gamma(20)$, $H_{4\tau}(11)$, $H_{2\omega}(10)$
	585	$K_{Cl}(58)$, $H_\Xi(17)$, $H_{2\omega}(14)$, $H_\gamma(13)$, $H_{1\omega}(10)$
	451	$H_\Xi(41)$, $H_{4\tau}(22)$, $K_{Cl}(14)$
	388	$H_\Xi(43)$, $K_{Cl}(29)$, $K_{RCl}(12)$, $H_\gamma(11)$
	268	$H_\theta(49)$, $H_\Xi(23)$, $K_{Cl}(11)$
	172	$H_\Xi(59)$, $H_{1\omega}(16)$, $H_{2\omega}(12)$
	148	$H_\Xi(127)$
	65	$H_{4\tau}(73)$, $H_\Xi(16)$
A''	2918	$K_d(100)$
	2854	$K_d(99)$
	1405	$H_\gamma(57)$, $H_{2\delta}(30)$, $K_{RCl}(23)$
	1373	$H_{2\delta}(51)$, $H_\gamma(46)$, $K_{RCl}(12)$
	1320	$H_\gamma(91)$
	1124	$K_{RCl}(49)$, $H_\gamma(20)$, $K_{Cl}(11)$
	1075	$K_{RCl}(101)$, $H_\Xi(18)$
	871	$H_\gamma(52)$, $K_{RCl}(19)$, $H_\Xi(15)$
	790	$H_\gamma(25)$, $H_\Xi(25)$, $K_{Cl}(22)$, $H_{2\omega}(19)$, $H_{1\omega}(10)$
	699	$K_{Cl}(96)$, $H_\Xi(44)$
	523	$K_{Cl}(54)$, $H_\Xi(19)$
	392	$H_\Xi(84)$, $H_{4\tau}(31)$
	283	$H_\theta(33)$, $H_\Xi(28)$, $K_{Cl}(20)$
	254	$H_\Xi(44)$, $H_\theta(20)$, $K_{Cl}(14)$
	168	$H_\Xi(73)$
	90	$H_{4\tau}(51)$, $H_\Xi(35)$, $H_\gamma(13)$

[a] Only diagonal force constants whose contribution is equal to or greater than 10% are included. Owing to contributions from interaction force constants the PED among the diagonal force constants may be greater than 100%.

frequencies should depend predominantly on the **G** matrix. A comparison of the calculated frequencies for the TGTG′ and TXTX′ (X = ±32.5°) conformational models appears to substantiate this assumption (60).

The results of the normal coordinate calculations for the TXTX′ (X = ±32.5°) conformational model of PVDC are given in Tables 16.13 and 16.14.

Table 16.14 A Comparison Between the Observed and Calculated Frequencies for the TXTX′ Chain Conformation of PVDC

Raman (cm^{-1})	Infrared (cm^{-1})	Calc. Freq. (cm^{-1})	Approx. Assignment[a]
		65	$\tau(\text{skel})_i\ A'$
99(m)	102(vw)[b]	90	$\tau(\text{skel})_o\ A''$
	113(vw)[b]		
143(w)	—	148	$t(CCl_2)_i\ A'$
161(s)	162(vw–sh)[b]	168	$w(CCl_2)_o\ A''$
184(ms)	185(w)π[b]	172	$\delta(CCC)_i\ A'$
242(s)	245(w)σ[b]	254	$r(CCl_2)_o\ A''$
290(vs)	291(m)π[b]	268	$s(CCl_2)_i\ A'$
—	307(w)[b]	283	$s(CCl_2)_o\ A''$
345(vw)	—		$\delta(CCC)_i + w(CCl_2)_o = 345?$
356(vs)	359(m)π[b]	388	$w(CCl_2)_i\ A'$
—	382(vw)σ[b]	392	$t(CCl_2)_o\ A''$
	430(vw)π[b]		Amorphous?
453(s)	454(m)π[b]	451	$r(CCl_2)_i\ A'$
531(m)	530(s) σ	523	$\nu^+(CCl_2)_o\ A''$
569(w)	568(vw)		Amorphous
	590(w–sh)		Amorphous
600(s)	602(s) σ	585	$\nu^+(CCl_2)_i\ A'$
653(s)	658(ms) σ	670	$\nu^-(CCl_2)_i\ A'$
688(ms)	688(vvw?)	699	$\nu^-(CCl_2)_o\ A''$
748(m)	752(m) σ	790	$\delta(CCC)_o\ A''$
879(m)	886(w) σ	871	$r(CH_2)_o\ A''$
	980(w–br)	920	$\nu^+(C{-}C)_i\ A'$
1042(w)	1041(vvs) σ	1075	$\nu^+(C{-}C)_o\ A''$
1071(w)	1070(s) σ	1053	$r(CH_2)_i\ A'$
1144(vw)	1142(w) π	1173	$\nu^-(C{-}C)_i\ A'$
	1180(vvw)	1124	$\nu^-(C{-}C)_o\ A''$
1273(w)	1265(vw–br)	1258	$t(CH_2)_i\ A'$
	1327(vw) σ	1320	$t(CH_2)_o\ A''$
	1342(vw–sh)	1373	$w(CH_2)_o\ A''$
1361(vvw?)	1358(w) π	1377	$w(CH_2)_i\ A'$
	1385(vvw)	1391	$s(CH_2)_i\ A'$
1403(m)	1403(mw), 1409(mw)	1405	$s(CH_2)_o\ A''$
2929(m)	2931(w) σ	2854	$\nu^+(CH_2)_o\ A''$
2944(m)	2947(w) σ	2853	$\nu^+(CH_2)_i\ A'$
2984(mw)	2982(mw), 2987(mw)	2920	$\nu^-(CH_2)_i\ A'$
3000(vw–sh)	3000(vvw?)	2918	$\nu^-(CH_2)_o\ A''$

[a]Based on the PED and cartesian displacement coordinates. Subscripts i and o denote in-phase and out-of-phase motions, respectively.
[b]From the data of Krimm (61).

The overall agreement between observed and calculated frequencies is very good; most frequencies are calculated to within 20 cm^{-1} of the observed data.

Details of the assignments of the normal modes of PVDC and a comparison of previous assignments based on the group frequency approach are presented by Wu et al. (55). However, the assignments of the normal modes in the 500–900 cm^{-1} region of the spectrum, where the four C—Cl stretching vibrations occur, deserves further comment. To elucidate the preferred conformation of crystalline PVDC from vibrational spectroscopic studies it is essential to identify the four C—Cl stretching vibrations. In early studies only polarized infrared spectra of PVDC copolymers (62) and the infrared spectra of deuterated PVDC and poly(vinylidene bromide) (63) were available. From this limited amount of data, attempts were made to assign relatively strong infrared bands to the four C—Cl stretching vibrations. This effort led to a debate in the literature concerning these assignments, and arguments were brought forth to rationalize individual interpretations. However, this is a classic case of a vibrational problem that is underdetermined by the experimental data. Raman spectra reported over a decade later revealed the presence in the spectrum of PVDC of a relatively strong line at 688 cm^{-1} that was either absent or of extremely weak intensity in the infrared spectrum of PVDC single crystals (64). A comparison of the infrared and Raman spectra in the 500–1000 cm^{-1} region is shown in Figure 16.14. The Raman line at 688 cm^{-1} is in the generally accepted range for a C—Cl stretching vibration and cannot be ignored. If one assumes that the 688 cm^{-1} line is indeed a C—Cl stretching vibration, it is obvious that a serious misinterpretation of the vibrational spectrum of PVDC arises from a study of the infrared spectra alone, and this explains the controversy in assignments of the four C—Cl infrared bands mentioned previously. Nonetheless, even with the additional Raman data the vibrational problem is still underdetermined. Ideally, it would be optimal to obtain polarized infrared and Raman data from a large perfect crystal of PVDC. However, this is practically impossible, and to date no one has succeeded in obtaining a highly oriented fiber or film of pure PVDC that can be used for infrared or Raman measurements. In fact, Coleman et al. (64) initially suggested, upon studying the infrared and Raman spectra of PVDC single crystals, that the preferred conformation of the PVDC chain was more consistent with a cis-planar model. Central to this interpretation was the fact that a cis-planar model has a line group symmetry isomorphous to the point group C_{2v}, which predicts that the asymmetric out-of-phase C—Cl stretching vibration occurs in the A_2 mode and is infrared inactive. At first glance, this interpretation appears to fit the experimental data, the 688 cm^{-1} Raman line being assigned to the A_2 mode. There are, however, several problems with this interpretation; the major one is that there is no obvious π band that can be assigned to the symmetric in-phase C—Cl stretching vibration.

Subsequently, these same authors examined the different conformations in detail (58) and considered geometric factors, symmetry, and predicted polarization and intensity estimations for the four C—Cl stretching vibrations. A

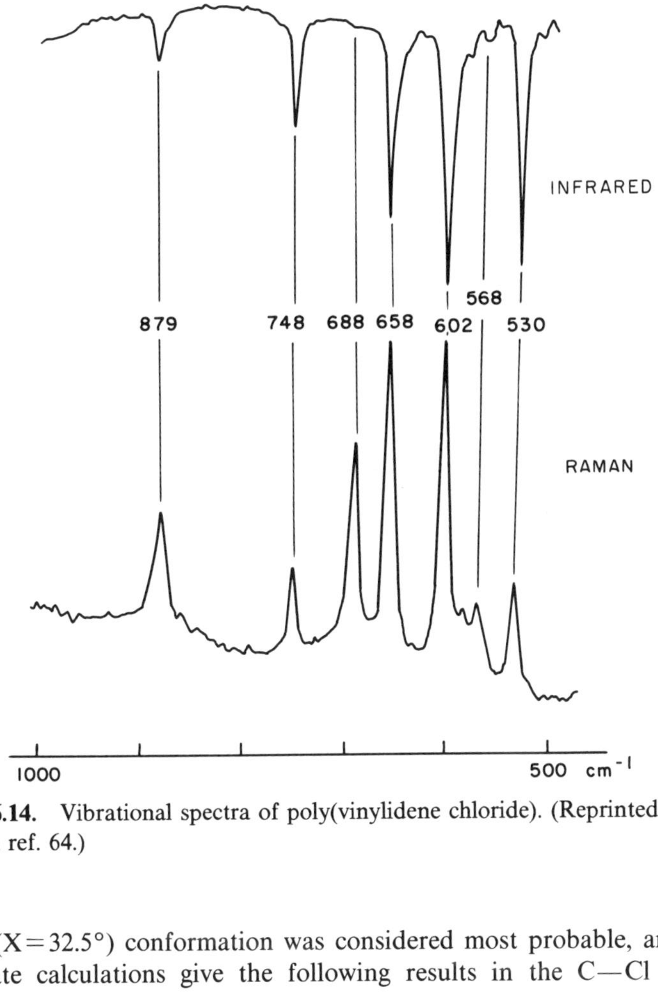

Figure 16.14. Vibrational spectra of poly(vinylidene chloride). (Reprinted by permission from ref. 64.)

TXTX′ (X=32.5°) conformation was considered most probable, and normal coordinate calculations give the following results in the C—Cl stretching region:

Observed (cm^{-1})			
Raman	Infrared	Calc. (cm^{-1})	Assignment
531(m)	530(s) σ	523	$\nu^+(CCl_2)_o\ A''$
600(s)	602(s) σ	585	$\nu^+(CCl_2)_i\ A'$
653(s)	658(ms) σ	670	$\nu^-(CCl_2)_i\ A'$
688(ms)	688(vvw?)	699	$\nu^-(CCl_2)_o\ A''$
748(m)	752(m) σ	790	$\delta(CCC)_o\ A''$
879(m)	886(w) σ	871	$r(CH_2)_o\ A''$

As mentioned previously, a symmetry analysis of the TXTX′ models indicated that the A'' modes are completely σ-polarized while the A' modes have a mixed σ and π character. Additionally, the relative intensities of three of the C—Cl stretching vibrations were estimated to be strong, while the other is weak (specifically the $\nu^-(CCl_2)_o$ mode). The $\nu^-(CCl_2)_o$ mode is calculated at 699 cm^{-1}, which is consistent with the observed data. The A' modes, which are considered to have both σ and π character, can be assigned to the vibrations occurring at 602 and 658 cm^{-1}, which from the experimental results (62) appear predominantly σ. The remaining totally σ A'' band is calculated at 523 cm^{-1} and therefore is assigned to the band at 530 cm^{-1}. The remaining infrared bands in this region at 752(m) and 886(w) are both σ-polarized, which is consistent with the calculated A'' (totally σ) modes associated predominantly with $\delta(CCC)_o$ and $r(CH_2)_o$ vibrations, respectively. Thus this region of the spectrum agrees well with the predictions based on the symmetry analysis and estimation of relative intensities.

Finally, it must be stated that although these results tend to substantiate a TXTX′ preferred conformation for crystalline PVDC, and although this conformation appears to be a pleasing compromise between the steric and repulsive factors involved, the vibrational spectroscopic studies to date cannot be considered to be definitive. The problem is still underdetermined.

Poly(vinylidene fluoride)

Normal coordinate calculations of poly(vinylidene fluoride) (PVDF) have been reported by Boerio and Koenig (65) and Kobayashi et al. (66). The former authors performed calculations employing a modified valence force field (VFF) on isolated chain models in two preferred conformations, a planar (all-trans) zigzag conformation representative of the so-called phase I (or form I) crystalline form and a TGTG′ conformation associated with phase II. Kobayashi et al. extended this work and included a third polymorphic crystalline form (phase III). In addition, these authors calculated the normal frequencies for the crystal lattices by assuming an intramolecular VFF together with van der Waals and electrostatic intermolecular forces. There is some controversy, however, concerning the crystal structure of phase III. This crystalline form was first suggested by Natta et al. (67) and Cortili and Zerbi (68). Doll and Lando (69) showed that phase III could be obtained by the crystallization of PVDF under high pressure. As noted by Kobayashi et al. (66), the result for the crystalline structure of phase III was not as definitive as that for phase I or II. At that time it was assumed that phase III has a monoclinic crystal system with a space group of $C121—C_2^3$ and a fiber axis repeat distance of 2.58 Å. Recent work by Weinhold et al. (70) indicates that the fiber axis repeat distance is actually 9.18 Å rather than 2.58 Å. This finding opens up the possibility for a number of different preferred conformations for phase III including one, a TTTGTTTG′ chain conformation, suggested by Bachmann et al. (71). In the following discussion we will initially concentrate on the normal coordinate calculations of Boerio and Koenig and then consider those of Kobayashi et al.

Boerio and Koenig considered two isolated chain conformations for PVDF. A planar zigzag (phase I) and a TGTG′ (phase II) conformation. The authors assumed the following molecular parameters:

$$\text{C—H} = 1.09\ \text{Å} \qquad \text{C—}\hat{\text{F}}\text{—C} = 109°\ 28'$$

$$\text{C—F} = 1.34\ \text{Å} \qquad \text{H—}\hat{\text{C}}\text{—H} = 109°\ 28'$$

$$\text{C—C} = 1.54\ \text{Å}$$

$$\text{C—}\hat{\text{C}}\text{—C for planar zigzag} = 112°$$

$$\text{C—}\hat{\text{C}}\text{—C for TGTG}' = 115°\ 36'$$

The isolated chain for the planar zigzag model of PVDF has a line group symmetry isomorphous to the point group C_{2v}. The optically active normal modes are distributed among the symmetry species as five A_1, three B_1, two A_2, and four B_2. All the normal modes are Raman active but the A_2 modes are infrared inactive. Additionally, the B_1 modes have parallel infrared dichroism while the A_1 and B_2 modes exhibit perpendicular infrared dichroism. In the case of the TGTG′ model there are two chemical repeats per translational repeat unit. The line group of the isolated chain is isomorphous to the point group C_s. The 32 optically active normal modes are split between the A' and A'' species. The A' modes, in which the two monomers in the translational repeat unit vibrate in phase, exhibit a mixture of perpendicular and parallel infrared dichroism. Conversely, the A'' modes, where the two monomer units vibrate out of phase, are purely perpendicular infrared dichroic.

Boerio and Koenig initially transferred the force constants associated with the methylene group from the VFF derived for polyethylene (34) and those for the difluoromethylene group from polytetrafluoroethylene (72). Preliminary assignments were made for the A_1, B_1, and B_2 modes of phase I and for some of the more prominent bands of phase II. New force constants were defined, and a final VFF was obtained by using a damped least-squares corefinement method in which the calculated frequencies were compared with the observed frequencies for both phases. This force field is shown in Table 16.15. The observed and calculated frequencies together with the final assignments for phases I and II of PVDF are given in Tables 16.16 and 16.17, respectively. The agreement between the calculated and observed frequencies is very good and a complete discussion of the assignments is given in (65).

As mentioned previously, Kobayashi et al. (66) subsequently studied the vibrational spectra of all three polymorphic forms of PVDF, taking into account not only intramolecular but also intermolecular forces. Furthermore, it has been established that a significant amount ($\simeq 10\%$) of sequence isomerism (head-to-head placements) occurs in the free radical polymerization of vinylidene fluoride. Accordingly, these authors also studied an alternating copolymer of ethylene and tetrafluoroethylene as a model for the head-to-head structural irregularities in PVDF.

Table 16.15 Valence Force Constants for PVDF

Constant	Coordinate(s) Involved	Common Atom(s)	Calc. Value[a]
1	CH		4.902
2	CC		4.413
3	CC, CF	C	0.740
4	CC, CC	C	0.148
5	CC, CCH	CC	0.206
6	CC, CCC	CC	0.273
7	CC, CCF	CC	0.567
8	CH, CH	C	0.058
9	CF		5.960
10	CF, CF	C	0.621
11	CF, CFF	CF	0.674
12	CCH		0.615
13	CCH, CCH	CC	0.105
14	CCH, CCH	CH	0.074
15	CHH		0.481
16	CCC		1.248
17	CCC, CCC(T)	CC	−0.036
18	CCF		1.262
19	CCF		1.280
20	τ	CC	0.050[b]
21	CF, CCF	CF	0.500
22	CCF, CCF	CC	0.178
23	CCF, CCF	CF	0.143
24	CCC, CCC(G)	CC	−0.061
25	CCC, CCH(G)	CC	0.106
26	CCC, CCH(T)	CC	0.207
27	CCC, CCF(G)	CC	−0.085
28	CCC, CCF(T)	CC	0.239
29	CCH, CCF(T)	CC	0.063
30	CCH, CCF(G)	CC	0.055

[a]Stretching constants have units of mdyn/Å; stretch-bend interactions have units of mdyn/rad; bending constants have units of mdyn-Å/rad^2.
[b]Torsional force constant constrained to this value.

X-ray studies by Hasegawa et al. (73) formed the basis for the calculations of Kobayashi et al. The crystallographic data are reproduced in Table 16.18. The intramolecular force constants employed were essentially the same as those obtained by Boerio and Koenig (Table 16.15). The only variations were as follows: For phase II, force constant 19 was changed from 1.28 to 1.50 mydn-Å/rad^2. Similarly, for phases I and III, force constants 3, 15, 19, 21, and 25 were changed from 0.740 to 0.403, 0.481 to 0.441, 1.280 to 1.50, 0.50 to 0.62, and 0.106 to 0.138, respectively, and in the appropriate units. Two types of

Table 16.16 Calculated and Observed Frequencies and Potential Energy Distribution (PED) for Planar PVDF[a,b]

Species	Frequency (cm^{-1}) Obs.	Calc.	PED (%)
A_1	2984	2980	CH(100)
	1435	1434	CHH(72)+CCH(17)
	1180	1185	CC(43)+CCC(47)+CF(45)
	879	886	CF(59)
	509	516	CFF(77)
A_2		990	CCH(144)
	(265)	270	CCF(131)
B_1	1400	1398	CCH(71)+CC(47)
	1070	1063	CC(71)+CCH(27)+CCF(29)
	480	484	CCF(79)
B_2	3020	3020	CH(100)
	1274	1256	CF(93)+CCF(27)
	840	844	CCH(58)+CF(26)
	490	483	CCF(67)

[a] Because of contributions from interactions, the PED among the diagonal force constants may be greater than 100%.
[b] Observed frequencies in parentheses were not used to compute force constants.

intermolecular forces were considered. The first is the van der Waals force acting between nonbonded atoms given by a Lennard-Jones type potential function, namely,

$$V(r)=\varepsilon\left[\left(\frac{r_{\min}}{r}\right)^{12}-2\left(\frac{r_{\min}}{r}\right)^{6}\right]$$

where ε is the depth of the potential energy minimum and $r_{\min}$ is the position of the minimum potential. The corresponding force constant was obtained by double differentiation of V with respect to r. Only H---F atomic pairs were considered, and values of $\varepsilon=0.3913$ kJ/mol and $r_{\min}=2.78$ Å were assumed. The other intermolecular force due to electrostatic interaction between the polar groups was calculated from

$$F(r)=\frac{2Q_1Q_2}{Dr^3}$$

where Q_1 and Q_2 are the values of the electric charges on the respective atoms, and D is the dielectric constant, assumed to be 4.0. Partial charges of 2.104×10^{-10} esu and -1.052×10^{-10} esu were assumed on the carbon and fluorine atoms, respectively, of the CF_2 group; these values were deduced from

Table 16.17 Calculated and Observed Frequencies and Potential Energy Distribution (PED) for TGTG′ PVDF[a, b]

	Frequency (cm^{-1})		
Species	Obs.	Calc.	PED (%)
A′	3022	3020	CH(100)
	2984	2980	CH(100)
	1428	1424	CHH(73) + CCH(17)
	1404	1404	CCH(76) + CC(33)
	1294	1285	CF(77) + CCF(24) + CC(13)
	1149	1161	CC(62) + CCH(22) + CCF(22) + CF(21)
	1070	1071	CF(48) + CCH(41) + CC(26)
	976	961	CCH(102)
	876	870	CC(31) + CF(21) + CCC(13)
	845	825	CCH(54) + CF(18)
	611	601	CFF(24) + CCF(24)
	488	498	CCF(45) + CFF(30)
	413	420	CCF(52)
	289	299	CCF(84)
	214	218	CCF(79) + CCC(34)

		63	CCC(37) + τ(36) + CCF(25)
A''		3020	CH(100)
		2980	CH(100)
	(1440)	1432	CHH(60) + CCH(26)
	1384	1383	CCH(53) + CC(18) + CF(17)
	1213	1251	CF(68) + CCH(36)
	1182	1175	CF(41) + CCH(38) + CC(24) + CCC(21)
	1059	1058	CC(91)
	(943)	948	CCH(103)
	(907)	909	CC(47) + CCH(18)
	799	811	CCH(72)
	768	757	CCF(35) + CF(29) + CCC(23)
	538	517	CFF(66)
		400	CCF(84)
	357	365	CCF(68)
		272	CCF(71)
		113	CCC(63) + CCF(41)

[a] Because of contributions from interactions, the PED among the diagonal force constants may be greater than 100%.
[b] Observed frequencies in parentheses were not used to compute force constants.

Table 16.18 Crystallographic Data of Three Forms of Poly(Vinylidene fluoride)

	Form I	Form II	Form III
Crystal system	Orthorhombic	Monoclinic	Monoclinic
Space group	$Cm2m\text{-}C_{2v}^{14}$	$P2_1/c\text{-}C_{2h}^{5}$	$C121\text{-}C_2^3$
Lattice constants	$a=8.58$ Å $b=4.91$ Å c(f.a.)$^a=$ 2.56 Å	$a=4.96$ Å $b=9.64$ Å c(f.a.)$=$ 4.62 Å $\beta=90°$	$a=8.66$ Å $b=4.93$ Å c(f.a.)$=$ 2.58 Å $\beta=97°$

[a] f.a. = fiber axis.

a CF bond moment of 1.41 D and a CF bond length of 1.34 Å. The values of the intermolecular force constants range from -0.002 to 0.0163 mdyn/Å. [A complete description of the individual force constants is given in (66).]

The number of normal modes and the selection rules for the three polymorphic crystalline forms of PVDF are shown in Table 16.19. The observed and calculated frequencies and the potential energy distribution for phases I and III are given in Table 16.20. Overall, the results of Kobayashi et al. and Boerio and Koenig for the intramolecular normal modes are in good agreement. However, the former authors have assigned infrared bands at 1273 and 1176 cm^{-1} to the A_1 and B_2 modes, respectively. These were assigned in the reverse manner by the latter authors. Additionally, the B_2 librational lattice mode was calculated by Kobayashi et al. to be at 72 cm^{-1}, which is in excellent agreement with the observed perpendicularly polarized infrared band at 70 cm^{-1}. At liquid nitrogen temperatures, very slight splittings of the librational lattice mode and other bands in the infrared spectrum were observed. The authors point out that these splittings are in conflict with the regular Cm2m crystal structure, since the spectroscopic unit cell contains only one chain, but they suggest that some kind of disordered structure may be the origin of this splitting.

As mentioned previously, there is considerable controversy concerning the crystal structure and chain conformation of phase III. Kobayashi et al. proposed a monoclinic cell with a space group of $C121\text{-}C_2^3$ upon which their calculations are based. The results, shown in Table 16.20, are in good agreement with the observed infrared and Raman data. The librational lattice mode is calculated at 106 cm^{-1} (somewhat higher than the similar mode calculated for phase I at 72 cm^{-1}, which is suggested to reflect stronger intermolecular forces in phase III) and is assigned to an infrared band observed at 84 cm^{-1}. Recent studies using computer-assisted infrared spectroscopy by Bachmann et al. (71) suggest that the number of “crystalline” infrared bands attributable to phase III is far greater than that observed previously and is inconsistent with

Table 16.19 Number of the Normal Modes and Selection Rules of the Polymorphic Forms of PVDF

Species	Molecular Modes	Lattice Modes[a]	Selection Rules Infrared[b]	Selection Rules Raman
		Phase I with Space Group $Cm2m\text{-}C_{2v}^{14}$		
A_1	5	T_b	Active ($\perp$)	Active
A_2	2		Forbidden	Active
B_1	3	T_c	Active ($\parallel$)	Active
B_2	4	T_a, $L(R_c)$	Active ($\perp$)	Active
		Phase III with Space Group $C121\text{-}C_2^3$		
A	7	T_b	Active ($\perp$)	Active
B	7	T_a, T_c, $L(R_c)$	Active ($\perp, \parallel$)	Active
		Phase II with Space Group $P2_1/c\text{-}C_{2h}^5$		
A_g	16	$L(T_a)$, $L(T_c)$	Forbidden	Active
B_g	16	$L(T_b)$, $L(R_c^0)$	Forbidden	Active
A_u	16	T_b, $L(R_c^\pi)$	Active ($\perp$)	Forbidden
B_u	16	T_a, T_c	Active ($\perp, \parallel$)	Forbidden

[a]T_a, T_b, T_c denote pure translation along the respective crystal axis; $L(R_c)$, librational lattice mode around fiber axis; $L(R_c^0)$, $L(R_c^\pi)$, librational lattice modes around fiber axis with phase difference of 0 and π between two chains in the unit cell, respectively.
[b]Parenthetical $\perp$ and $\parallel$ denote infrared polarization for a uniaxial specimen.

the results of Kobayashi et al. It should be emphasized, however, that Kobayashi et al. point out in their manuscript that they observed many bands in their infrared and Raman spectra that are associated with the crystalline phase but that are not considered fundamental vibrations. The authors obtained dispersion curves and calculated the frequency distribution functions $g(\nu)$ for phases I and III [see (66) for complete details]. The additional bands, which are not assigned to optically active fundamentals of the regular lattice, are attributed to a disordered crystal lattice. Four types of disorder were considered:

1 ***Disorder in Molecular Structure*** It was noted that most of the peaks of $g(\nu)$ corresponding to weak adsorptions appear at the positions close to the molecular modes with a phase angle of π. These modes are inactive for the infinite fully extended zigzag chain but will be activated if the molecular chains twist alternately from the planar structure.

2 ***Disorder in Molecular Packing*** The splitting of some infrared bands, especially at low temperatures, can be ascribed to this effect.

3 ***Distortions in Molecular Conformation*** Bands associated with the amorphous phase (e.g., gauche conformations) are present in the spectrum of semicrystalline PVDF. The authors were able to identify several of these bands by their increased intensity in the molten state. The advent of

Table 16.20 Observed and Calculated Frequencies and Potential Energy Distribution (PED) of Crystal Forms I and III of Poly(vinylidene fluoride)

	Form I						Form III						
	Frequency (cm^{-1})							Frequency (cm^{-1})					
	Observed							Observed					
Species	Infrared		Raman		Calc.	Species	Infrared		Raman		Calc.	PED[a] (%)
A_1	2980 (⊥)[b]	vw[c]	2984	s	2980	A	2980 (⊥)	vw	2984	s	2987	$\nu_s(CH_2)$[d] (99)
	1428 (⊥)	w	1436	s	1423		1427 (⊥)	w	1434	vs	1430	$\delta(CH_2)$ (81)
	1273 (⊥)	s	1283	m	1286		1269 (⊥)	w	1270	m	1287	$\nu_s(CF_2)$ (40) − $\nu_s(CC)$ (22) + $\delta(CCC)$ (15)
	884 (⊥)	s	886	s	879		882 (⊥)	s	884	s	880	$\nu_s(CF_2)$ (54) + $\nu_s(CC)$ (18)
	508 (⊥)	s	514	m	508		510 (⊥)	s	516	m	510	$\delta(CF_2)$ (98)
A_2	Inactive		980	w	983		950 (−)	vw	942	w	982	$t(CH_2)$ (100)
			268	m	262		270 (−)	vw	268	m	262	$t(CF_2)$ (100)
B_1	1398 (‖)	s	1400	w	1396	B	1400 (‖)	s	1397	w	1396	$w(CH_2)$ (58) − $\nu_a(CC)$ (35)
	1071 (‖)	m	1078	m	1065		1073 (‖)	w	1078	m	1065	$\nu_a(CC)$ (54) − $w(CF_2)$ (22) + $w(CH_2)$ (24)
	468 (‖)	s	475	w	470		483 (‖)	vs	487	m	473	$w(CF_2)$ (92)
B_2	3022 (⊥)	vw	3020	vs	3029		3020 (⊥)	vw	3020	vs	3036	$\nu_a(CH_2)$ (99)
	1176 (⊥)	s	1175	w	1182		1175 (⊥)	s	1178	m	1182	$\nu_a(CF_2)$ (64) − $r(CF_2)$ (21) + $r(CH_2)$ (15)
	840 (⊥)	s	845	vs	825		838 (⊥)	m	843	vs	825	$r(CH_2)$ (60) − $\nu_a(CF_2)$ (31)
	442 (⊥)	w	445	w	443		440 (⊥)	w	437	m	458	$r(CF_2)$ (74) + $r(CH_2)$ (26)
	70 (⊥)	s	77	w	72		84 (−)	s			106	Librational lattice mode

[a] The values obtained by the normal coordinate treatment for a single chain.

[b] Infrared dichroism; ⊥, electric vector perpendicular to the orientation direction; ‖, electric vector parallel to the orientation direction.

[c] Relative intensity; vs, very strong; s, strong; m, medium; w, weak; vw, very weak.

[d] Symmetry coordinates; ν_s, symmetric stretching; ν_a, antisymmetric stretching; δ, bending; w, wagging; t, twisting; r, rocking. The + or − denotes the phase relation among the symmetry coordinates.

computer-assisted vibrational spectroscopy should further aid in the elucidation of amorphous contributions by digital subtraction techniques, as indicated by Bachmann et al. (71).

4 ***Disorder in Chain Structure*** The amount of sequence isomerism (i.e., head-to-head placements) in commercial PVDF is substantial. From NMR studies it has been found that about 10% of the monomer units are incorporated "backward" into the chain, leading to a typical chain structure:

$$—CH_2CF_2CH_2CF_2CF_2CH_2CH_2CF_2CH_2CF_2—$$

Additionally, there is x-ray evidence (69) to suggest that these structural irregularities can be accommodated within the crystalline regions. In order to investigate the effect of these head-to-head sequences on the vibrational spectra of PVDF, Kobayashi et al. studied the infrared and Raman spectra of an alternating ethylene–tetrafluoroethylene copolymer. This is an excellent model for the head-to-head sequence. Furthermore, the authors performed normal coordinate calculations, assuming a planar zigzag conformation [details of the force field and a comparison of the observed and calculated frequencies together with the PED are given in (66)]. Strong infrared absorptions observed in the copolymer at 1453, 1323, and 666 cm^{-1} appear to be localized modes characteristic of head-to-head and tail-to-tail structures, since there are no optically active fundamentals occurring at these frequencies in any of the three crystalline forms of PVDF. In addition, the authors performed normal coordinate calculations, assuming a planar zigzag conformation, of sequences of the type $\text{+}(CH_2CF_2)_mCF_2CH_2\text{+}$, where m varied from 2 to 4. Attention was focused on the three localized modes mentioned earlier. The CH_2 bending mode (1453 cm^{-1}) was found not to depend on the value of m and is clearly an "in-band" localized defect mode characteristic of tail-to-tail units. Similarly, the CH_2 wagging mode is also an in-band localized defect mode due to $CH_2—CH_2$ groups with a frequency in the range of 1320–1335 cm^{-1}, depending on the number of tail-to-tail units involved. The CF_2 wagging mode (head-to-head units) is more complex. The frequency of this mode is strongly dependent on the number of head-to-head units in PVDF. As m increases, the atomic displacements tend to concentrate on the $CF_2—CF_2$ group and the frequency rises asymptotically to about 670 cm^{-1}. Thus the infrared band observed at 678 cm^{-1} in commercial PVDF is considered to be a "pseudolocalized" band characteristic of head-to-head units.

To return to the controversy concerning the crystal structure and chain conformation of phase III PVDF, there is obviously more work to be performed. Bachmann et al. (71) have cast doubt on the interpretation of

Kobayashi et al. (66). The basis for their arguments is more recent x-ray data, which suggest a much larger translational repeat unit, and their statement that "the number of allowed modes is inconsistent with the 45 bands which are observed in the crystalline spectrum of phase III." This argument is somewhat simplistic, however, especially in the light of the discussion earlier of the effects of disorder.

Finally, we turn our attention to the results obtained by Kobayashi et al. on phase II of PVDF. As anticipated, the intramolecular vibrations calculated by these authors do not differ significantly from those of Boerio and Koenig (65). This is not surprising since the intramolecular force fields employed are almost identical. However, Kobayashi et al. considered the normal vibrations associated with the space group $P2_1/C\text{-}C_{2n}^5$ and included their intermolecular force constants. In these circumstances, the intramolecular vibrations of each mode of the isolated chain belonging to the species A' and A'' (C_s site symmetry) are split into two modes in the crystal (A' to A_g and B_u and A'' to A_u and B_g, respectively, for C_{2h} site symmetry). The A_g and B_g modes are Raman active, while the A_u and B_u modes are infrared active. However, the frequency splittings due to the intermolecular interactions were too small to be detectable in the Raman and infrared spectra, even at liquid nitrogen temperatures. Five lattice modes were calculated, corresponding to the one A_u species (librational and infrared active with perpendicular dichroism), two A_g and one B_g species (translational and Raman active), and one B_g species (librational and Raman active). The A_u librational mode was assigned to the sharp perpendicular infrared band at 53 cm^{-1} (which shifts to 60 cm^{-1} at liquid nitrogen temperatures) and was calculated at 51 cm^{-1}. The three translational lattices modes calculated at 84, 59, and 11 cm^{-1} were assigned to Raman lines at 99, 52, and 29 cm^{-1}, respectively. The final librational mode calculated at 62 cm^{-1} has not yet been observed.

Polytetrafluoroethylene

Normal coordinate calculations of polytetrafluoroethylene (PTFE) were first reported in 1956 by Liang and Krimm (74) based on infrared data and a planar zig-zag chain model. Hannon, Boerio, and Koenig (75), who had obtained additional Raman data, calculated the normal vibrations of PTFE assuming a 15_7 helical chain model and employing a simple valence force field containing nine "assumed" and seven "adjusted" force constants. Subsequently, Boerio and Koenig (72) improved their force field by employing a damped least-squares method to refine a 19-parameter force field to the experimental PTFE data available from infrared, Raman, and neutron scattering experiments. Piseri et al. (76) studied the Boerio-Koenig force field and pointed out that although application of this force field yielded calculated frequencies that were in satisfactory agreement with the observed data, some imaginary frequencies (specifically the ν_8 branch of the dispersion curve) were calculated. This indicates that the chain is unstable under the force field. A

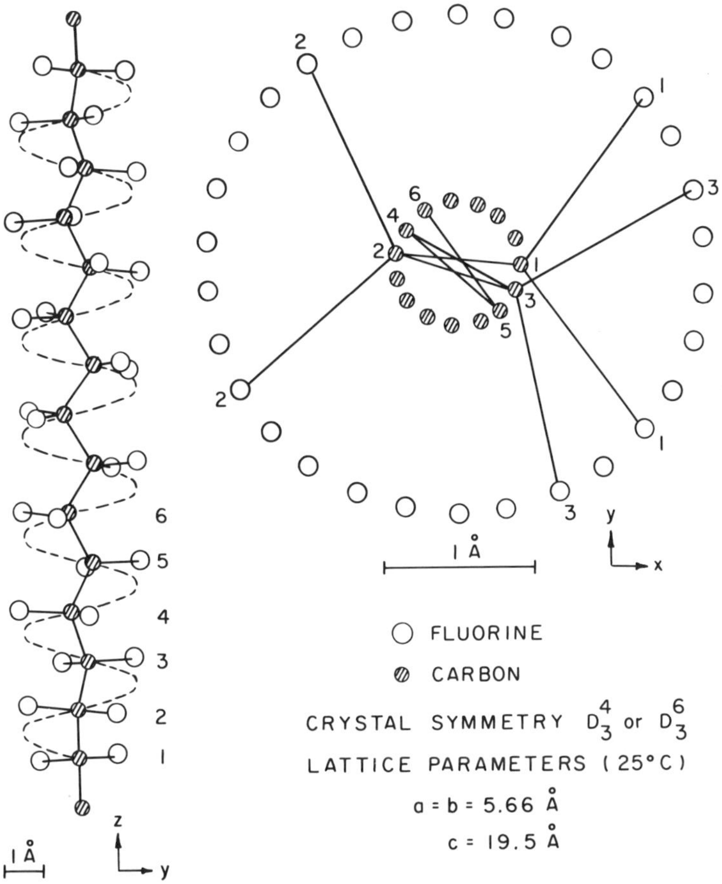

Figure 16.15. Elevation and plan views of one unit cell of the 15_7 conformation of PTFE at room temperature. Note that the scales of the two views are different. (Reprinted by permission from ref. 76.)

new 17-parameter force field was developed by Piseri et al. from primarily a least-squares fit of neutron scattering data. Zerbi and co-workers (77, 78) critically reviewed the two force fields of Boerio and Koenig and Piseri et al. and concluded that although there are problems associated with the former, it gives a better agreement to the optical spectra. Accordingly, Zerbi et al. calculated the phonon dispersion curves, density of states, and $\mathbf{k}=0$ phonon frequencies for several chain models of PTFE. Additionally, these authors studied the polymer chain as a disordered system. In the following discussion we will concentrate primarily on the results of Boerio and Koenig (72) and Zerbi et al. (77, 78).

PTFE is a homopolymer containing regular difluoromethylene units $(CF_2)_n$ along the chain and exists as a largely crystalline material in the solid state.

Two solid state transitions at approximately 30°C and 19°C have been observed and the exact source of these transitions is still open to some question. However, it is generally agreed that between 19 and 30°C, the PTFE polymer chain has a helical conformation in which 15 CF_2 groups are arranged in seven turns of the helix, as illustrated in Figure 16.15. The unit cell is believed to be trigonal with the chains forming a hexagonal array. Below 19°C, it has been suggested that the helix tightens to a 13_6 conformation in which the chains form an almost regular monoclinic or triclinic structure. Above 30°C the conformation may become more irregular, although the hexagonal arrangement of chains is maintained (79–82). Several interpretations of the solid state transitions have been advanced; see the summaries given by Tadokoro (83) and Zerbi and Sacchi (77) for further details.

The 15_7 helical conformation of PTFE was assumed by Boerio and Koenig (72) in their calculations. An isolated chain of PTFE has a line group isomorphous to the point group $D(14\pi/15)$. The optically active normal modes are distributed among the symmetry species in the following manner:

$4\,A_1$	Raman active (polarized)
$3\,A_2$	Infrared active (parallel dichroism)

Table 16.21 Force Field for PTFE[a]

Symbol	Value	Dispersion	Environment	Coordinate (S) Coupled
1. tt	6.095	0.107	$C_1(F_1,F_2)$	C_1,F_1
2. $\psi\psi$	1.628	0.156	$C_1(F_1,F_2)$	F_1,C_1,F_2
3. $\alpha\alpha$	1.086	0.023	$C_1—C_2(F_1,F_2)—C_3$	C_1,C_2,F_1
4. $\Phi\Phi$	1.375	0.610	$C_1—C_2—C_3$	C_1,C_2,C_3
5. rr	5.057	2.261	$C_1—C_2$	C_1,C_2
6. rt	0.702	0.104	$C_1—C_2(F_1,F_2)—C_3$	$(C_1,C_2)(C_2,F_1)$
7. $t\psi$	0.488	0.064	$C_1(F_1,F_2)$	(C_1,F_1), (F_1,C_1,F_2)
8. $t\alpha$	0.673	0.023	$C_1—C_2(F_1,F_2)—C_3$	(C_1,C_2,F_1), (C_2,F_1)
9. $r\alpha$	0.551	0.468	$C_1—C_2(F_1,F_2)—C_3$	(C_1,C_2), (C_1,C_2,F_1)
10. $r\Phi$	0.516	0.207	$C_1,C_2—C_3$	(C_1,C_2), (C_1,C_2,C_3)
11. $\Phi\Phi_1$	0.532	0.327	$C_1—C_2—C_3—C_4$	(C_1,C_2,C_3), (C_2,C_3,C_4)
12. rr_1	0.116	0.447	$C_1—C_2—C_3$	(C_1,C_2), (C_2,C_3)
13. $\alpha\gamma$	−0.114	0.117	$C_1—C_2(F_1,F_2)—C_3$	(C_1,C_2,F_1), (C_3,C_2,F_1)
14. $\gamma\alpha_1$	−0.025	0.121	$F_1—C_1—C_2—F_2$	(C_2,C_1,F_1), $(C_1,C_2,F_2)t$
15. $\gamma\beta_1$	0.126	0.121	$F_1—C_1—C_2—F_2$	(C_2,C_1,F_1), $(C_1,C_2,F_2)g$
16. $\alpha\Phi$	0.387	0.147	$C_1—C_2(F_1,F_2)—C_1$	(C_1,C_2,C_3), (C_1,C_2,F_1)
17. $\alpha\delta$	−0.118	0.116	$C_1—C_2(F_1,F_2)—C_3$	(C_1,C_2,F_1), (C_3,C_2,F_2)
18. ts	0.093	0.109	$C_1(F_1,F_2)$	(C_1,F_1), (C_1,F_2)
19. $\tau\tau$[b]	0.050			

[a] t, trans; g, gauche, bond stretching constants in mdyn/Å;
[b] Torsional force constant constrained to this value. Angle bending constants in mdyn-Å/rad^2.

Table 16.22 Comparison of Observed and Calculated Fundamentals

	Frequency (cm^{-1})	
Species	Calc.	Obs.
A_1	304	291
	387	383
	731	729
	1379	1379
A_2	522	516
	636	636
	1212	1210
E_2	14	*a*
	145	140
	300	308
	385	385
	523	524
	675	676
	741	741
	1216	1215
	1347	*a*
E_1	6	*a*
	188	198
	271	277
	271	277
	322	323
	552	553
	1150	1150
	1241	1242
	1298	1298

[a]Fundamental not observed experimentally.

8 E_1 Raman active (depolarized), infrared active (perpendicular dichroism)

9 E_2 Raman active (depolarized)

(The reader is referred to Chapter 11 where we discussed the normal modes of a helical polymer.) For the PTFE chain with a helix angle of $14\pi/15$, or 168°, the only optically active modes are those with phase angles of 0° (A_1 and A_2), 168° (E_1), and 24° (E_2). It will be seen from calculations of the dispersion curves that the E_2 modes lie close to the A modes in frequency.

After careful review of the experimental infrared, Raman, and neutron scattering results of PTFE and its oligomers, the authors concluded that they could confidently assign 21 optically active modes to the various symmetry species (four A_1, three A_2, seven E_1, and seven E_2) and six optically inactive modes from neutron scattering data. The 19-parameter force field reproduced

Table 16.23 Potential Energy Distribution of Optically Active Modes[a]

Species	Calc. Freq. (cm^{-1})	Force Constant (%)					
		tt	$\psi\psi$	$\alpha\alpha$	$\Phi\Phi$	rr	$\tau\tau$
$A_1(0)$	304	0	0	87	1	1	0
	387	0	68	21	0	9	0
	731	76	3	0	2	2	0
	1379	39	19	2	45	44	0
$A_2(0)$	522	6	0	87	0	0	0
	636	3	0	85	0	0	0
	1212	112	0	55	0	0	0
$E_2(24)$	14	0	0	2	0	0	99
	145	3	3	23	26	38	0
	300	0	0	88	0	1	0
	385	0	65	22	9	10	0
	523	8	0	83	0	2	1
	675	27	0	62	2	13	0
	741	49	3	12	2	13	0
	1216	111	0	52	0	0	1
	1347	42	20	4	43	40	0
$E_1(168)$	6	0	0	18	62	0	13
	188	7	0	99	0	0	0
	271	0	1	115	0	0	0
	322	0	1	54	0	9	0
	552	17	68	10	0	0	0
	1150	82	21	3	0	0	0
	1241	108	0	2	0	0	0
	1298	0	0	39	0	118	0

[a]Contributions to the potential energy distribution are given to nearest percent.

in Table 16.21 was developed. The torsional force constant, $\tau\tau$, was not included in the refinement procedure and was constrained to a value of 0.05 mdyn-Å/rad^2. The remaining force constants were employed in the refinement procedure and allowed to vary in order to obtain the best fit of observed to calculated frequencies. A comparison of the observed and calculated frequencies together with the potential energy distribution in terms of the force constants involved is given in Tables 16.22 and 16.23, respectively. A good agreement between the observed and calculated frequencies is apparent. Boerio and Koenig also indicated that their major force constant values are reasonably consistent with those reported by previous workers for model fluorocarbons. Dispersion curves were calculated by the authors and several of the branches exhibited considerable dispersion, indicating that significant coupling of the normal modes occurs. [Similar calculations of the 15_7 helix by Zerbi and Sacchi (77) yield identical results and are shown in Figure 16.16.]

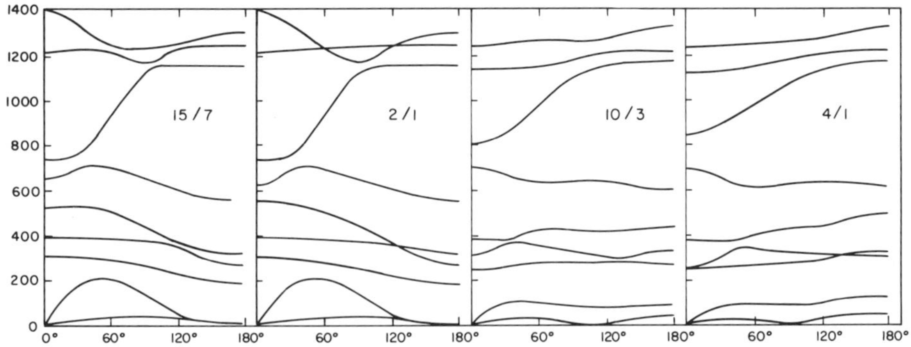

Figure 16.16. Dispersion curves for single-chain poly(tetrafluoroethylene): (*a*) 2/1 helix; (*b*) 15/7 helix; (*c*) 10/3 helix; (*d*) 4/1 helix. (Reprinted by permission from ref. 77.)

Zerbi and Sacchi (77), after an extensive review of the previously published studies of the vibrational spectra of PTFE, employed the Boerio-Koenig force field to calculate the normal vibrations of several chain models. The authors fully realized the limitations of this force field (in terms of large statistical dispersion and calculation of imaginary values of frequencies for the lowest acoustic branch for a phase shift between neighboring chemical units near 180°) but pointed out that it produces a superior fit to the optical spectrum when compared to the force field developed by Piseri et al. (76). The four models chosen, together with their relevant point groups and distribution of the $\mathbf{k}=0$ optical modes, are summarized in Table 16.24. The major goal of these studies was to predict spectral variations between these models, and it was tacitly assumed that most of the spectral changes could be ascribed to variations in the geometry (**G** matrix) of the isolated polymer chain. This assumption is justified by the fact that slight changes in the force constants do not radically change the calculated frequencies. In contrast, the dynamics of hydrocarbon polymers is strongly affected by both force field and geometry. Dispersion curves, density of states, and the calculation of the $\mathbf{k}=0$ phonon frequencies for the four models are reproduced in Figures 16.16 and 16.17 and Table 16.25, respectively. From Table 16.25 it is clear that while changes in the geometry yield minimal frequency changes in some of the fundamentals, others are markedly affected. This is more readily observed in the schematic diagram shown in Figure 16.18. The phonon dispersion curves (Figure 16.16) for the four models indicate extensive coupling of the phonons of different branches. Strong repulsions between optical branches of the same symmetry species are evident. A comparison of the density of states $g(\omega)$ of the planar zigzag 2_1 and the 15_7 helical models is informative (Figure 16.17). Since there is only a small change in the geometry of the chain between those two models, the position of calculated singularities, with the exception of the distribution of the phonons in the 500–600 cm^{-1} range, are very similar. For the 15_7 helix, two singularities at 520 and 552 cm^{-1} coalesce into a single peak at $\sim$545 cm^{-1} for the 2_1

Table 16.24 Models of Infinite One-Dimensional Lattices of Poly(tetrafluoroethylene) (PTFE) and Structure of the Irreducible Representations for Spectroscopically Active Phonons at $k=0$

Model	θ^a(deg)	τ^b(deg)	No. of CF_2 in Identity Period	No. of Turns	Point Group	Distribution of $k=0$ Optical Modes $\varphi=0$	$\varphi=\theta$	$\varphi=2\theta$
1	180	180	2	1	D_{2h}	3 A_g(R) 2 B_{3g}(R) 1 A_u(inactive) 1 B_{3u}(IR ∥)	2 B_{1g}(R) 1 B_{2g}(R) 2 B_{1u}(IR, ⊥) 2 B_{2u}(IR, ⊥)	
2	168	165.66	15	7	$D_{14\pi/15}$	4 A_1(R) 3 A_2(IR ∥)	8 E_1(IR ⊥, R)	9 E_2(R)
3	108	90.89	10	3	$D_{6\pi/10}$	4 A_1(R) 3 A_2(IR ∥)	8 E_1(IR ⊥, R)	9 E_2(R)
4	90	64.87	4	1	D_4	4 A_1(R) 3 A_2(IR ∥)	8 E(IR ⊥, R)	5 B_1(R) 4 B_2(R)

[a] Rototranslational angle.
[b] Torsional angle.

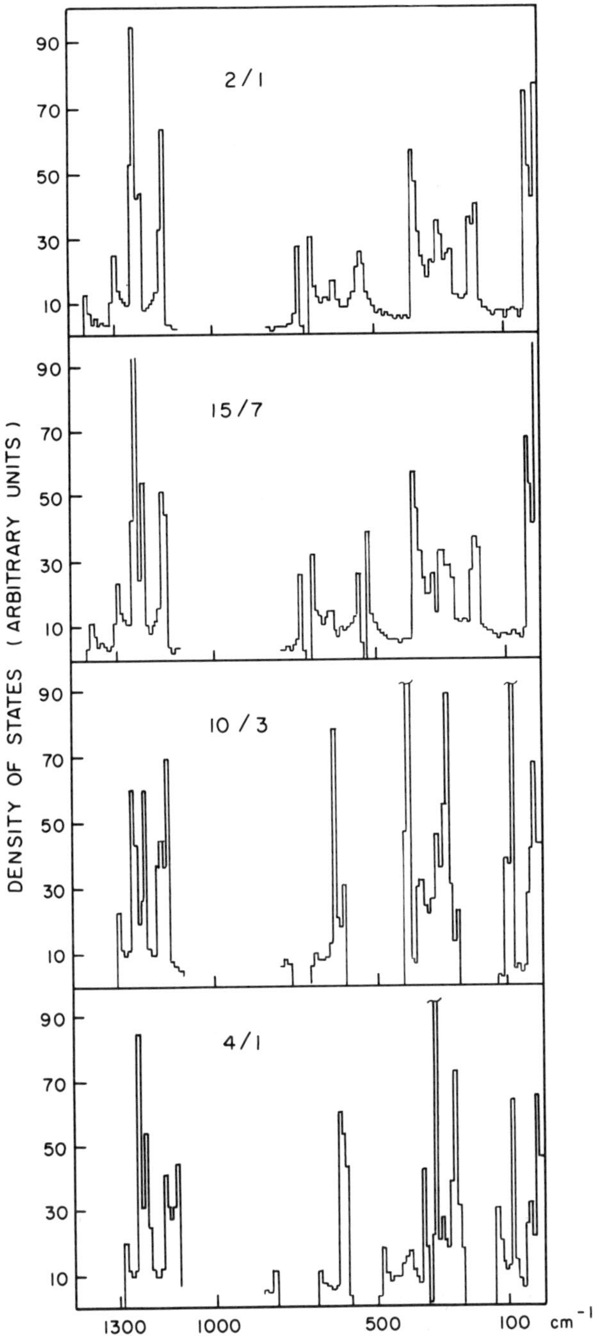

Figure 16.17. Density of vibrational states $g(\omega)$ for single-chain poly(tetrafluoroethylene): (*a*) 2/1 helix; (*b*) 15/7 helix; (*c*) 10/3 helix; (*d*) 4/1 helix. (Reprinted by permission from ref. 77.)

Table 16.25 Spectroscopically Active $k=0$ Phonon Frequencies (cm^{-1}) for Various Models of One-Dimensional PTFE Polymer Chains

	Model 1 $\theta=180°$		Model 2 $\theta=168°$		Model 3 $\theta=108°$		Model 4 $\theta=90°$	
$\varphi=0$	A_g(R)	1388	A_1(R)	1381	A_1(R)	1142	A_1(R)	1123
		731		731		701		687
		387		386		379		376
	B_{3g}(R)	1214		304		248		251
		542						
	A_u(inactive)	305	A_2(IR, ∥)	1213	A_2(IR, ∥)	1245	A_2(IR, ∥)	1236
				640		802		841
	B_{3u}(IR∥)	619		519		313		246
$\varphi=\theta$	B_{1g}(R)	1300	E_1(IR⊥,R)	1298	E_1(IR⊥,R)	1267	E(IR⊥,R)	1257
		316		1241		1198		1175
	B_{2g}(R)	265		1150		1134		1065
	B_{1u}(IR ⊥)	1243		552		635		617
		182		322		417		421
	B_{2u}(IR⊥)	1149		271		311		316
		548		187		277		272
				6.2		82		81
$\varphi=2\theta$			E_2(R)	1348	E_2(R)	1304	B_1(R)	1324
				1216		1217		1216
				741		1163		493
				677		615		301
				522		421		41
				385		308		
				300		278	B_2(R)	1169
				145		84		612
				14		21		325
								123

helix. The calculated density of states for the 10_3 and 4_1 helical models indicates that specific singularities can be taken as being characteristic of the models. Given that the force field can be accepted with reasonable confidence, the following calculated characteristic singularities may be used to differentiate between the models:

2_1 helix ~ 545 cm^{-1}
10_3 helix ~ 630, 405, 285, and 80 cm^{-1}
4_1 helix ~ 365, 325, and 260 cm^{-1}

Zerbi and Sacchi then turned their attention to the interpretation of the infrared and Raman spectra of PTFE. Particular attention was given to the phase transition of PTFE where the possibility of a transformation of the 15_7

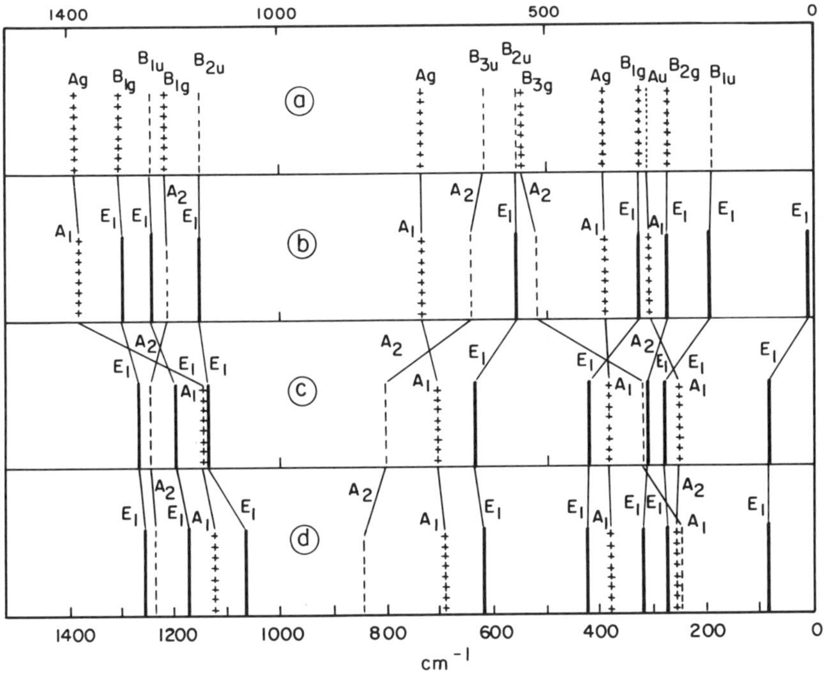

Figure 16.18. Correlation between $k=0$ phonon frequencies of various models of poly(tetrafluoroethylene). For each spectrum the symmetry species and Raman and infrared activities are indicated: (a) 2/1 helix; (b) 15/7 helix; (c) 10/3 helix; (d) 4/1 helix; _____, infrared and Raman active; + + +, Raman active; (---) IR active; (---) inactive. (Reprinted by permission from ref. 77.)

to the 2_1 helical chain conformation might occur. The predictions, based on optical selection rules and frequency calculations, suggest that the E_1 modes at 1298, 322, and 271 cm^{-1} and the A mode at 1213 cm^{-1} should disappear (become inactive) in the infrared spectrum upon transformation to the 2_1 helix. Furthermore, the A_2 mode at 640 cm^{-1} shifts to 625 cm^{-1} and the E_1–A_2 doublet at 552–519 cm^{-1} narrows at 548–542 cm^{-1} to a B_{2u}–B_{3g} doublet, of which one component is infrared inactive. Good experimental evidence (78) for these predictions has been observed for the 19°C phase transition and it was concluded that at about 19°C a mixture of 15_7 and 2_1 helices exist. The concentration of 2_1 helices increases significantly at this transition temperature. For evidence of 10_3 and 4_1 helical conformations the authors suggest that the relatively strong A_2 modes at ~800 and ~850 cm^{-1} would be the easiest to observe. Some supporting experimental evidence is observed in the infrared spectrum of PTFE. Specifically, the band at 780 cm^{-1}, which increases in intensity with increasing temperature and has been previously assigned to "amorphous" contributions, could be associated with the A_2 mode of the 10_3 helix. The experimental details are fully discussed in the paper of Masetti et al. (78).

Finally, in a manner similar to that discussed for PVC (see Section 16.2.), Zerbi and Sacchi calculated the lattice dynamics of a conformationally disordered PTFE chain. They considered finite chains of 200 CF_2 units with the following disordered structures:

1 $-(165.66°)_m—X—(165.66°)_n—$ where $m=n$. The defect X corresponds to internal rotational angles τ of 60°, 90°, and 180° distributed as singlets, doublets, or triplets (e.g., X=60°, −60° −90°, etc.).

2 $—(165.66°)_m—X—(165.66°)_n—$ where $m=n$ (random distribution).

3 A randomly coiled chain containing a statistical distribution of torsional angles whose population is dictated by a Boltzmann distribution at room temperature, as suggested by the potential energy calculations of DeSantis et al. (84).

The density of vibrational states $g(\omega)$ for these disordered chains was calculated using the negative eigenvalue theory (see Chapter 15). It was postulated that the vibrational spectrum of disordered PTFE chains should reveal (a) $k=0$ phonons of the host lattice; (b) the activation of inactive $k=0$ phonons of the perfect lattice and the activation of $k \neq 0$ phonons due to the lack of translational symmetry; (c) out-of-band, gap, or band modes; and (d) additional features arising from coupling between defects or with phonons of the host lattices. An example of the calculated density of states for the randomly coiled chain containing a distribution of torsional angles is shown in Figure 16.19. Zerbi and Sacchi concluded from their studies that in contrast to polyethylene, the vibrational spectrum of PTFE is not greatly changed by the

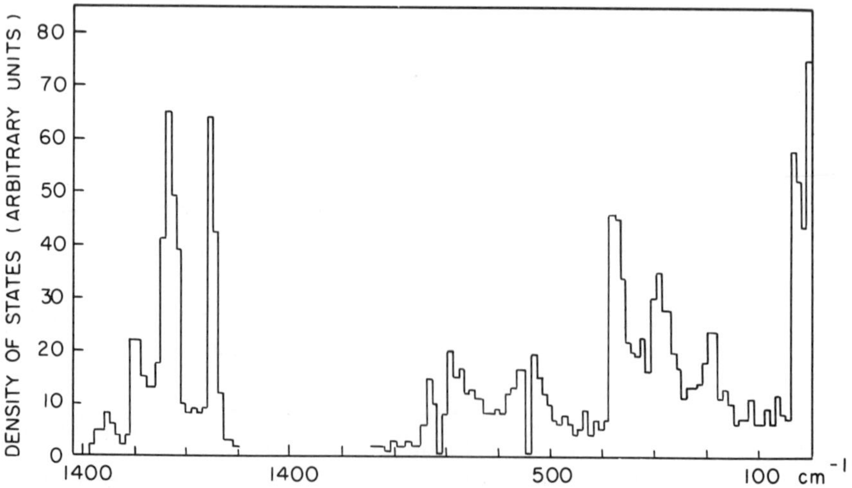

Figure 16.19. Density of vibrational states of conformationally disordered chain of poly(tetrafluoroethylene) consisting of 200 CF_2 units in a randomly kinked chain with a population of conformations given by a Boltzmann distribution. When the distribution percentage is 32%, $\tau=165.66°$; 32%, $\tau=-165.66°$; 22%, $\tau=180°$; 6%, $\tau=90°$; 6%, $\tau=-90°$; 1%, $\tau=-60°$. (Reprinted by permission from ref. 77.)

introduction of conformational defects. Furthermore, there is no clear evidence for gap or resonance modes characteristic of a given defect. The clearest evidence of the activation of $k \neq 0$ phonons from a breakdown of the selection rules due to lack of translational symmetry is the presence of an infrared band at 384 cm^{-1} that corresponds to the very strong A_1 Raman-active mode in the PTFE 15_7 helix at 383 cm^{-1}.

16.3 Polydienes and Polyalkenylenes

In this section we will consider the results of normal coordinate calculations performed on a general class of homopolymers that contain unconjugated olefinic double bonds in the chemical repeat unit of the polymer chain. The term polydiene, which is traditional and is still used extensively, is somewhat misleading; it was adopted to describe polymers synthesized from butadiene and its derivatives ($CH_2{=}CR{-}CR'{=}CH_2$). Many commercially significant polymers fall into this category, including the polybutadienes (R=R′=H), polyisoprenes (R=CH_3, R′=H), poly(2,3-dimethyl butadiene) (R=R′=CH_3), polychloroprenes (R=Cl, R′=H), and poly(2,3-dichlorobutadiene) (R=R′=Cl). More recently, the ring-opening polymerization of cycloolefins of the type

$$\overline{(CH_2)_n CH{=}CH}$$

has led to a class of polymers, called polyalkenylenes, that contain the general chemical repeat unit ($-CH{=}CH-(CH_2)_n-$).

Returning to the polymers synthesized from butadiene and its derivatives, we must be aware that many different polymers can be synthesized from a single diene monomer. By varying the polymerization conditions (e.g., method of polymerization, catalyst type, temperature, and solvent) it is feasible to synthesize polymers containing a distribution of *trans*-1,4 and *cis*-1,4, -1,2, and −3,4 structural units. Another complication arises from the fact that the *trans*-1,4 and *cis*-1,4 units may be incorporated into a chain "backward" leading to head-to-head and tail-to-tail placements. (See Figure 15.1.) However, in many cases, homopolymers can be synthesized (or obtained from natural sources such as Hevea (natural rubber) and balata) that contain chains made up predominantly of a single structural unit. Naturally, normal coordinate studies have been restricted to these microstructurally pure polymers. In the following subsections we will describe the results obtained for *trans*-1,4, *cis*-1,4, and syndiotactic 1,2-polybutadienes; *trans*-1,4-polychloroprene; *trans*-1,4-poly(2,3-dichlorobutadiene); *trans*-1,4-polyisoprene; *trans*-1,4-poly(2,3-dimethyl butadiene); poly(1-*trans*-pentenylene) and poly(1-*trans*-heptenylene).

Syndiotactic 1,2-Polybutadiene

Normal coordinate calculations of syndiotactic 1,2-polybutadiene were performed by Zerbi and Gussoni (85) in 1966. To our knowledge, this was the first published normal coordinate analysis of a polydiene.

The crystalline structure and chain conformation of the polymer were determined from x-ray studies by Natta and Corradini (45). The results indicate that syndiotactic 1,2-polybutadiene crystallizes in a rhombic lattice containing four chemical repeat units per unit cell with an identity period of 5.1 Å. Zerbi and Gussoni assumed a planar zigzag chain conformation for the isolated chain (see Figure 16.20) with the structural parameters as given below:

C—C=	1.53 Å	C—C	1.54 Å
C=C	1.337 Å	C—H	1.092 Å
=C—H	1.086 Å	Angles about the saturated C atom all tetrahedral	
H / =C \ H	120°		
H / =C \ C	120°		

$=CH_2$ and $=CCH$ groups are coplanar

The factor group of the isolated and infinite polymer chain is isomorphous with the point group C_{2v}. The translational repeat unit contains two chemical repeat units and the 56 optically active vibrational modes are distributed among 17 A_1, 11 A_2, 11 B_1, and 17 B_2 species. The A_1 and B_2 modes are predicted to be infrared active with perpendicular polarization; the B_1 modes, infrared active with parallel polarization; and the A_2 modes inactive in the infrared. The internal coordinates are depicted in Figure 16.21. Particular care was taken by Zerbi and Gussoni in the removal of redundancies and the definition of torsional coordinates.

Zerbi and Gussoni state that one of the major goals in their work is to test the concept of transferability of valence force constants between chemically and structurally similar molecules. Thus they directly transferred the valence force constants determined by the overlay method for branched paraffins and for methyl derivatives of ethylene that were reported by Snyder and Schachtschneider (11) and Schachtschneider (86), respectively. These are given in Tables 16.26 and 16.27.

The results of the calculation are presented in Table 16.28. As pointed out by the authors, the number of infrared-active fundamentals expected in the region from 3500 to 650 cm^{-1} is much larger than the number of bands observed by Morero et al. (87). Nevertheless, after some slight modification to the original assignments of Morero et al., which were based on the dichroism data and the group frequency approach, the calculated results are in good agreement with the experimentally observed frequencies. Zerbi and Gussoni

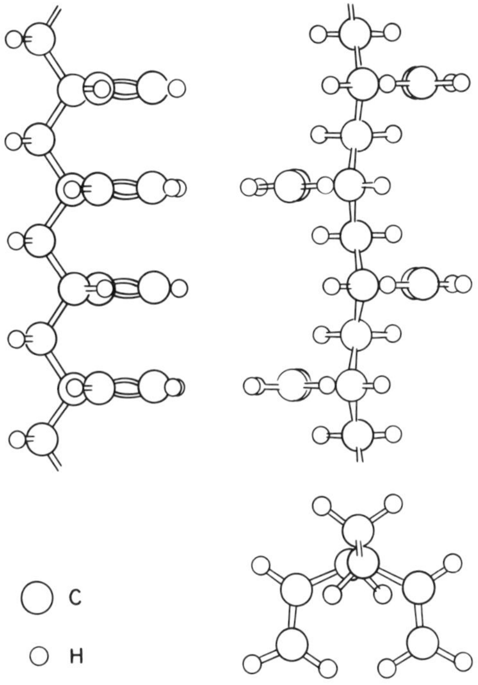

Figure 16.20. Structure of syndiotactic 1,2-polybutadiene. (Reprinted by permission from ref. 85.)

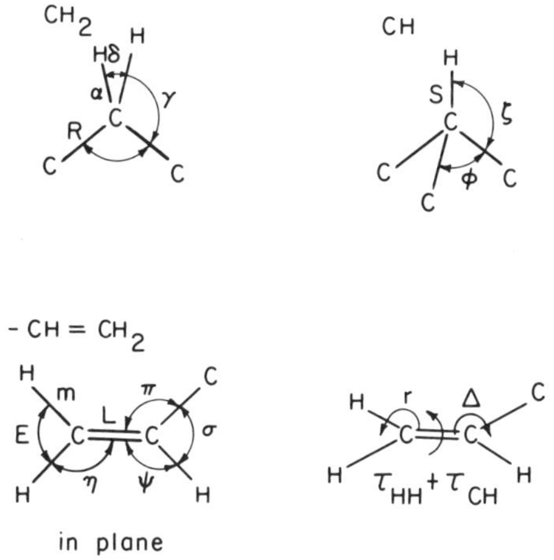

Figure 16.21. Definition of internal coordinates used in the calculations of the normal coordinates of syndiotactic 1,2-polybutadiene. (Reprinted by permission from ref. 85.)

Table 16.26 Valence Force Constants for Substituted Ethylenes[a]

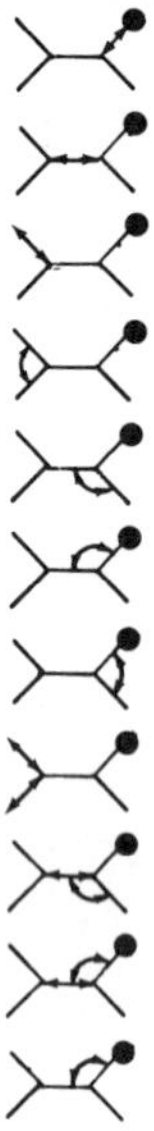

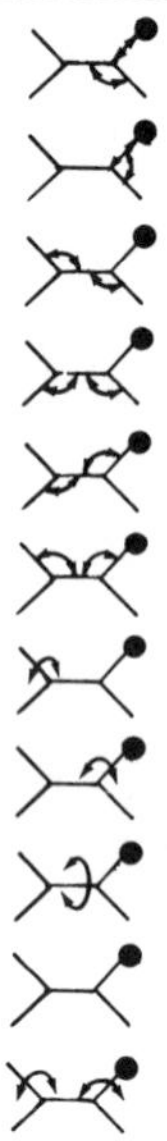

$K_p=4.47$	$F_{p\psi}=0.018$
$K_L=9.70$	$F_{P\sigma}=0.446$
$K_m=5.05$	$F^t_{\eta\psi}=0.122$
$H_\epsilon=0.378$	$F^c_{\eta\psi}=-0.042$
$H_\psi=H_\eta=0.553$	$F^t_{\eta\pi}=0.122$
$H_\pi=0.947$	$F^c_{\eta\pi}=0.056$
$H_\sigma=0.534$	$H_\Gamma=0.251$
$F_m=0.06$	$H_\Delta=0.297$
$F_{L\psi}=0.34$	H_τ } $=0.441$
$F_{L\pi}=1.16$	H'_τ
$F_{P\pi}=0.447$	$F_{\Gamma\Delta}=-0.012$

[a]For complete description see ref. 85.

Table 16.27 Valence Force Constants from Branched Paraffins Used for the Calculation of a Linear Hydrocarbon-like Polymer with a Vinyl Substituent[a]

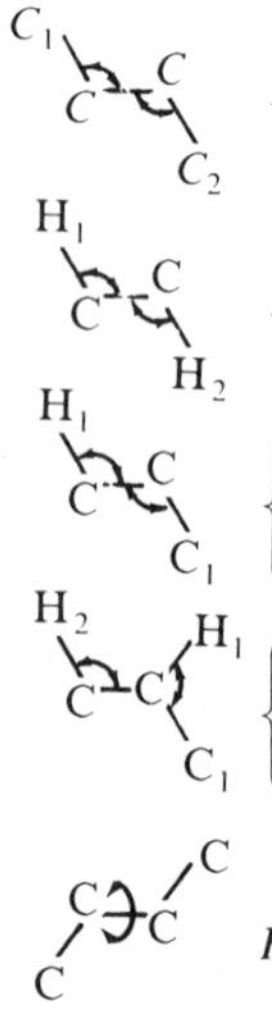

K_d 4.55	K_S 4.588	
K_R 4.387	H_ζ 0.657	
H_δ 0.550	H_φ 1.084	
H_γ 0.656	F'_ζ 0.012	
H_ω 1.13	F_φ 0.041	F^t_ω −0.011; F^γ_ω 0.11
F_d 0.01	$F_{\gamma\omega}$ −0.031	F^t_γ 0.127; F^g_γ 0.005
F_R 0.101	$F'_{R\gamma}$ 0.079	
$F_{R\omega}$ 0.328		$F^t_{\gamma\omega}$ 0.049; $F^g_{\gamma\omega}$ 0.052
$F_{R\beta}$ −0.021		F'^t_γ 0.044; F'^g_γ 0.09
F_γ 0.012		
F'_γ		H_τ

[a]For complete description see ref. 85.

Table 16.28 Syndiotactic 1,2-Polybutadiene

	Obs.[a]	Calc.	Error (%)	Description from PED[b]					
A_1	(3080)s	3085	−0.17	$=CH_2$	s				
		3047	—	$=CH$	s				
	(3012)shd	3007	−0.15	$=CH_2$	s				
	(2913)m	2905	0.26	$-CH$	s				
	(2847)m	2856	−0.31	CH_2	s				
	1643 s	1642	0.04	C=C	s	(86)			
	1452 m	1457	−0.39	CH_2	d_i	(90)			
	1419 s	1424	−0.37	$=CH$	d_i	(44),	$=CH_2$	d_i	(44)
	1333 s	1329	0.27	$=CH$	d_i	(50)			
	1265 shd	1258	0.54	CH_2	t_o	(41)			
	—	1205	—	CH	d_i	(40)			
	1131 w	1131	0.00	CC	s	(42),	CH	d_i	(37)
	1045 m	1054	−0.89	$=CH$	d_i	(43),	CH_2	t_o	(35)
	815 w	814	0.13	=C—C		(57),	CH_2	r_o	(30)
	—	470[c]	—	C=C—C	d	(35)			
	—	321[c]	—	$=CH_2$	r_o	(39),	C=C—C	d	(34)
	—	—[c]	—						
B_2	(3080) s	3085	−0.17	$=CH_2$	s				
	—	3047	—	$=CH$	s				
	(3012) shd	3007	0.15	$=CH_2$	s				
	(2913) m	2927	−0.47	CH_2	s				
	—	2902	—	CH	s				
	1643 s	1642	0.04	C=C	s	(86)			
	1419 s	1424	−0.4	$=CH_2$	d_o	(88)			
	1333 s	1334	−0.09	CH	d_o	(31)			
	1294 m	1290	0.25	Mixing					
	—	1194	—	C—C=	s	(32)			
	—	1117	—	CH_2	w_o	(72)			
	—	1085	—	C—C	s	(64)			
	—	1021	—	$=CH_2$	r_o	(45)			
	768 w	790	−0.27	C—C=	s	(52)			
	—	579	—	C—C—C=	d	(36),	C=C—C	d	(27)
	—	356	—	C—C—C	d	(57)			
	—	295	—	Skeletal deformation					
B_1	—	2927	—	CH_2	s				
	1343 m	1332	0.77	CH	w_i	(63)			
	1213 m	1200	1.09	CH_2	w_i	(62)			
	—	1131	—	$=CH_2$	t_o	(60),	$=CH$	w_i	(34)
	1070 w	1076	−0.57	CH_2	w_i	(57)			
	991 m	985	0.61	$=CH_2$	w_i	(92)			
	908 m	907	0.07	CH_2	r_o	(63)			
	664 s	664	−0.05	$=CH$	w_i	(50),	$=CH_2$	t_o	(27)
	—	349[c]	—	C—C—C	d	(73)			
	—	—	—	Torsional					

Table 16.28 Continued

	Obs.[a]	Calc.	Error (%)	Description from PED[b]
A_2	(Inactive)	2855		CH_2 s
		1447		CH_2 d_i (75)
		1433		CH_2 t_i (46)
		1131		$=CH_2$ t_i (61)
		1074		C—C s (84)
		984		$=CH_2$ w_o (90)
		897		CH_2 t_i (95)
		679		=CH w_o (33)
		516		C—C—C d (47)
		197		C—C—C d (41)

[a] The experimental frequencies and polarization have been taken from ref. 87. The assignment as proposed there has been rearranged.
[b] Only the most significant contribution of PED has been reported. The symbols i and o refer to in-phase and out-of-phase motions, respectively.
[c] These frequencies are somewhat affected by the value of the torsional force constant when it is introduced in the calculation.

discuss the assignments in more detail in their paper and further discussion would be redundant. It is emphasized, however, that this work does substantiate the concept of transferability of valence force constants to similar molecules.

trans-1,4-Polybutadiene

Normal coordinate calculations of *trans*-1,4-polybutadiene (TPB) were performed in 1967 by Neto and DiLauro (88) and in 1975 by Hsu et al. (89). The former authors published the results of calculations based on the isolated chain of TPB and its symmetrically deuterated derivatives. More recently, Hsu and co-workers extended this work and calculated the normal vibrations of crystalline TPB by combining interchain atom-atom potentials with the intrachain force field of Neto and DiLauro.

The conformation of an isolated chain of TPB in its preferred crystalline conformation has been determined from x-ray studies (90, 91). It has been shown that the translational repeat unit contains one chemical repeat unit, which is consistent with x-ray results that yield a 4.85 Å repeating distance. A schematic drawing of the model employed by Neto and DiLauro, together with their nomenclature for the internal coordinates, is shown in Figure 16.22. In their calculations they employed the following molecular parameters:

$$\mathrm{C{=}C} = 1.34\ \text{Å} \qquad \mathrm{C{-}\hat{C}{=}C} = 125^\circ$$
$$\mathrm{{=}C{-}C{-}{=}} = 1.52\ \text{Å} \qquad \mathrm{C{-}\hat{C}{-}C} = 110^\circ$$

—C—C— = 1.52 Å	H—Ĉ—H = 109.5°
—C—H = 1.10 Å	Internal rotation angle about =C—C—bond = 120°
=C—H = 1.09 Å	Internal rotation angle about —C—C—bond = 180°

However, as noted by Hsu et al., there is no agreement as to the correct value of the torsional angle of the $=CH—CH_2—$ group. Neto and DiLauro report that they did study the effect of slightly changing the valence and internal torsional angles on the calculated frequencies. Apparently, only a few calculated frequencies are appreciably altered by small changes in the $=CH—CH_2$ — torsional angles, whereas small changes in the valence and other torsional angles do not significantly affect the results.

The isolated chain model of TPB has a line group symmetry isomorphous to the point group C_i. There are 26 optically active fundamental vibrations distributed between the A_u and A_g species. Mutual exclusion is predicted, and the 12 A_u modes are only infrared active while the 14 A_g modes are only Raman active.

The valence force field used by Neto and DiLauro is reproduced in Table 16.29 and was transferred directly, without refinement, from a force field derived from the model compound *trans-trans-trans*-1,5,9-cyclododecatriene (92). The authors point out that there is a distinct geometric similarity between the $—CH_2—CH=CH—CH_2—$ monomer unit of TPB and the model compound. In fact, this work demonstrates the transferability of valence force fields of model compounds to structurally similar polymers. Table 16.30 summarizes the observed and calculated frequencies together with the PED for TPB. Calculations of the symmetrically deuterated derivatives were also performed, and the reader is referred to ref. 88 for details. The results are in fairly

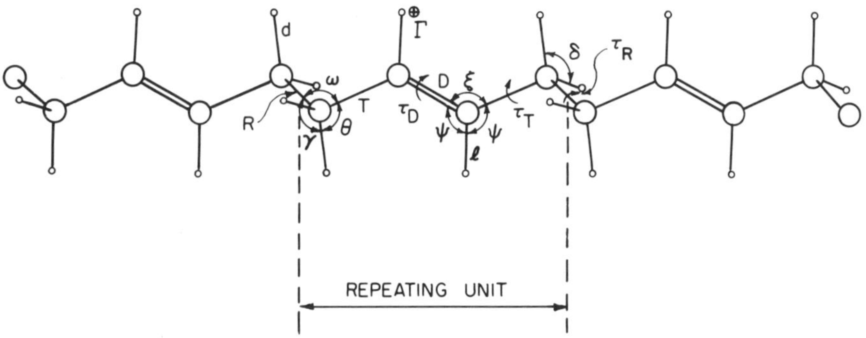

Figure 16.22. A schematic drawing of an isolated chain of *trans*-1,4-polybutadiene, showing the nomenclature of the internal coordinates. (Reprinted by permission from ref. 88.)

Table 16.29 Valence Force Constants for *trans*-1, 4-Polybutadiene

Force Constant[a]	Value	Force Constant[a]	Value
K_l	4.947	$F_{T\theta}$	0.341
K_d	4.524	$F_{R\gamma}$	0.348
K_D	8.702	$F_{T\phi}$	0.075
K_T	4.384	$F_{T\gamma}$	0.066
H_ω	1.049	$F_{T\omega}$	0.419
H_δ	0.540	$F_{\theta\theta}$	−0.032
H_ε	0.910	$F_{\theta\gamma}$	0.015
H_ϕ	0.500	$F_{\phi\psi}$	0.027
H_ψ	0.480	$F_{\varepsilon\phi}$	−0.029
H_θ	0.668	$F_{\omega\theta}$	−0.028
H_γ	0.664	$f_{\phi\phi}$	0.121
H_Γ	0.199	$f^{t}_{\gamma\gamma}$	0.120
τ_D	0.328	$f^{g}_{\gamma\gamma}$	−0.012
τ_T	0.021	$f^{t'}_{\theta\gamma}$	0.008
τ_R	0.024	$f^{g'}_{\theta\gamma}$	0.029
F_{dd}	0.004	$f^{g''}_{\theta\theta}$	−0.006
F_{DT}	0.094	$f^{g}_{\omega\gamma}$	−0.045
F_{TR}	0.122	$f^{t}_{\omega\omega}$	0.080
$F_{D\phi}$	0.355	$f^{g}_{\varepsilon\omega}$	0.012
$F_{T\psi}$	0.333		
$K_T = K_R$		$f^{t}_{\gamma\gamma} = f^{t}_{\psi\theta}$	
$F_{T\phi} = F_{D\psi}$		$f^{g}_{\gamma\gamma} = f^{g}_{\psi\theta}$	
$F_{T\gamma} = F_{R\theta}$		$f^{t'}_{\theta\gamma} = f^{t'}_{\phi\gamma} = f^{t'}_{\phi\theta} = f^{t'}_{\phi\psi}$	
$F_{T\omega} = F_{R\omega} = F_{T\varepsilon} = F_{D\varepsilon}$		$f^{g'}_{\theta\gamma} = f^{g'}_{\psi\gamma} = f^{g'}_{\phi\theta}$	
$F_{\theta\theta} = F_{\gamma\gamma}$		$f^{g''}_{\theta\theta} = f^{g''}_{\phi\gamma} = f^{g''}_{\psi\psi}$	
$F_{\varepsilon\phi} = F_{\varepsilon\psi}$		$f^{g}_{\gamma\omega} = f^{g}_{\varepsilon\theta} = f^{g}_{\varepsilon\phi} = f^{g}_{\omega\psi}$	
$F_{\omega\theta} = F_{\omega\gamma}$		$f^{t}_{\omega\omega} = f^{t}_{\varepsilon\varepsilon}$	

[a]Stretching constants are in units of mdyn/Å; Stretch-bend interaction in units of mdyn/rad; bending constants in units of mdyn-Å/rad^2.

good agreement with the observed experimental infrared frequencies and it must be reemphasized that no refinement of the force constants to obtain a better fit to the observed frequencies was attempted. Although Neto and DiLauro calculated the frequencies and PED of the A_g mode of TPD, they did not have Raman data with which to compare their results. Subsequently, Cornell and Koenig (93) obtained Raman spectra of TPB and compared the experimental frequencies with the normal coordinate calculations of Neto and DiLauro. The calculated results, which are included in Table 16.30, are in good agreement with the experimental Raman data. A more detailed discussion of the assignments of the normal modes of an isolated chain of TPB is included in the paper of Neto and DiLauro and will not be discussed further here.

Table 16.30 Calculated and Observed Frequencies and Approximate Potential Energy Distribution of *trans*-1, 4-Polybutadiene

Species	Obs.	Calc.	Δ_ν	K_l	K_d	K_D	K_T	H_ω	H_δ	H_ε	H_ϕ	H_ψ	H_θ	H_γ	H_Γ	τ_D	τ_T	τ_R
A_g	3005[a]	3027	24	98														
	2932	2923	−9		98													
	2846	2848	2		99													
	1664	1667	3			76	15											
	1431	1433	2						77				10	12				
	1324	1331	7				11				10		26	43				
	1301	1304	3										25	56				
	1267	1259	−8			10	17				29	13	16	16				
	1124	1141	17				31			16		11	27					
	1011	1039	28				79	11										
	969	962	−7				17						39		12			
	761	759	−2										13		66			
	537	555	18					32		31		16	17					
	221	223	2				17										57	
A_u	3018	3009	−9	99														
	2915	2914	−1		99													
	2840	2842	2		100													
	1453	1426	−27						80				14					
	1312	1302	−10								16	22	36	20				
	1235	1245	10										53	43				
	1075	1066	−9				37				21	20	25	21				
	1054	1035	−19				58				12	12		10				
	978	986	8												32	59		
	773	769	−4										22	74				
	440	426	−14					56		10	11			13	13	11		
	—	301	—							52	13				12	10		16

[a] Raman data of Cornell and Koenig (93) for A_g species.

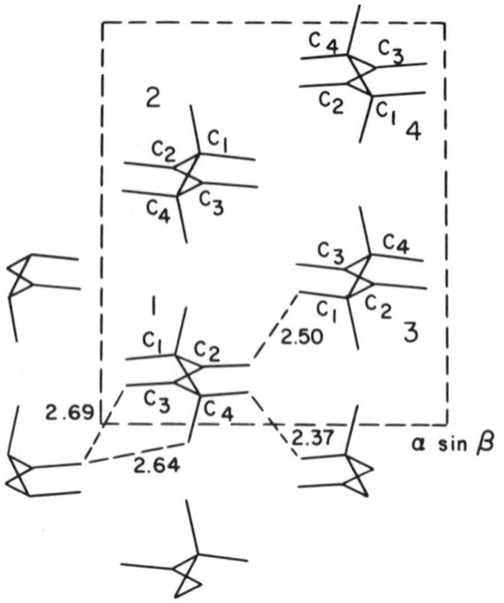

Figure 16.23. Crystal structure of TPB projected along the *c* axis. Short H · · · H contact distances are given. (Reprinted by permission from ref. 89.)

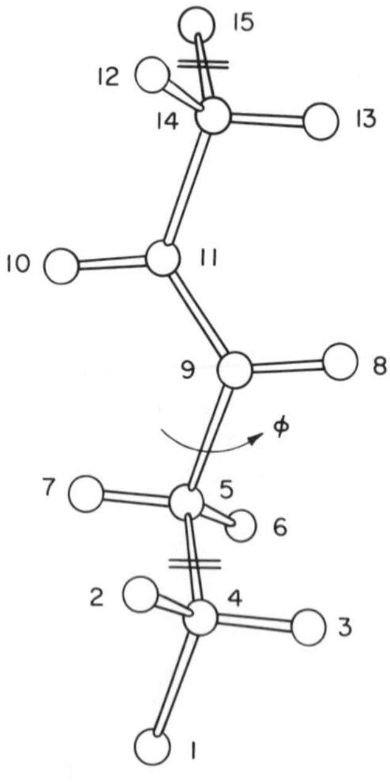

Figure 16.24. Single chain of TPB. Atoms numbered in accordance with the definition of the internal coordinates. (Reprinted by permission from ref. 89.)

Hsu, Moore, and Krimm (89) in a subsequent study obtained infrared and Raman spectra of an essentially all-*trans* TPB polymerized in a urea complex. Spectra were obtained from the as-polymerized polymer as well as from single-crystal preparations. The authors noted that the spectra of TPB exhibit explicit evidence of interchain interactions in both the splitting of intramolecular modes and the presence of lattice modes. Thus they conclude that the interchain interactions must be taken into account in order to interpret the vibrational spectrum of TPB satisfactorily.

Iwayanagi et al. (91) determined the crystal structure of TPB and proposed a monoclinic unit cell containing four molecules with a $P2_1/a$ space group, as shown in Figure 16.23. This space group corresponds to the factor group C_{2h}. The optically active normal modes are distributed between the Raman-active A_g and B_g species and the infrared-active A_u and B_u species as follows: 30 A_g, 30 B_g, 29 A_u, and 28 B_u. In contrast to the isolated chain (where the center of inversion lies midway between the olefinic bond), in the case of the crystal the center of inversion lies between the chains. Thus, in theory, mutual exclusion should not now apply to the isolated chain modes. However, Hsu et al. point out that since the perturbation of the interaction potential by the interchain interactions is very small, mutual exclusion of the intrachain modes will not be observed, except in those cases involving large contributions from hydrogen motions. A model of the single chain of TPB that shows the atoms numbered in accordance with the definition of the internal coordinates is shown in Figure

Table 16.31 Definitions of Internal Coordinates for TPB

	Atoms		Atoms
$R_1=r(\mathrm{C{-}C})$	(4,5)	$R_{18}=<(\mathrm{CCC})$	(5,9,11)
$R_2=r(\mathrm{C{-}C})$	(5,9)	$R_{19}=<(\mathrm{CCH})$	(11,9,8)
$R_3=r(\mathrm{C{=}C})$	(9,11)	$R_{20}=<(\mathrm{CCH})$	(9,11,10)
$R_4=r(\mathrm{C{-}C})$	(11,14)	$R_{21}=<(\mathrm{CCC})$	(9,11,14)
$R_5=r(\mathrm{C{-}H})$	(14,12)	$R_{22}=<(\mathrm{CCH})$	(14,11,10)
$R_6=r(\mathrm{C{-}H})$	(14,13)	$R_{23}=<(\mathrm{CCC})$	(11,14,15)
$R_7=r(\mathrm{C{-}H})$	(11,10)	$R_{24}=<(\mathrm{CCH})$	(11,14,12)
$R_8=r(\mathrm{C{-}H})$	(9,8)	$R_{25}=<(\mathrm{CCH})$	(11,14,13)
$R_9=r(\mathrm{C{-}H})$	(5,7)	$R_{26}=<(\mathrm{HCH})$	(12,14,13)
$R_{10}=r(\mathrm{C{-}H})$	(5,6)	$R_{27}=<(\mathrm{CCH})$	(15,14,12)
$R_{11}=<(\mathrm{CCH})$	(4,5,6)	$R_{28}=<(\mathrm{CCH})$	(15,14,13)
$R_{12}=<(\mathrm{CCH})$	(4,5,7)	$R_{29}=$CH opb	(11,10,9,14)
$R_{13}=<(\mathrm{HCH})$	(6,5,7)	$R_{30}=$CH opb	(9,8,5,11)
$R_{14}=<(\mathrm{CCH})$	(9,5,6)	$R_{31}=\mathrm{C{-}C}(R)$ tor	(4,5)
$R_{15}=<(\mathrm{CCH})$	(9,5,7)	$R_{32}=\mathrm{C{-}C}(T)$ tor	(5,9)
$R_{16}=<(\mathrm{CCC})$	(4,5,9)	$R_{33}=\mathrm{C{=}C}$ tor	(9,11)
$R_{17}=<(\mathrm{CCH})$	(5,9,8)	$R_{34}=\mathrm{C{-}C}(T)$ tor	(11,14)

Table 16.32 Definitions of Local Symmetry Coordinates for TPB

C—C(*R*) str	$S_1=R_1$
C—C(*T*) str	$S_2=R_2$
C=C(*D*) str	$S_3=R_3$
C—C(*T*) str	$S_4=R_4$
CH sym str	$S_5=(R_5+R_6)/2^{1/2}$
CH asy str	$S_6=(R_5-R_6)/2^{1/2}$
=C—H	$S_7=R_7$
=C—H	$S_8=R_8$
CH sym str	$S_9=(R_9+R_{10})/2^{1/2}$
CH asy str	$S_{10}=(R_9-R_{10})/2^{1/2}$
CH_2 wagging	$S_{11}=(R_{11}+R_{12}-R_{14}-R_{15})/2$
CH_2 bending	$S_{12}=(4R_{13}-R_{11}-R_{12}-R_{14}-R_{15})/20^{1/2}$
CH_2 twisting	$S_{13}=(R_{11}-R_{12}-R_{14}+R_{15})/2$
CH_2 rocking	$S_{14}=(R_{11}-R_{12}+R_{14}-R_{15})/2$
CCC def	$S_{15}=(5R_{16}-R_{11}-R_{12}-R_{13}-R_{14}-R_{15})/30^{1/2}$
CCC′ def	$S_{16}=(2R_{18}-R_{17}-R_{19})/6^{1/2}$
CH ipb	$S_{17}=(R_{17}-R_{19})/2^{1/2}$
CCC′ def	$S_{18}=(2R_{21}-R_{20}-R_{22})/6^{1/2}$
CH ipb	$S_{19}=(R_{20}-R_{22})/2^{1/2}$
CH_2 wagging	$S_{20}=(R_{24}+R_{25}-R_{27}-R_{28})/2$
CH_2 bending	$S_{21}=(4R_{26}-R_{24}-R_{25}-R_{27}-R_{28})/20^{1/2}$
CH_2 twisting	$S_{22}=(R_{24}-R_{25}-R_{27}-R_{28})/2$
CH_2 rocking	$S_{23}=(R_{24}-R_{25}+R_{27}-R_{28})/2$
CCC def	$S_{24}=(5R_{23}-R_{24}-R_{25}-R_{26}-R_{27}-R_{28})/30^{1/2}$
CH opb	$S_{25}=R_{29}$
CH opb	$S_{26}=R_{30}$
C—C(*R*) tor	$S_{27}=R_{31}$
C—C(*T*) tor	$S_{28}=R_{32}$
C=*C* tor	$S_{29}=R_{33}$
C—C(*T*) tor	$S_{30}=R_{34}$

16.24. Tables 16.31 and 16.32 give the definitions of the internal coordinates and local symmetry coordinates, respectively.

Hsu et al. employed the basic intrachain potential function derived by Neto and DiLauro (the only variation being the value of the hydrogen out-of-plane force constant, H_Γ). Interchain potentials attributed to Williams (94) that have the form $Ad^{-6}+B\exp(-Cd)$, where d is the interatomic distance, were employed with the following constants:

	A (kcal/mole-Å^6)	*B* (kcal/mole)	*C* (Å^{-1})
H---H	−27.3	2654	3.74
C---H	−125	8766	3.67
C---C	−568	83630	3.60

Table 16.33 Structural Parameters of Proposed Geometries of a Single Chain of *trans*-1,4-Polybutadiene.

	A^a	B^b	C^c	D^d
l(=C—H) (Å)	1.09	1.09	1.09	1.09
l(—C—H) (Å)	1.10	1.10	1.10	1.10
l(—C=C—) (Å)	1.34	1.34	1.34	1.34
l(=C—C—) (Å)	1.52	1.54	1.54	1.54
l(—C—C—) (Å)	1.52	1.54	1.54	1.54
<(—C—C=C—) (deg)	125	125	125	125
<(—C—C—C—) (deg)	110	109.5	109.5	109.5
<(H—C—H) (deg)	109.5	109.5	109.5	109.5
τ(=C—C—) (deg)	60	60	71	100
τ(—C—C—) (deg)	0	0	0	0
Repeat (Å)	4.85	4.89	4.83	4.66

[a]From Neto and di Lauro (88).
[b]From Natta and Corradini (90).
[c]From Iwayanagi et al. (91).
[d]High-temperature form.

The attraction-repulsion center for the H atom was assumed to be located 0.07 Å toward the C atom along the C—H bond and radii of interaction of 3.5, 4.0, and 4.3 Å were assumed for the H---H, C---H, and C---C interactions, respectively. Force constants were obtained by the double differentiation of the potential with the first derivative terms included in their evaluation.

Initially, Hsu et al. decided to study the effect of conformation, specifically the effect of changing the torsional angle ψ about the =C—C—C group, on the calculated frequencies of the isolated chain. Values of ψ were employed corresponding to the different crystalline chain conformations suggested in the literature, as shown in Table 16.33. The results are displayed in Table 16.34. It is apparent that some of the modes, especially the low frequency Raman-active vibrations, are particularly sensitive to changes in ψ. Calculations of the crystalline TPB were performed based on the unit cell of Iwayanagi et al., in which the value of ψ for the chain is 60°. These results are listed in Tables 16.35 and 16.36.

Hsu et al. discuss the results in detail, and we will present only the major conclusions here. The use of single-crystal preparations and annealing studies was useful in assigning noncrystalline contributions to the spectra. As mentioned previously, the effect of interchain interactions in the crystal is predicted to be observed by the splitting of intrachain vibrations and the appearance of lattice modes. Such splittings are observed for the infrared bands in the 1450 and 770 cm^{-1} region of the spectrum. Additionally, in the low temperature Raman spectra similar splittings are noted for lines appearing

Table 16.34 Calculated Normal Vibrations of a Single Chain of *trans*-1,4-Polybutadiene

Structure[a]			
B	*C*	*D*	Assignment and Potential Energy Distribution[b]
3026	3026	3026	A_g: CH str (98,98,98)
3009	3009	3009	A_u: CH str (100,100,100)
2922	2923	2923	A_g: CH_2 asy str (98,98,98)
2914	2914	2914	A_u: CH_2 asy str (100,100,100)
2848	2848	2848	A_g: CH_2 sym str (100,100,100)
2842	2842	2842	A_u: CH_2 sym str (100,100,100)
1665	1666	1665	A_g: C=C str (76,75,75), CH ipb (16,16,16), C—C(*T*) str (14,14,16)
1436	1436	1436	A_g: CH_2 bend (98,98,98)
1428	1429	1432	A_u: CH_2 bend (104,102,102)
1327	1324	1322	A_g: CH_2 wag (42,44,38), CH_2 twist (18,18,20), CH ipb (14,14,16)
1301	1300	1299	A_g: CH_2 twist (57,58,56), CH_2 wag (19,16,16), CH ipb (4,10,12)
1298	1296	1286	A_u: CH ipb (36,38,44), CH_2 twist (28,32,44), CH_2 wag (22,14,0)
1257	1257	1260	A_g: CH ipb (36,36,32), CH_2 wag (34,36,42), C=C str (10,10,9)
1241	1244	1245	A_u: CH_2 wag (76,82,100), CH_2 twist (16,10,0)
1139	1138	1136	A_g: C—C(*T*) str (30,28,24), CH_2 rock (24,26,30), CCC′ def (22,10,20)
1062	1061	1059	A_u: CH_2 twist (48,46,42), CH ipb (42,38,30), C—C(*T*) str (34,38,48)
1033	1032	1026	A_g: C—C(*R*) str (79,83,79), CCC def (12,14,18)
1032	1031	1028	A_u: C—C(*T*) str (62,56,48), CH ipb (22,24,30), CH_2 twist (14,18,22)
975	975	976	A_u: C=C tor (62,61,61), CH opb (34,33,34)
960	944	898	A_g: CH_2 rock (38,36,34), C—C(*T*) str (14,14,18), CH opb (12,12,9), CCC def (10,12,18)
774	773	772	A_u: CH_2 rock (94,94,92)
748	756	775	A_g: CH opb (70,66,58), CH_2 rock (12,12,6), C—C(*R*) str (4,6,14)
547	561	598	A_g: CCC′ def (44,42,34), CCC def (36,34,28), CH opb (8,10,18)
423	419	416	A_u: CCC def (64,68,70), CCC′ def (20,10,3), CH opb (14,18,20), C=C tor (10,13,15)
296	303	307	A_u: CCC′ def (68,84,96), C—C(*R*) tor (16,16,15)
224	234	243	A_g: C—C(*T*) tor (60,56,56), C—C(*T*) str (12,14,13), CCC def (10,10,12)

[a]See Table 16.33 for structural parameters.
[b]Numbers in parentheses refer to structures *B*, *C*, and *D*, respectively.

Table 16.35 Raman-Active Modes of Crystalline *trans*-1,4-Polybutadiene

Frequency				
Observed		Calculated		
110°K	RT	A_g	B_g	Assigment and Potential Energy Distribution[a]
3002	2999	3030	3030	CH str (98,98)
			3013	CH str (99)
		3012		CH str (99)
2948	2948			CH_2 asy str (100,99) } plus combinations and overtones
2922				
2920	2920	2926	2926	CH_2 asy str (99) } plus combinations and overtones
2899	2900	2916	2917	CH_2 asy str (99) } plus combinations and overtones
2883	2880			
2869				
2838	2841	2852		CH_2 sym str (98)
			2851	CH_2 sym str (98)
		2845	2845	CH_2 sym str (99,99)
1665	1665	1665	1665	C=C str (75,75), CH ipb (16,16), C—C(*T*) str (14,14)
1456	1456		1448	CH_2 bend (96)
1436	1436	1447		CH_2 bend (97)
1433		1434		CH_2 bend (102)
1427	1427		1434	CH_2 bend (92)
			1331	CH_2 wag (41), CH_2 twist (21), CH ipb (15)
1329	1331	1330		CH_2 wag (34), CH_2 twist (21), CH ipb (15)
			1320	CH_2 twist (51), CH_2 wag (26), CH ipb (10)
1307	1310	1318		CH_2 twist (35), CH_2 wag (26), CH ipb (10)
		1305		CH ipb (33), CH_2 twist (31), CH_2 wag (17)
			1303	CH_2 twist (34), CH ipb (32), CH_2 wag (19)
1268	1269	1265		CH_2 wag (36), CH ipb (36), C—C(*T*) str (12)
			1263	CH_2 wag (38), CH ipb (35), C—C(*T*) str (11)
		1245	1245	CH_2 wag (73,73), CH_2 twist (15,14)
1151	1153	1143	1143	C—C(*T*) str (29,30), CH_2 rock (24,24), CCC′ def (22,22)
			1073	CH ipb (47), CH_2 twist (30), CH_2 bend (26), C—C(*T*) str (22)
		1070		CH ipb (47), CH_2 twist (29), CH_2 bend (24), C—C(*T*) str (23)
1014	1016	1037	1037	C—C(*R*) str (66,76), C—C(*T*) str (17,17)
			1035	C—C(*T*) str (71), CH ipb (11)
		1034		C—C(*T*) str (56), C—C(*R*) str (14), CH ipb (14)
997		988		C=C tor (60), CH opb (32)
			982	C=C tor (61), CH opb (33)
			965	CH_2 rock (37), C—C(*T*) str (14), CCC def (12), CH opb (11)
965	967	964		CH_2 rock (38), C—C(*T*) str (14), CCC def (12), CH opb (11)
		796		CH_2 rock (89)
			792	CH_2 rock (88)
764	766		758	CH opb (68), CH_2 rock (13)
759		755		CH opb (69), CH_2 rock (12)
539				noncrystalline
528	530	550	550	CCC′ def (43,44), CCC def (36,36), CH opb (8,12)
		434		CCC def (61), CCC′ def (20), CH opb (12)
			431	CCC def (62), CCC′ def (20), CH opb (12)

Table 16.35 Continued

Frequency				
Observed		Calculated		
110°K	RT	A_g	B_g	Assigment and Potential Energy Distribution[a]
		306		CCC′ def (63), C—C(R) tor (15), CH opb (14)
			304	CCC′ def (64), C—C(R) tor (15), CH opb (14)
251	239		243	C—C(T) tor (48), C—C(T) str (12)
237	234	239		C—C(T) tor (51), C—C(T) str (12)
145	118		127	R:$H_{1,8}H_{3,7}$(12), $H_{1,8}H_{3,10}$(11), $H_{1,8}H_{4,13}$(13)$H_{1,13}H_{4,13}$(21)
124	104	108		R:$H_{1,8}H_{4,13}$(19), $H_{1,13}H_{4,13}$(33), $H_{1,10}H_{4,10}$(10)
			88	T:$H_{1,13}H_{4,13}$(10), $H_{1,10}H_{4,12}$(13)
93	70	68		T: $H_{1,10}H_{4,12}$(19)
		59		T:$H_{1,10}H_{4,10}$(22), $H_{1,10}H_{4,12}$(12)
			58	T:$C_{1,9}H_{2,12}$(14), $H_{1,10}H_{4,10}$(26)
56	48		45	T:$H_{1,6}H_{3,7}$(10), $H_{1,8}H_{3,7}$(18), $H_{1,13}H_{4,13}$(14)
		36		T:$H_{1,6}H_{2,12}$(12), $H_{1,8}H_{4,13}$(21), $H_{1,13}H_{4,13(11)}$

[a]R rotatory lattice mode, T translatory lattice mode. Interacting chains (first subscript, see Figure 16.23) and atoms (second subscript, see Figure 16.24) are given. For these modes, there is essentially no contribution to the PED from intrachain coordinates.

Table 16.36 Infrared-Active Modes of Crystalline *trans*-1,4-Polybutadiene

Frequency			
	Calculated		
Observed	A_u	B_u	Assignment and Potential Energy Distribution
3030		3030	CH (98)
	3029		CH (98)
2995	3012	3012	CH (100, 100)
2957			CH_2 asy str (98)
2928	2925	2924	CH_2 asy str (98) } plus combinations
2922	2916	2916	CH_2 asy str (100, 100) } and overtones
2905	2850	2850	CH_2 sym str (99,99)
2844	2844	2844	CH_2 sym str (99,99)
	1665	1665	C═C str (76,75), CH ipb (16,16), C—C(T) str (14,14)
1456	1442		CH_2 bend (97)
1447		1440	CH_2 bend (97)
1438	1433	1433	CH_2 bend (102, 103)
1350			Noncrystalline
		1329	CH_2 wag (38), CH_2 twist (21), CH ipb (15)
1336	1328		CH_2 wag (45), CH_2 twist (14), CH ipb (17)
	1306		CH_2 twist (60), CH_2 wag (14)
1312		1304	CH_2 twist (44), CH_2 wag (25), CH ipb (15)
		1302	CH_2 twist (37), CH ipb (30), CH_2 wag (17)

Table 16.36 Continued

Frequency			
	Calculated		
Observed	A_u	B_u	Assignment and Potential Energy Distribution
	1300		CH ipb (36), CH_2 wag (22), CH_2 twist (29)
		1261	CH ipb (36), CH_2 wag (34), C=C str (10), C—C(*T*) str (10)
	1259		CH_2 wag (37), CH ipb (34), C=C str (10), C—C(*T*) str (12)
	1244		CH_2 wag (75), CH_2 twist (15)
1235		1242	CH_2 wag (76), CH_2 twist (16)
1124	1141	1141	C—C(*T*) str (28,30), CH_2 rock (24,24), CCC′ def (22,22)
	1067		CH_2 twist (52), CH ipb (46), C—C(*T*) str (27)
1075			Noncrystalline
		1064	CH_2 twist (48), CH ipb (42), C—C str (32)
		1035	C—C(*R*) str (66), C—C(*T*) str (12)
	1035		C—C(*T*) str (63), CH ipb (14), CH_2 twist (10)
	1034		C—C(*R*) str (72), CCC def (11)
1053		1033	C—C(*T*) str (52), CH_2 twist (12), CH ipb (17)
982			Noncrystalline
970		985	C=C tor (60), CH opb (32)
963	979		C=C tor (61), CH opb (34)
		963	CH_2 rock (38), C—C(*T*) str (14), CH opb (11), CCC def (10)
	962		CH_2 rock (38), C—C(*T*) str (14), CH opb (11), CCC def (10)
773		785	CH_2 rock (91)
768	781		CH_2 rock (92)
	756		CH opb (68)
		754	CH opb (69)
		551	CCC′ def (42), CCC def (35)
	550		CCC′ def (43), CCC def (36)
443		427	CCC def (63), CCC′ def (20), CH opb (12)
	426		CCC def (63), CCC′ def (20), CH opb (13)
330	306		CCC′ def (66), C—C(*R*) str (15), CH opb (13)
		305	CCC′ def (63), C—C(*R*) tor (15), CH opb (14)
	240		C—C(*T*) tor (49), C—C(*T*) str (12)
		233	C—C(*T*) tor (54), C—C(*T*) str (12)
104	102		T: $H_{1,3}H_{3,7}$(11), $H_{1,8}H_{3,7}$(22); $H_{1,8}H_{3,10}$(17), $H_{1,10}H_{4,12}$(12)
64		76	T: $H_{1,3}H_{3,6}$(13), $H_{1,6}H_{3,7}$(10), $H_{1,8}H_{4,12}$(18), $H_{1,10}H_{4,12}$(20)
		58	R: $C_{1,9}H_{2,12}$(15), $H_{1,8}H_{3,7}$(16), $H_{1,8}H_{3,10}$(13)
	54		R: $H_{1,8}C_{3,5}$(10), $H_{1,6}H_{3,7}$(25)
	35		T: $H_{1,6}H_{2,12}$(15), $H_{1,8}H_{3,7}$(25)

at approximately 1435, 760, and 240 cm^{-1}. Bands observed below 200 cm^{-1} are assigned to lattice modes. We find this somewhat suprising, since torsional modes are often calculated below 200 cm^{-1} for many polymers.

In summary, the addition of interchain atom-atom potentials to the intrachain force field of TPB appears to account for the band splittings and low frequency modes observed in the vibrational spectra of this polymer.

cis-1,4-Polybutadiene

Normal coordinate calculations of an isolated chain of *cis*-1,4-polybutadiene (CPB) were reported in 1974 by Coleman et al. (95). A model was employed that contained two chemical repeat units per translational repeat distance and was based on the x-ray studies of Natta and Corradini (96). The bond angles and lengths employed are shown schematically in Figure 16.25. Figure 16.26 shows the translational repeat unit. The factor group of the line group of the isolated molecule is isomorphous to the point group C_{2h}. Active line group normal modes are distributed among the symmetry species as 14 A_g, 15 B_g, 14 A_u, and 13 B_u. Mutual exclusion occurs, and only the $A_u(\pi)$ and $B_u(\sigma)$ modes are predicted to be infrared active. Conversely, Raman activity is predicted for only the A_g (polarized) and B_g (depolarized) modes.

A valence force field was employed for the calculations. Table 16.37 lists the initial force constant values Φ_0, which were transferred, where applicable,

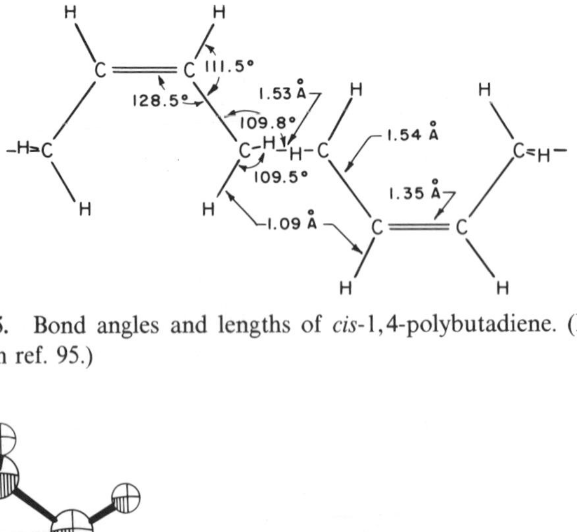

Figure 16.25. Bond angles and lengths of *cis*-1,4-polybutadiene. (Reprinted by permission from ref. 95.)

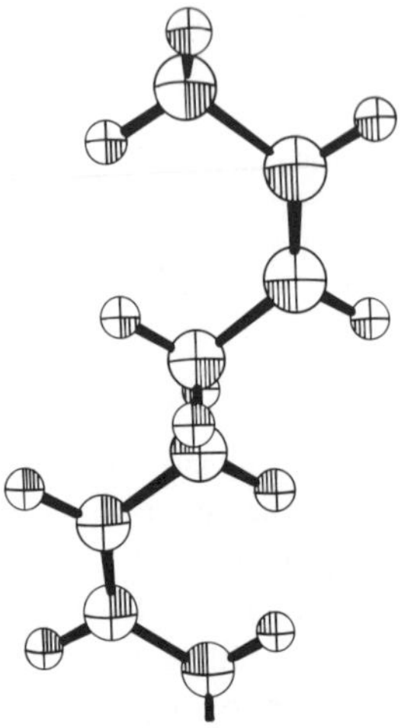

Figure 16.26. Model of repeat unit for isolated chain of *cis*-1,4-polybutadiene. (Reprinted by permission from ref. 95.)

Table 16.37 Force Constant Data for CPB[a]

	Φ_0	Φ_r	$\delta(\Phi_r)^b$
K_l	5.07	5.00	0.05
K_d	4.55	4.63	0.04
K_D	8.70	8.65	0.46
$K_T = K_R$	4.38	4.32	0.30
H_ε	0.92	1.07	0.19
H_ω	1.05	1.19	0.22
H_δ	0.54	0.51	0.02
H_ϕ	0.50	0.39	0.16
H_ψ	0.48	0.62	0.13
H_θ	0.67	0.68	0.07
H_γ	0.66	0.63	0.07
H_Γ	0.22	0.25	0.02
τ_D	0.22	0.15	0.03
F_{RT}	0.12	0.18	0.32
F_{DT}	0.10	0.32	0.34
$F_{\theta T}$	0.36	0.26	0.34
$F_{\gamma R}$	0.34	0.31	0.20
$F_{\psi T}$	0.32	0.20	0.21
$F_{\phi D}$	0.37	0.67	0.32
$F_{T\gamma'} = F_{R\theta'}$	0.05	−0.01	0.33
$F_{T\phi'} = F_{D\psi'}$	0.08	0.21	0.18
$F_{R\omega} = F_{T\varepsilon} = F_{D\varepsilon} = F_{T\omega}$	0.42	0.60	0.14
$F_{\gamma\gamma}(t) = F_{\psi\theta}(t)$	0.14	0.12	0.04
$F_{\phi\phi}$	0.12	0.12	0.06

[a] Not refined: $\tau_T = \tau_R = 0.02$; $F_{\phi\psi'}(\mathrm{tr}) = F_{\gamma\theta'}(\mathrm{t}) = \mathrm{F}_{\psi\psi} = \mathrm{F}_{\theta\theta''}(\mathrm{t}) = F_{dd} = 0.01$; $F_{\gamma\gamma}(\mathrm{g}) = F_{\theta\gamma}(\mathrm{g}) = F_{\psi\theta}(\mathrm{g}) = F_{\theta\theta''}(\mathrm{g}) = F_{\gamma\phi''}(\mathrm{g}) = -0.01$; $F_{\gamma\theta'} = 0.02$; $F_{\gamma\gamma} = F_{\theta\theta} = F_{\varepsilon\varepsilon}(\mathrm{g}) = -0.02$; $F_{\phi\psi} = F_{\theta\phi'}(\mathrm{g}) = F_{\gamma\theta'}(\mathrm{g}) = 0.03$; $F_{\gamma\omega} = F_{\theta\omega} = -0.03$; $F_{\varepsilon\phi} = F_{\varepsilon\psi} = -0.04$; $F_{\phi\varepsilon}(\mathrm{t}) = 0.06$; $F_{\gamma\omega}(\mathrm{g}) = F_{\theta\varepsilon}(\mathrm{g}) = F_{\omega\omega}(\mathrm{t}) = F_{\psi\omega}(\mathrm{g}) = -0.07$; $F_{\omega\varepsilon}(\mathrm{t}) = 0.07$. Force constant units as in Table 16.29.

[b] Standard error in Φ.

from the refined values obtained by Neto et al. (97) for cyclohexene. A few minor interactions, not applicable to cyclohexene, were given reasonable values based on the results of Neto and DiLauro (88) for *trans*-1,4-polybutadiene. The internal coordinate nomenclature is identical to that used by Neto et al., and is illustrated in Figure 16.27. Also included in Table 16.37 are the refined force constants Φ_r, which were obtained from a damped least-squares method (72) so as to obtain the best fit with the observed experimental data. Force constants associated with the methylene part of the molecule were restrained around the limits indicated by the results of Snyder and Schachtschneider (11, 21). Minor interaction force constants, shown at the bottom of Table 16.37, were held constant during the refinement to reduce computation time.

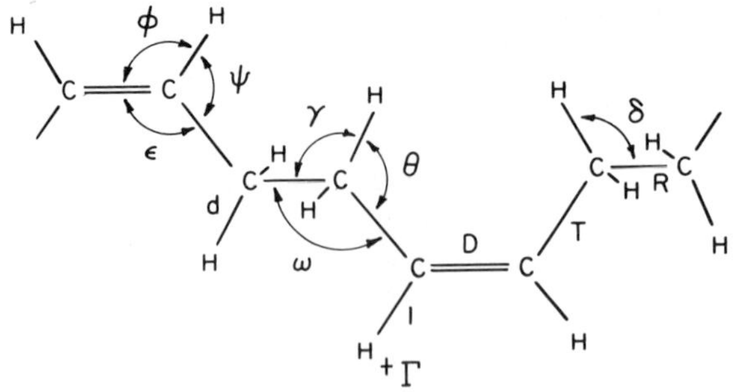

Figure 16.27. Internal coordinates for *cis*-1,4-polybutadiene. (Reprinted by permission from ref. 95.)

Fourier transform infrared and polarized Raman spectra of CPB were obtained by Coleman et al. (95). In addition, low temperature Raman studies of CPB have also been reported (98). In Tables 16.38 and 16.39 the observed frequencies from the infrared and Raman spectra are compared with those calculated. The agreement is quite satisfactory; the total squared deviation of the calculated compared to the observed frequencies resulting from the refinement of all four species is 0.067. The theoretical band assignments were made from the potential energy distribution (PED) and are also given in Tables 16.38 and 16.39. Contributions of less than 10% were omitted.

In the completely amorphous state, the model for CPB will have only local symmetry of C_s around the double bond. As the temperature is reduced toward the crystalline melting point of 2°C (99, 100), more units will assume the preferred conformation. At room temperature, it is evident from the polarized Raman spectra that a considerable number of translational repeat units are in the preferred conformation. The Raman polarization data were particularly useful for the assignment of the observed lines to the A_g and B_g modes, and this increases the confidence of the refinement procedure because errors can be introduced by a mismatching of the observed to the calculated frequencies.

With the exception of one Raman line at 706 cm^{-1}, the polarization data are entirely consistent with the calculated modes and their respective frequencies. In the Raman spectrum at room temperature the 706 cm^{-1} line is observed to be polarized, indicating an A_g mode. However, the normal coordinate calculations predict a normal mode in this region belonging to the B_g species and hence depolarized. Polarization data for the 710–699 cm^{-1} doublet observed in the low temperature spectrum were not available. A rationale for this apparent anomaly is as follows. The Raman line at 706 cm^{-1} observed in the room temperature spectrum is uncharacteristically broad and strongly suggests an amorphous line that would be polarized. The two lines observed at 710 and 699 cm^{-1} in the low temperature spectrum are probably masked at room temperature by the broad 706 cm^{-1} amorphous line. This contention is supported by the similar appearance of the 410–398 cm^{-1} doublet in the low temperature spectrum. In going from 25° to −77°C one

Table 16.38 Infrared Data for CPB[a]

Observed[b] (cm^{-1})	Calc. (cm^{-1})	Group	Approximate PED[c]
3065 (w)	3034	B_u	99% K_l
3007 (vvs)	3034	A_u	99% K_l
2972 (w)	2945	B_u	99% K_d
2935 (vvs)	2945	A_u	99% K_d
2915 (vs)	2878	B_u	99% K_d
2850 (vs)	2878	A_u	99% K_d
1655 (m)	1659	A_u	74% K_D, 17% K_T, 11% H_ψ
1450 (s)	1453	B_u	68% H_δ, 25% H_θ
1433 (s)	1443	A_u	77% H_δ, 17% H_θ
1404 (m)	1373	B_u	10% K_T, 11% H_δ, 16% H_ϕ, 29% H_ψ, 16% H_θ, 19% H_γ
1310 (m)	1301	B_u	13% H_ϕ, 34% H_θ, 35% H_γ
1263 (w)	1300	A_u	32% H_θ, 38% H_γ
1240 (m)	1270	A_u	14% H_ϕ, 23% H_ψ, 31% H_θ, 19% H_γ
1212 (vw)			
1170 (vw)	1173	B_u	24% K_T, 10% H_ε, 55% H_θ, 21% H_γ
1045 (vvw)	1118	A_u	14% H_ϕ, 18% H_ψ, 55% H_θ, 26% H_γ
1015 (m-sh)			
994 (ms)	975	A_u	72% H_Γ, 18% τ_D
970 (m)	954	B_u	46% K_T, 13% H_ε, 33% H_ψ, 11% H_θ, 23% H_γ
913 (w)	872	A_u	12% K_D, 79% K_T, 13% H_γ
820 (vvw)	822	B_u	16% K_T, 18% H_ε, 10% H_ϕ, 43% H_γ
775 (m)	751	A_u	21% H_θ, 71% H_γ
740 (vvs)	728	B_u	20% H_γ, 61% H_Γ
690 (sh)	669	B_u	16% H_ε, 16% H_θ, 23% H_γ, 31% H_Γ
635 (w)			
515 (vvw)	491	A_u	30% H_ω, 10% H_γ, 48% τ_D
	287	A_u	59% H_ε, 10% H_ϕ
	285	B_u	84% H_ω, 12% H_γ
	103	A_u	10% H_ε, 50% H_ω, 10% H_Γ, 18% τ_T
	62	B_u	92% τ_R

[a]Key: s = strong; m = medium; w = weak; sh = shoulder; v = very.
[b]At 25°C.
[c]Because of contributions from interactions the PED among the diagonal force constants may be greater than 100%.

observes intensity changes and a sharpening of Raman lines, as evidenced by the separation of specific normal vibrations into A_g and B_g modes. Furthermore, crystal field splitting was suggested by Cornell and Koenig (98) for the Raman lines at 829, 706, and 406 cm^{-1}. Since the analysis of the isolated crystalline chain predicts only one frequency in each of these regions, this prediction tends to substantiate the assumption of Cornell and Koenig.

Table 16.39 Raman Data for CPB

Observed[a] (cm^{-1})		Calc.		
25°C	−76°C[b]	(cm^{-1})	Group	Approximate PED[c]
3065		3034	B_g	99% K_l
3006		3034	A_g	99% K_l
2972		2954	B_g	99% K_d
2935		2954	A_g	99% K_d
2896		2883	B_g	99% K_d
2850		2883	A_g	99% K_d
1650 p	1650	1645	A_g	77% K_D, 15% K_T, 12% H_ψ
1438 d	1438	1456	B_g	75% H_δ, 15% H_θ
	1429	1453	A_g	76% H_δ, 12% H_θ, 10% H_γ
1399 d	1399	1382	B_g	15% K_T, 13% H_ϕ, 22% H_ψ, 27% H_θ, 22% H_γ
1323 ?	1337	1332	A_g	15% K_T, 34% H_θ, 51% H_γ
	1325	1330	B_g	13% K_T, 41% H_θ, 51% H_γ
1305 ?	1301	1316	A_g	37% H_θ, 49% H_γ
1277 ?	1282	1272	B_g	21% H_ϕ, 27% H_ψ, 37% H_γ
1258 ?	1257	1257	A_g	23% H_ϕ, 41% H_ψ, 16% H_γ
1206 d	1206	1141	B_g	37% K_T, 27% H_ε, 21% H_θ, 10% H_γ
1078	1078	1046	A_g	46% K_T, 21% H_θ, 19% H_Γ
	1028	1017	A_g	33% K_T, 34% H_Γ
1009 d	1013	1003	B_g	83% K_T
986 p	984	957	A_g	27% K_T, 36% H_θ, 16% H_Γ
971 d	950	947	B_g	30% K_T, 19% H_ω, 10% H_ψ, 36% H_θ
831 p	834, 819[d]	852	A_g	59% K_T, 31% H_ω, 10% H_θ, 13% τ_D
784 p	777			
706 p	710, 699[d]	693	B_g	77% H_Γ
561 d	537	605	B_g	36% K_T, 36% H_ε, 11% H_ω, 15% H_θ
466 d	461	448	B_g	11% K_T, 62% H_ω, 10% H_θ
	410, 398[d]	439	A_g	15% K_T, 13% H_ω, 40% τ_D
237	250	242	A_g	10% K_T, 19% H_ε, 42% H_ω, 11% H_θ, 13% H_Γ, 17% τ_D
155		182	A_g	44% H_ε, 10% τ_D, 15% τ_T
126		72	B_g	96% τ_T

[a] Key: p = polarized; d = depolarized.
[b] Data of Cornell and Koenig (98).
[c] Because of contributions from interactions the PED among the diagonal force constants may be greater than 100%.
[d] Crystal field splitting (98).

A tentative assignment for the Raman lines in CPB was advanced by Cornell and Koenig (98) based on the studies of *trans*-1,4- (88) and syndiotactic 1,2-polybutadiene (85) and the infrared results of Binder (101). The PED from this normal coordinate analysis gives a much more detailed assignment of the Raman lines, as is shown in Table 16.39. There are some discrepancies, the most notable of which is the reassignment of the 1305 and 1258 cm^{-1} lines.

As with the polarization data used in the Raman, infrared dichroic data would also have been valuable for assignment purposes in order to assist in the differentiation of A_u and B_u modes. Attempts to obtain dichroic data, however, were not successful. The PED from the vibrational analysis is given in Table 16.38. Most interest in the infrared spectra of polybutadienes has been centered around the hydrogen out-of-plane motions, which have been utilized for structure determination. The results are consistent with the hydrogen out-of-plane motion occurring at 740 cm^{-1} for CPB, and the overall assignments for the infrared spectrum compare favorably with those proposed by Silas et al. (102).

trans-1,4-Polychloroprene and *trans*-1,4-Poly(2,3-Dichlorobutadiene)

Normal coordinate calculations of *trans*-1,4-polychloroprene (TPC) were initially performed by Tabb and Koenig (103). Attempts to transfer the valence force field derived for TPC to the chemically and structurally similar *trans*-1,4-poly(2,3-dichlorobutadiene) (TPDCB) by Petcavich and Coleman (104) were not successful. An improved valence force field that satisfactorily fits both TPC and TPDCB was developed by the latter authors, and we will concentrate on this study in the discussion that follows.

The crystal structures of TPC and TPDCB have been studied, using x-ray diffraction, by Bunn (105) and Chatani and Nakatani (106), respectively. The preferred conformation of the isolated chain of both polymers is similar. This conformation is illustrated in Figure 16.28, and the molecular parameters of the models based on the X-ray data are listed below:

C=C	1.32 Å	C—Ĉ=C	129.3°
=C—C—	1.54 Å	C—Ĉ—C	108.0°
—C—C—	1.54 Å	H—Ĉ—H	109.5°
—C—H	1.09 Å	C—Ĉ—Cl	113.3°
=C—H	1.08 Å	C=Ĉ—Cl	117.4°

Internal rotation angle about =C—C—bond = 105°
Internal rotation angle about —C—C—bond = 180°

The line group of an isolated chain of TPDCB is isomorphous to the point group C_i. There are 26 optically active fundamental vibrations distributed

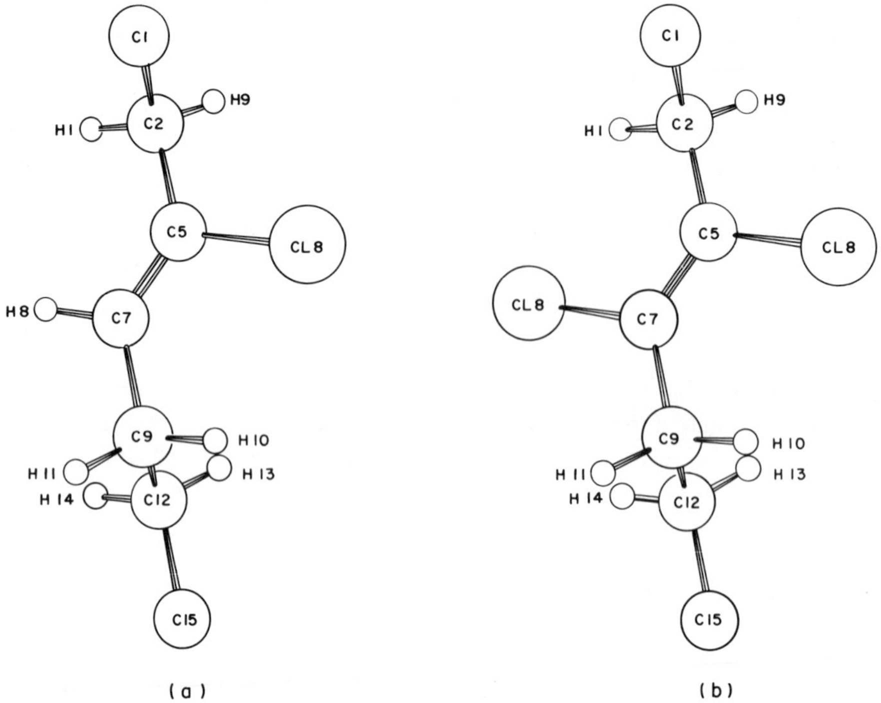

Figure 16.28. Model of the preferred conformation of (*a*) TPC and (*b*) TPDCB. (Reprinted by permission from ref. 104.)

between the two symmetry species A_u and A_g. The 12 A_u modes are infrared active and the 14 A_g are Raman active. There are no nontrivial symmetry elements associated with the line group for the isolated chain of TPC. Hence the 26 optically active fundamental vibrations are both infrared and Raman active.

Petcavich and Coleman emphasize that a unique force field cannot be defined for complex molecules such as TPDCB and TPC, since the number of force constants always exceeds the number of fundamental normal modes. (However, this has never stopped us before!) Nevertheless, both TPC and TPDCB have similar chemical structures and preferred conformations and it was postulated that if a common valence force field (with the exception of a few force constants applicable only to a specific molecule) could be derived that satisfactorily fits the experimental vibrational spectra of both molecules, this result would increase the confidence of the normal coordinate calculations. Furthermore, the authors decided to transfer force constants associated with the aliphatic part of the two molecules directly, without refinement, from the valence force field derived by Neto and DiLauro (88) for *trans*-1,4-polybutadiene. This force field was subsequently used successfully, with minimal revision, by Hsu et al. (89) in their studies of the same polymer (see

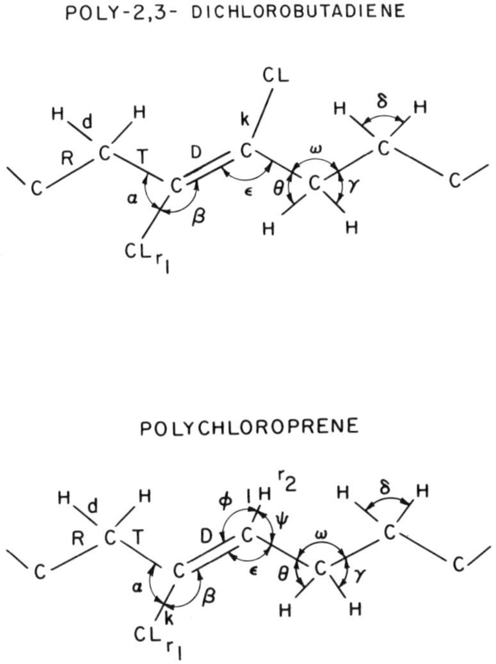

Figure 16.29. Internal coordinates of TPC and TPDCB. (Reprinted by permission from ref. 104.)

Section 16.3.). Initial force constant values involved with the chlorine atoms were transferred, where applicable, from vinyl chloride (107) and 2-chloro-1-butene (108).

Figure 16.29 shows the internal coordinates used for both polymers. The force constant notation of Tabb and Koenig (103) was employed. Table 16.40 lists the initial force constant values (ϕ_1) and those obtained by corefining (ϕ_2) both molecules together using a damped least-squares method (72), so as to obtain the best fit with the observed frequencies.

The theoretical band assignments were made from the potential energy distribution (PED), and the eigenvectors in terms of cartesian displacement coordinates were obtained from the **L** matrix. Dispersion curves were calculated for both molecules using phase angles at increments of 20° through the range $0-\pi$.

In order to obtain infrared spectra of the preferred conformation of TPC that were essentially free from spectral contributions attributable to nonpreferred (amorphous) conformations, Coleman et al. (109, 110) employed a digital subtraction technique. This is illustrated in Figure 16.30. Spectrum *A* is of a TPC polymerized at −150°C, which is essentially the all-trans-1,4-polymer (111) in the semicrystalline state at room temperature. Spectrum *B* is that of the identical sample in the amorphous (molten) state obtained at 75°C, which

Table 16.40 Force Constant Data for TPC and TPDCB

Refined				Unrefined			
Force Constant	ϕ_1^a	ϕ_2	$\sigma(\phi_2)^b$	Force Constant	ϕ_1^a	Force Constant	ϕ_1^a
K_T	4.384	4.424	0.129	K_D	8.503[c]	$F_{\gamma\gamma}$	−0.032
K_K	3.311[c]	2.961	0.185	K_d	4.523	$F_{\alpha\theta}$	0.015
H_ε	0.910	1.037	0.169	K_l	4.947	$F_{\omega\gamma}$	−0.028
H_β	0.832[c]	1.343	0.200	H_ω	1.049	$F_{T\gamma}$	0.066
Γ_1	0.344[c]	0.263	0.022	H_γ	0.664	$F_{T\varepsilon}$	0.419
Γ_2	0.199	0.160	0.019	H_θ	0.668	$F_{D\beta}$	0.340
$F_{T\omega}$	0.419	0.608	0.081	H_α	0.944[c]	$F_{D\phi}$	0.355
F_{DK}	0.0	−0.077	0.250	H_δ	0.540	$F_{T\phi}$	0.075
$F_{D\alpha}$	0.0	−0.490	0.146	H_ϕ	0.500	$F_{T\psi}$	0.333
$F_{T\alpha}$	0.0	0.057	0.183	H_ψ	0.480	$F_{\varepsilon\phi}$	−0.029
$F_{T\beta}$	0.0	0.389	0.161	τ_D	0.328	$f_{\gamma\gamma}^t$	0.120
$F_{K\alpha}$	0.0	−0.019	0.193	τ_T	0.024	$f_{\gamma\gamma}^g$	−0.012
$F_{K\varepsilon}$	0.0	−0.474	0.188	F_{dd}	0.004	$f_{\theta\gamma}^t$	0.008
$F_{K\beta}$	0.0	0.494	0.205	F_{TR}	0.122	$f_{\theta\gamma}^g$	0.029
$f_{\beta\phi}$	−0.476[d]	−0.216	0.124	F_{TD}	0.094	$f_{\theta\theta}^t$	0.010
$f_{\beta\beta}$	−0.476[d]	−0.728	0.174	F_{TK}	0.588[c]	$f_{\theta\theta}^g$	0.012
$f_{\varepsilon\beta}$	0.0	−0.100	0.125	$F_{R\gamma}$	0.348	$f_{\omega\omega}$	0.080
				$F_{T\theta}$	0.341	$f_{\omega\gamma}$	0.045

$K_T = K_R$	$F_{\gamma\gamma} = F_{\theta\theta}$	$F_{T\phi} = F_{D\psi}$	$f_{\omega\omega} = f_{\varepsilon\varepsilon}$	$f_{\theta\gamma}^g = f_{\psi\gamma} = f_{\phi\theta}$
$F_{T\gamma} = F_{R\theta}$	$\tau_T = \tau_R$	$F_{\varepsilon\phi} = F_{\varepsilon\psi}$	$f_{\gamma\gamma}^g = f_{\psi\theta}$	$f_{\theta\theta}^g = f_{\gamma\phi} = f_{\varepsilon\omega}$
$F_{T\varepsilon} = F_{D\varepsilon}$	$F_{\omega\gamma} = F_{\omega\theta}$	$f_{\gamma\gamma}^t = f_{\psi\theta}$	$f_{\theta\gamma}^t = f_{\psi\gamma} = f_{\phi\theta}$	$f_{\omega\gamma} = f_{\varepsilon\theta} = f_{\omega\psi}$

[a] Force constants obtained from *trans*-1,4-polybutadiene (88) except where otherwise noted. Force constant units are the same as in Table 16.29.
[b] Standard error in ϕ_2.
[c] Force constants obtained from 2-chloro-1-butene (108).
[d] Force constants obtained from vinyl chloride (107).

is above the crystalline melting point of the polymer. By subtracting spectrum *B* from *A*, using an appropriate weighting factor based on the elimination of the 603 cm^{-1} amorphous band, the spectrum of the preferred (crystalline) conformation was obtained. Raman spectra of TPCs have also been reported by Coleman et al. (112). In the case of TPDCB, infrared and Raman spectra (which are shown in Figures 16.31 and 16.32) were obtained from single-crystal preparations of the polymer. The polymer is highly crystalline and spectral contributions from nonpreferred (amorphous) conformations are minimized. The infrared bands and Raman lines are extremely sharp for a polymeric system, which indicates that the material is highly ordered. Furthermore, a

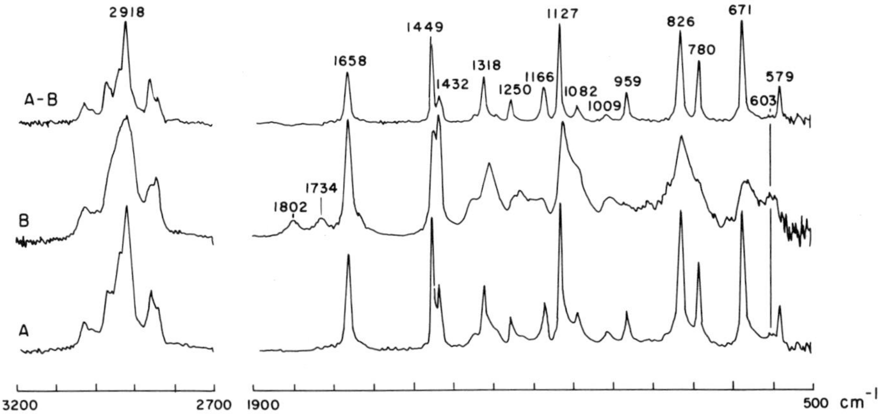

Figure 16.30. FT-IR spectra in the range of 500–3200 cm^{-1} of chloroprene polymerized at $-150°C$. *A*, spectrum recorded at room temperature; *B*, spectrum recorded at 80°C; *A* − *B*, difference spectrum obtained by subtracting *B* from *A*. (Reprinted by permission from ref. 110.)

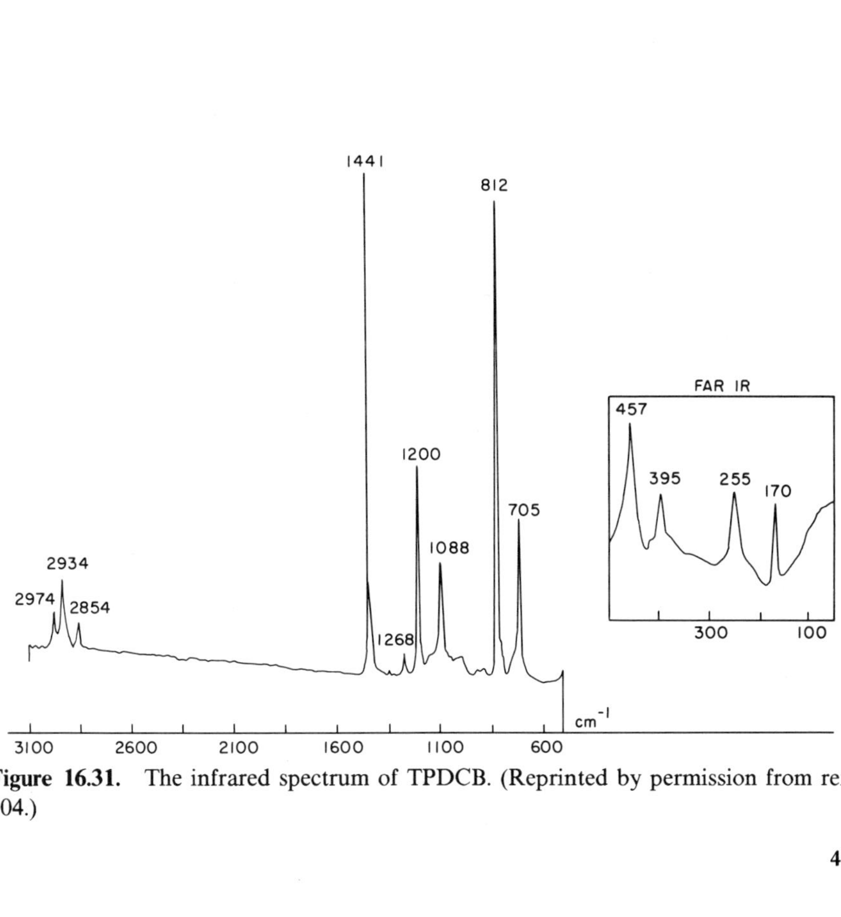

Figure 16.31. The infrared spectrum of TPDCB. (Reprinted by permission from ref. 104.)

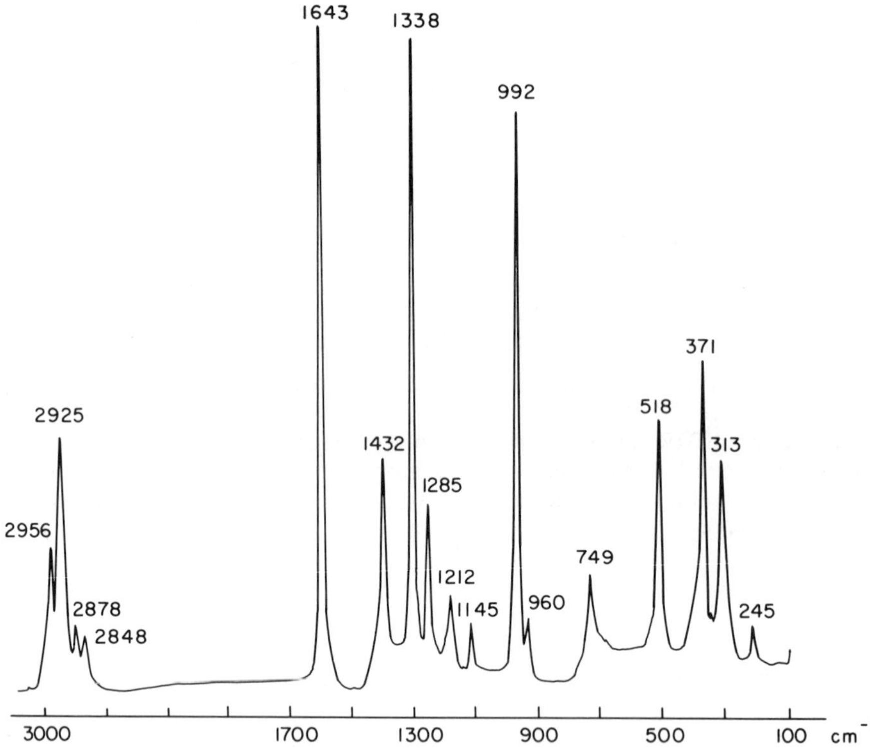

Figure 16.32. The Raman spectrum of TPDCB. (Reprinted by permission from ref. 104.)

comparison of the infrared and Raman spectra reveals that they exhibit an outstanding example of mutual exclusion.

The results of the normal coordinate calculations for TPC and the A_g and A_u modes of TPDCB are shown in Tables 16.41, 16.42, and 16.43, respectively. The initial unrefined calculated values are in good agreement with the observed frequencies for both polymer molecules. These results are an improvement on those of Tabb and Koenig (103) for TPC, especially in the region below 1000 cm^{-1}. Also included in these tables are the calculated frequencies and PED after corefinement. The agreement between observed and calculated frequencies is excellent; the total squared deviation of the calculated compared to the observed frequencies for both molecules is 0.024. The dispersion curves for TPDCB and TPC are shown in Figure 16.33. Figure 16.34 illustrates the cartesian displacement coordinates for several normal modes occurring below 900 cm^{-1} in TPC.

A few comments concerning the assignments of the normal modes of TPC are in order. Petcavich and Coleman's assignments of the normal modes below 900 cm^{-1} are significantly different from those of Tabb and Koenig. A

Table 16.41 ***trans*-1,4-Polychloroprene**

Frequency (cm^{-1})			
	Calculated		
Observed	Initial	Refined	Approx. Potential Energy Distribution(%)
3023	3016	3017	100 K_l
2966	2924	2924	100 K_d
2918	2914	2915	100 K_d
2857	2848	2846	99 K_d
2840	2842	2842	100 K_d
1660	1668	1653	19 K_T, 79 K_D
1449	1459	1458	14 H_θ, 76 H_δ
1432	1451	1451	11 H_γ, 14 H_θ, 79 H_δ
1342	1345	1343	17 K_T, 38 H_γ, 32 H_θ, -11 $F_{T\gamma}$
1318	1322	1317	10 K_T, 38 H_γ, 32 H_θ, -11 $F_{T\gamma}$, 12 $f_{\gamma\gamma}$
1287	1298	1298	54 H_γ, 16 H_θ
1250	1210	1216	50 H_γ, 56 H_θ, -12 $F_{T\theta}$
1167	1180	1177	17 K_T, 17 H_γ, 38 H_θ, 18 H_ε
1127	1097	1131	13 K_T, 18 H_γ, 36 H_θ, 13 H_ε, 26 H_ϕ
1083	1049	1071	80 K_T
1007	1037	1002	86 K_T, -16 $F_{R\omega}$
958	964	942	21 K_T, 17 H_ω, 35 H_θ, -16 $F_{R\omega}$
826	870	826	10 H_θ, 27 τ_D, 36 Γ_2
780	795	794	79 H_γ, 17 H_θ, -15 $f_{\gamma\gamma}$
671	736	684	12 K_K, 34 H_ω, 13 H_ε, 10 H_β, 14 τ_D
577	566	572	71 K_K
477	483	471	16 K_T, 11 H_ω, 57 Γ_1
407	421	409	46 H_ω, 24 τ_D, 17 Γ_2
339	324	349	31 H_ε, 41 H_α
253	201	225	34 H_ε, 35 H_β
153	102	103	11 K_T, 12 H_ω, 42 τ_R

summary of the assignments in the 450–900 cm^{-1} spectral region is given in Table 16.44. It can be seen that the assignments of the modes at 826, 671, 577, and 477 cm^{-1} are markedly different. Previous assignments by Mochel and Hall (113) on TPC and *trans*-1,4-polybromoprene (TPBP), using the group frequency approach, suggested that the 826 cm^{-1} infrared band did not appreciably involve the C—Cl bond, since it occurred within 2 cm^{-1} of the identical band in TPBP. Furthermore, they inferred that the 671 cm^{-1} and especially the 577 cm^{-1} bands involve the C—Cl bond. The results of Petcavich and Coleman are in complete agreement with these findings. The 826 cm^{-1} is assigned predominantly to the CH out-of-plane bend, which is consistent with the generally accepted range (114) of 800–840 cm^{-1} for this type of structure. There is no debate concerning the 780 cm^{-1} band assigned to the CH_2 rocking mode. The 671 cm^{-1} band is a highly mixed mode with

Table 16.42 ***trans*-1,4-Poly-2,3-Dichlorobutadiene (A_g Mode)**

Frequency (cm^{-1})			
	Calculated		
Observed	Initial	Refined	Approx. Potential Energy Distribution (%)
2925	2923	2924	99 K_d
2848	2848	2846	99 K_d
1643	1707	1651	19 K_T, 81 K_D
1432	1459	1457	14 H_θ, 76 H_δ
1338	1329	1332	19 K_T, 57 H_γ, 40 H_θ, -19 $F_{R\gamma}$
1285	1309	1307	63 H_γ, 20 H_θ, 11 $f_{\gamma\gamma}$
1145	1178	1145	32 K_T, 16 K_K, 26 H_θ, 33 H_ε, 23 H_β, -12 $F_{T\varepsilon}$, -12 $F_{K\varepsilon}$
992	1043	1005	82 K_T, -14 $F_{R\omega}$
960	966	945	27 K_T, 17 H_ω, 37 H_θ, -20 $F_{R\omega}$
749	817	748	14 K_K, 28 H_ω, 11 H_θ, 15 H_ε, 22 H_β, 27 Γ_1
518	606	529	22 K_T, 12 H_β, 53 Γ_1, -14 $F_{R\omega}$
371	326	363	75 K_K, 10 $F_{K\varepsilon}$
245	224	245	56 H_α, 45 H_β, -24 $f_{\beta\beta}$
70 ?	74	73	12 K_T, 14 H_α, 27 H_β, 42 τ_R, -15 $f_{\beta\beta}$

Table 16.43 ***trans*-1,4-Poly-2,3-Dichlorobutadiene (A_u Mode)**

Frequency (cm^{-1})			
	Calculated		
Observed	Initial	Refined	Approx. Potential Energy Distribution (%)
2934	2914	2914	100 K_d
2854	2842	2842	100 K_d
1441	1451	1451	11 H_γ, 13 H_θ, 80 H_δ
1268	1227	1238	21 K_T, 38 H_γ, 51 H_θ, -16 $F_{T\theta}$
1200	1198	1193	36 H_γ, 68 H_θ
1088	1083	1110	64 K_T, 13 H_γ, 10 H_θ
812	862	824	25 K_K, 66 H_γ, -12 $f_{\gamma\gamma}$
705	728	699	67 K_K, 19 H_γ, 11 H_θ
457	472	445	50 H_ω, 33 Γ_1, -11 $F_{R\omega}$
395	378	397	15 K_K, 13 H_ε, 36 H_α
260	202	236	34 H_ε, 43 H_β, 23 $f_{\beta\beta}$
170	162	156	28 Γ_1, 69 τ_D

contributions from the ν(C—Cl), CCC bending, and C=C twisting vibrations. It is a complex mode (see Figure 16.34) but does involve a contribution from the C—Cl stretching force constant (see Table 16.41). The 577 cm^{-1} band may be characterized as predominantly the C—Cl stretching vibration, which is also consistent with the observations of Mochel and Hall (113). Finally, the out-of-plane C—Cl bending vibration is assigned to the band occurring at 477

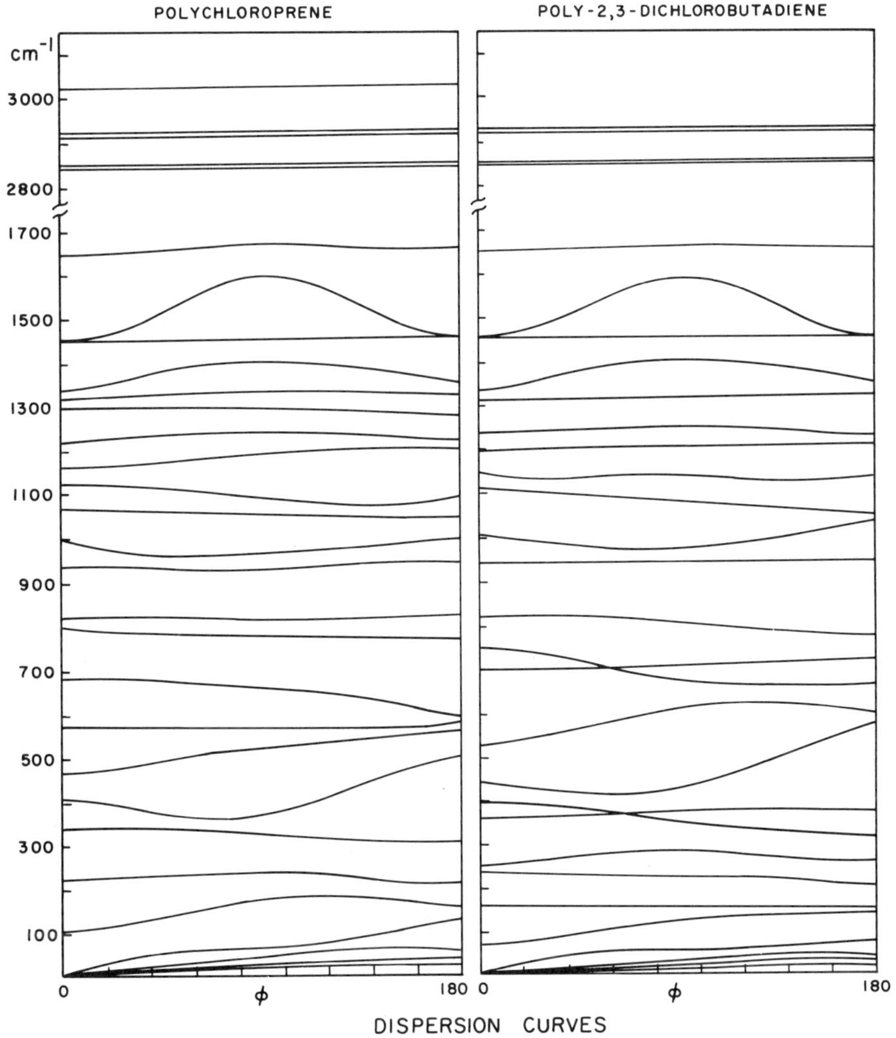

Figure 16.33. Dispersion curves of TPC and TPDCB. (Reprinted by permission from ref. 104.)

cm^{-1}, which is close to that assigned by Crowder et al. for 2,3-dichloropropene (115) and 2-chloro-1-butene (108). Each of these modes is drawn schematically from the cartesian displacement coordinates and is shown in Figure 16.34.

Tabb and Koenig (103) also published a dispersion curve of TPC based on their initial and refined valence force fields. They concluded that since the branches exhibited little curvature as a function of phase angle, there was very little coupling between adjacent translational repeat units of TPC. The results of Petcavich and Coleman (see Figure 16.33) are significantly different and several branches show considerable curvature, indicating that specific normal

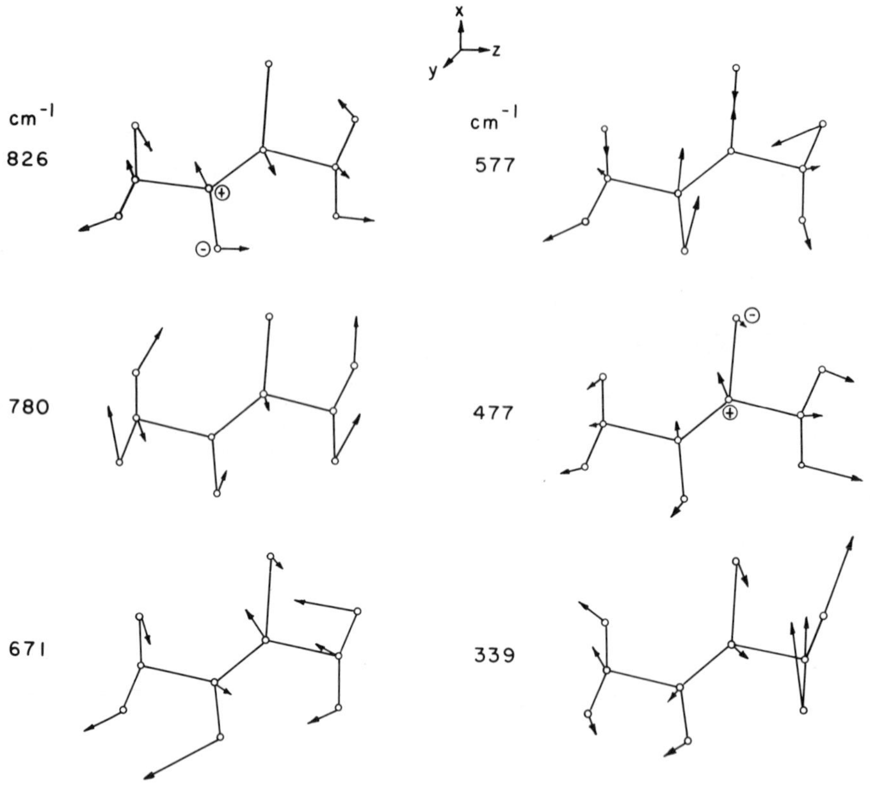

Figure 16.34. Cartesian displacement coordinates for the normal modes of TPC. (Reprinted by permission from ref. 104.)

modes are coupled to adjacent units down the chain. For example, the branches ν_5, ν_8, ν_{11}, ν_{17}, and ν_{24} exhibit distinct curvature (branches are numbered sequentially starting from 0 cm^{-1} at a phase angle equal to zero).

A detailed discussion of the assignments of the normal modes of TPC and TPDCB is given in the publication of Petcavich and Coleman.

trans-1,4-Polyisoprene and *trans*-1,4-Poly(2,3-Dimethyl butadiene)

Normal coordinate calculations of *trans*-1,4-polyisoprene (TPI) and *trans*-1,4-poly(2,3-dimethyl butadiene) (TPDMB) were recently reported by Petcavich and Coleman (16). The approach taken in this work was similar to that described in Section 16.3. for the chlorodiene polymers (104).

TPI can exist in two distinct crystalline forms, the alpha (α-TPI) and beta (β-TPI) polymorphic forms. The crystal structure of β-TPI has been determined by Bunn (105) from x-ray diffraction studies. The unit cell was found to be orthorhombic with a space group of $P_{2_1 2_1 2_1}$ and a c-axis fiber repeat

Table 16.44 Comparison of the Results for TPC

Frequency[a] (cm^{-1})		Approx.	Frequency[b] (cm^{-1})	Approx.
Obs.	Calc.	Assignment[a]	Calc.	Assignment[b]
826	826	CH opb	848	ν(C—Cl)
780	794	$r(CH_2)$	779	$r(CH_2)$
671	684	Mixed mode containing ν(C—Cl), δ(C—C—C), and δ(C=C—C)	677	CH opb
577	572	ν(C—Cl)	576	Mixed mode containing ν(C—Cl), δ(C—C—C), and δ(C=C—C)
477	471	CCl opb	427	Mixed mode containing δ(C=C—C), δ(C—C—C), and δ(C=C—Cl)

[a] From Petcavich and Coleman (104).
[b] From Tabb and Koenig (103).

Table 16.45 Molecular Parameters

	β-TPI and TPDMB		
C=C	1.33 Å	C—Ĉ=C	125°
=C—C—	1.54 Å	C—Ĉ—C	108°
—C—C—	1.54 Å	H—Ĉ—H	109.5°
—C—H	1.09 Å	C—Ĉ—CH_3	115°
=C—H	1.08 Å	C=Ĉ—CH_3	120°
Internal rotation angle about =C—C—bond = 105°			
Internal rotation angle about —C—C—bond = 180°			
	α-TPI		
C=C	1.38 Å	—C—Ĉ=C	125°
=C—C—	1.54 Å	C—Ĉ—C	109.5°
—C—C—	1.54 Å	H—Ĉ—H	109.5°
—C—H	109 Å	C—Ĉ—CH_3	115°
		C=Ĉ—CH_3	120°
Internal rotation angle about =C—C—bond = 120°			
Internal rotation angle about —C—C—bond = 180°			

distance of 4.70 Å. Chantani and Nakatani (106) reported similar results for TPDMB. In this case the unit cell was found to be orthorhombic with a space group of $Pna2_1$ and a c-axis fiber repeat distance of 4.70 Å. From the molecular parameters shown in Table 16.45, the cartesian coordinates for an isolated chain in the preferred conformation of both β-TPI and TPDMB were calculated. The conformations of the two polymers are illustrated in Figure 16.35. This conformation is similar to those of several other diene polymers, such as *trans*-1,4-polybutadiene (88), *trans*-1,4-polychloroprene (105), and *trans*-1,4-poly(2,3-dichlorobutadiene) (106). It contains one chemical repeat unit per translational repeat unit with torsional angles of 180° between adjacent CH_2 units and ±105° between the CH_2 and

$$-\overset{|}{C}=C$$

units.

The crystal structure of α-TPI has been studied by several investigators. In an early study, van der Wyk and Misch (117) proposed a simple planar model for the molecular structure of α-TPI. Thereafter, Bunn (105) and Natta et al. (118) proposed different molecular models. Recent work by Shimanouchi and Abe (119) and Takahashi et al. (120) suggests that the molecular model proposed by Natta et al. is more reasonable than Bunn's. In this study the authors used the results of Takahashi et al. From their x-ray diffraction studies the unit cell of α-TPI was determined to be monoclinic with a space group of

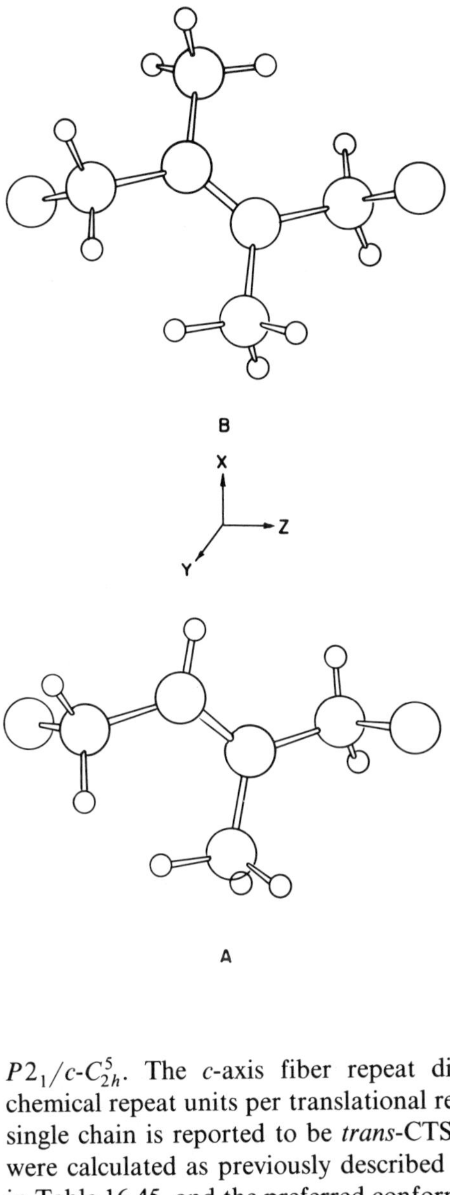

Figure 16.35. Molecule structure of (*A*) β-TPI, (*B*) TPDMB. (Reprinted by permission from ref. 116.)

$P2_1/c$-C_{2h}^5. The *c*-axis fiber repeat distance is 8.77 Å and contains two chemical repeat units per translational repeating unit. The conformation of the single chain is reported to be *trans*-CTS-*trans*-CT$\bar{\text{S}}$. The cartesian coordinates were calculated as previously described using the molecular parameters listed in Table 16.45, and the preferred conformation is illustrated in Figure 16.36.

The line group of an isolated chain of TPDMB is isomorphous to the point group C_i. There are 44 optically active fundamental vibrations distributed between two symmetry species A_u and A_g. The 23 A_g modes are Raman active and the 21 A_u are infrared active. There are no nontrivial symmetry elements associated with the line groups for the isolated chains of α- and β-TPI. Therefore there are 35 optically active fundamentals for β-TPI and 74 optically active fundamentals for α-TPI, which are all infrared and Raman active.

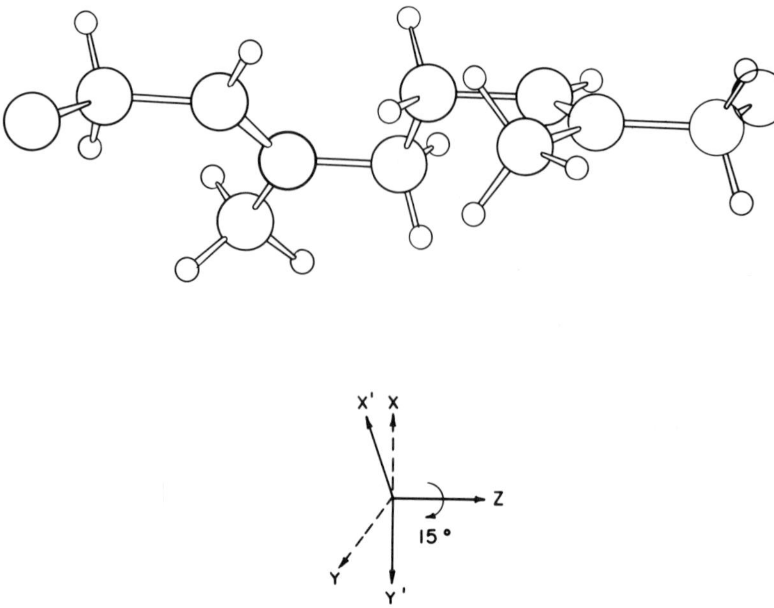

Figure 16.36. Molecular structure of α-TPI. (Reprinted by permission from ref. 116.)

Since β-TPI and TPDMB have very similar chemical structures and preferred conformations, Petcavich and Coleman decided to apply a common valence force field (VFF) to both molecules in order to increase the confidence of the normal coordinate calculations. The force constants were employed directly from the VFF of Neto and Dilauro (88). The initial force constants associated with the methyl group were transferred from Snyder and Schachtschneider (121), where applicable. Figure 16.37 shows the internal coordinates employed for β-TPI and TPDMB. Those employed for α-TPI are similar to those presented for β-TPI.

Experimentally, Petcavich and Coleman, using Fourier transform infrared spectroscopy, were able to obtain spectra characteristic of the preferred conformations of both α- and β-TPI by digitally subtracting the amorphous component from semicrystalline samples of TPI in their respective polymorphic forms. This is illustrated in Figure 16.38 in the range of 500–2000 cm^{-1}. Additionally, the authors obtained far infrared and Raman spectra of α- and β-TPI and the infrared and Raman spectra of TPDMB. Table 16.46 lists the initial force constant values (ϕ_1) and those obtained by corefining (ϕ_2) both molecules (i.e., β-TPI and TPDMB) together using a damped least-squares method (72) so as to obtain the best fit with the observed frequencies. The theoretical band assignments were made from the potential energy distribution (PED) of each normal vibration. An approximate description of the normal modes was made from a consideration of the cartesian displacement coordinates (**L** matrix).

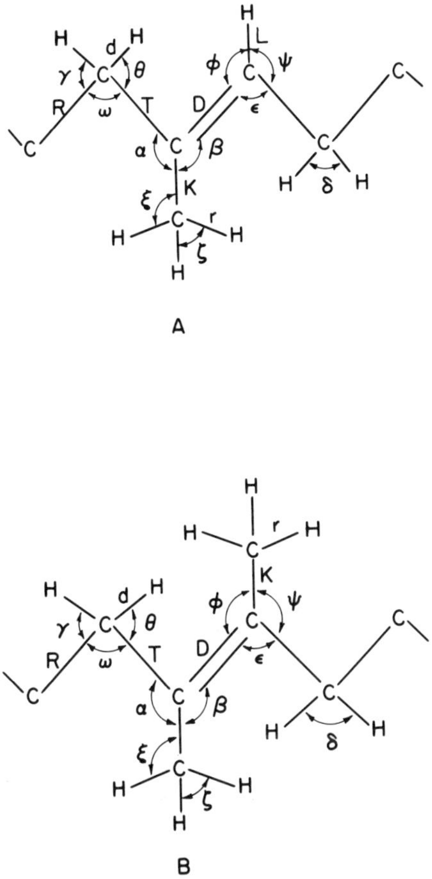

Figure 16.37. Internal coordinates for (*A*) β-TPI and (*B*) TPDMB. (Reprinted by permission from ref. 116.)

Tables 16.47 to 16.49 compare the results obtained from the normal coordinate calculations with the experimentally observed infrared and Raman frequencies for β-TPI and TPDMB. Included in these tables are the frequencies calculated for the unrefined VFF transferred directly from the force fields of Neto and DiLauro (88) and Snyder and Schachtschneider (121), together with the frequencies calculated from the VFF obtained by corefining both molecules. It can be seen that even in the case of the unrefined VFF the calculated results are in surprisingly good agreement with the observed spectral data. After corefinement, in which only the force constants associated with the double bond and corresponding interactions were allowed to vary (see Table 16.46), the total square deviation of the calculated compared to the observed frequencies for both molecules is 0.047. The PED and approximate description of the normal modes obtained from the refined force field are also included in these tables. Dispersion curves for β-TPI and TPDMB were also reported that,

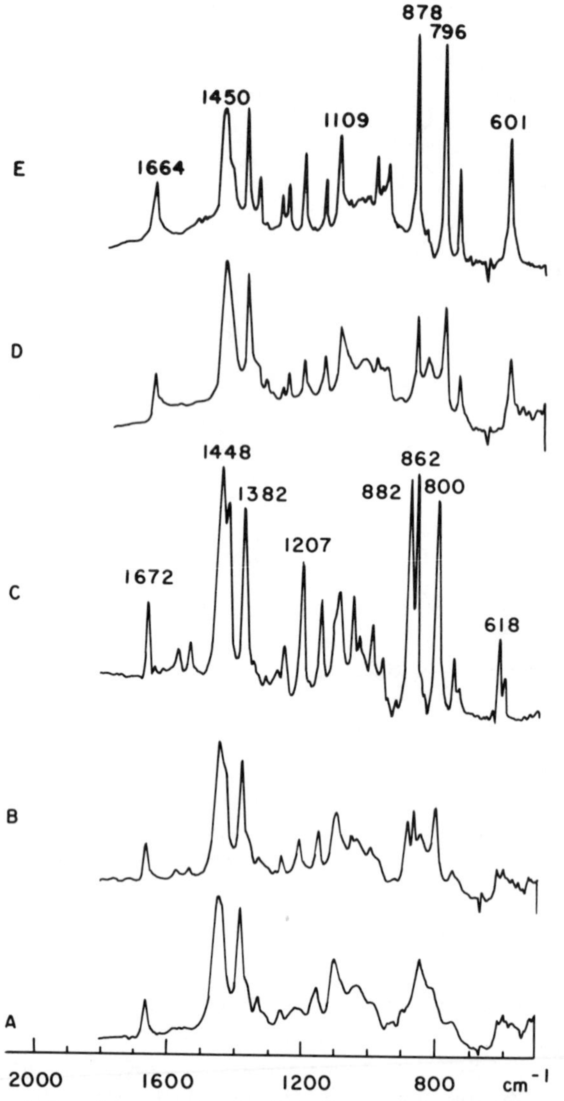

Figure 16.38. Infrared spectra in the 500–3200 cm^{-1} range. (*A*) Amorphous spectrum of TPI recorded at 80°C. (*B*) Semicrystalline spectrum of TPI in predominantly α form recorded at room temperature. (*C*) Difference spectrum $B - A$. (*D*) Semicrystalline spectrum of TPI in predominantely β form recorded at room temperature. (*E*) Difference spectrum $D - A$. (Reprinted by permission from ref. 116.)

Table 16.46 Force Constant Data for TPI and TPDMB

Refined				Unrefined			
Force Constant	ϕ_1[a]	ϕ_2	$\sigma(\phi_2)$[b]	Force Constant	ϕ_1[a]	Force Constant	ϕ_1[a]
H_ε	0.910	0.783	082	K_T	4.384	$F_{T\gamma}$	0.066
H_β	0.944	0.882	0.105	K_D	8.702	$F_{T\phi}$	0.075
Γ_1	0.344	0.288	0.023	K_l	4.947	$F_{T\omega}$	0.419
Γ_2	0.199	0.197	0.035	K_d	4.524	$F_{K\xi}$	0.174[c]
τ_D	0.328	0.348	0.077	K_r	4.699[c]	$F_{\theta\theta}$	−0.032
$F_{T\omega}$	0.419	0.632	0.087	H_ω	1.049	$F_{\theta\gamma}$	0.015
$F_{D\alpha}$	0.0	−0.384	0.132	H_γ	0.664	$F_{\phi\psi}$	0.027
$F_{T\alpha}$	0.0	0.363	0.198	H_θ	0.668	$F_{\varepsilon\psi}$	−0.029
$F_{T\beta}$	0.0	0.042	0.196	H_ζ	0.539[c]	$F_{\omega\theta}$	−0.028
$F_{K\alpha}$	0.0	−0.006	0.190	H_ξ	0.618[c]	$F_{\xi\xi}$	−0.031
$F_{K\varepsilon}$	0.0	0.058	0.179	H_δ	0.540	$f_{\theta\theta}^{t}$	0.010
$F_{K\beta}$	0.0	−0.052	0.183	H_ϕ	0.560	$f_{\theta\theta}^{g}$	−0.006
$f_{\varepsilon\beta}$	0.0	−0.120	0.166	H_ψ	0.480	$f_{\gamma\gamma}^{t}$	0.120
$f_{\beta\beta}$	0.0	−0.204	0.109	τ_T	0.024	$f_{\gamma\gamma}^{g}$	−0.02
$f_{\beta\phi}$	0.0	−0.050	0.162	F_{DT}	0.094	$f_{\gamma\theta}^{t}$	0.008
				F_{TR}	0.122	$f_{\gamma\theta}^{t}$	0.029
				F_{dd}	0.004	$f_{\omega\omega}^{t}$	0.080
				F_{rr}	0.032[c]	$f_{\omega\gamma}^{g}$	−0.045
				$F_{D\phi}$	0.355	$f_{\omega\gamma}^{t}$	0.073[c]
				$F_{T\theta}$	0.341	f_{ω}^{t}	−0.011[c]
				$F_{T\psi}$	0.333	f_{ω}^{g}	0.012
				$F_{T\gamma}$	0.348		

$K_T = K_R = K_K$	$F_{\gamma\gamma} = F_{\theta\theta}$	$F_{\varepsilon\phi} = F_{\varepsilon\psi}$	$f_{\theta\gamma}^{t} = f_{\psi\gamma} = f_{\phi\theta}$
$H_\alpha = H_\beta$	$\tau_T = \tau_R = \tau_K$	$f_{\gamma\gamma}^{t} = f_{\psi\theta}$	$f_{\theta\gamma}^{g} = f_{\psi\gamma} = f_{\phi\theta}$
$F_{T\gamma} = F_{R\theta}$	$F_{\omega\gamma} = F_{\omega\theta}$	$f_{\omega\omega} = f_{\varepsilon\varepsilon}$	$f_{\theta\theta}^{g} = f_{\gamma\phi} = f_{\varepsilon\omega}$
$F_{T\varepsilon} = F_{D\varepsilon}$	$F_{T\phi} = F_{D\psi}$	$f_{\gamma\gamma}^{g} = f_{\psi\theta}$	$f_{\omega\gamma} = f_{\varepsilon\theta} = f_{\omega\psi}$

[a] Force constants obtained from *trans*-1,4-polybutadiene (88) except where otherwise noted.
[b] Standard error in ϕ_2.
[c] Force constants obtained from Snyder and Schachtschneider (121).
[d] Stretch constants are in units of mdyn/Å; stretch-bend interaction constants are in units of mdyn/rad; bending constants are in units of mdyn-Å/rad^2.

not surprisingly, closely resemble those calculated for the analogous chlorobutadiene polymers (see Figure 16.33). A complete discussion of the assignments of the normal vibrations of β-TPI and TPDMB is given in the publication of Petcavich and Coleman (116). However, a few interesting points deserve further comment. The infrared bands occurring in the spectrum of β-TPI at 878, 796, and 751 cm^{-1} have been used as characteristic frequencies

Table 16.47 β-Polyisoprene

Observed Frequency[a] (cm^{-1})		Calc. Frequency (cm^{-1})		Approx. Potential Energy Distribution	Type of Vibration[b]
IR	Raman	Initial	Refined		
3025(m)	3020(w)	3016	3014	99% K_l	$\nu(=C-H)$
2980(s)		2964	2964	100% K_r	
2962(s)	2962(sh)	2963	2963	100% K_r	$\nu_{as}(CH_3)$
2941(s)	2930(sh)	2923	2923	99% K_d	
2916(s)	2912(s)	2914	2914	100% K_d	$\nu_{as}(CH_2)$
	2880(sh)	2876	2876	98% K_r	$\nu_s(CH_3)$
2852(s)		2848	2846	99% K_d	$\nu_s(CH_2)$
2845(s)	2845(ms)	2842	2842	100% K_d	
1664(m)	1666(vs)	1706	1657	17% K_T, 79% K_D	$\nu(C=C)$
	1460(m-sh)	1466	1466	92% H_ζ	
1450(m)		1462	1462	86% H_ζ	$\delta_{as}(CH_3)$
	1445(m-sh)	1454	1460	12% H_θ, 67% H_δ	
1430(m)	1432(s)	1453	1454	10% H_γ, 15% H_θ, 78% H_δ	$\delta_s(CH_2)$
1384(m)	1386(mw)	1397	1394	28% K_T, 34% H_ζ, 39% H_ξ, -12% $F_{R\beta}$	$\nu(CC)$, $\delta_s(CH_3)$
1350(w)	1350(vw)	1335	1331	14% K_T, 40% H_γ, 33% H_θ, -13% $F_{T\gamma}$	$\gamma_\omega(CH_2)+\gamma_t(CH_2)$
1326(w)	1325(m)	1314	1314	53%, H_γ, 26% H_θ, 12% $f_{\gamma\gamma}$	$\gamma_t(CH_2)$
1280(w)	1280(m)	1312	1301	19% K_T, 21% H_γ, 27% H_θ, 12% H_ϕ	$\gamma_\omega(CH_2)+\gamma_t(CH_2)$

1260(w)		1260	1260	38% K_T, 20% H_γ, 12% H_θ	$\gamma_\omega(CH_2)+\gamma_t(CH_2)$
1215(s)	1210(vw-br)	1213	1215	48% H_γ, 49% H_θ, −10% $f_{\gamma\gamma}$	$\gamma_\omega(CH_2)$
1151(m)	1151(w)	1118	1115	29% H_γ, 47% H_θ, 19% H_ϕ, 15% H_ψ	$\gamma_t(CH_2)+\delta(=C-H)$i.p.
1109(m)	1109(vw)	1093	1077	57% K_T, 15% H_θ, 12% H_ξ	ν(CC)
	1038(w)	1056	1033	74% H_ξ	$\gamma_r(CH_3)$
998(m)	1000(ms)	1015	998	76% K_T, 11% H_θ	ν(CC)
981(m)	980(w-sh)	976	972	20% K_T, 12% H_θ, 49% H_ξ	$\gamma_r(CH_3)+\gamma_r(CH_2)$
965(m)		953	951	32% K_T, 20% H_θ, 24% H_ξ	
878(s)	878(w)	877	883	12% K_T, 22% τ_D, 36% Γ_2	$\delta(=C-H)$o.p.
796(s)	798(m)	817	800	15% K_T, 52% H_γ, 13% H_θ, −11% $F_{T\omega}$	
751(m)	751(w)	760	743	39% K_T, 26% H_ω, 25% H_γ, 12% H_θ, −27% $F_{T\omega}$	$\gamma_r(CH_2)$
601(m)	602(m)	636	618	16% K_T, 12% H_ω, 24% H_ε, 17% H_β	δ(CCC)
474(w)	475(w)	492	468	13% K_T, 57% Γ_1	$\delta(CH_3)$o.p.
425(vw-sh)		424	417	10% K_T, 47% H_ω, 25% τ_D, 13% Γ_2, −14% $F_{T\omega}$	
325(vw)	380(w)	379	361	26% H_ε, 58% H_β	$\delta(C-C-C)+\delta(C=C-C)+\tau$
225(w)		219	240	43% H_ε, 27% H_β	
	165(w)	158	158	97% τ_T	
153(vw)		128	130	10% K_T, 15% H_ω, 43% τ_T, 12% $F_{T\omega}$	

[a]Intensities in parentheses: s=strong, m=medium, w=weak, sh=shoulder, br=broad, and v=very.
[b]Type of vibration: ν=stretch, δ=bending, γ_ω=wagging, γ_t=twisting, γ_r=rocking, τ=torsion, i.p.=in plane, o.p.=out of plane; s=symmetric, and as=asymmetric.

Table 16.48 Poly-2,3-Dimethylbutadiene (A_g Mode)

Observed Frequency[a] (cm^{-1})	Calculated Frequency (cm^{-1})		Approx. Potential Energy Distribution	Type of Vibration[a, b]
Raman	Initial	Refined		
2980(m)	2964	2964	100% K_r	$\nu_{as}(CH_3)_i$
2950(sh)	2963	2963	100% K_r	
2910(s)	2919	2920	99% K_d	$\nu_{as}(CH_2)_i$
2861(m)	2877	2877	98% K_r	$\nu_s(CH_3)_i$
2855(sh)	2850	2849	99% K_d	$\nu_s(CH_2)_i$
2714(w)				$1331(R)+1377(R)=2708?$
1665(s)	1764	1673	18% K_T, 78% K_D	$\nu(C{=}C)$
	1472	1476	12% K_T, 13% H_θ, 55% H_δ	$\delta_s(CH_3)_i+\delta(CH_2)_i$
1462(m)	1464	1462	92% H_ζ	
	1451	1455	93% H_ζ	$\delta_{as}(CH_3)_i$
1438(m)	1424	1432	36% K_T, 14% H_ζ, 17% H_ξ, 21% H_δ	
1377(ms)	1350	1350	32% H_ζ, 37% H_ξ	$\delta_s(CH_3)+\delta(CH_2)_i$

1331(ms)	1308	1299	28% K_T, 50% H_γ, 51% H_θ, -15% $F_{T\theta}$, -20% $F_{T\gamma}$	$\gamma_\omega(CH_2)_i$
	1142	1139	60% H_γ, 37% H_θ, -11% $f_{\gamma\gamma}$	$\gamma_t(CH_2)_i$
1036(m)	1105	1075	34% K_T, 13% H_ω, 32% H_ξ, 11% Γ_1	
1018(m)	1025	1017	50% K_T, 43% H_ξ	$\gamma_r(CH_3)_i+\gamma_\omega(CH_2)_i+\nu(C—C)_i$
978(w)	960	966	65% H_ξ, 11% H_γ	$\gamma_r(CH_3)_i+\gamma_r(CH_2)_i$
943(w)	950	946	13% K_T, 35% H_γ, 27% H_θ, 22% H_ξ	$\gamma_r(CH_2)_i+\gamma_r(CH_3)_i$
782(mw)	796	779	44% K_T, 33% H_ω, -16% $F_{T\omega}$	$\gamma_r(CH_2)_i+\gamma_r(CH_3)_i$, highly mixed
750(w)				?
688(w-br)				Amorphous
580(m)	595	560	26% K_T, 60% Γ_1	$\delta(CH_3)$o.p.
489(mw)	502	495	32% K_T, 20% H_ε, 22% H_β, 10% $F_{\varepsilon\beta}$	
415(vw-br)?	303	287	16% H_ε, 81% H_β	$\delta(CCC)$, $\tau(CCCC)$
150(w-sh)?	158	158	98% τ_t	$\tau(C—C=C—C)$
125(w-sh)?	119	121	15% H_ω, 52% τ_t	

[a]Notation as in Table 16.47.
[b]Subscript i=in phase.

Table 16.49 Poly 2,3-Dimethylbutadiene (A_u Mode)

Observed Frequency[a] (cm^{-1})	Calculated Frequency (cm^{-1})		Approx. Potential Energy Distribution	Type of Vibration[a,b]
Infrared	Initial	Refined		
2970(m-sh)	2964	2964	100% K_r	$\nu_{as}(CH_3)_o$
2948(s)	2963	2964	100% K_r	
2918(vs)	2913	2913	100% K_d	$\nu_{as}(CH_2)_o$
	2876	2877	98% K_r	$\nu_s(CH_3)_o$
2860(s)	2843	2843	100% K_d	$\nu_s(CH_2)_o$
2720(w)				1331(R)+1375(IR)=2706?
1463(m)	1465	1464	92% H_ξ	$\delta_{as}(CH_3)_o$
	1459	1459	12% H_θ, 17% H_ζ, 65% H_δ	$\delta(CH_2)_o+\delta_{as}(CH_3)_o$
1445(m)	1457	1457	78% H_ζ, 14% H_δ	
1376(s)	1393	1393	16% K_T, 44% H_ζ, 50% H_ξ, −10% $F_{R\beta}$	$\delta_s(CH_3)_o+\delta(CH_2)_o$
1260(w)	1316	1316	34% H_γ, 46% H_θ	$\gamma_t(CH_2)_o$
1223(m)	1231	1228	13% K_T, 53% H_γ, 53% H_θ, −15% $F_{T\theta}$	$\gamma_\omega(CH_2)_o$
1155(m)	1180	1163	73% K_T, 14% H_ξ	$\nu(C—C)_o$
1100(w-br)				Amorphous?
1095(w-br)	1082	1074	37% K_T, 13% H_γ, 28% H_ξ	$\nu(C—C)_o+\gamma_r(CH_3)_o$
1050(w)	1032	1031	84% H_ξ	$\gamma_r(CH_3)_o$
896(w)	926	925	32% K_T, 46% H_ξ	$\gamma_r(CH_3)_o+\gamma_r(CH_2)_o$
880(w-br)				
803(w)				Amorphous?
760(w)	825	825	22% K_T, 27% H_γ, 38% H_θ	$\gamma_r(CH_2)$
454(w)	491	474	36% H_ω, 34% Γ_1, 21% τ_D	$\delta(CH_3)$o.p.
407(vw?)	435	419	12% H_ε, 74% H_β	
260(vvw?)	232	242	31% H_ε, 35% H_β, −10% $F_{T\omega}$, 16% $F_{\varepsilon\beta}$	δ(CCC), τ(CCCC)
215(vw)	201	206	30% H_ω, 45% τ_D	τ(C—C=C—C)
	157	157	94% τ_T	

[a]Notation as in Table 16.47.
[b]Subscript o=out of phase.

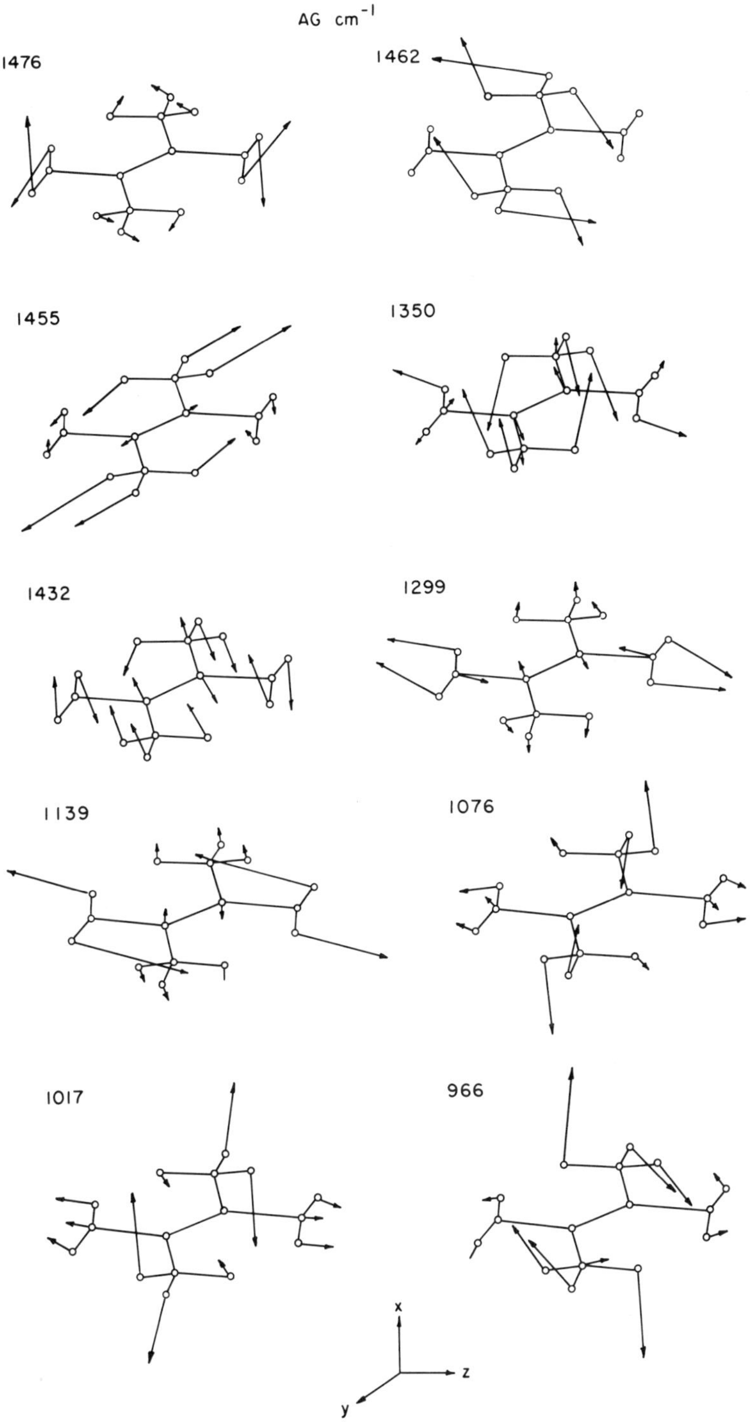

Figure 16.39. Cartesian displacements coordinates for the A_g normal modes of TPDMB. (Reprinted by permission from ref. 116.)

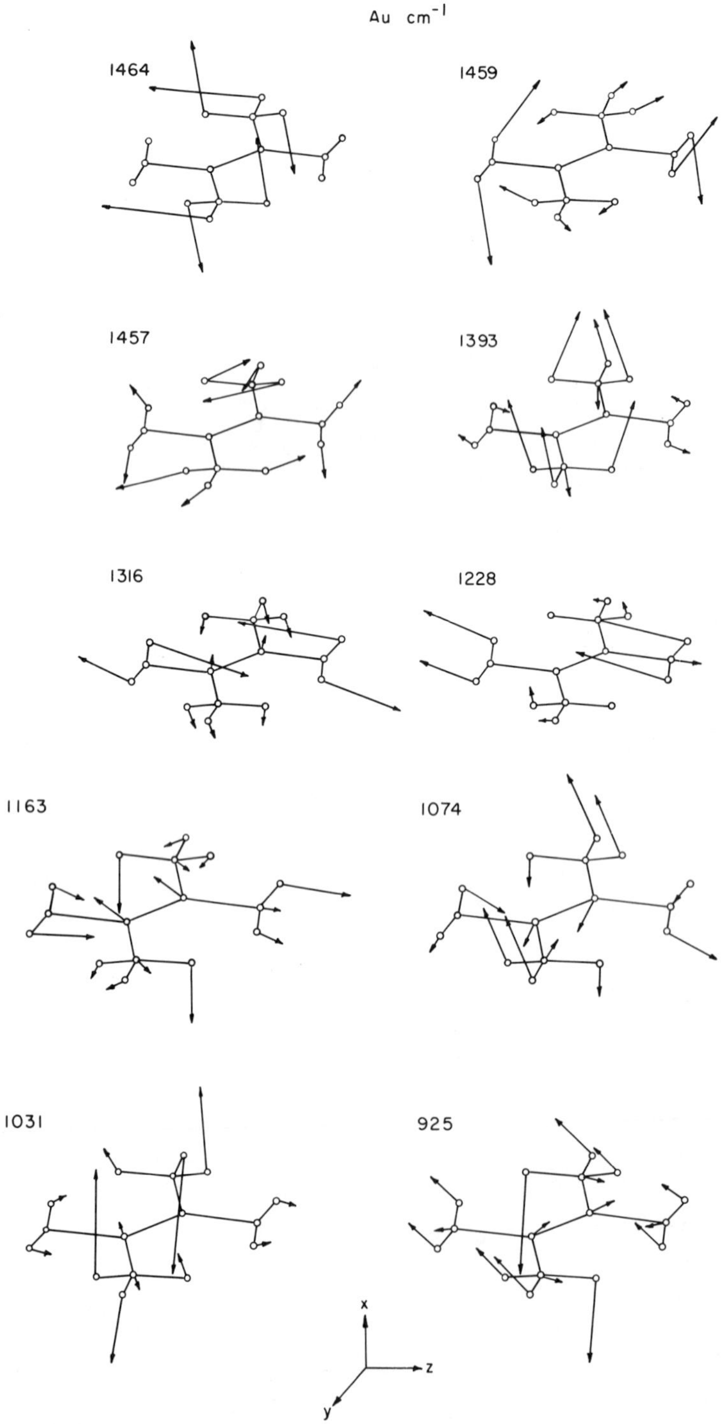

Figure 16.40. Cartesian displacements coordinates for the A_u normal modes of TPDMB. (Reprinted by permission from ref. 116.)

for this polymorphic form. From the normal coordinate calculations these characteristic bands are assigned to predominantly CH out-of-plane and two CH_2 rocking modes, respectively. A comparison with the assignments of the characteristic bands of α-TPI will be discussed later. Schematic representations of the cartesian displacement coordinates for the A_g (Raman-active) and A_u (infrared-active) normal modes of TPDMB occurring between 900 and 1500 cm^{-1} are reproduced in Figures 16.39 and 16.40, respectively, and are excellent examples of the difference between the relative displacements of the totally symmetric A_g and the asymmetric A_u modes. Furthermore, these figures superbly illustrate that only a few of the normal vibrations can be considered pure modes (i.e., the 1462 and 1455 cm^{-1} A_g and the 1464 and 1228 cm^{-1} A_u modes), while the remainder are highly mixed modes characteristic of CH_3 deformations and rocking, CH_2 scissoring, wagging, twisting, and rocking, and C—C skeletal vibrations.

Petcavich and Coleman also transferred their force field derived from β-TPI and TPDMB to the preferred conformational model of α-TPI. The authors stressed the assumptions and severe limitations that are inherent in this approach. In effect, they were assuming that any changes in the calculated frequencies for α-TPI relative to β-TPI were primarily dependent on corresponding changes in the **G** matrix. The conformation of α-TPI is markedly different from that of the other two molecules, and certain interaction force constants obviously are not entirely transferable. However, the CH_2 groups are still in the *trans* conformation and the authors anticipated that the force constants involved with these groups would not alter significantly. The C=C—C internal rotation angle is 120° in the α-TPI, whereas it is 105° in β-TPI and TPDMB. Neto and DiLauro (88) have stated, and the authors have verified, that reasonable changes in these internal rotational angles have only minimal effects on the calculated frequencies. The translational repeat unit contains two chemical repeat units and consequently there are 74 optically active modes that are both infrared and Raman active. There is always the temptation in this type of calculation to match the observed and calculated frequencies, but it was realized that this is a spurious exercise. The prime objective was to elucidate qualitatively the assignments of the vibrational modes occurring in the spectra of α-TPI that serve to differentiate between this polymorphic form and that of the β form of TPI. The results, described earlier for β-TPI (for which there was a high degree of confidence), permitted the authors to compare the frequencies and approximate PED of similar modes in both polymers.

Infrared bands occurring at 882, 863, and 800 cm^{-1} have been previously used to identify the α form of TPI. The calculations suggest that the former two bands are mixed modes containing predominantly out-of-plane =CH vibrations. Since we have two chemical repeat units, two such modes were calculated at 882 and 871 cm^{-1}. These are in very good agreement with the observed frequencies. The 800 cm^{-1} band, calculated at 819 cm^{-1}, was assigned to a predominantly CH_2 rocking mode. For comparison, the out-of-plane $=CH$ vibration is observed at 878 cm^{-1} and the CH_2 rocking mode vibration at 796 cm^{-1} in the β-TPI. Another CH_2 rocking mode is calculated at 811 cm^{-1} in α-TPI and was assigned to the weak Raman line at 785 cm^{-1}. The other normal modes of α-TPI occurring below 750 cm^{-1} could not be assigned with confidence because of the overriding assumptions made in the calculations.

Poly(1-*trans*-Pentenylene) and Poly(1-*trans*-Heptenylene)

Normal coordinate calculations of poly(1-*trans*-pentenylene) (PTP) and poly(1-*trans*-heptenylene) (PTH) have been performed by Holland-Moritz and van Werden (122). These polymers have the general chemical structure $\leftarrow (CH_2)_nCH{=}CH \rightarrow$ where n equals 3 and 5 for PTP and PTH, respectively. The geometric structures of the poly(alkenylenes) are well established from the x-ray studies of Natta and Bassi (123, 124), and are summarized in Table 16.50. The internal coordinates for a chemical repeat unit of PTP are shown in Figure 16.41. An analogous model was employed for PTH.

The ideal, infinite, isolated polymer chain in the preferred conformation has a line group symmetry isomorphous to the point group C_{2h} for both PTP

Table 16.50

Polymer	Crystal Structure	Unit Cell Dimensions (nm) *a*	*b*	*c*	Space Group
Poly(1-*trans*-pentenylene)	Orthorhombic	0.728	0.497	1.190	P_{nam}
Poly(1-*trans*-heptenylene)	Orthorhombic	0.74	0.5	1.71	P_{nam}

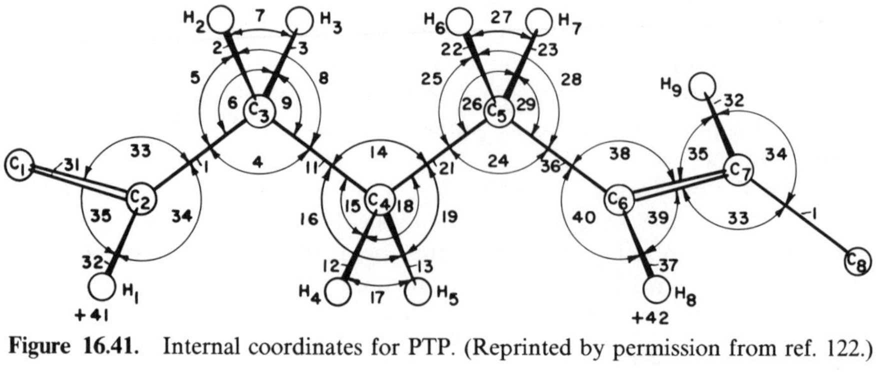

Figure 16.41. Internal coordinates for PTP. (Reprinted by permission from ref. 122.)

Table 16.51 Force Constant Data for PTP

Force Constant Symbol	Internal Coordinates[a]	Force Constant[b]	Force Constant Symbol	Internal Coordinates[a]	Force Constant[b]
		Diagonal Elements			
K_T	1, 31	4.384	H_ε	33, 38	0.91
K_R	11, 21	4.384	H_ϕ	35, 39	0.5
K_D	31	8.71	H_Ψ	34, 40	0.48
K_d	2, 3, 22, 23	4.524	H_θ	5, 6, 28, 29	0.668
K'_d	12, 13	4.524	H_γ	8, 9, 25, 26	0.664
K_e	32, 37	4.947	H'_T	15, 16, 18, 19	0.66
H_ω	14	0.901	H_Γ	41, 42	0.1999
H'_ω	4, 24	1.049	τ_D	43	0.3
H_δ	7, 27	0.53	τ_T	10, 44	0.21
H'_δ	17	0.53	τ_R	20, 30	0.24
		Off-diagonal Elements			
F_{dd} (α—CH_2)		0.004	$F_{\Psi\varepsilon}$		−0.029
F_{dd} (β—CH_2)		0.016	$F_{\Gamma\omega}$		0.419
F_{DT}		0.094	$F_{\Gamma D}$		0.094
F_{TR}		0.122	$f_{\phi\phi}$		0.121
F_{RR}		0.064	$f^{t}_{\gamma\gamma}$		0.106
$F_{D\phi}$		0.355	$f^{g}_{\gamma\gamma}$		−0.024
$F_{\Psi T}$		0.333	$f^{t}_{\omega\omega}$		0.093
$F_{D\Psi}$		0.075	$f^{t}_{\varepsilon\varepsilon}$		0.8
$F_{T\phi}$		0.075	$f^{g}_{\gamma\omega}$		−0.058
$F_{T\theta}$		0.4	$f^{g}_{\varepsilon\theta}$		−0.045
$F_{R\gamma}$ (α—CH_2)		0.348	$f^{g}_{\varepsilon\phi}$		−0.045
$F_{R\gamma}$ (β—CH_2)		0.261	$f^{g}_{\Psi\omega}$		−0.045
$F_{T\omega}$		0.419	$f'^{g}_{\gamma\gamma}$		0.002
$F_{D\varepsilon}$		0.419	$f'^{t}_{\theta\gamma}$		−0.002
$F_{R\omega}$ (α—CH_2)		0.419	$f'^{g}_{\theta\gamma}$		0.002
$F_{R\omega}$ (β—CH_2)		0.351	$f'^{t}_{\Psi\gamma}$		0.008
$F_{\theta\theta}$		−0.032	$f'^{g}_{\Psi\gamma}$		0.029
$F_{\gamma\gamma}$ (α—CH_2)		−0.032	$f'^{t}_{\phi\theta}$		0.008
$F_{\theta\omega}$		−0.028	$f'^{g}_{\phi\theta}$		0.029
$F_{\gamma\omega}$ (α—CH_2)		−0.028	$f'^{t}_{\phi\Psi}$		0.008
$F_{\gamma\theta}$		0.015	$f''^{t}_{\phi\gamma}$		−0.001
$F_{\gamma\gamma}$ (β—CH_2)		−0.016	$f''^{g}_{\theta\gamma}$		0.001
$F'_{\gamma\gamma}$		0.040	$f''^{g}_{\phi\gamma}$		−0.006
$F_{\gamma\omega}$ (β—CH_2)		−0.124	$f''^{t}_{\Psi\Psi}$		−0.006
$F_{\phi\varepsilon}$		0.029	$f^{t}_{\varepsilon\omega}$		0.012

[a]See Figure 16.41.
[b]Units of force constants are stretch (mdyn/Å), angle deformation and torsions (mdyn-Å/rad^2), and angle-stretch (mdyn/rad).

Table 16.52 Poly(1-trans-pentenylene) $\leftarrow CH=CH-(CH_2)_3 \rightarrow$

Calculated Frequency				Observed Frequency		Assignment and PED	Type of
A_g	B_g	A_u	B_u	A_g, B_g	$A_u(\parallel), B_u(\perp)$	(%)	Vibration
		46				$57\ \tau_R, 25\ \tau_T$	τ
	74					$88\ \tau_T$	τ
77						$67\ \tau_R$	τ
			84			$62\ \tau_R, 26\ \tau_T$	τ
		88				$37\ H_\Gamma, 21\ H_\varepsilon, 20\ \tau_D$	$\tau, \delta(C-C-C)$
	99					$86\ \tau_R$	τ
135						$32\ K_T, 32\ H'_\omega$	$\nu(C-C), \delta(C-C-C)$
167						$45\ H_\omega, 22\ \tau_T$	$\delta(C-C-C), \tau$
			242			$33\ H_\varepsilon, 30\ H_\Gamma$	$\delta(C-C=C), \tau$
		264			260 w $\parallel$	$52\ H_\varepsilon$	$\delta(C-C=C)$
			311		300 m $\perp$	$40\ H_\varepsilon, 29\ H'_\omega$	$\delta(C-C=C), \delta(C-C-C)$
	374					$72\ H_\omega$	$\delta(C-C-C)$
			425			$20\ H_\omega, 20\ K_R$	$\delta(C-C-C), \nu(C-C)$
	439					$43\ H_\varepsilon, 35\ K_T$	$\delta(C=C-C), \nu(=C-C)$
		519			531 m $\parallel$	$65\ H_\omega$	$\delta(C-C-C)$
594				585 w		$32\ H_\varepsilon, 22\ H'_\omega$	$\delta(C=C-C), \delta(C-C-C)$
713						$33\ H_\Gamma, 27\ H_\gamma, 25\ H'_\gamma$	op(CH), $r(CH_2)$
			736		734 s B_{2u} $\perp$ 744 m B_{3u} $\perp$	$55\ H_\gamma, 39\ H'_\gamma$	$r(CH_2)$

	741			758 vw B_g		62 H_Γ	op(CH)
768				780 w		52 H_Γ, 28 H_γ	op(CH), $t(\alpha\text{-CH}_2)$, $r(\beta\text{-CH}_2)$
		838			838 m ∥	55 H'_γ, 28 H_θ, 21 H_γ	$r(\alpha\text{-CH}_2)$, $t(\beta\text{-CH}_2)$
	851			863 m B_g		42 H'_γ, 26 H_Γ	$t(\text{CH}_2)$, op(CH)
		968			963 vs ?	55 τ_D, 34 H_Γ	
			971			58 τ_D, 35 H_Γ	τ(H—C=C—H), op(CH)
990				974		57 K_R	
		999			999 w ∥	74 K_T, 21 K_R	ν(C—C)
			1013		1015 w ⊥	69 K_R	
			1040		1041 m ⊥	51 H_θ, 21 H_γ	$t(\text{CH}_2)$
1040				1035 vs		35 H_θ, 30 K_R	
	1057					78 K_R	ν(C—C)
	1068			1070 vw B_g		34 K_T, 26 H_γ	ν(=C—C), $t(\beta\text{-CH}_2)$
		1071			1072 s ∥	48 K_R, 24 K_T	ν(C—C)
			1086		1084 w ⊥	84 K_T	ν(=C—C)
		1134				40 H_θ, 31 H_γ	
	1169					53 H_θ, 31 H_γ, 21 H'_γ	$t(\text{CH}_2)$
1174				1175 vw		21 K_T, 21 H_Ψ	ν(C=C—C), δ(C=C—H)
	1193			1190 m B_g		31 H_θ, 20 H_γ, 18 H_Ψ	$w(\alpha\text{-CH}_2)$, $t(\beta\text{-CH}_2)$, δ(C=C—H)
		1197		(1200 m)	1218 m ∥	50 H_θ, 39 H_γ	$w(\text{CH}_2)$
			1211			35 H_Ψ, 28 H_ϕ, 27 H_θ, 22 H_γ	δ(C=C—H), $w(\alpha\text{-CH}_2)$
1241				1260 s		36 H'_γ, 35 H_θ, 20 K_T	$w(\alpha\text{-CH}_2)$, ν(=C—C)
		1251				36 H_Ψ, 30 H_ϕ, 22 H_γ	δ(C=C—H), $t(\text{CH}_2)$
			1291		1290 vw ⊥	48 H'_γ, 26 H_θ	
1293				1282 s A_{1g} 1292 s B_{1g}		54 H'_γ, 21 H_θ	$t(\alpha\text{-CH}_2)$, $r(\beta\text{-CH}_2)$

Table 16.52 Continued

Calculated Frequency				Observed Frequency		Assignment and PED	Type of
A_g	B_g	A_u	B_u	A_g, B_g	$A_u(\parallel)$, $B_u(\perp)$	(%)	Vibration
	1297			1308 sh B_g		49 H_γ, 28 H'_γ	$t(CH_2)$
			1312			38 H_γ, 27 H_θ	$t(\alpha\text{-}CH_2)$, $r(\beta\text{-}CH_2)$
		1301			1310 w	42 H_γ, 29 H'_γ	$t(CH_2)$
	1327					74 H_γ, 23 K_R	$w(\beta\text{-}CH_2)$, $r(C—C)$
		1338			1350 w ?	65 H_γ, 25 K_R	$w(CH_2)$, $r(C—C)$
1363	1361					24 H_ϕ	$\delta(C{=}C—H)$
			1430		1436 s $\perp$	50 H'_δ, 27 H_δ	
				1416 m A_{1g}			
1433						43 H'_δ, 31 H_δ	$\delta(CH_2)$
				1435 s B_{1g}			
		1439				79 H'_δ	
	1441					74 H'_δ	$\delta(\alpha\text{-}CH_2)$
					1452 s B_{2u} $\perp$		
			1454			49 H_δ, 28 H'_δ	
					1458 s B_{3u} $\perp$		
1454						46 H_δ, 31 H'_δ	$\delta(CH_2)$
1653	1653			1668 vs ?		70 K_D	$\nu(C{=}C)$
2840			2840	2838 m	2823 sh	70 K'_d, 30 K_d	
	2841	2841			2852 vs	99 K'_d	
2848			2848	2852 m	2847 sh	69 K_d, 29 K'_d	$\nu(CH_2)$
2913			2913	2888 vs	2905 m	60 K_d, 41 K'_d	
	2922	2922		2905 sh	2924 vs	99 K'_d	
2927			2921	2922 sh	2927 sh	59 K'_d, 40 K_d	
		3010	3010		3002 w	100 K_l	
3029	3029			2988 m		98 K_l	$\nu(=C—H)$

and PTH. Hence, the normal modes (excluding pure translations and rotations) can be classified into the following irreducible representations:

PTP	20 A_g, 18 B_g, 17 A_u, and 19 B_u
PTH	29 A_g, 27 B_g, 26 A_u, and 28 B_u

The A_g and B_g modes are only Raman active while the A_u and B_u modes are only infrared active.

Holland-Moritz and van Werden initially transferred the well-established force constants derived for *trans*-1,4-poly(butadiene) by Neto and DiLauro (88) for the olefinic part of the molecule and those obtained by Schachtschneider and Snyder (21) for the corresponding methylene portion. A minor modification to some of the force constants was reported after a refinement procedure involving a least-squares best fit of the observed and calculated frequencies. The final set of force constants is listed in Table 16.51.

The authors obtained polarized infrared and Raman data on stretched films of the polymers. Thus they were able to separate the infrared-active A_u (parallel polarization) modes from the B_u (perpendicular polarization) modes, using computer-assisted infrared spectroscopy. By ratioing the parallel and perpendicular polarized spectra the different dichroic bands were readily identified. Similarly, polarized Raman experiments led to a reasonable separation of the A_g and B_g modes. The results of the normal coordinate calculations of PTP are summarized in Table 16.52. Included in this table is a comparison of the observed and calculated frequencies together with assignments from the potential energy distribution. In general, there is an excellent correspondence between the observed and calculated frequencies. The authors also present calculations for PTH in their publication and the reader is referred to ref. 122 for a detailed discussion of these results.

In a discussion similar to that presented by Hsu et al. (89) for crystalline *trans*-1,4-poly(butadiene) (see Section 16.3), Holland-Moritz and van Werden present arguments concerning the effects of a perturbation from an interchain potential to explain some anomalies in the experimental Raman data. The assignments of the normal modes of PTP and PTH are discussed in detail in the authors' publication. It is sufficient to restate that these results once again indicate the value of transferring well-established valence force fields derived basically from model compounds to analogous polymer systems.

16.4 Polymers Containing Aromatic Rings

In this section we will restrict ourselves to two polymers containing aromatic groups: isotactic polystyrene (IPS) and poly(ethylene terephthalate) (PET).

The infrared spectrum of polystyrene is perhaps the most easily recognized of any polymeric material since films of this polymer have traditionally been used to standardize conventional dispersive instruments. Polystyrenes are polymers containing essentially head-to-tail chemical repeating units

$$\text{--}(\text{CH}_2\text{CH})\text{--}_n$$
$$|$$
$$\text{C}_6\text{H}_5$$

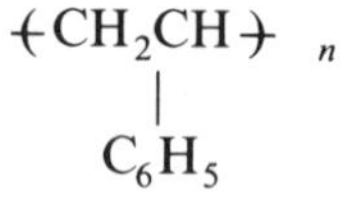

along the chain. Owing to the presence of an asymmetric α-carbon atom, polystyrenes may contain a complex distribution of stereoisomeric placements. In fact, free radical polymerization of styrene usually results in random placements of syndio- and isotactic isomers yielding a basically atactic polymer. Polymers containing significantly different sequence distributions of the stereoisomeric placements may be obtained by cationic and anionic polymerization. However, IPS is synthesized predominantly by Zeigler-Natta catalysis. Because of the stereoregularity of this polymer, IPS crystallizes and has a crystalline melting point of 230°C.

PET is synthesized by the condensation of ethylene glycol with terephthalic acid and has a regular chemical repeat unit of

$$\text{--}\!\!\!+ \text{ O--CH}_2\text{--CH}_2\text{--O--}\overset{\displaystyle O}{\overset{\displaystyle \|}{C}}\text{--}\langle\!\langle O \rangle\!\rangle\text{--}\overset{\displaystyle O}{\overset{\displaystyle \|}{C}} \text{ }+\!\!\!\text{--}_n$$

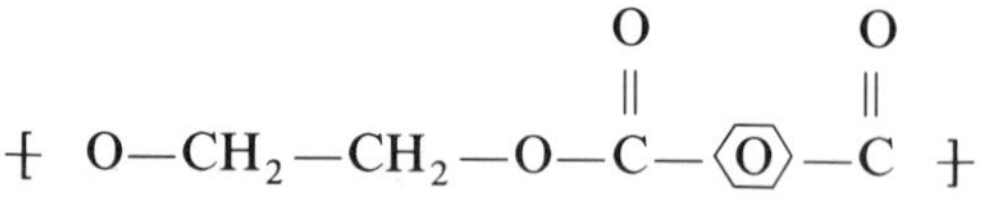

along the chain. Because of its structural regularity PET readily crystallizes and has a crystalline melting point of 265°C.

In the following subsections we will consider the normal coordinate calculations that have been performed on models of IPS and PET.

Polystyrene

Normal coordinate calculations of isotactic polystyrene (IPS) were initially performed in 1977 by Painter and Koenig (125). These authors employed a valence force field that was basically derived from the force constants determined by R. G. Snyder (34) for the aliphatic chain and those determined by La Lau and R. G. Snyder (126) for alkyl benzenes. Recently, Painter and R. W. Snyder (127) have revised the Painter-Koenig force field to account for discrepancies in the value of an important interaction force constant that should be zero under symmetry considerations. Additionally, a revised definition of the C—C—H in-plane bending coordinate was suggested. In the following discussion we will concentrate on the results obtained from the more recent studies.

The preferred conformation of the isolated chain of crystalline IPS, as determined by the x-ray studies of Natta et al. (128), is a 3_1 helix. The line group of the infinite perfect 3_1 helical chain is isomorphous to the point group

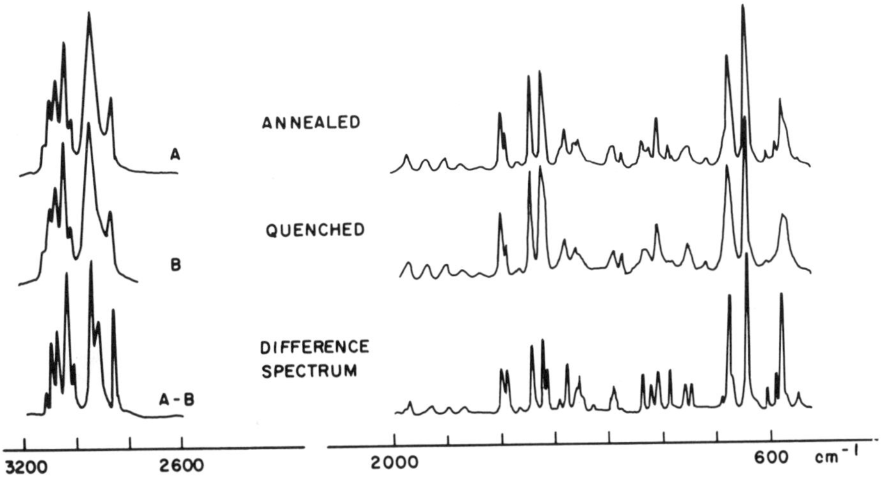

Figure 16.42. Infrared spectrum of (A) annealed IPS and (B) quenched IPS; ($A-B$) the difference spectrum obtained by subtracting the quenched from the annealed spectrum. (Reprinted by permission from ref. 125.)

C_3, which predicts that all the normal modes are both Raman and infrared active. However, the local C_{2v} symmetry of the monosubstituted benzene group must also be considered.

Experimentally, Painter and Koenig (125) obtained infrared spectra of the preferred conformation of IPS by digitally subtracting the spectrum of a quenched from that of an annealed sample. Examples are shown in Figures 16.42 and 16.43. These authors also obtained infrared spectra as a function of annealing time, polarized infrared spectra of an oriented film of IPS, and Raman spectra of quenched and annealed samples of the polymer. More recently, Painter and R. W. Snyder have obtained infrared and Raman spectra of the low frequency region.

The internal coordinates of a monosubstituted benzene ring of IPS are illustrated in Figure 16.44. Those coordinates associated with the aliphatic part of the molecule are identical to those defined by Snyder and Schachtschneider (3, 11) for polyalkanes. As mentioned previously, the original force field of Painter and Koenig (125) perpetuated two significant errors. Painter and R. W. Snyder, in an attempt to obtain a general force field for monosubstituted benzenes, determined that the C—C—H in-plane bending of the phenyl ring had been inadequately defined. (This finding was discussed in greater detail in Chapter 9.) Furthermore, the interaction force constant $F_{R(X)\Phi(X)}$ should have been restrained to a value of zero based on symmetry considerations, but was refined to give a relatively large value in the work of La Lau and R. G. Snyder (126) for alkyl benzenes and by Painter and Koenig (125) for polystyrene. Painter and R. W. Snyder thus returned to the problem of the alkyl benzenes and obtained a new force field by corefining toluene and ethyl benzene using

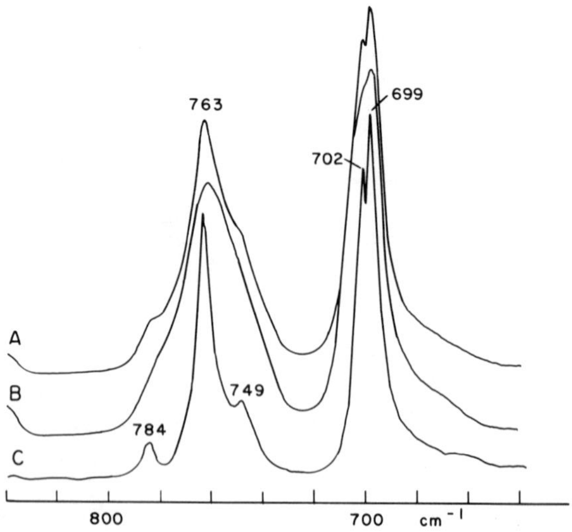

Figure 16.43. Infrared spectrum of (*A*) annealed IPS, (*B*) quenched IPS, and (*C*) annealed minus quenched; frequency-expanded scale between 850 and 650 cm^{-1}. (Reprinted by permission from ref. 125.)

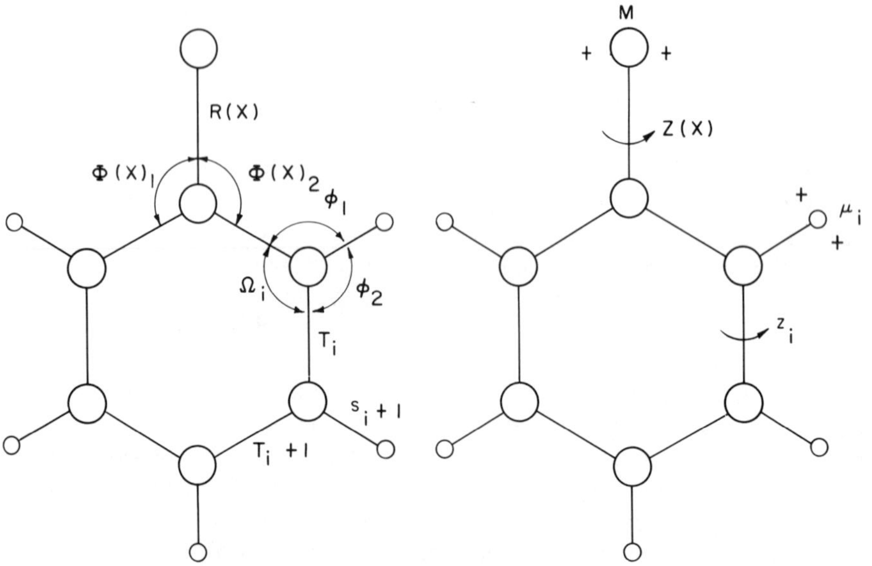

Figure 16.44. Internal coordinates of the phenyl ring of IPS.

Table 16.53 Force Constants for Isotactic Polystyrene

Force Constant	Coordinate Involved	ϕ_i [a]
	Phenyl Group, In Plane	
K_T	T	7.084
K_S	s	5.022
H_Ω	Ω	1.461
H_{ϕ_o}	ϕ	0.496
F_T^0	T_i, T_{i+1}	0.693
F_T^m	T_i, T_{i+2}	−0.510
T_T^p	T_i, T_{i+3}	0.604
$F_{T\phi}^0$	T_i, ϕ_i	−0.271
$F_{T\phi}'^0$	T_i, ϕ_{i+1}	0.271
$F_{T\phi}^m$	T_i, ϕ_{i-1}	−0.057
$F_{T\phi}'^m$	T_i, ϕ_{i+2}	0.057
$F_{T\phi}^p$	T_i, ϕ_{i-2}	0.102
$F_{T\phi}'^p$	T_i, ϕ_{i+3}	−0.102
$F_{T\Omega}$	$T_i, \Omega_i; T_i, \Omega_{i+1}$	0.344
F_{Ts}	$T_i, s_i; T_i, s_{i+1}$	−0.115
F_s^0	s_i, s_{i+1}	0.013
F_Ω	Ω_i, Ω_{i+1}	0.085
$F_{\Omega\phi}^0$	Ω_i, ϕ_{i-1}	−0.101
$F_{\Omega\phi}'^0$	Ω_i, ϕ_{i+1}	0.101
F_ϕ^0	ϕ_i, ϕ_{i+1}	0.013
F_ϕ^m	ϕ_i, ϕ_{i+2}	−0.001
F_ϕ^p	ϕ_i, ϕ_{i+3}	−0.019
	Phenyl Group, Out of Plane	
H_μ	μ	0.324
H_z	z	0.119
F_μ^0	μ_i, μ_{i+1}	0.018
F_μ^m	μ_i, μ_{i+2}	−0.024
F_μ^p	μ_i, μ_{i+3}	−0.023
F_z^0	z_i, z_{i+1}	−0.039
$F_{z\mu}$	z_i, μ_i	0.027
$F_{z\mu}^0$	z_i, μ_{i+1}	−0.026
	Phenyl Group–Alkyl Group Interactions	
$K_{R(X)}$	$R(X)$	4.746
$H_{\Phi(X)}$	$\Phi(X)$	0.581 (0.717)
$H_{Z(X)}$	$Z(X)$	0.085 (0.010)
H_M	M	0.370 (0.383)
$F_{TR(X)}^0$	$T_{i-1}, R(X); T_{i-2}, R(X)$	0.173
$F_{\Omega\Phi(X)}^0$	$\Omega_i, \Phi(X)$	0.076
$F_{\Omega\Phi(X)}'^0$	$\Omega_{i-2}, \Phi(X)$	−0.076
$F_{T\Phi(X)}$	$T_{i-1}, \Phi(X)$	−0.375
$F_{T\Phi(X)}'$	$T_{i-2}, \Phi(X)$	0.375
$F_{R(X)\Omega}$	$R(X), \Omega_{i-1}$	−0.526

Table 16.53 Continued

Force Constant	Coordinate Involved	ϕ_i [a]
$F^0_{R(X)\Omega}$	$R(X), \Omega_i;\ R(X), \Omega_{i-2}$	0.123
	Chain	
$K_{s'}$	s'	4.588
K_d	d	4.544 (4.538)
K_R	R	4.337 (4.532)
H_δ	δ	0.550 (0.533)
H_Γ	γ	0.656
H_ζ	ζ	0.657
H_ω	ω	1.130
H_Λ	Λ	1.118 (1.086)
H_τ	τ	0.024
F_d	d, d	0.006
F_R	R, R	0.101
$F_{RR(X)}$	$R, R(X)$	0.274 (0.083)
$F_{R\gamma}\ (F_{R\zeta})$	R, γ	0.328
$F'_{R\gamma}\ (F'_{R\zeta})$	R, γ	0.079
$F_{R(X)\zeta}$	$R(X), \zeta$	0.268 (0.328)
$F'_{R(X)\zeta}$	$R(X), \zeta$	0.062 (0.079)
$F_{R\omega}$	R, ω	0.417
F_γ	γ, γ	−0.021
$F'_\gamma\ (F'_\zeta)$	γ, γ	0.012
$F_{\gamma\omega}\ (F_{\zeta\omega})$	γ, ω	−0.031
$F^{t}_{\gamma\zeta}$	γ, ζ	0.127
$f^{g}_{\gamma\zeta}$	γ, ζ	−0.005
$F'^{t}_{\gamma\zeta}$	γ, ζ	0.002
$f'^{g}_{\gamma\zeta}$	γ, ζ	0.009
$f''^{g}_{\gamma\zeta}$	γ, ζ	−0.015
$F''^{g}_{\gamma\zeta}$	γ, ζ	−0.025
$f^{t}_{\gamma\omega}\ (f^{T}_{\zeta\omega})$	γ, ω	0.049
$f^{g}_{\gamma\omega}\ (f^{G}_{\zeta\omega})$	γ, ω	−0.052
f^{t}_{ω}	ω, ω	−0.231 (0.011)
f^{g}_{ω}	ω, ω	−0.139 (0.011)
F_Λ	Λ, Λ	0.308
$F_R\ (F_{R(X)})$	R, Λ	0.219 (0.303)
F'_ζ	ζ, ζ	0.012

[a]Force constant units: stretching constants, mdyn/Å; bending constants, mdyn-Å/rad^2; stretch-bend interactions, mdyn/rad.

Table 16.54 Observed and Calculated Frequencies and Approximate Potential Energy Distribution of Isotactic Polystyrene

Frequency (cm^{-1})				
Observed		Calculated		
Crystalline Phase				Potential Energy Distribution
Infrared	Raman	Initial	Final	(Final)
		A Modes		
(3105)		3062	3062	98% K_s
(3085)		3054	3053	98% K_s
(3063)		3049	3048	98% K_s
(3019)		3042	3042	99% K_s
(3000)		3037	3037	99% K_s
2929		2925	2924	99% K_d
2904		2908	2907	98% K_s'
2845		2851	2851	99% K_d
1604	1602	1598	1597	93% K_T, 15% H_Ω
1584	1583	1590	1587	95% K_T, 13% H_Ω
1494		1507	1506	55% H_ϕ, 40% K_T
1453	1440	1464	1462	76% H_δ, 23% H_γ
1438		1449	1446	62% H_ϕ, 39% K_T
1387		1422	1391	51% H_ζ, 29% H_γ, 34% K_R, (−30% $F_{R\gamma}$)
1364	1367	1374	1363	42% H_ζ, 26% H_γ, 17% $K_{R(X)}$
1325	1329	1336	1336	71% K_T, 65% H_ϕ, (−14% F_T^0, −10% F_T^m)
1303		1305	1301	48% K_T, 42% H_ϕ, 14% H_ζ, 13% H_γ
1298		1292	1288	73% H_γ, 17% K_T
1198	1195	1217	1212	34% H_ϕ, 21% $K_{R(X)}$, 15% H_γ, 11% K_T
1188	1192	1188	1187	43% H_ζ, 32% H_ϕ, 14% H_γ
1181	1184	1169	1168	69% H_ϕ, 15% H_Ω, (9% $K_{R(X)}$)
1156	1156	1151	1152	57% H_ϕ, 36% K_T
1113	1101	1126	1114	54% K_R, 16% H_γ, 11% H_Λ, (17% $F_{R\gamma}$)
1083		1071	1067	50% K_T, 32% H_ϕ
1026	1029	1039	1039	59% K_T, 20% H_ϕ, 13% H_Ω
	1004	1008	1008	79% H_Ω, 23% K_T
981		990	991	90% H_μ, 29% H_z, (−16% F_z^0)
983 (1004?)		998	982	52% K_R, 18% H_γ, 11% H_Λ, (−11% $F_{R\omega}$)
965		966	966	92% H_μ, 27% H_z, (−12% $F_{z\mu}$)
899		915	913	89% H_μ, 20% H_z
	842	878	872	44% H_γ, 12% H_ζ, 10% K_T
840		855	854	81% H_μ, 21% H_z
784	785	795	787	34% H_γ, 23% $K_{R(X)}$, 17% H_Ω, 17% K_T
764	764	772	768	48% H_μ, 23% H_M, 14% H_z
702		703	702	65% H_μ, 19% H_z
620	622	626	626	95% H_Ω, 12% H_ϕ
586		584	584	72% H_Ω, 14% H_ϕ
567	565	546	546	37% H_Λ, 17% H_M, 16% H_z, 16% H_μ
426		429	439	16% H_Λ, 15% $H_{\Phi(X)}$, 15% H_ω, 12% H_M
	411	408	407	51% H_z, 18% H_μ (17% F_z)
313	320	307	303	36% H_Λ, 16% $H_{\Phi(X)}$, 16% H_z
225	225	237	240	80% H_Λ (−22% F_Λ)
	123	153	146	21% H_z, 15% H_M, 13% $H_{\Phi(X)}$, 12% H_Λ, 10% H_ω
	123	140	137	47% $H_{\Phi(X)}$, 15% H_Λ
	57	45	43	39% H_Λ, 25% H_ω, 16% H_M, 13% H_τ, (−10% F_Λ)
		42	36	74% $H_{Z(X)}$ (9% H_M)

Table 16.54 Continued

Frequency (cm^{-1})				
Observed		Calculated		
Crystalline Phase				Potential Energy Distribution
Infrared	Raman	Initial	Final	(Final)
		E Modes		
(3105)		3062	3062	98% K_s
(3085)		3054	3054	98% K_s
(3063)		3049	3049	98% K_s
(3019)		3042	3042	99% K_s
(3000)		3037	3037	99% K_s
2929		2924	2928	99% K_d
2904		2907	2905	88% K'_s, 11% K_d
2845		2852	2851	99% K_d
1604	1602	1598	1598	93% K_T, 15% H_Ω, 10% H_ϕ
1584	1583	1590	1587	95% K_T, 13% H_Ω
1494		1508	1506	54% H_ϕ, 40% K_T
1453	1454	1466	1463	73% H_δ, 21% H_γ
		1449	1446	61% H_ϕ, 38% K_T
1387		1415	1385	50% H_ζ, 31% H_γ, 21% K_R (−20% $F_{R\gamma}$)
1364	1367	1380	1365	39% H_γ, 32% H_ζ, 15% $K_{R(X)}$
1325	1329	1336	1335	75% K_T, 65% H_ϕ (−14% F_T^0, −11% F_T^m)
1315		1310	1307	47% K_T, 37% H_ϕ, 18% H_ζ, 16% H_γ
1262		1272	1267	36% H_γ, 23% K_T, 18% H_ϕ, 15% $K_{R(X)}$, 12% $K_{S'}$
1198	1195	1211	1208	26% H_ϕ, 19% H_ζ, 18% H_γ, 11% $K_{R(X)}$
1188	1192	1192	1187	33% H_ϕ, 27% H_γ, 20% H_ζ, 14% K_R
1181	1184	1170	1170	68% H_ϕ, 12% K_T
1156	1157	1152	1152	57% H_ϕ, 35% K_T
1113	1101	1100	1104	37% K_R, 32% H_γ
1083		1078	1071	36% K_T, 26% H_ϕ, 21% K_R
1052		1055	1044	37% K_R, 22% K_T, 15% H_γ (9% H_ϕ)
1026	1029	1039	1038	54% K_T, 17% H_ϕ, 12% H_Ω
	1004	1008	1007	77% H_Ω, 25% K_T
981		991	991	91% H_μ, 30% H_z (−14% $F_{z\mu}$, −14% $F'_{z\mu}$)
965		966	966	92% H_μ, 28% H_z
922		925	922	26% H_μ, 19% H_γ, 16% K_R, 10% K_T
922		911	909	63% H_μ, 15% H_z (8% H_γ)
840		855	854	81% H_μ, 21% H_z
764	764	771	769	28% H_μ, 14% H_M, 10% H_γ
749		765	760	20% H_μ, 13% H_γ, 12% $K_{R(X)}$, 12% H_Ω
702		703	702	65% H_μ, 19% H_z (11% F_z^0)
620	622	626	627	92% H_Ω, 12% H_ϕ
586		596	594	57% H_Ω, 14% H_Λ, 12% H_ϕ, 10% H_ω
567	565	545	552	29% H_Λ, 12% H_γ (9% H_M)
497		496	495	18% H_Λ, 15% H_M, 11% H_z
	411	408	407	51% H_z, 18% H_μ (17% F_z^0)
407 (?)		372	373	19% H_ω, 17% H_Λ, 12% H_z, 12% H_μ
225	225	239	229	47% H_Λ, 35% $H_{\Phi(X)}$
167	160	192	188	25% H_z, 22% $H_{\Phi(X)}$, 13% H_Λ, 11% H_ω
	123	134	132	55% H_Λ, 24% $H_{\Phi(X)}$ (−13% F_Λ)
	50	51	41	60% H_z, 23% H_Λ
		33	31	48% H_Λ, 23% $H_{Z(X)}$, 22% H_M, 13% H_ω
		13	13	83% H_τ

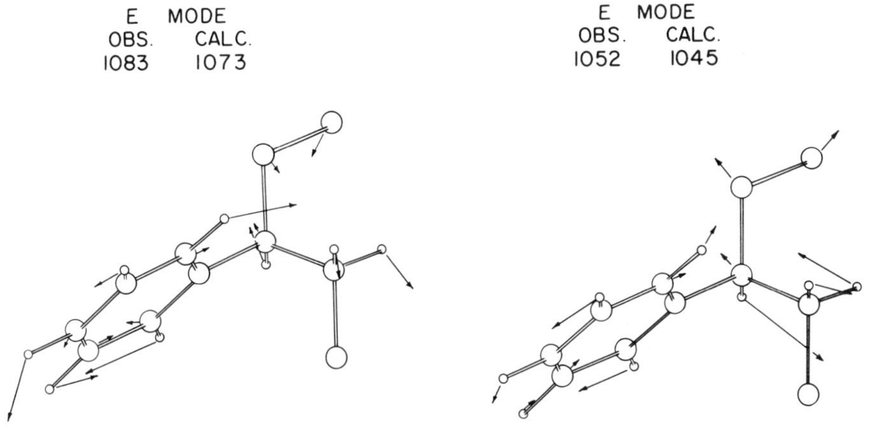

Figure 16.45. Cartesian displacement coordinates of the 1083 and 1052 cm^{-1} modes in IPS.

the overlay technique (129). From this work several of the interaction force constants associated with the phenyl and alkyl groups were determined. This new force field was then transferred to the model of IPS. Only those force constants associated with the interactions involving the phenyl group and the main chain were refined to the observed frequencies of IPS. The force field is reproduced in Table 16.53. The calculated frequencies together with their associated potential energy distributions are compared to the observed frequencies in Table 16.54. The total squared deviation for both the A and E modes was determined to be 0.023. A complete discussion of the assignments of the normal modes of IPS is detailed in the publication of Painter and R. W. Snyder (127), to which the interested reader is referred. However, one of the more important conclusions of this work concerns the assignments of the

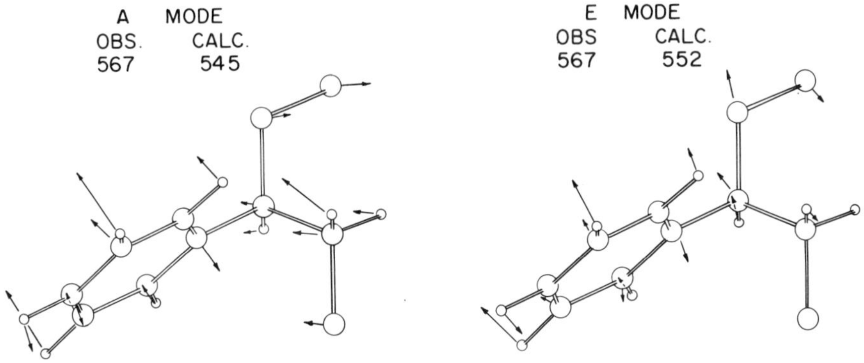

Figure 16.46. Internal coordinate displacements of the 567 cm^{-1} A and E modes of IPS.

conformationally sensitive modes, and this deserves further comment. Four normal modes at 1083, 1052, 567, and 225 cm^{-1} are particularly sensitive to conformation. The doublet at 1083–1052 cm^{-1} is characteristic of the 3_1 helical conformation of IPS, whereas a broad singlet is observed at 1069 cm^{-1} in atactic polystyrene. Previously, Onishi and Krimm (130) suggested that rotational disordering of the benzene ring and the resultant local disappearance of C_3 symmetry could result in the splitting of the doubly degenerate E mode at 1069 cm^{-1} into two components. This suggestion was rejected by Kobayashi et al. (131, 132), who observed a rapid decrease in the splitting of the two bands upon deuteration of the skeletal α-hydrogen atoms. Accordingly, it was concluded that both the 1083 and 1052 cm^{-1} bands must be due to backbone vibrational modes, since the splitting was associated with regularity of the skeletal chain. The results of Painter and R. W. Snyder indicate that the 1083 and 1052 cm^{-1} bands are due to a mixture of a backbone and in-plane ring mode, making them conformationally sensitive. A schematic diagram of the cartesian displacements of these two normal modes is reproduced in Figure 16.45 and the highly coupled nature of these modes is immediately apparent. The 567 and 225 cm^{-1} bands have previously been assigned to ring modes based on the group frequency approach. The calculations of Painter and R. W. Snyder indicate once again that these modes are in fact due to ring modes that are highly coupled to backbone modes, as illustrated in Figures 16.46 and 16.47.

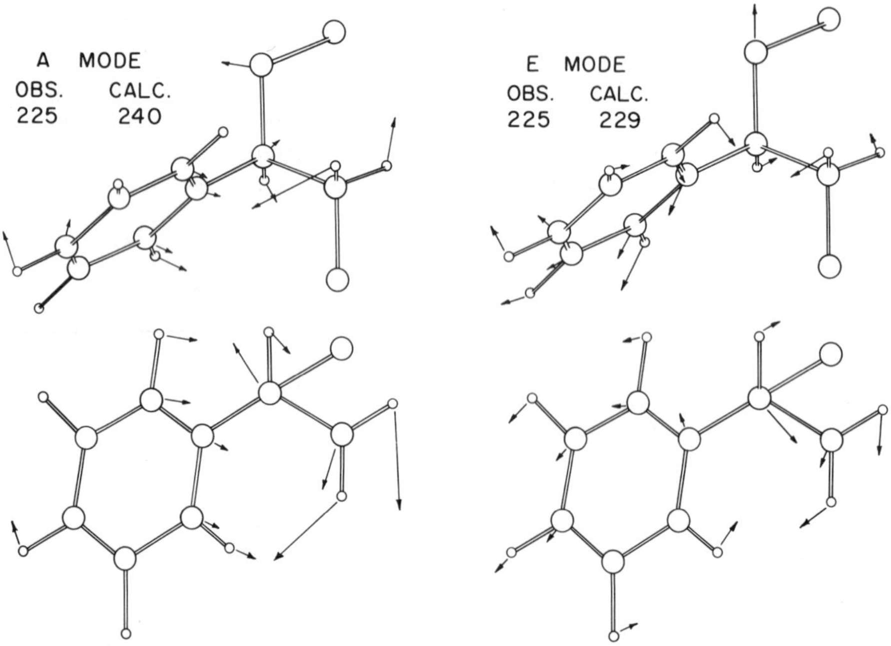

Figure 16.47. Internal coordinate displacements of the 225 cm^{-1} A and E modes of IPS.

Poly(ethylene terephthalate)

Normal coordinate calculations of poly(ethylene terephthalate) (PET) have been performed by Boitsov and Gotlib (133), Danz et al. (134), and Boerio and co-workers (135, 136). The two former groups based their calculations on approximate models and force fields, and their conclusions, which were considered to support the contentions of Ward (137) concerning the observed changes in the infrared intensities upon crystallization of PET, can only be considered tentative. Ward related these intensity changes upon crystallization to rotational isomerism of the ethylene glycol segments in PET. Alternatively, Liang and Krimm (138) suggest that symmetry and resonance characteristics of the benzoid ring framework are responsible for the changes. Boerio and co-workers undertook a more complete normal coordinate study of PET and two deuterated analogues. In this case a valence force field was derived by the overlay technique from vibrational studies of a series of low molecular weight

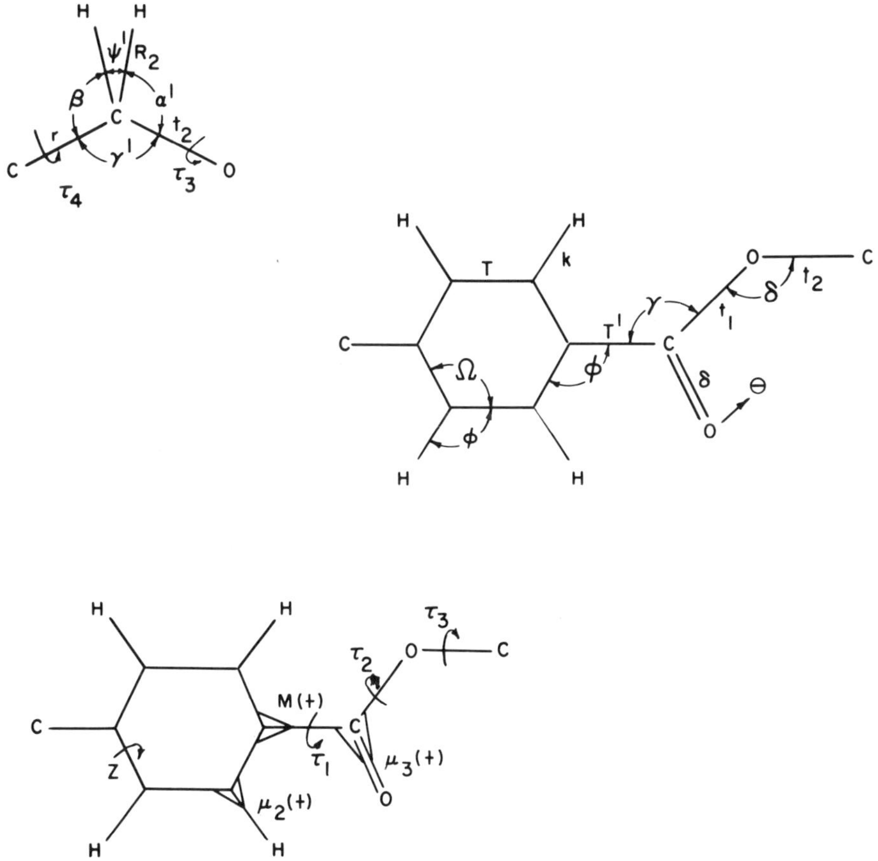

Figure 16.48. Internal coordinates for PET. (Reprinted by permission from ref. 136.)

Table 16.55 Valence Force Constants for Polyethylene Terephthalate[a]

Force Constant	Coordinate(s) Involved	Atom(s) Common	Value
	Ring, in Plane[b]		
K_k	k		5.1265
K_T	T		6.3816
H_ϕ	ϕ		0.5050
H_Φ	Φ		0.8375
H_Ω	Ω		1.0157
F_{TT}^0	T, T	C	0.7716
F_{TT}^m	T, T		−0.3189
F_{TT}^p	T, T		0.2895
$F_{TT'}$	T, T'		0.2506
$F_{T\phi}$	T, ϕ	CC	0.0918
$F_{T\Phi}$	T, Φ	CC	0.1860
$F_{T\Omega}$	T, Ω	CC	0.1187
$F_{T'\Phi}$	T', Φ	CC	0.7261
$K_{T'}$	T'		4.7299
F_ϕ^m	ϕ, ϕ		−0.0052
F_ϕ^p	ϕ, ϕ		0.0044
	Ring Out of Plane[b]		
H_{u_2}	u_2(C<u>H</u>H)		0.4643
H_M	M		0.4936
H_Z	Z		0.2763
f_{MZ}^0	M, Z		0.1589
f_{MZ}^m	M, Z		0.0281
f_u^0	u, u		−0.0723
f_u^m	u, u		−0.0003
f_u^p	u, u		−0.0143
f_M^p	M, M		−0.0003
f_{uZ}^0	u, Z		−0.1589
f_{uZ}^m	u, Z		0.0281
f_{uM}^0	u, M		−0.0723
f_{uM}^m	u, M		−0.0003
f_Z^0	Z, Z		−0.0578
	Substituent		
K_S	S		11.4763
K_{t_1}	t_1		6.7147
K_{t_2}	t_2		4.2902
H_θ	θ		1.3484
H_γ	γ		1.9670
H_δ	δ		1.5792

Table 16.55 Valence Force Constants for Polyethylene Terephthalate[a]

Force Constant	Coordinate(s) Involved	Atom(s) Common	Value
$H_{\alpha'}$	α'		0.9320
H_{β}	β		0.7429
$H_{\psi'}$	ψ'		0.4754
K_{R_2}	R_2		4.6535
K_r	r		4.0475
$H_{\gamma'}$	γ'		1.2174
H_{u_3}	u_3		0.6987
H_{τ_1}	τ_1		0.024
H_{τ_2}	τ_2		0.026
H_{τ_3}	τ_3		0.026
H_{τ_4}	τ_4		0.024
$F_{T'\theta}$	T', θ	CC(O)	0.0648
$F_{t_1\theta}$	t_1, θ	OC(O)	−0.7720
$F_{T'\gamma}$	T', γ	CC(O)	0.6932
$F_{r\gamma'}$	r, γ'	CC	0.0815
$F_{t_1\gamma}$	t_1, γ	OC(O)	1.1975
$F_{t_2\gamma'}$	t_2, γ	CO	0.5112
$F_{t_1\delta}$	t_1, δ	OC(O)	0.6252
$F_{t_2\alpha'}$	t_2, α'	OC	0.387
$F_{t_2\delta}$	t_2, δ	CO	0.450
$F_{r\beta}$	r, β	CC	0.478
$F_{T'S}$	T', S	C(O)	1.2197
F_{t_1S}	t_1, S	C(O)	0.8086
$F_{T't_1}$	T', t_1	C(O)	0.8446
F_{t_2r}	t_2, r	C	0.1538
$F_{t_1t_2}$	t_1, t_2	O	0.8286
$F_{\alpha'\alpha'}$	α', α'	OC	−0.005
$F^{t}_{\gamma'\gamma'}$	γ', γ'	CC(trans)	−0.011
$F_{\gamma\delta}$	γ, δ	OC(O)	−0.0924
$F^{g}_{\delta\alpha'}$	δ, α'	CO(gauche)	0.004
$F^{g}_{\beta\beta}$	β, β	CC(gauche)	0.004
$F^{g}_{\gamma'\beta}$	γ', β	CC(gauche)	−0.113
$F^{t}_{\beta\beta}$	β, β	CC(trans)	0.121
$F_{\beta\beta}$	β, β	CC	0.0475
$F_{\delta\gamma'}$	δ, γ'	CO	−0.011
$F_{\alpha'\beta}$	α', β	CH	0.115
$F_{\gamma'\alpha'}$	γ', α'	CO	0.031
$F_{\gamma'\beta}$	γ', β	CC	−0.031

[a]Units are mdyn/Å for stretching force constants, mdyn/rad for stretch -bend interactions, and mdyn-Å/rad^2 for bending constants.

[b]For a more complete description of these force constants, see ref. 126.

Table 16.56 Calculated and Observed Frequencies and Potential Energy Distribution for Polyethylene Terephthalate[a, b]

Symmetry Species	ν_{obs} (cm^{-1}) IR	ν_{obs} (cm^{-1}) Raman	ν_{calc} (cm^{-1})	PED (%)
A_g		3085	3074	$k(99)$
		3085	3072	$k(99)$
		2912	2889	$R_2(99)$
		1730	1723	$S(90)$
		1615	1606	$T(74)$, $\phi(25)$
			1576	$T(82)$, $\phi(13)$
		1462	1452	$\psi'(58)$, $\alpha'(44)$
		1418	1395	$\beta(62)$, $\alpha'(28)$
		1310	1316	$\phi(71)$
		1295	1273	$T'(37)$, $t_1(32)$, $\phi(18)$, $\theta(18)$
		1192	1178	$\phi(72)$, $T(15)$
		1119	1118	$t_1(38)$, $r(35)$, $\gamma'(15)$
		1096	1100	$T(31)$, $r(28)$, $t_1(10)$
		1000	1002	$t_2(80)$, $r(21)$
		857	844	$T(27)$, $t_1(15)$, $\theta(12)$, $T'(11)$, $r(10)$
		701	693	$\Omega(36)$, $\theta(22)$, $T'(14)$
		626	628	$\Omega(60)$, $\phi(15)$
			559	$\gamma(37)$, $\Phi(22)$
			373	$\gamma'(23)$, $\Phi(21)$, $\Omega(10)$
		278	282	$\theta(33)$, $\Phi(28)$, $\delta(21)$
			151	$\delta(44)$, $r(13)$, $\Phi(11)$
B_g		2968	2967	$R_2(99)$
			1285	$\alpha'(62)$, $\beta(48)$
			1174	$\beta(43)$, $\alpha'(39)$
			972	$\mu_2(136)$
			859	$\mu_2(114)$
		800	799	$Z(48)$, $\mu_3(48)$, $M(40)$
			673	$Z(86)$, $\mu_3(39)$
			245	$M(77)$
			119	$\tau_3(89)$
			63	$\tau_1(60)$, $\tau_2(37)$
			37	$\tau_2(60)$, $\tau_1(34)$
$A_u(\perp)$	2962		2957	$R_2(100)$
			1271	$\alpha'(94)$
			991	$\mu_2(126)$
	875		871	$\mu_2(81)$, $M(24)$, $Z(15)$
	845		837	$\beta(98)$
	727		723	$\mu_3(60)$, $\mu_2(43)$
			470	$Z(89)$, $M(79)$
	430		411	$Z(135)$
			141	$\tau_4(64)$, $\tau_2(18)$
			80	$Z(29)$, $M(27)$, $\mu_2(17)$
			54	$\tau_3(49)$, $\tau_1(32)$

Table 16.56 Continued

Symmetry Species	ν_{obs} (cm^{-1}) IR	Raman	ν_{calc} (cm^{-1})	PED (%)
B_u (‖, ⊥)	3081		3074	$k(99)$
	3067		3072	$k(99)$
	2889		2882	$R_2(100)$
	1727		1721	$S(93)$
	1504		1504	$\phi(54)$, $T(39)$
	1475		1465	$\psi'(62)$, $\alpha'(35)$
	1410		1407	$T(46)$, $\phi(39)$
	1337		1334	$\beta(55)$, $\alpha'(41)$
			1293	$T(138)$
	1263		1254	$t_1(50)$, $T'(38)$, $\theta(23)$
	1126		1126	$\Omega(26)$, $t_1(26)$, $\phi(19)$, $T(12)$
	1109		1102	$\phi(56)$, $T(33)$
	1018		1019	$\Omega(37)$, $T(28)$, $\phi(27)$
	973		971	$t_2(89)$
			868	$t_1(29)$, $\delta(23)$, $\theta(14)$
	502		506	$T'(37)$, $\theta(19)$
	438		459	$\delta(31)$, $\gamma(27)$, $\theta(16)$
	382		384	$\gamma(29)$, $\gamma'(18)$, $\theta(17)$
	145		135	$\Phi(46)$, $\gamma'(30)$

[a] Because of contributions from off-diagonal force constants, the potential energy distribution as listed may be greater than or less than 100%.
[b] Observed frequencies in parentheses may be assigned to more than one calculated mode.

aromatic esters (139). In the following discussion we will restrict ourselves to a description of this work.

A planar model for PET was assumed and the majority of the bond lengths and angles were obtained from previous work on aliphatic esters (140) and alkyl benzenes (126) as shown below.

Benzene ring	C—H = 1.084 Å, C—C = 1.397 Å
	C—C—C angle = 120°
Ring ester	C—C = 1.51 Å
Ester group	C—H = 1.096 Å, C—O = 1.41 Å
	C=O = 1.24 Å, C—C = 1.54 Å
	C—C=O and O—C=O angle = 125°16′
	All other angles were tetrahedral.

The factor group of the line group is isomorphous to the point group C_{2h} for this planar model. The optically active normal modes are distributed among the symmetry species as 21 A_g (Raman active), 11 B_g (Raman active), 11 A_u (infrared active, perpendicular polarization), and 19 B_u (infrared active, parallel or perpendicular polarization). The internal coordinates and valence force constants for PET are reproduced in Figure 16.48 and Table 16.55, respectively. As mentioned in Section 16.4, Painter and R. W. Snyder (127) have pointed out that the interaction force constant, $F_{R(X)\Phi(X)}$ should have a value of zero based on symmetry considerations, whereas Boerio and co-workers have employed a relatively large value of 0.7261 mdyn/rad. It has also been shown that the force constant value of $F_{R(X),\Omega}$ is significant and should be included in the force field of molecules containing aromatic rings. However, it turns out that inclusion of an $F_{R(X),\Omega}$ interaction force constant and restraining $F_{R(X)\Phi(X)}$ to zero does not appear to affect greatly the form of the normal modes (141).

The results of the normal coordinate analysis of PET are presented in Table 16.56. Boerio et al. also calculated the normal modes of two deuterated PETs. In general, the agreement between the observed and calculated frequencies for PET is satisfactory (the average difference between them is 9 cm^{-1} for 110 observed frequencies) but the authors note that the results for the deuterated polymers is not quite as good. (See ref. 136 for a complete discussion of the assignment and form of the normal modes.)

16.5 Polyamides, Polypeptides, and Proteins

Polyamides are a general class of synthetic polymers, commonly called nylons, that are usually prepared by condensation polymerization from aliphatic and aromatic diamines and diacids; from the self-condensation of amino acids, or from ring opening of lactams. All polyamides have in common the amide chemical grouping

$$
\begin{array}{c}
O \\
\parallel \\
(-C-N-) \\
\mid \\
H
\end{array}
$$

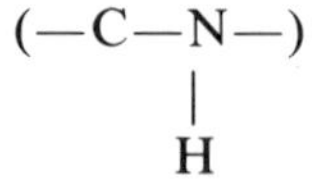

present in the polymer backbone. Nylons of particular commercial significance include nylon-6, $(-HN(CH_2)_5CO-)_n$, nylon-6,6 $(-OC(CH_2)_4CO-NH(CH_2)_6NH-)_n$, and the aromatic nylons

$$(-HN-\langle\bigcirc\rangle-NH-CO-\langle\bigcirc\rangle-CO-)_n$$

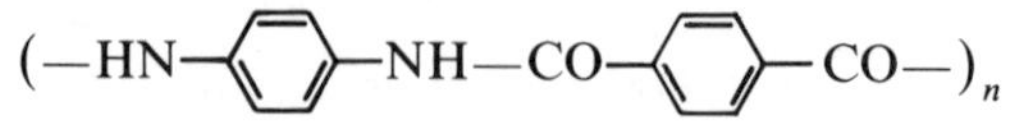

which are synthesized from phthalic acid and diaminobenzene.

Proteins are naturally occurring polyamides that can be considered to be copolymers of amino acids (and/or imino acids) of the general formula

$$NH_2\text{—}\underset{}{\overset{\displaystyle R\\ |}{CH}}\text{—}COOH$$

(the exceptions being proline and hydroxyproline). In general, the microstructure of the polymer chain in proteins consists of a complicated sequence of up to 20 different amino acids. The precise sequence of amino and imino acids in the protein molecule is responsible for the conformation and its biological function and activity. Polypeptides are synthetic polymers prepared from one or more of the amino acids generally found in proteins. Thus, synthetic homopolymers have been prepared from, for example, glycine, alanine, and valine, to yield

$$\underset{\text{polyglycine}}{(\text{—}HN\text{—}CH_2\text{—}CO\text{—})_n} \qquad \underset{\text{polyalanine}}{(\text{—}HN\text{—}\overset{\displaystyle CH_3\\ |}{CH}\text{—}CO\text{—})_n}$$

$$\underset{\text{polyvaline}}{(\text{—}HN\text{—}\overset{\displaystyle C_3H_7\\ |}{CH}\text{—}CO\text{—})_n}$$

In the following we will briefly describe the normal coordinate calculations that have been performed on models for nylons and homopolypeptides. An example of the normal coordinate calculations of polyglycine will then be discussed in more detail.

Theoretical studies of the vibrational spectra of polyamides and polypeptides date back to the 1950s when Miyazawa and co-workers (142) performed normal coordinate calculations on a model compound, *N*-methylacetamide (NMA), containing the amide group. These authors employed a Urey-Bradley force field and the methyl groups were taken as point masses. Subsequently, this force field was refined to a complete molecular model of NMA and analogues containing deuterated methyl groups (143, 144). The first complete valence force field for the R—CO—NH—R′ unit, where R and R′ are CH_3 or CH_2 groups, was reported by Jakeš and Krimm (145). The overlay technique was employed and the force constants were refined, using a least-squares procedure, to the observed infrared frequencies of NMA, 7 nylons, and 13 deuterated analogues of NMA and the nylons. This resultant force field was then transferred to molecular models of three nylons, some *N*-alkyl amides, and C-deuterated derivatives of nylon-6,6 that were not used in the original refinement. In general, the agreement between the observed and calculated frequencies was good and a reasonable assignment of the infrared frequencies was established.

Abe and Krimm (146) then turned their attention to the normal vibrations of crystalline polyglycine, which is known to exist in two distinct polymorphic crystalline forms, denoted PGI and PGII, respectively. Normal coordinate calculations of PGI had been performed previously by Fukushima et al. (147), Gupta et al. (148), and Fukushima and Miyazawa (149), but in all cases the polymer chain was assumed to be in a planar conformation and the CH_2 groups were approximated by a point mass. In contrast, Abe and Krimm performed calculations based on the chain conformation and crystalline unit cell determined by x-ray diffraction and did not replace the CH_2 groups with a single mass. At that time the x-ray evidence suggested that crystalline PGI contains a unit cell consisting of two pleated sheet antiparallel chains and four peptide units. The space group of the crystal is isomorphous to the point group D_2. Accordingly, there are four symmetry species, A (Raman active) and B_1, B_2, and B_3 (infrared and Raman active). A natural starting point for the determination of a valence force field for polyglycine was the previously obtained amide force field of Jakeš and Krimm. The only interchain coordinate considered was the one between the C=O stretching and the O---H stretching coordinates. From their calculations, the authors reported that the agreement between the observed and calculated frequencies of PGI was good except for the amide I modes. In Chapter 14 we described the important concept of transition dipole coupling in polypeptides, and the interested reader is referred to Section 14.5 for an in-depth discussion. It is emphasized, however, that it was the work of Abe and Krimm (146) on PGI and PGII that first suggested that the amide I splittings could not be rationalized on the basis of a large D_{10} term in the Miyazawa perturbation theory (i.e., no reasonable force field predicts a large enough D_{10} term). These authors suggested that the amide I splittings could be explained by including additional direct interaction force constants between C=O groups in neighboring chains through transition dipole coupling. In a subsequent publication, Krimm and Abe (150) examined the transition dipole coupling approach in depth and suggested that the amide I vibrations in β-polypeptides could be satisfactorily resolved by inclusion of a D_{11} term in the perturbation expression, and that the D_{10} interaction constant is effectively zero. More recently, Krimm and co-workers (151–154) in a series of papers have presented normal coordinate calculations incorporating all transition dipole coupling terms over an effective "sphere of influence" for a series of polypeptides. As an example of this work we will briefly describe the results obtained for crystalline polyglycine I.

Polyglycine I

In an initial study of crystalline PGI, Abe and Krimm (146) had assumed an antiparallel chain pleated sheet (APPS) structure. However, in subsequent

Table 16.57 Structural Parameters of Crystalline Polyglycine I

Chain Parameters	
Bond Length (Å)	Bond Angles
$r(C^\alpha—C)=1.53$	$C^\alpha CN=114°$
$r(C—N)=1.32$	$CNC^\alpha=123°$
$r(N—C^\alpha)=1.47$	$C^\alpha CO=121°$
$r(C=O)=1.24$	$CNH=123°$
$r(C^\alpha—H)=1.07$	
$r(N—H)=1.00$	
$NC^\alpha C=NC^\alpha H=CC^\alpha H=$tetrahedral	

electron diffraction studies Lotz (155) proposed an antiparallel chain rippled sheet (APRS) structure, which has also been suggested for other β-polypeptides (156). Moore and Krimm (152) presented a detailed spectroscopic analysis of PGI that also tends to favor the APRS structure. Accordingly, this APRS structure was employed in the normal coordinate calculations and it was determined that the resultant force field was transferable to other β-polypeptides. A 2_1 helix is adequate to describe the conformation of the isolated chain, and the parameters employed are reproduced in Table 16.57. However, the relationship between the two antiparallel chains in the unit cell is

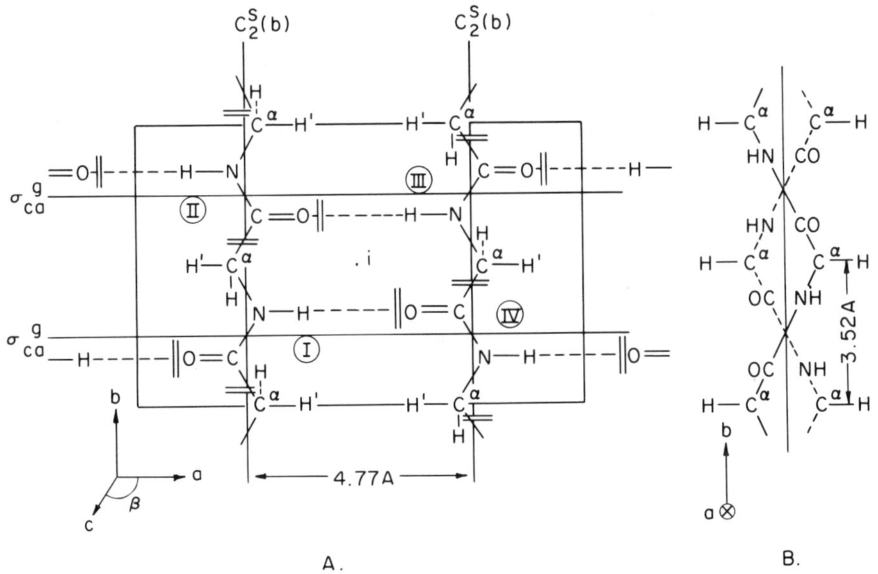

Figure 16.49. Unit cell of the rippled sheet structure of polyglycine. (Reprinted by permission from ref. 152.)

Table 16.58 Force Constants for Polyglycine I

1. $f(\mathrm{N{-}C^{\alpha}})$	5.043
2. $f(\mathrm{C^{\alpha}{-}C})$	4.509
3. $f(\mathrm{C{-}N})$	6.299
4. $f(\mathrm{C{=}O})$	9.720
5. $f(\mathrm{N{-}H})$	5.840
6. $f(\mathrm{C^{\alpha}{-}H})$	4.564
7. $f(\mathrm{CNC^{\alpha}})$	0.787
8. $f(\mathrm{NC^{\alpha}C})$	0.819 (0.032)[a]
9. $f(\mathrm{C^{\alpha}CN})$	1.400
10. $f(\mathrm{C^{\alpha}C{=}O})$ }	
11. $f(\mathrm{NC{=}O})$ }	1.246
12. $f(\mathrm{CNH})$ }	
13. $f(\mathrm{C^{\alpha}NH})$ }	0.527 (0.015)
14. $f(\mathrm{NC^{\alpha}H})$	0.715
15. $f(\mathrm{CC^{\alpha}H})$	0.684
16. $f(\mathrm{HC^{\alpha}H})$	0.584 (0.013)
17. $f(\mathrm{C{=}O\ op})$	0.587 (0.019)
18. $f(\mathrm{N{-}H\ op})$	0.129 (0.006)
19. $f(\mathrm{N{-}C^{\alpha}\ tor})$	0.037
20. $f(\mathrm{C^{\alpha}{-}C\ tor})$	0.037
21. $f(\mathrm{C{-}N\ tor})$	0.680 (0.014)
22. $f(\mathrm{N{-}C^{\alpha}, C^{\alpha}{-}C})$	0.300
23. $f(\mathrm{C^{\alpha}{-}C, C{-}N})$	0.300
24. $f(\mathrm{C{-}N, N{-}C^{\alpha}})$	0.300
25. $f(\mathrm{C^{\alpha}{-}C, C{=}O})$	0.500
26. $f(\mathrm{C{-}N, C{=}O})$	0.500
27. $f(\mathrm{C{-}N, CNC^{\alpha}})$	0.300
28. $f(\mathrm{N{-}C^{\alpha}, CNC^{\alpha}})$	0.300
29. $f(\mathrm{N{-}C^{\alpha}, NC^{\alpha}C})$	0.300
30. $f(\mathrm{C^{\alpha}{-}C, NC^{\alpha}C})$	0.300
31. $f(\mathrm{C^{\alpha}{-}C, C^{\alpha}CN})$	0.300
32. $f(\mathrm{C{-}N, C^{\alpha}CN})$	0.300
33. $f(\mathrm{C^{\alpha}{-}C, C^{\alpha}C{=}O})$	0.200
34. $f(\mathrm{C{-}N, NC{=}O})$	0.200
35. $f(\mathrm{C{=}O, C^{\alpha}C{=}O})$	0.450
36. $f(\mathrm{C{=}O, NC{=}O})$	0.450
37. $f(\mathrm{C{-}N, CNH})$ }	
38. $f(\mathrm{N{-}C^{\alpha}, C^{\alpha}NH})$ }	0.294
39. $f(\mathrm{N{-}C^{\alpha}, NC^{\alpha}H})$ }	
40. $f(\mathrm{N{-}C^{\alpha}, CC^{\alpha}H})$ }	0.517
41. $f(\mathrm{C^{\alpha}{-}C, NC^{\alpha}H})$	0.026
42. $f(\mathrm{C^{\alpha}{-}C, CC^{\alpha}H})$	0.205
43. $f(\mathrm{CNC^{\alpha}, NC^{\alpha}C})$	0.000

Table 16.58 Continued

44.	$f(NC^{\alpha}C, C^{\alpha}CN)$	0.000
45.	$f(C^{\alpha}CN, CNC^{\alpha})$	0.000
46.	$f(NC^{\alpha}C, C^{\alpha}C{=}O)$	0.000
47.	$f(CNC^{\alpha}, NC{=}O)$	0.000
48.	$f(C^{\alpha}CN, CNH)$	0.200
49.	$f(NC^{\alpha}C, C^{\alpha}NH)$	−0.100
50.	$f(CNC^{\alpha}, NC^{\alpha}H)$	0.000
51.	$f(CNC^{\alpha}, NC^{\alpha}H')$	0.000
52.	$f(C^{\alpha}CN, CC^{\alpha}H)$	0.000
53.	$f(C^{\alpha}CN, CC^{\alpha}H')$	0.000
54.	$f(NC^{\alpha}H, NC^{\alpha}H')$	0.026
55.	$f(CCH, CCH')$	−0.023
56.	$f(NC^{\alpha}H, CC^{\alpha}H)$	0.033
57.	$f(NC^{\alpha}H, HC^{\alpha}H)$	0.0767
58.	$f(CC^{\alpha}H, HC^{\alpha}H)$	0.0215
59.	$f(NC{=}O, CNH)$	0.251
60.	$f(C^{\alpha}C{=}O, CC^{\alpha}H)$	0.100
61.	$f(C^{\alpha}C{=}O, CC^{\alpha}H')$	0.000
62.	$f(C^{\alpha}NH, NC^{\alpha}H)$	0.058
63.	$f(C^{\alpha}NH, NC^{\alpha}H')$	0.077
64.	$f(C{=}O \text{ op}, NC^{\alpha}C)$	0.020 (0.015)
65.	$f(C{=}O \text{ op}, CCH)$	0.100
66.	$f(C{=}O \text{ op}, CCH')$	0.000
67.	$f(N{-}H \text{ op}, NC^{\alpha}C)$	0.000
68.	$f(N{-}H \text{ op}, NC^{\alpha}H)$	0.000
69.	$f(N{-}H \text{ op}, NC^{\alpha}H')$	0.100
70.	$f(C{=}O \text{ op}, N{-}H \text{ op})$	0.114 (0.009)
71.	$f(C{=}O \text{ op}, C{-}N \text{ tor})$	0.111 (0.02)
72.	$f(N{-}H \text{ op}, C{-}N \text{ tor})$	−0.130
73.	$f(H{-}{-}{-}O)$	0.125
74.	$f(C{=}O{-}{-}{-}H \text{ ip})$	0.010
75.	$f_{\tau}(C{=}O)$	0.001
76.	$f(N{-}H{-}{-}{-}O \text{ ip})$	0.030
77.	$f_{\tau}(N{-}H)$	0.0015
78.	$f(H^{\alpha}{-}{-}{-}H^{\alpha})$	0.007
79.	F^{I}_{10}	0.102
80.	F^{I}_{01}	−0.244
81.	F^{I}_{11}	0.160
82.	F^{II}_{10}	−0.036
83.	F^{II}_{01}	−0.0095
84.	F^{II}_{11}	0.0091

[a]Figure in parentheses is dispersion for force constants that were allowed to refine.

Table 16.59 Observed and Calculated Frequencies of Crystalline Polyglycine I

Frequency (cm^{-1})						
Observed		Calculated				
Raman	IR	A_g	A_u	B_g	B_u	Potential Energy Distribution[a] (%)
	3272[b]	3272	3272	3272	3272	NH str(98)
2932	2929	2933	2931	2932	2932	$C^{\alpha}H$ asym str(99)
2869	2869	2860	2860	2860	2860	$C^{\alpha}H$ sym str(99)
				1690		C=O str(77), CN str(14), $C^{\alpha}CN$ def(12)
	1685		1685			C=O str(76), CN str(16), $C^{\alpha}CN$ def(12)
1674		1674				C=O str(76), CN str(16), $C^{\alpha}CH$ def(12)
	1636				1637	C=O str(77), CN str(16), $C^{\alpha}CN$ def(12)
				1588	1563 }	NH ipb(42), CN str(26), $C^{\alpha}C$ str(18), NC^{α} str(11), C=O ipb(11)
1515	1517	1515	1516		}	NH ipb(43), CN str(31), C=O opb(13) $C^{\alpha}C$ str(12)
1460		1460				CH_2 bend(97)
			1460			CH_2 bend(97)
				1432		CH_2 bend(98)
	1432				1432	CH_2 bend(99)
1410	1408	1393	1391		}	CH_2 wag(60), NH ipb(19), $C^{\alpha}C$ str(17)
1341	1338			1350	1353 }	CH_2 wag(75)
1255		1252	1252		}	CH_2 twist(79)
				1300		NH ipb(41), CH_2 twist(14), $C^{\alpha}C$ str(13), C=O ipb(12)
	1295				1286	NH ipb(43), $C^{\alpha}C$ str(13), C=O ipb(12), CH_2 twist(11)
1234				1250		CH_2 twist(76), NH ipb(12)
	1236				1245	CH_2 twist(79)
1220	1214	1220	1219		}	NH ipb(28), CH_2 wag(29), CN str(16), NC^{α} str(14), CH_2 twist(13)
1162	1054	1159	1158		}	NC^{α} str(62), $C^{\alpha}C$ str(14)
1034						PGII impurity (?)
1021	1016			1020	1024 }	NC^{α} str(67), $C^{\alpha}C$ str(11)
		1001				CH_2 rock(51), C=O str(10)
			998			CH_2 rock(56)
				973		CH_2 rock(70)
	987				992	CH_2 rock(59), CN str(11)
		944				CH_2 rock(24), CN str(13), $C^{\alpha}C$ str(12), C=O str(10), $NC^{\alpha}C$ def(10)
	936		941			CH_2 rock(18), CN str(15), $C^{\alpha}C$ str(12), C=O str(12), $NC^{\alpha}C$ def(11)
884				887		$C^{\alpha}C$ str(27), CN str(27), C=O str(14)
	888?				893	$C^{\alpha}C$ str(27), CN str(26), C=O str(14)
				759	760 }	C=O ipb(22), $C^{\alpha}C$ str(21), NC^{α} str(15), CNC^{α} def(15), $NC^{\alpha}C$ def(14)

Table 16.59 Continued

Frequency (cm^{-1})						
Observed		Calculated				
Raman	IR	A_g	A_u	B_g	B_u	Potential Energy Distribution[a] (%)
			730			CN tor(73), NH opb(35), H---O (17), NH---O ipb(14)
		722				CN tor(71), NH opb(37), NH---O ipb(15), H---O (14)
				724		CN tor(66), NH opb(21), H---O (15), CH_2 rock(14), NH---O ipb(12)
708 (720)[c]					716	CN tor(75), NH opb(24), H---O (13), NH---O ipb(13)
	628		629			C=O ipb(35), C=O opb(25), CH_2 rock(11)
		628				C=O ipb(37), C=O opb(22), CH_2 rock(11)
	614				618	C=O opb(52), C^{α}CN def(18), NH opb(14), CH_2 wag(10)
				627		C=O opb(67), NH opb(19)
599				589		C^{α}CN def(58)
					585	C^{α}CN def(45), C=O opb(18)
	589		585			C=O opb(39), C=O ipb(30), C^{α}C str(12)
589		584				C=O opb(40), C=O ipb(28), C^{α}C str(12)
327				318		NC^{α}C def(25), NH opb(23), C=O ipb(17)
	321				317	NC^{α}C def(25), NH opb(23), C=O ipb(17)
	285		290			C^{α}CN def(58), NC^{α}C def(18), NC^{α} str(12)
		286				C^{α}CN def(54), NC^{α}C def(21), NC^{α} str(12)
260				257		CNC^{α} def(38), C=O ipb(28), NH opb(12)
					257	CNC^{α} def(38), C=O ipb(27)
	217		230			CNC^{α} def(67), C=O ipb(10)
211		217				CNC^{α} def(75)
170 (175)		181				NH opb(69), C=O opb(14)
			179			NH opb(71), C=O opb(18)
	140				145	H---O str(52), CN tor(28), NH opb(24)
		134				H---O str(30), CN tor(23), NC^{α}C def(12)
112 (114)				110		H---O str(79), CN tor(19)
			110			NC^{α}C def(38), CN tor(15), C^{α}C tor(14)
82 (87)		96				H^{α}---H^{α} str(31), NC^{α}C def(22), NH opb(15), H---O str(14)
				69		C^{α}C tor(33), NC^{α} tor(32), NH---O ipb(17)
			34			NH---O ipb(42), C=O---H ipb(32), H---O str(17)
				29		NH---O ipb(45), C=O---H ipb(35), CN tor(11)
		12				NH tor(54), C=O tor(37)

[a]Only contributions of 10% or more are included.
[b]Unperturbed frequency.
[c]Frequencies at 120°K in parentheses.

complex and must also be specified; see ref. 152 for specific details. It should be noted that unlike the APPS structure, which is characterized by short linear hydrogen bonds, the APRS structure contains long and nonlinear hydrogen bonds. This necessitated the replacement of the in- and out-of-plane bending coordinates used in the APPS model with angle bending and torsional coordinates for the APRS model. The rippled sheet structure has a symmetry isomorphous to the point group C_{2h}; it is illustrated in Figure 16.49. The optically active symmetry coordinates are thus classified into four symmetry species: A_g and B_g (Raman active) and A_u and B_u (infrared active with parallel and perpendicular dichroism, respectively).

The force field of Abe and Krimm (146) was employed as a starting point, and 9 of the 72 intrachain force constants were refined to fit the observed frequencies of PGI, β-polyalanine, and β-poly(L-alanyl glycine). Adjustments were made to the N—H stretching force constant to make the calculated frequencies appear close to the unperturbed N—H stretching frequency. An H^{α}---H^{α} interaction force constant was introduced consistent with a Williams potential and it was also necessary to adjust two interaction force constants in order to account for the observed splitting of the infrared and Raman-active CH_2 bending modes. As mentioned previously, Krimm and his co-workers had shown that an f(C=O, O---H) interaction force constant was not applicable and that transition dipole coupling satisfactorily accounts for the amide I splittings. Accordingly, values for F_{ij}^{I} and F_{ij}^{II} interaction force constants (where I and II relate to the amide I and II modes, respectively) were determined by transition dipole coupling. The final valence force field is reproduced in Table 16.58 and the results of the normal coordinate calculations are shown in Table 16.59. In general, there is excellent agreement between the observed and calculated frequencies for PGI. The force field, which incorporates the transition dipole coupling concept, satisfactorily accounts for the important splittings observed in the amide I, II, and III modes. In addition, the authors have demonstrated that the force field is applicable to other β-polypeptides and with slight modification is also applicable to α-helical structures.

References

1 G. Natta, I. Pasquon, and A. Zambelli, *J. Am. Chem. Soc.* **84**, 1488 (1962).
2 R. G. Snyder and J. H. Schachtschneider, *Spectrochim. Acta* **20**, 853 (1964).
3 J. H. Schachtschneider and R. G. Snyder, *Spectrochim. Acta* **21**, 1527 (1965).
4 G. Gramberg, *Kolloid-Z.* **175**, 119 (1960).
5 T. Miyazawa, Y. Ideguchi, and K. Fukushima, *J. Chem. Phys.* **38**, 2709 (1963).
6 T. Miyazawa and Y. Ideguchi, *Bull. Chem. Soc. Japan* **36**, 1125 (1963); **37**, 1065 (1964).
7 H. Tadokoro, M. Kobayashi, M. Ukita, K. Yasufuku, S. Murahashi, and T. Torii, *J. Chem. Phys.* **42** (4), 1432 (1965).
8 G. Natta, P. Corradini, and M. Ceseari, *Atti Accad. Naz. Lincei Rend. Cl. Sci. Fis. Mat. Nat.* **21**, 365 (1956).

9 G. Natta, L. Pasquon, P. Corradini, M. Peraldo, and A. Zambelli, *Rend. Accad. Naz. Lincei* **28** (8), 539 (1960).

10 G. Natta, M. Peraldo, and G. Allegra, *Makromol. Chem.* **75**, 215 (1964).

11 R. G. Snyder and J. H. Schachtschneider, *Spectrochim. Acta* **21**, 169 (1965).

12 M. P. McDonald and S. M. Ward, *Polymer* **2**, 341 (1961).

13 K. Fukushima, Y. Ideguchi, and T. Miyazawa, Abstracts of the International Symposium on Molecular Structure and Spectroscopy, Tokyo (1962).

14 M. C. Tobin, *J. Chem. Phys.* **64**, 216 (1960).

15 M. Peraldo and M. Cambini, *Spectrochim. Acta* **21**, 1509 (1963).

16 G. Zerbi, M. Gussoni, and F. Ciampelli, *Spectrochim Acta* **23A**, 301 (1967).

17 T. Miyazawa, *J. Polymer Sci. Polymer Symp.* **C7**, 59 (1963).

18 M. Ukita, *Bull. Chem. Soc. Japan* **39**, 742 (1966).

19 S. W. Cornell and J. L. Koenig, *J. Polymer Sci.* **A2** (7), 1965 (1969).

20 K. Holland-Moritz and E. Sausen, *J. Polymer Sci. Polymer Phys. Ed.* **17**, 1 (1979).

21 J. H. Schachtschneider and R. G. Snyder, *Spectrochim Acta* **19**, 117 (1963).

22 G. Natta, P. Pino, P. Corradini, F. Danusso, E. Mantica, G. Mazzanti, and G. Moraglio, *J. Am. Chem. Soc.* **77**, 1708 (1955).

23 F. Danusso and G. Gianotti, *Makromol. Chem.* **6**, 139 (1963).

24 R. Zannetti, P. Manaresi, and G. C. Buzzoni, *Chim. Ind.* **43**, 735 (1961).

25 G. Natta, P. Corradini, and I. Bassi, *Makromol. Chem.* **21**, 240 (1956).

26 R. L. Miller and V. F. Holland, *J. Polymer Sci. Polymer Lett. Ed.* **2**, 519 (1964).

27 G. Geacintov, R. S. Schotland, and R. B. Miles, *J. Polymer Sci.* **B1**, 587 (1963).

28 A. Turner Jones, *J. Polymer Sci. Polymer Lett. Ed.* **1**, 445 (1963).

29 V. F. Holland and R. L. Miller, *J. Appl. Phys.* **35** 241 (1964).

30 C. Cojazzi, V. Malta, G. Celotti, and R. Zanelli, *Makromol. Chem.* **177**, 915 (1976).

31 G. Goldbach and G. Peitscher, *J. Polymer Sci. Polymer Lett. Ed.* **6**, 783 (1968).

32 J. P. Luongo and R. Salovey, *J. Polymer Sci.* **B3**, 513 (1965).

33 G. Zerbi, L. Piseri, and F. Cabassi, *Mol. Phys.* **22**, 241 (1971).

34 R. G. Snyder, *J. Chem. Phys.* **47**, 1316 (1967).

35 K. Holland-Moritz, E. Sausen, P. Djudovic, M. M. Coleman, and P. C. Painter, *J. Polymer Phys. Polymer Phys. Ed.* **17**, 25 (1979).

36 S. Krimm, A. R. Berens, V. L. Folt, and J. J. Shipman, *Chem. Ind.* (*London*) 1512 (1958); 433 (1959).

37 S. Krimm, *Pure Appl. Chem.* **16**, 369 (1968).

38 C. G. Opaskar, Ph.D. Thesis, University of Michigan, Ann Arbor (1966).

39 W. H. Moore and S. Krimm, *Makromol. Chem. Suppl.* 1, **491**, (1975).

40 A. Rubĉić and G. Zerbi, *Macromolecules* **6**, 751 (1974).

41 A. Rubĉić and G. Zerbi, *Macromolecules* **6**, 759 (1974).

42 T. Shimanouchi and M. Tasumi, *Bull. Chem. Soc. Japan* **34**, 359 (1961).

43 M. Tasumi and T. Shimanouchi, *Polymer J.* **2**, 62 (1971).

44 V. G. Boitsov and Y. Y. Gotlib, *Opt. Spectrosc. USSR Suppl.* **2**, 65 (1966).

45 G. Natta and P. Corradini, *J. Polymer Sci.* **20**, 251 (1956).

46 C. G. Opaskar and S. Krimm, *Spectrochim. Acta* **23A**, 2261 (1967).

47 W. H. Moore and S. Krimm, *Spectrochim. Acta* **29A**, 2025 (1973).

48 A. V. R. Warrier and S. Krimm, *J. Chem. Phys.* **52**, 4316 (1970).

49 A. V. R. Warrier and S. Krimm *Macromolecules* **3**, 709 (1970).

50 S. Krimm, V. L. Folt, J. J. Shipman, and A. R. Berens, *J. Polymer Sci.* **A1**, 2621 (1963).

51 J. L. Koenig and D. Druesedow, *J. Polymer Sci* **A2** (7), 1075 (1969).

52 S. Krimm, *J. Polymer Sci. Polymer Symp.* **C7**, 3 (1964).

53 C. G. Opaskar and S. Krimm, *J. Polymer Sci.* **A2**(7), 57 (1969).

54 T. Miyazawa and Y. Ideguchi, *J. Polymer Sci.* **B3**, 541 (1965).

55 M. S. Wu, P. C. Painter, and M. M. Coleman, *J. Polymer Sci. Polymer Phys. Ed.* **18**, 95 (1980).

56 S. Narita and K. Okuda, *J. Polymer Sci.* **38**, 270 (1959).

57 C. S. Fuller, *Chem. Rev.* **26**, 143 (1940).

58 M. S. Wu, P. C. Painter, and M. M. Coleman, *J. Polymer Sci. Polymer Phys. Ed.* **18**, 111 (1980).

59 M. S. Wu, M. M. Coleman, and P. C. Painter, *Spectrochim. Acta* **35A**, 823 (1979).

60 M. S. Wu, Ph.D. Thesis, Pennsylvania State University, University Park, (1979).

61 S. Krimm, *Fortschr. Hochpolym.-Forsch.* **2**, 51 (1960).

62 S. Krimm and C. Y. Liang, *J. Polymer Sci.* **22**, 95 (1956).

63 S. Narita, S. Ichinohe, and S. Enomoto, *J. Polymer Sci.* **37**, 251 (1959); **37**, 263 (1959).

64 M. M. Coleman, M. S. Wu, I. R. Harrison, and P. C. Painter, *J. Macromol. Sci. Phys. Ed.* **B15** (3), 463 (1978).

65 F. J. Boerio and J. L. Koenig, *J. Polymer Sci.* **A2**(9), 1517 (1974).

66 M. Kobayashi, K. Tashiro, and H. Tadokoro, *Macromolecules* **8**(2), 158 (1975).

67 G. Natta, G. Allegra, I. W. Bassi, D. Sianesi, G. Caporiccio, and E. Torti, *J. Polymer Sci.* **A3**, 4263 (1965).

68 G. Cortili and G. Zerbi, *Spectrochim. Acta* **23A**, 2216 (1967).

69 W. W. Doll and J. B. Lando, *J. Macromol. Sci.* **B2**, 219 (1968).

70 S. Weinhold, M. J. Litt and J. B. Lando, *Macromolecules*, **13**, 1178 (1980).

71 M. A. Bachmann, W. L. Gordon, J. L. Koenig, and J. B. Lando, *J. Appl. Phys.* **50**, 6106 (1979).

72 F. J. Boerio and J. L. Koenig, *J. Chem. Phys.* **52**, 4826 (1970).

73 R. Hasegawa, Y. Takahashi, Y. Chatani, and H. Tadokoro, *Polymer J.* **3**, 600 (1972).

74 C. Y. Liang and S. Krimm, *J. Chem. Phys.* **25**, 563 (1956).

75 M. J. Hannon, F. J. Boerio, and J. L. Koenig, *J. Chem. Phys.* **50** 2829 (1969).

76 L. Piseri, B. M. Powell, and G. Dolling, *J. Chem. Phys.* **57**, 158 (1973).

77 G. Zerbi and M. Sacchi, *Macromolecules* **6** (5), 692 (1973).

78 G. Masetti, F. Cabassi, G. Morelli, and G. Zerbi, *Macromolecules* **6** (5), 700 (1973).

79 E. S. Clark and L. T. Muus, *Kristallografiya* **117**, 119 (1962).

80 C. W. Bunn and E. R. Howells, *Nature* (*London*) **174**, 549 (1954).

81 H. G. Kilian, *Kolloid-Z* **185**, 13 (1962).

82 D. W. McCall, D. C. Douglass, and D. R. Falcone, *J. Phys. Chem.* **71**, 998 (1967).

83 H. Tadokoro, "Structure of Crystalline Polymers," Wiley-Interscience, New York, 1979.

84 P. DeSantis, E. Giglio, A. M. Liguori, and A. Ripamonti, *J. Polymer Sci.* **A1**, 1383 (1963).

85 G. Zerbi and M. Gussoni, *Spectrochim. Acta* **22**, 2111 (1966).

86 J. H. Schachtschneider, private communication to G. Zerbi.

87 D. Morero, F. Ciampelli, and E. Mantica, in "Advances in Molecular Spectroscopy," Vol. 2, A. Mangini, Ed., Pergamon Press, Oxford, 1962, p. 898.

88 N. Neto and C. DiLauro, *Eur. Polymer J.* **3**, 645 (1967).

89 S. L. Hsu, W. H. Moore, and S. Krimm, *J. Appl. Phys.* **46** (10), 4185 (1975).

90 G. Natta and P. Corradini, *Nuovo Cimento Suppl.* **1**, 9 (1960).

91 S. Iwayanagi, I. Sakurai, T. Sakurai, and T. Seto, *J. Macromol. Sci. Phys. Ed.* **B2**, 163 (1968).

92 C. DiLauro, N. Neto, and S. Califano, *Spectrochim Acta* **24A**, 385 (1968).

93 S. W. Cornell and J. L. Koenig, *Macromolecules* **2**, 540 (1969).

94 D. E. Williams, *J. Chem. Phys.* **45**, 3770 (1966).

95 M. M. Coleman, D. L. Tabb, B. L. Farmer, and J. L. Koenig, *J. Polymer Sci. Polymer Phys. Ed.* **12**, 445 (1974).

96 G. Natta and P. Corradini, *Nuovo Cimento Suppl.* **15**, 111 (1960).

97 N. Neto, C. DiLauro, E. Castellucci, and S. Califano, *Spectrochim. Acta* **23A**, 1763 (1967).

98 S. W. Cornell and J. L. Koenig, *J. Polymer Sci.* **B8**, 137 (1970).

99 M. Dannis, *J. Appl. Polymer Sci.* **1**, 121 (1959).

100 G. Natta, SPE *Trans.* **71**, 99 (1963).

101 J. L. Binder, *J. Polymer Sci.* **A1**, 47 (1963).

102 R. J. Silas, J. Yates, and V. Thorton, *Anal. Chem.* **31**, 529 (1959).

103 D. L. Tabb and J. L. Koenig, *J. Polymer Sci. Polymer Phys. Ed.* **13**, 1159 (1975).

104 R. J. Petcavich and M. M. Coleman, *J. Macromol. Sci. Phys. Ed.* **B18** (1), 47 (1980).

105 C. W. Bunn, *Proc. Roy. Soc. (London)* **A180**, 40 (1942).

106 Y. Chatani and S. Nakatani, *Macromolecules* **5** (5), 597 (1972).

107 M. Z. El-Sabban and B. J. Zwolinski, *J. Mol. Spect.* **27**, 1 (1968).

108 G. A. Crowder and N. Smyrl, *J. Mol. Struct.* **10**, 373 (1971).

109 M. M. Coleman, P. C. Painter, D. L. Tabb, and J. L. Koenig, *J. Polymer Sci. Polymer Lett. Ed.* **12**, 577 (1974).

110 M. M. Coleman and P. C. Painter, *J. Macromol. Sci. Revs. Macromol. Chem.* **C16** (2), 197 (1978).

111 M. M. Coleman, D. L. Tabb, and E. G. Brame, Jr., *Rubber Chem. Technol.* **50**, 49 (1977).

112 M. M. Coleman, P. C. Painter and J. L. Koenig, *J. Raman Spect.* **5**, 417 (1976).

113 W. E. Mochel and M. B. Hall, *J. Am. Chem. Soc.* **71**, 4082 (1949).

114 D. Pasto and C. Johnson, "Organic Structure Determination," Prentice-Hall, Englewood Cliffs, New Jersey, 1969, p. 115.

115 G. A. Crowder, *J. Mol. Spect.* **20**, 430 (1966).

116 R. J. Petcavich and M. M. Coleman, *J. Polymer Sci. Polymer Phys. Ed.* **18**, 2097 (1980).

117 A. J. van der Wyk and L. Misch, *J. Chem. Phys.* **8**, 127 (1940).

118 G. Natta, P. Corradini, and L. Porri, *Atti Accad. Naz. Lincei. Rend. Cl. Sci. Fis. Mat. Nat.* **20**, 728 (1956).

119 T. Shimanouchi and Y. Abe, *J. Polymer Sci.* **A2** (6), 1419 (1968).

120 Y. Takahashi, T. Sato, H. Tadokoro, and Y. Tanaka, *J. Polymer Sci. Polymer Phys. Ed.* **11**, 233 (1973).

121 R. G. Snyder and J. H. Schachtschneider, *Spectrochim. Acta* **22**, 2111 (1966).

122 K. Holland-Moritz and K. van Werden, *J. Polymer Sci. Polymer Phys. Ed.* **18**, 1753 (1980).

123 G. Natta and I. W. Bassi, *Eur. Polymer J.* **3**, 33 (1967).

124 G. Natta and I. W. Bassi, *Atti Accad. Naz Lincei Rend. Cl. Sci. Fis. Mat. Nat.* **38**, 315 (1965).

125 P. C. Painter and J. L. Koenig, *J. Polymer Sci. Polymer Phys. Ed.* **15**, 1885 (1977).

126 C. La Lau and R. G. Snyder, *Spectrochim Acta* **27A**, 2073 (1971).

127 P. C. Painter and R. W. Snyder, *Polymer* (accepted).

128 G. Natta, P. Corradini, and I. W. Bassi, *Nuovo Cimento Suppl.* **15**, 68 (1960).

129 P. C. Painter and R. W. Snyder, *Polymer* (accepted).

130 T. Onishi and S. Krimm, *J. Appl. Phys.* **32**, 2320 (1961).

131 M. Kobayashi, *Bull. Chem. Soc. Japan* **34**, 560 (1961).

132 M. Kobayashi, K. Akita, and H. Tadokoro, *Makromol. Chem.* **118**, 324 (1968).

133 V. B. Boitsov and Y. Y. Gotlib, *Opt. Spekt.* **15** (2), 216 (1963).

134 R. Danz, J. Dechant, and C. Rusher, *Faserforsch. Textiltech.* **21**, 503 (1970).

135 S. K. Bahl, D. D. Cornell, F. J. Boerio, and G. E. McGraw, *J. Polymer Sci. Polymer Lett. Ed.* **12**, 13 (1974).

136 F. J. Boerio, S. K. Bahl, and G. E. McGraw, *J. Polymer Sci. Polymer Phys. Ed.* **14**, 1029 (1976).

137 I. M. Ward, *Chem. Ind.* (*London*) 905 (1956); *Ibid.*, 1102 (1957).

138 C. Y. Liang and S. Krimm, *J. Mol. Spect.* **3**, 554 (1959).

139 S. K. Bahl, Ph.D. Thesis, University of Cincinnati (1974).

140 R. G. Snyder and G. Zerbi, *Spectrochim. Acta* **23A**, 391 (1967).

141 P. C. Painter, unpublished data.

142 T. Miyazawa, T. Shimanouchi, and S. Mizushima, *J. Chem. Phys.* **24**, 408 (1956); **29**, 611 (1958).

143 C. D. Needham, Ph.D. Thesis, University of Minnesota (1965).

144 J. Jakeš, *Coll. Czech. Chem. Commun.* **33**, 643 (1968).

145 J. Jakeš and S. Krimm, *Spectrochim. Acta* **27A**, 19 (1971); **27A**, 35 (1971).

146 Y. Abe and S. Krimm, *Biopolymers* **11**, 1817 (1972); **11**, 1841 (1972).

147 K. Fukushima, Y. Ideguchi, and T. Miyazawa, *Bull. Chem. Soc. Japan* **36**, 1301 (1963).

148 V. D. Gupta, S. Trevino, and H. Boutin, *J. Chem. Phys.* **48**, 3008 (1968).

149 T. Miyazawa, in "Poly-α-Amino Acids," G. D. Fasman, Ed., Marcel Dekker, New York, 1967.

150 S. Krimm and Y. Abe, *Proc. Natl. Acad. Sci.* **69**, 2788 (1972).

151 W. H. Moore and S. Krimm, *Proc. Natl. Acad. Sci.* **72**, 4933 (1975).

152 W. H. Moore and S. Krimm, *Biopolymers* **15**, 2439 (1976); **15**, 2465 (1976).

153 S. L. Hsu, W. H. Moore, and S. Krimm, *Biopolymers* **15**, 1513 (1976).

154 J. F. Rabolt, W. H. Moore, and S. Krimm, *Macromolecules* **10**, 1065 (1977).

155 B. Lotz, *J. Mol. Biol.* **87**, 169 (1974).

156 L. Pauling and R. B. Corey, *Proc. Natl. Acad. Sci.* **39**, 253 (1953).

AUTHOR INDEX

SUBJECT INDEX